Columnar Cacti and Their Mutualists

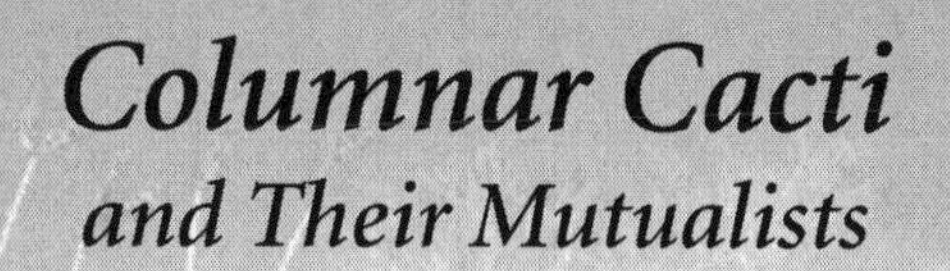

Columnar Cacti and Their Mutualists

Evolution, Ecology, and Conservation

EDITED BY

Theodore H. Fleming

AND

Alfonso Valiente-Banuet

THE UNIVERSITY OF ARIZONA PRESS
Tucson

The University of Arizona Press
www.uapress.arizona.edu

First paperback edition 2019

ISBN-13: 978-0-8165-2204-0 (cloth)
ISBN-13: 978-0-8165-4021-1 (paper)

Cover design by Sara Thaxton
Cover photo by Corey Taratuta, Organ Pipe National Monument, Southern Arizona

Library of Congress Cataloging-in-Publication Data
Columnar cacti and their mutualists : evolution, ecology, and conservation / edited by
Theodore H. Fleming and Alfonso Valiente-Banuet.
p. cm.
Includes bibliographical references and index.
ISBN 0-8165-2204-9 (cloth : alk. paper)
1. Cactus—Southwest, New. 2. Cactus—Mexico. 3. Cactus—South America.
4. Mutualism (Biology)—Southwest, New. 5. Mutualism (Biology)—Mexico.
6. Mutualism (Biology)—South America. I. Fleming, Theodore H. II. Valiente-Banuet, Alfonso.
QK495.C11 C64 2002
583′.56—dc21
2002003840

Printed in the United States of America
♾ This paper meets the requirements of ANSI/NISO Z39.48-1992 (Permanence of Paper).

We dedicate this book to two scientists who have made major contributions to our understanding of the evolution of Mexico's columnar cacti:

Dra. Helia Bravo-Hollis and Dr. Arthur C. Gibson

CONTENTS

PREFACE

Columnar cacti (Cactaceae, subfamily Cactoideae) are the most conspicuous and, arguably, the ecologically most important plants over vast areas of arid to semiarid land in North, Central, and South America and the Caribbean. Occurring in at least four tribes of the Cactoideae (Browningieae, Cereeae, Pachycereeae, and Trichocereeae), these impressive plants are represented by about 25 genera and about 170 species. Many species are the dominant elements of arid and semiarid ecosystems, where they sometimes attain densities of 1,200 to 1,800 adult cacti per ha. Because of this dominance, many species of animals, ranging from cactophilic *Drosophila* to nectar-feeding birds and bats, as well as humans, rely on columnar cacti for food and/or shelter. Recent studies indicate that many of these plants are pollinated primarily by bats in central Mexico and Venezuela, whereas extratropical cacti in northern Mexico and Peru are pollinated by a wider variety of species, including diurnal birds and bees. Other studies have revealed specialized relationships between cacti and animals, including the third example ever detected of active pollination in an obligate pollination mutualism—the relationship between the senita cactus (*Lophocereus schottii*) and the pyralid moth (*Upiga virescens*) in the Sonoran Desert. Careful studies of the breeding systems of columnar cacti have also revealed the occurrence of rare breeding systems—trioecy and androdioecy—in Mexican *Pachycereus pringlei* and *Neobuxbaumia mezcalaensis,* respectively, as well as examples of autotetraploidy in Mexico, Venezuela, and Peru. These and other studies indicate that we have made much progress recently in understanding how abiotic and biotic factors interact to influence the evolution and distribution and abundance of columnar cacti and their animal mutualists.

In addition to their intrinsic biological interest, columnar cacti and their mutualists are of considerable interest from a conservation viewpoint. The biota of the earth is currently being altered at an unprecedented rate by humans with unknown consequences in many geographic areas. Only by determining the functional roles of individual species and their web of interactions within ecosystems will we be able to mitigate these disturbances through enlightened management and conservation.

This book focuses on the evolution, ecology (sensu lato), and conservation needs of columnar cacti and their vertebrate mutualists. Our goal is to summarize much of the current knowledge of the biology of these two groups of organisms—from their physiological ecology to their ecological interactions in different geographic areas. We also explore the ways in which humans, through the process of domestication, have modified these plants for their economic benefit. We review phylogenetic relationships within cacti and nectar-feeding bats in an effort to understand how bat-plant interactions have influenced the evolution of diversity and ecological specialization within both groups of organisms. Finally, we make conservation recommendations aimed at preserving fully functional ecosystems in arid portions of the New World tropics and subtropics.

This book is the result of a conference held in Tehuacan City, Puebla, Mexico, on June 28–July 3, 1998. The conference was sponsored by the Instituto de Ecología of the Universidad Nacional Autónoma de México (UNAM) and the University of Miami, Coral Gables, Florida. Financial support was provided by the Consejo Nacional de Ciencia y Tecnología (CONACyT) and the U.S. National Science Foundation (NSF). At the conference and afterwards, participants reviewed each other's chapters. The entire book was thoroughly reviewed by Peter Scott, to whom we give special thanks. We thank our funding agencies and all of the participants for making the conference and its summary volume a scientific success.

THEODORE H. FLEMING
ALFONSO VALIENTE-BANUET

CONTRIBUTORS

SALVADOR ARIAS-MONTES, Jardin Botánico, Universidad Nacional Autónoma de México, Ciudad Universitaria, Mexico, D.F. 04510, Apartado Postal 70-614

HÉCTOR T. ARITA, Instituto de Ecología, Universidad Nacional Autónoma de México, Ciudad Universitaria, Mexico, D.F. 04510, Apartado Postal 70-275

MARÍA DEL CORO ARIZMENDI, Unidad de Biología, Tecnología y Prototipos, FES-Iztacala, Universidad Nacional Autónoma de México, Avenida de los Barrios s.n., Los Reyes Iztacala, Tlalnepantla, Estado de Mexico 54090, Apartado Postal 314

JAVIER CABALLERO, Jardin Botánico, Universidad Nacional Autónoma de México, Ciudad Universitaria, Mexico, D.F. 04510, Apartado Postal 70-614

ALEJANDRO CASAS, Instituto de Ecología, Universidad Nacional Autónoma de México, Ciudad Universitaria, Mexico, D.F. 04510, Apartado Postal 70-275

JAIME CAVELIER, 2828 Connecticut Avenue NW, Apartment 608, Washington, D.C. 20008

MARTIN L. CODY, Department of Biology, University of California–Los Angeles, Los Angeles, California 90095-1606

PATRICIA DÁVILA-ARANDA, Unidad de Biología, Tecnología y Prototipos, FES-Iztacala, Universidad Nacional Autónoma de México, Avenida de los Barrios s.n., Los Reyes Iztacala, Tlalnepantla, Estado de Mexico 54090, Apartado Postal 314

THEODORE H. FLEMING, Department of Biology, University of Miami, Coral Gables, Florida 33124

HÉCTOR GODÍNEZ-ALVAREZ, Instituto de Ecología, Universidad Nacional Autónoma de México, Ciudad Universitaria, Mexico, D.F. 04510, Apartado Postal 70-275

J. L. HAMRICK, Departments of Botany and Genetics, University of Georgia, Athens, Georgia 30602

RAFAEL LIRA-SAADE, Unidad de Biología, Tecnología y Prototipos, FES-Iztacala, Universidad Nacional Autónoma de México, Avenida de los Barrios s.n., Los Reyes Iztacala, Tlalnepantla, Estado de Mexico 54090, Apartado Postal 314

Sofía Loza-Cornejo, Programa de Botánica, Colegio de Postgraduados, Montecillo, Estado de México 56230

John D. Nason, Department of Botany, Iowa State University, Ames, Iowa 50011-1020

Jafet Nassar, Instituto Venezolano de Investigaciones Científicas, Biblioteca "Marcel Roche," 8424 NW 56th Street, Suite CCS 00206, Miami, Florida 33166

Park S. Nobel, Department of Biology, University of California–Los Angeles, Los Angeles, California 90095-1606

Alberto Rojas-Martínez, Instituto de Ecología, Universidad Nacional Autónoma de México, Ciudad Universitaria, Mexico, D.F. 04510, Apartado Postal 70-275

Adriana Ruiz, Postgrado en Ecología Tropical, ICAE, Facultad de Ciencias, Universidad de Los Andes, Mérida 5101, Venezuela

Mery Santos, Instituto de Ecología, Universidad Nacional Autónoma de México, Morelia, C.P. 58089, Michoacán, Mexico, Apartado Postal 27-3 (Xangari)

Carlos Silva, Instituto de Ecología, Universidad Nacional Autónoma de México, Ciudad Universitaria, Mexico, D.F. 04510, Apartado Postal 70-275

Nancy B. Simmons, Department of Mammalogy, American Museum of Natural History, Central Park West at 79th Street, New York, New York 10024

Pascual J. Soriano, Departamento de Biología, Facultad de Ciencias, Universidad de Los Andes, Mérida 5101, Venezuela

Vinicio J. Sosa, Departamento de Ecología y Conservación de Ecosistemas Templados, Instituto de Ecología, Xalapa 91000, Veracruz, Mexico

Teresa Terrazas, Programa de Botánica, Colegio de Postgraduados, Montecillo, Estado de Mexico 56230

Alfonso Valiente-Banuet, Instituto de Ecología, Universidad Nacional Autónoma de México, Ciudad Universitaria, Mexico, D.F. 04510, Apartado Postal 70-275

Thomas R. Van Devender, Arizona-Sonora Desert Museum, 2021 North Kinney Road, Tucson, Arizona 85743

José Luis Villaseñor, Instituto de Biología, Universidad Nacional Autónoma de México, Ciudad Universitaria, Mexico, D.F. 05410, Apartado Postal 70-233

Robert S. Wallace, Department of Botany, Iowa State University, Ames, Iowa 50011-1020

Andrea L. Wetterer, Department of Mammalogy, American Museum of Natural History, Central Park West at 79th Street, New York, New York 10024

ABOUT THE EDITORS

THEODORE H. FLEMING was born and raised in Detroit, Michigan. He received a B.A. degree from Albion College in 1964 and M.S. and Ph.D. degrees from the University of Michigan in 1968 and 1969, respectively. From 1969 to 1978, he was an assistant and then an associate professor of biology at the University of Missouri–St. Louis. In 1978, he joined the Department of Biology, University of Miami, where he became a full professor in 1980. He is a Fulbright fellow, a fellow of the American Association for the Advancement of Science, and a National Science Foundation Mid-Career fellow, and he has served as president of the Association for Tropical Biology. He has written one book on tropical bat-plant interactions and has co-edited two books on frugivory and seed dispersal. His research interests deal with ecological interactions between vertebrate frugivores and pollinators and their food plants. He has conducted field work in Panama, Costa Rica, England, Australia, Mexico, and the United States.

ALFONSO VALIENTE-BANUET received his Ph.D. degree from the Universidad Nacional Autónoma de México (UNAM) in 1991. His research contributions are related to the study of Mexican deserts, including the role of biotic interactions in the maintenance of biological diversity and the reconstruction of paleoenvironments during the Late Quaternary. Dr. Valiente-Banuet is presently head of the Community Ecology Laboratory and head of the Department of Functional and Applied Ecology of the Institute of Ecology, UNAM. Research in his lab focuses on three areas. One is the study of biotic interactions and their effect on plant demography and plant diversity in communities dominated by cacti at different field sites in Mexico. The second area is the study of evergreen-sclerophyllous vegetation located in non-Mediterranean climates in Mexico, focusing on the comparative community ecology of chaparral vegetation in Mediterranean climates. Finally, his lab is interested in the paleoclimatic reconstruction of the Chihuahuan and Tehuacán-Cuicatlán Deserts to provide a historical framework for the analysis of present ecological conditions, the demography of long-lived species, and vegetation dynamics.

PART I

Geology and Evolution

CHAPTER 1

Environmental History of the Sonoran Desert

Thomas R. Van Devender

Introduction

The Sonoran Desert is considered to be the most "tropical" of the North American deserts, partly because of its climate—mild, virtually frost-free winters and summer monsoonal rainfall from tropical oceans—and partly because of its physical connections with more tropical communities to the south (Shreve, 1964; Turner and Brown, 1982). Its structurally diverse vegetation is dominated by columnar cacti and leguminous trees, and certainly differs from the various shrub-dominated desertscrub communities of the Great Basin, Mohave, and Chihuahuan deserts. In northwestern Mexico, the transition from temperate biomes to the New World tropics occurs in Sonora between 31° and 28° north latitude, where winter freezes decrease in frequency and intensity, and absolute amounts and percentages of summer precipitation increase (Van Devender et al., 1994b). The subdivisions of the Sonoran Desert from Arizona to Guaymas on the Gulf of California in Sonora are Arizona Upland, Plains of Sonora, and Central Gulf Coast. South of Guaymas, the Sonoran Desert is replaced on the coast by coastal thornscrub and in wetter areas inland by foothills thornscrub and tropical deciduous forest (Búrquez et al., 1999). On rocky slopes in central Sonora, the Arizona Upland dominants *Carnegiea gigantea* (saguaro) and *Cercidium microphyllum* (foothills palo verde) are joined by such trees as *Fouquieria macdougalii* (tree ocotillo, jaboncillo), *Bursera fagaroides,* and *B. laxiflora* (torotes), as desertscrub becomes denser and taller, gradually changing into foothills thornscrub.

However, this remarkable vegetational gradient does not really explain our perception of the Arizona Upland as "tropical." The ranges of a few Sonoran Desert dominants such as *Cercidium praecox* (palo brea), *F. macdougalii, Guaiacum coulteri* (guayacán), and *Stenocereus thurberi* (organpipe cactus, pitahaya) extend into dry tropical forests near Alamos (Van Devender et al., 2000) but not those of *C. gigantea, C. microphyllum, Larrea divaricata* (creosotebush), or *Olneya tesota* (desert ironwood). Surprisingly, most of the visual dominants in the saguaro "forests" of the Tucson Mountains of Arizona do not occur in tropical deciduous forests near Alamos in southern Sonora. Most of the shared species are widespread herbs and grasses, and not the Arizona Upland trees, shrubs, and cacti. Moreover, many tropical species such as *Erythrina flabelliformis* (coral bean), *Manihot davisae* (manihot), and *Oxybelis aeneus* (brown vine snake) reach their northern limits in southern Arizona in desert grassland at 1,220–1,525 m elevation above the Sonoran Desert (Van Devender et al., 1994b).

To understand the "tropicalness" of the Sonoran Desert, we must look for its evolutionary roots in the distant past. Changes in the distributions of plants and animals in the fossil record provide evidence of more or less tropical climates, with profound implications for the northern limits of the New World tropics and the development of the Sonoran Desert. Here I explore the fossil record in search of insights into the deep history of the region.

Early Tertiary Tropical Forests

Seasonality in Tropical Forests

In the Paleocene epoch, soon after the extinction of the dinosaurs (65 mya [million years ago]), temperate evergreen and tropical rainforests were widespread across North America with little regional differentiation (Wolfe, 1977; Graham, 1993). The climates of North America were warm with humid forests with strong Asian affinities and primitive ferns (*Anemia*), cycads *(Dioon, Zamia*), and palms occurring as far north as Alaska. Palms grew at 70° north latitude in Greenland.

In the Eocene (54 to 35 mya), deciduous trees became increasingly more common, providing the first evidence of tropical deciduous forests (Wolfe and Hopkins, 1967; Axelrod and Bailey, 1969; Dilcher, 1973; Graham, 1993). Middle Eocene (45 mya) leaf floras from Kentucky and Tennessee are interpreted to reflect dry tropical forests analogous to those of the Pacific coast of Mexico (Dilcher, 1973). There were evolutionary radiations in many plants and animals as they adapted to new heat and moisture regimes as more sunlight penetrated the forest canopy. The result was a dramatic shift in the biota as more advanced tropical forms displaced archaic ones. Although cacti are essentially absent from the sedimentary fossil record, their evolutionary roots likely took hold in the dry seasons of these early tropical forests.

Polar Tropical Forests

Tropical communities, with their great diversities and mixtures of archaic and advanced species, traditionally have been thought to be important evolutionary arenas. However, the speciation mechanisms creating this diversity are not so clear, considering that climatic fluctuations that isolate populations are more extreme at high latitudes. Recent paleomagnetic dating of fossil-bearing sediments in northern Canada indicates that some plants first appeared as much as 18 million years earlier, and some mammals two to four million years earlier than at lower latitudes (Hickey et al., 1983). If these relative ages are not artifacts of a fragmentary fossil record, important biotic innovations may have evolved in the Arctic, with its unusual combination of mild climate and months-long polar day-night cycle, and then moved southward into more tropical latitudes!

The remarkable similarities of Eocene mammal faunas from western North America and western Europe led to a search for a land route in the Arctic that connected the now-sundered landmass of Euramerica (Dawson et al., 1976; Graham, 1993). The early Eocene Eureka Sound Formation on Ellesmere Island in the Canadian Arctic Archipelago (78° N) not only yielded typical mammals but also *Allognathosuchus* (an extinct alligator), *Geochelone* (tortoise), *Trionyx* (softshell turtle), a varanid lizard (monitor lizards and relatives), and a ground boa (Estes and Hutchison, 1980). Eocene plant fossils from Alaska are a surprising mixture of evergreens, including *Magnolia* (magnolia), *Phoenicites,* and *Sabalites* (palms), and a mangrove; deciduous plants, including *Caesalpinites* (a legume) and *Ginkgo* (ginkgo); and temperate forest and riparian trees, including *Alnus* (alder), *Juglans* (walnut), *Platanus* (sycamore), *Prunus* (cherry), *Salix* (willow), and *Sorbus* (mountain ash). This Paratropical rainforest also had a number of species with lowland Malaysian affinities (see discussion in Graham, 1993).

Fossil records of tropical plants and animals from Arctic latitudes indicate equable climates with winters that rarely suffered freezing temperatures (Estes and Hutchison, 1980). The presence of tropical plants and animals at latitudes with six-month-long nights raises important questions about how they survived the dark. Today reptiles spend cold winters in hibernation and hot, dry periods in estivation. Deciduous plants shed their leaves for long periods, triggered by photoperiodic responses to shorter days, the onset of cold temperatures in temperate latitudes, or the aridity of the dry season in the tropics where winter temperatures are mild. Unlike temperate areas where plants are mostly inactive in the dormant season, many tropical trees flower during the dry season. The Arctic fossils suggest that deciduousness in plants and hibernation and estivation in reptiles could have arisen as responses to the polar night and later shifted to other stimuli.

Miocene Revolution

Rising Mountains

Geologic factors have been extremely important in the evolution of plants, animals, and communities throughout earth history (Tiffany, 1985). A series of enormous volcanic eruptions from the late Oligocene to the middle Miocene (about 30 to 15 mya) changed climates and established the modern biogeographic provinces of North America (Axelrod, 1979). The Rocky Mountains were uplifted at least 1,525 m near Florissant, Colorado, and a layer more than 2,135 m thick of volcanic rocks was deposited in the Jackson Hole area of west-central Wyoming (Leopold and MacGinitie, 1972). In the Sierra Madre Occidental in northwestern Mexico, extensive Oligocene (35 to 22 [14] mya) rhyolitic ash-flow tuffs were deposited on top of Laramide (90 to 40 mya) andesites in combined layers as much as 2 km thick, forming the modern high plateaus of the Sierra Madre Occidental (Swanson and Wark, 1988; Cochemé and Demant, 1991; Roldán and Clark, 1992). Basaltic volcanism predominated in the northern Sierra Madre (24 to 12 mya; Cochemé and Demant, 1991), accompanied by basin and range faulting and extension (Henry and Aranda, 1992), especially from 15 to 5 mya (Menges and Pearthree, 1989). As the mountains were uplifted higher, they interrupted the upper flow of the atmosphere for the first time in the Tertiary. Tropical moisture from both the Pacific Ocean and the Gulf of Mexico was blocked from the mid-continent, drying out the modern Great Plains and Mexican Plateau. Harsher climates segregated drought- and cold-tolerant species into new, environmentally limited biomes, including tundra, modern conifer forests, and grasslands that were distributed along elevational and latitudinal environmental gradients (Van Devender, 1995). The "Miocene Revolution" also initiated major evolutionary radiations in many of today's successful groups. For example, pollen of the Compositae first appeared in abundance at the Oligocene-Miocene boundary (Graham, 1993).

In the Miocene, tropical forests were restricted by the rising Sierra Madres to lowland ribbons along the coasts from about modern Sonora and Tamaulipas southward. Evidence for tropical deciduous forests on the eastern side of the continent (southern Mexico and Central America) was not found in the pollen record until the middle Pliocene (4 mya; Graham and Dilcher, 1995), eight to ten million years later than postulated for the Pacific coast. Although in a general sense these tropical communities were more "ancient" than the "new" pine-oak forests of the Sierra Madre Occidental, the lowland forests were also derived descendants of their Eocene-Oligocene precursors. The flora of the tropical deciduous forests of southern Sonora today are mixtures of archaic and more recently evolved species. Apparently, thornscrub was also a new vegetation type in the early Miocene that formed on the lower, drier edges of tropical deciduous forest. Thornscrub may well have been the regional vegetation covering the drier areas to the north that are now Sonoran Desert.

Biogeography of Columnar Cacti

The distributions and phylogenetic relationships of the columnar cacti in the Sonoran Desert Region provide additional biogeographic insight. There are ten species in the genera *Stenocereus* (five), *Lophocereus* (two), *Pachycereus* (two), and *Carnegiea* in the Pachycereeae tribe of the Cactaceae in Sonora and Baja California (Gibson and Horak, 1978; Cornejo, 1994). Two species (*L. gatesii, S. eruca*) are narrow endemics in southern Baja California. The closest relatives of most of the other species occur to the southeast in Mexico. The only species whose distributions extend south of central Sinaloa are the tropical deciduous forest species *P. pecten-aboriginum* (hecho) and *S. montanus* (saguira). The thornscrub/desertscrub *S. alamosensis* (sina) and *S.*

thurberi are northwestern species in separate series of closely related species along the Pacific coast (Cornejo, 1994). Several Sonoran endemics are closely related to—and are likely derived from—columnar cacti in the Río Balsas Basin, Sierra Madre Sur, Trans-Volcanic Range, and Valle de Tehuacán, generally from Michoacán southeast to Oaxaca in south central Mexico. Species pairs include *C. gigantea/Neobuxbaumia* species, *L. schottii/P. marginatus, Myrtillocactus cochal/M. geometrizans,* and *P. pringlei* (sahueso, cardón)/*P. grandis.* Only the Baja California endemics *L. gatesii* and *S. eruca* (creeping devil) evolved from other Sonoran species, *L. schottii* (senita) and *S. gummosus* (pitahaya agria), respectively. At least for the columnar cacti, the floristic connections between the Sonoran Desert Region and tropical deciduous forests are with areas to the southeast, presumably from a time when the Sierra Madre Occidental and the Western Mexican Volcanic Belt were not such formidable geographic barriers. Cladistic theory aside, there is no reason to believe that the ancestors of the Sonoran Desert columnar cacti are extinct (Van Devender et al., 1992).

Paleoelevation

A new method of estimating paleoelevations has challenged the Miocene Revolution scenario developed above. Wolfe (1971) correlated leaf morphology in fossil floras with modern floras to estimate the history of climatic temperatures in the Tertiary. He has recently expanded his analyses to include regressions between foliar morphology and various temperature and precipitation parameters, and to provide a method of estimating the elevations of fossil floras at the time of deposition (Wolfe, 1993). In this approach, the physiological tolerances and limits of the living populations or closest relatives of fossil taxa are considered irrelevant or erroneous because extinct species might have had different environmental adaptations; the leaf morphology-climate relationships based on worldwide floras are thought to be better indicators of climate. The first studies using this methodology reached dramatically different paleoelevation estimates than previous studies based on floristic affinities. For example, MacGinitie (1953) inferred a paleoelevation of 915 m for the latest Eocene-early Oligocene (34–35 mya) Florissant Beds in Colorado based on a paleoflora closely allied with the highlands of northeastern Mexico. The flora was a mixture of plants now found in lowland and montane tropical areas and *Sequoia affinis.* In contrast, Gregory (1994), using Wolfe's (1993) multivariate climate analysis techniques, estimated that the Florissant Beds were at 2,300–3,300 m elevation; the site is now at 2,500 m. The climatic implications of the additional 1,385–2,385 m elevation in the late Eocene of the Rocky Mountains are profound: All of the ecological, evolutionary, and biogeographic changes in the biota discussed for the Oligocene–middle Miocene in the Axelrod model should have occurred earlier. It is difficult to accept this in light of tropical Eocene floras in the Arctic, the presence in the Florissant flora of such tropical plants as *Cedrela* (cedar), palms, and *Trichilia* (piocha), or the gradual modernization of the Rocky Mountain flora from the Oligocene to the early Miocene (Leopold and MacGinitie, 1972; Graham, 1993).

A reexamination of Gregory's (1994) study of the Florissant flora is enlightening. The living relatives of at least 34% of the 29 taxa used in the paleoclimatic analysis are restricted today to elevations and latitudes lower than Florissant. *Cedrela, Dodonaea* (hopbush), *Prosopis* (mesquite), and *Sapindus* (soapberry) are genera with tropical affinities whose extant species live in areas far to the south with higher mean annual temperatures than the upper elevations of the Rockies. Other tropical or Sonoran Desert genera previously reported from the Florissant include *Bursera* (torote, copal), *Cardiospermum* (balloon vine), and *Colubrina* (a rhamnaceous shrub; Leopold and MacGinitie, 1972; Graham, 1993). *Trichilia* in particular is a tropical genus in the Meliaceae that reaches its northern distributional limit at 27° north latitude in southern Sonora; both *T. americana* and *T. hirta* live in tropical deciduous forest below about 1,000 m elevation on the Sierra de Alamos. The inescapable conclusions of a paleoelevation of 2,300–3,300 m for the Florissant Beds are that a third of the flora had greater cold tolerances 35 mya than their living relatives, and that the cold-tolerant lineages did not survive the general cooling trend from the Eocene to the Miocene. If the Wolfe multivariate climate analyses of leaf floras systematically underestimate mean annual temperatures, then paleoelevations are overestimated, resulting in questionable landscape and paleoclimatic reconstructions. I feel that the important paleoecological signals from the plant taxa in these floras must be considered. For the present, the Miocene Revolution model of landscape evolution is the most useful.

EVOLUTION OF THE SONORAN DESERT

The Sonoran Desert itself is the result of a drying trend in the middle Miocene (8 to 15 mya; Axelrod, 1979). Some of the species in the new desertscrub communities such as *Carnegiea gigantea, Cercidium microphyllum, C. praecox, Fouquieria macdougalii, Guaiacum coulteri, Lophocereus schottii, Olneya tesota,* and *Stenocereus thurberi* were segregated out of thornscrub. However, *Carnegiea gigantea* is not closely related to the columnar cacti of the tropical deciduous forests of Sinaloa and Sonora (*Pachycereus pecten-aboriginum* or *Stenocereus montanus*) or adjacent thornscrub (*L. schottii* or *S. thurberi*) but was derived from the central Mexican *Neobuxbaumia.*

Another important chapter in the history of the Sonoran Desert is the evolution of Baja California, which was attached to the mainland in the Miocene. About twelve million years ago, the proto-Gulf of California opened and several large islands stocked with tropical plants and animals drifted in splendid isolation northwestward to meet California. The actual timing of the formation of the present Gulf of California along the San Andreas Fault, and the biological and evolutionary importance of the proto-Gulf, have been controversial, depending on interpretations of the geological record (Axelrod, 1979; Morafka et al., 1992; Grismer, 1994). Natural selection shaped the plants isolated on Baja California into many unique endemics, including *Fouquieria columnaris* (boojum tree, cirio), *Lophocereus gatesii,* and *Stenocereus eruca.*

Thus, the Sonoran Desert was in existence by the late Miocene (5–8 mya)—one of the youngest biotic communities of North America. The geologic events that shaped the landscape and altered regional climates occurred mostly in the early-middle Miocene, suggesting that the speciation of many of the prominent desert-scrub plants occurred prior to the formation of the Sonoran Desert.

Post-Miocene Environments

High Sea Levels and Iguanas

After the Miocene, changing climates and immigration rather than evolution were the major impacts on the biota, dramatically shifting species ranges and community compositions. During the Pliocene (1.8–5 mya), there was a reversal to more tropical climates.

During the latest Miocene–early Pliocene (centered on about 5 mya), sea level rose enough that the Gulf of California expanded into the Salton Trough of southeastern California to deposit the marine sediments of the Imperial Formation (see discussion and references in Spencer and Patchett, 1997). The extensive sediments of the roughly contemporaneous Bouse Formation in the Lower Colorado River Valley Basin in Arizona and California have been interpreted as estuarine, based on invertebrate fossils including barnacles or as freshwater lakes, based on strontium isotopic analyses.

In Anza Borrego Desert State Park in southern California, the Pliocene–early Pleistocene terrestrial sediments of the Palm Springs Formation overlie the marine Imperial Formation. A fossil lizard skull from the Pliocene (ca. 2.5–4.3 mya) Vallecito Creek local fauna was described as *Pumilia novaceki* (Norell, 1989), although it could have easily been placed in the extant *Iguana.* The modern *Iguana iguana* (green iguana) is a tropical lizard that today occurs no farther north than southern Sinaloa, about 1,500 km to the southeast of Vallecito Creek. Tropical species in the Sonoran Desert and much higher sea levels would indicate warmer global temperatures and ocean water, enhanced monsoonal summer rainfall, and the northward expansion of tropical thornscrub and deciduous forests.

Ice Age Environments

The warmth of the Pliocene ended with the advent of the Pleistocene, as the earth entered a new climatic era that far surpassed the middle Miocene in cool, continental conditions. Traditionally, four ice ages or glacial periods based on terrestrial sedimentary deposits were recognized in North America and were widely correlated with similar periods in Europe and South America. However, recent studies of isotopic climatic indicators in continuous sediment cores from the ocean floors document 15 to 20 glacial periods in the past 2.4 million years, with ice ages being about five to ten times longer than the 10,000- to 20,000-year interglacials (Imbrie and Imbrie, 1979).

In the last glacial period (the Wisconsin), the massive Laurentide ice sheet covered most of Canada and extended as far south as New York and Ohio. Boreal forest with *Picea spp.* (spruces) and *Pinus banksiana* (jack pine) moved southward, displacing mixed deciduous forest in much of the eastern United States (Delcourt and Delcourt, 1993). Glaciers covered the tops of the Rocky Mountains and the Sierra Nevada in the western United States and the Sierra Madre del Sur in south-central Mexico. Now-dry playa lakes in the Great Basin were full. Enough water was tied up in ice on land to lower sea level about 100 m.

Unlike paleoclimatic reconstructions for most of the southwestern United States and northwestern Mexico during the late Wisconsin (Thompson et al., 1993), tropical lowlands from the Yucatán Peninsula to Venezuela were apparently much drier than today (Leyden 1984; Leyden et al., 1994, and references therein). Water levels in lakes in the karst terrains of northern Guatemala were 30–40 m lower than today. The modern semideciduous tropical forests in areas with 900–1,600 mm/yr annual rainfall were replaced by more xeric, temperate forests. Dominants in pollen assemblages were pine, juniper, and temperate trees with few, if any, tropical species. However, considering that relatively few tropical trees can be detected in pollen analyses in modern tropical forests, the pollen spectra could have been overwhelmed by temperate, wind-pollinated forests on the modest uplands while the lowlands were dominated by tropical trees that left no pollen trace. The implications of greater aridity than today 12,000 years ago—the time that Porter (1989) suggested was typical of the entire Pleistocene—is that columnar cacti and nectar-feeding bats probably were more widely distributed across the now-wet tropical lowlands of Central America for much of the last two million years. This is of particular interest in the discussion of the taxonomic status of *Leptonycteris* in Mexico. Simmons and Wetterer (chapter 5) concluded that *L. curasoae* and *L. yerbabuenae* should be considered separate species rather than the latter a subspecies of the former because of non-overlapping ranges, minor morphological differences, and mitochondrial DNA evidence indicating that the clades diverged approximately 540,000 years ago. This conclusion might warrant reevaluation considering (1) the probability that the non-overlapping ranges is an ephemeral interglacial situation not typical of the last 540,000 years; (2) the minor morphological differences are found in different geographic ranges, and thus reflect environmental conditions; and (3) the difficulty of using evidence of molecular evolution (mitochondrial DNA with characters randomly mutating, independent of climate) to support taxonomic levels (morphological characters produced by natural selection in response to climate). The increasing weight placed on geographic ranges as taxonomic characters rather than any measure of phenetic divergence is questionable, considering the fossil evidence that distributions have changed so dramatically and so often (Van Devender et al., 1992).

Tropical Interglacials

The general environmental history of the Sonoran Desert region ranges from tropical deciduous forests in the early Tertiary to more temperate ice age woodlands and interglacial deserts in the Pleistocene. However, vertebrate fossils record that some interglacials were more tropical than the Holocene, suggesting that the Sonoran Desert and its columnar cacti may have extended farther north than they do today.

El Golfo, Sonora. The first fossil record of *Myrmecophaga tridactyla* (giant anteater) in North America was in early Pleistocene (Irvingtonian Land Mammal Age, beginning 1.8 mya) sediments from El Golfo de Santa Clara in northwestern Sonora (Shaw and McDonald, 1987; Lindsay, 1984). Today, the nearest populations of this large tropical mammal are 3,000 km to the southeast in the humid, tropical lowlands of Central America! As for many large mammals, its historical distribution may not accurately reflect the potential range because of human predation in the past 11,000 years (Martin, 1984). The El Golfo fauna included many extinct large mammals including antelope (*Tetrameryx* sp.), a bear (*Tremarctos* cf. *floridanus*), camels, cats, horses (*Equus* spp.), proboscidians, and a tapir (*Tapirus* sp.). Extant species in the fauna included *Bufo alvarius* (Sonoran Desert toad), *Trachemys scripta* (slider), *Boa constrictor* (boa constrictor), *Castor* cf. *C. californicus* (extinct beaver), and *Felis* cf. *F. onca* (jaguar). *Bufo alvarius* is a regional endemic, but *T. scripta* and *C. constrictor* occur today in Sonora in wetter, more tropical areas to the southeast. El Golfo is at the head of the Gulf of California in the Lower Colorado River Valley subdivision of the Sonoran Desert (Turner and Brown, 1982). Today, this hyperarid desert is too dry to support any of these animals, although historically the delta of the Colorado River was a very wet area with abundant beaver (*Castor canadensis*) and extensive *Populus fremontii* (cottonwood) gallery forests (Davis, 1982). This fauna is significant because it reflects an early Pleistocene interglacial with a climate much more tropical than is true today—frost-free, much greater rainfall in the warm season, and higher humidity.

Rancho La Brisca, Sonora. The late Pleistocene (Rancholabrean Land Mammal Age) Rancho La Brisca fossil locality is located in a riparian canyon north of Cucurpe, north-central Sonora (Van Devender et al., 1985b). The fauna was dominated by *Kinosternon sonoriense* (Sonoran mudturtle) and fish, reflecting a wet ciénega paleoenvironment. The presence of *Leptodactylus melanonotus* (sabinal frog) 240 kilometers north of its northernmost population on the Río Yaqui indicates that the climates were more tropical than today. The presence of a large extinct *Bison,* which only immigrated to North America from Siberia in the late Pleistocene between about 170,000 and 150,000 years ago (C. A. Repenning, pers. com., 1984) limits the age of the deposit. The coexistence of bison and tropical species helped establish that sediments near Rancho La Brisca in north-central Sonora were deposited about 80,000 years ago during the last interglacial (the Sangamon), which was considerably more tropical than the Holocene.

Ice Ages in the Desert

Plant remains in ancient packrat (*Neotoma* spp.) middens document the expansion of woodland trees and shrubs into desert elevations from 45,000 to 11,000 yr B.P. (radiocarbon years before 1950; Betancourt et al., 1990; Van Devender et al., 1987). Woodlands with *Pinus monophylla* (singleleaf pinyon), *Juniperus* spp. (junipers), *Quercus turbinella* (shrub live oak), and *Yucca brevifolia* (Joshua tree) were widespread in the present Arizona Upland Sonoran Desert (Van Devender, 1990b). Ice age climates with greater winter rainfall from Pacific and reduced summer monsoonal rainfall from the tropical oceans likely favored woody cool-season shrubs with northern affinities (Neilson, 1986) rather than the summer-rainfall trees, shrubs, and cacti of tropical forests and subtropical deserts. Warm desertscrub communities dominated by *Larrea divaricata* were restricted to below 300 m elevation in the Lower Colorado River Valley in the Sonoran Desert and to the southern Chihuahuan Desert (Van Devender, 1990a, 1990b).

Although *Carnegiea gigantea* and *Encelia farinosa* (brittlebush) returned to Arizona soon after the beginning of the present interglacial (the Holocene) about 11,000 years ago, Sonoran desertscrub did not form until about 9,000 years ago, when displaced woodland retreated upslope. However, relatively modern community composition was not achieved until *Cercidium microphyllum*, *Olneya tesota*, and *Stenocereus thurberi* arrived about 4,500 years ago. Similar successional stages likely occurred during each of 15 to 20 interglacials. Although the late Holocene desertscrub communities probably resembled the original late Miocene Sonoran Desert, relatively modern communities were only developed for about 5–10% of the 2.4 million years of the Pleistocene (Porter, 1989; Winograd et al., 1997); ice age woodlands were in desert lowlands for about 90% of this period.

Cactus Fossils

Cacti are mostly plants of seasonally dry tropical communities, semi-arid grasslands, and dry deserts, some of the most unlikely environments for the preservation of macrofossils or pollen. Considering their diversity and huge distribution in the New World, cacti have a very poor fossil record. Seeds and stems of cacti preserved in packrat middens are notable exceptions. Packrats and other mammals and birds readily eat the fruit and disperse the seeds to dry rockshelters, where the middens are preserved (see chapter 15). Packrats readily eat cacti, one of their principal sources of water in dry, desert habitats. Cacti are reasonably common fossils in Sonoran Desert middens, especially *Echinocereus* sp. (hedgehog cactus), *Ferocactus cylindraceus* (California barrel cactus), *Mammillaria grahamii* (fishhook cactus), *Opuntia acanthocarpa* (buckhorn cholla), *O. chlorotica* (pancake cactus), and *O. phaeacantha* (variable prickly pear). *Echinocactus polycephalus* (many-headed barrel cactus) is present in two Wisconsin (15,680 and >37,000 yr B.P.) samples from the Tinajas Altas Mountains, and four Holocene (3,820 to 10,360 yr B.P.) samples from the Butler Mountains. Various other species have been encountered as well, especially in the younger Holocene samples. A few seeds of *Echinocactus horizonthalonius* var. *nicholii* (a restricted Sonoran Desert variety of a widespread Chihuahuan Desert species) found with *Pinus monophylla* needles and *Artemisia tridentata*-type leaves in middens from the Waterman Mountains (dated at 11,470 and 22,380 yr B.P.) are possible examples of past associations that no longer occur today (Van Devender, 1990b).

The abundances of plant macrofossils in packrat middens are often expressed on a relative abundance scale, with 1 = a single specimen and 5 = the most common taxon; other species are ranked between these two extremes (2, 3, 4; Van Devender, 1990b). Tiny columnar cactus seeds with relative abundances of 1 or 2 (two to ca. ten seeds) could be younger contaminants adhering to the surfaces of older samples. In some cases, radiocarbon dates using an accelerator mass spectrometer (AMS) can be run on the seeds themselves. For example, *Carnegiea gigantea* seeds were found in a middle Wisconsin midden (actually an *Erethizon dorsatum* [porcupine] deposit) from the Castle Mountains of Arizona dated at 25,210 yr B.P., apparently associated with *Pinus monophylla, Juniperus* sp., and *Quercus turbinella.* An AMS date directly on the seeds themselves yielded a middle Holocene age of 5,630 yr B.P. (Van Devender et al., 1985a). Other AMS dates on *C. gigantea* seeds are 2,410 yr B.P. from the Tucson Mountains and 8,590 and 9,230 yr B.P. from the Ajo Mountains, both in Arizona. Probable contaminants are indicated in Table 1.1.

Only *Carnegiea gigantea* has an extensive midden record (Table 1.1). *Lophocereus schottii* and *Pachycereus pringlei* were only found in a 320-yr B.P. sample from the Sierra Bacha on the coast of the Gulf of California near Puerto Libertad, Sonora (Van Devender et al., 1994a). The absence of seeds from eight other samples dating as far back as 9,970 yr B.P. is probably insignificant, because pollen of *Carnegiea*-type cacti (includes *Carnegiea, Lophocereus,* and *Pachycereus*) was identified in them (Anderson and Van Devender, 1995). Additional records of *P. pringlei* may be found in middens from Cataviña and San Fernando in the modern *Fouquieria columnaris-P. pringlei* desertscrub in the northern Viscaino subdivision in Baja California when they are analyzed (Lanner and Van Devender 1998; Peñalba and Van Devender, 1998).

Excluding probable contaminants in a 7,580-yr B.P. sample, *Stenocereus thurberi* apparently arrived in the Puerto Blanco Mountains (in Organ Pipe Cactus National Monument, Arizona) at the beginning of the late Holocene (after 5,240 and before 3,480 yr B.P.; Van Devender, 1990b). It was more common than today at the midden sites dated at 980, 990, 3,220, and 3,400 yr B.P.

In contrast, *Carnegiea gigantea* seeds have been identified in middens from 14 different study areas in Arizona and Sonora (Table 1.1). For the most part, *C. gigantea* was absent from the middle and late Wisconsin woodlands in the modern Sonoran Desert in Arizona and California, although probable contaminant seeds were found in the Ajo (14,500 to 32,000 yr B.P., five samples), Kofa (13,400 yr B.P.), Waterman

TABLE 1.1

Fossil Records of Columnar Cacti in Sonoran Desert Packrat Middens[a]

Species	*Location*	*Elevation (m)*	*Dates and Abundances (B.P.)[b]*	*Source*
Carnegiea gigantea	Tucson Mountain, Pima County, Az. (32° N)	700–800	2,410 (3, AMS)[c,d], 3,550 (5), 3,940 (1), 7,985 (3), *12,430,* 21,000 (1)	Rondeau et al. (1996)
	Picacho Peak, Pima County, Az. (32°38′ N)	655	*9,420, 10,280,* 11,100 (1), *13,170*	Van Devender et al. (1991)
	Waterman Mountains, Pima County, Az. (32°21′ N)	975	1,200 (3), 1,320 (4), 2,600 (5), 3,880 (3), 5,190 (5), 6,195 (5), 8,310 (2), 8,910 (2), 9,920 (2), *9,920, 11,470, 11,510,* *12,530 (2), *12,690, 19,270,* *22,380 (2), *22,450*	Anderson and Van Devender (1991)
	Wolcott Peak, Pima County, Az. (32°27′ N)	860	5,020 (2), 5,350 (5), *12,130 (2)	Van Devender and Weins (1993)
	Castle Mountains, Pima County, Az. (32°24′ N)	790	1,300 (3), 1,670 (5), 5,630 (2; AMS), 6,440 (2), 8,720 (3) 8,990 (2)	
	Ajo Mountains, Pima County, Az. (32°05′ N)	915–975	1,150 (2), 8,130 (3), 8,590 (3; AMS), 9,230 (2; AMS), *13,500,* *14,500 (1), *17,830,* *20,490 (2), *21,840 (1), *29,110 (1), *32,000 (2)	
	Puerto Blanco Mountains, Pima County, Az. (31°58′ N)	535–605	30 2(2), 130 (2), 980 (4), 990 (3), 1,910 (3), 2,160 (3), 2,340 (3), 3,220 (2), 3,400 (2), 3,440 (4), 3,480 (5), 5,240 (5), 7,560 (3), 7,580 (5), 7,970 (4), 8,790 (2), 9,070 (2), 9,860 (2), 10,540 (2), *14,120*	Van Devender (1990b)
	Tinajas Altas Mountains, Yuma County, Az. (32°19′ N)	365–580	1,230 (2), 4,010 (2), 5,080 (1), *5,860, 5,940,* 7,860 (3), 8,255 (2), *8,910,* 8,970 (3), *9,230 to >43,200* (14 samples)	
	Butler Mountains, Yuma County, Az. (32°21′ N)	250	*740,* 3,820 (1), 8,160 (2), *8,570, 10,360,* 11,060 (1), 11,250 (2)	

	Wellton Hills, Yuma County, Az. (32°36′ N)	160–180	*3,520, 8,150, 8,750, 10,750*	
	Harquahala Mountains, Maricopa County, Az. (33°53′ N)	800–825	1,595 (1), 1,885 (3), 2,980 (2), 3,340 (2), 4,040 (5), 4,540 (3), 5,130 (3), 5,335 (5), *6,425,* 10,440 (2), *22,140*	McAuliffe and Van Devender (1998)
	Kofa Mountains, La Paz County, Az. (33°26′ N)	550	3,620 (1), 9,750 (1), 11,450 (1)	
	New Water Mountains, La Paz County, Az. (33°36′ N)	605–615	*10,880, 11,000, 12,090*	
	Hornaday Mountains, Sonora (33°59′ N)	240–260	*1,720,* 1,850 (3), 1,930 (2), 2,320 (5), 4,430 (5), 6,065 (5), 8,660 (2), 8,910 (3), 9,370 (3), *10,000*	Van Devender et al. (1990)
	Picacho Peak, San Bernardino County, Calif. (32°58′ N)	240–300	*modern to 12,730* (21 samples)	Cole (1986)
	Whipple Mountain, San Bernardino County, Calif. (34°16′ N)	490–525	*470 to 13,810* (25 samples)	Van Devender (1990b)
Lophocereus schottii	Sierra Bacha, Sonora (29°50′ N)	100	320 (2), *2,330 to 9,970* (8 samples)	Van Devender et al. (1994a)
Pachycereus pringlei	Sierra Bacha, Sonora (29°50′ N)	100	320 (2), *2,330 to 9,970* (8 samples)	Van Devender et al. (1994a)
Stenocereus thurberi	Puerto Blanco Mountains, Pima County, Az. (31°58′ N)	535–605	30 (3), 130 (3), 980 (4), 990 (5), 1,910 (2), 2,160 (2), 2,340 (1), 3,220 (3), 3,400 (4), 3,440 (2), 3,480 (1), *5,240, 7,560,* *7,580 (2), *7,970, 8,790, 9,070, 9,860, 10,540, 14,120*	Van Devender (1990b)

Note: Time periods: middle Wisconsin: ca. 22,000 to > 43,200 yr B.P.; late Wisconsin = 11,000 to 22,000 yr B.P.; early Holocene = 8,900 to 11,00 yr B.P.; middle Holocene = 4,000 to 8,900 yr B.P.; late Holocene = present to 4,000 yr B.P.

[a] Sites arranged from east to west.

[b] Dates in radiocarbon years before 1950. * = possible contaminant. Dates in italics for samples without columnar cacti.

[c] AMS = accelerator mass spectrometer date on cactus seeds.

[d] Numbers in parentheses are internal relative abundances: 1 = a single sepcimen; 5 = the most common taxon; 2, 3, and 4 = varying numbers depending on the total number of specimens of a taxon in the sample.

mountains (12,130 and 22,380 yr B.P.), and Wolcott Peak (12,130 yr B.P.). All of these are represented by so few seeds (one to five) that an AMS radiocarbon date is not possible. The ice age "refugium," or relictual area, for *C. gigantea* was likely in areas of central and southern Sonora that continued to have sufficient summer rainfall. Unfortunately, with the exception of a single middle Wisconsin sample (>33,000 yr B.P.) reported to contain *Juniperus* sp. only from near Hermosillo (Wells and Hunziker, 1976), ancient middens from this area have not been found.

Carnegiea gigantea apparently migrated into the northern part of its present range in the early Holocene, soon after the end of the Wisconsin glacial period (about 11,000 yr B.P.) (Van Devender, 1990b). The Puerto Blanco Mountains record is especially complete, with *C. gigantea* present in 19 samples from 30 to 10,540 yr B.P. *C. gigantea* was present but uncommon with *Juniperus californica* (California juniper) in transitional early Holocene (9,860–10,540 yr B.P.) communities. The Holocene record of *C. gigantea* is reasonably good in sites in the modern Arizona Upland, less so in Lower Colorado River Valley sites, and absent altogether in sites at the lowest elevations (Wellton Hills, 160–180 m elevation) and near its modern western distributional limits (Picacho Peak and the Whipple Mountains, California; Table 1.1). In a number of cases, *C. gigantea* was more common than today near the sites, especially in the middle Holocene of the Waterman Mountains (5,190 and 6,195 yr B.P.), Wolcott Peak (5,350 yr B.P.), and the Puerto Blanco (5,240, 7,560, 7,580, and 7,970 yr B.P.) and Harquahala (4,040, 4,540, 5,130, and 5,335 yr B.P.) mountains. *C. gigantea* was also more common at times in the late Holocene in the Tucson (2,410 yr B.P.), Waterman (1,320 and 2,600 yr B.P.), Castle (1,670 yr B.P.), Puerto Blanco (980, 3,440, and 3,480 yr B.P.), and Hornaday (990, 3,220, and 3,400 yr B.P.) mountains. In the Puerto Blanco Mountains sequence, the relative abundances of *C. gigantea* appeared to be at times inversely related to those of *Stenocereus thurberi*. For example, *C. gigantea* dominance in the 3,480- and 3,440-yr B.P. samples was reversed in the 3,400-yr B.P. sample. Interestingly the two columnars were codominants in samples dated at 980 and 990 yr B.P., a brief wet period when the desert Hohokam culture thrived as well (Van Devender, 1990b).

Thus the midden fossils document that *Carnegiea gigantea* was able to expand its range into Arizona quickly at the end of the Wisconsin glacial and that it was more common at times in the Holocene than it is today. Although similar range expansions likely occurred in each of the 15 or 20 interglacials, its range was contracted into a relatively small area for most of the Pleistocene.

The Baja California Connection

The Central Gulf Coast subdivision of the Sonoran Desert occurs along the coasts of the Gulf of California in Baja California and Sonora. To the north, many of the Central Gulf Coast species are limited by the hyperarid climates of the Lower Colorado

River Valley in northeastern Baja California and northwestern Sonora. Southern Sonora shares many coastal thornscrub species with Baja California. Of particular interest are disjunct populations of typical Baja California plants on the Sonoran coast, especially between Guaymas and Puerto Libertad. Notable examples include *Ambrosia chenopodifolia, A. camphorata, A. divaricata, A. magdalenae* (bursages), *Bursera hindsiana* (torote prieto), *Ebanopsis confinis* (ejotón), *Fouquieria columnaris, F. diguetii* (palo ádan), *Lysiloma candidum* (palo blanco), *Pachycereus pringlei, Pedilanthus macrocarpus* (candelilla), *Ruellia californica, R. peninsularis* (rama parda), *Senna polyantha, Stenocereus gummosus, Viguiera laciniata, V. microphylla* (goldeneyes), and *Viscainoa geniculata* (guayacán; Turner et al., 1995). These distributions may have come about in a number of ways.

The simplest explanation of the distributions is that they are vicariant populations whose ranges were split by the formation of the Gulf of California without dispersal. This seems unlikely, as the species mostly occur in winter-rainfall climates in Baja California, and probably evolved there in isolation. Although little studied, the Sonoran isolates do not appear to be much differentiated. Hamrick et al. (chapter 6) did find that Sonoran populations of *Lophocereus schottii, Pachycereus pringlei,* and *Stenocereus thurberi* had much less genetic diversity than Baja populations, indicating likely derivation from Baja California populations and shorter isolation times.

Another possibility is that the plants dispersed around the head of the Gulf of California. For summer-rainfall species, climatic conditions for circum-Gulf dispersal were likely reasonable during warm periods in the Pliocene or wetter Pleistocene interglacials equivalent to those recorded in the El Golfo (1.8 mya) or Rancho La Brisca (80,000 yr B.P.) deposits. High sea levels and the extensive wetlands of the Colorado River delta would presumably have been a barrier between Baja and Sonora. If late Holocene climates are any indication, hyperaridity would have made circum-Gulf dispersals difficult during drier portions of interglacials.

During Pleistocene glacials, sea level was 100 m or more lower than it is today (Bloom, 1983), expanding the lowlands around the head of the Gulf. In the Sierra Bacha, middens record that the early Holocene (9,270–9,970 yr B.P.) climates had greater winter rainfall and cooler summers (Van Devender et al., 1994a). *Fouquieria columnaris* was more common and widespread. Presumably, the climates of the late Wisconsin and earlier Pleistocene glacials along the coast of Sonora were even more like those of central Baja California. A contemporaneous early Holocene (8,910–10,000 yr B.P.) record from the Hornaday Mountains of the Pinacate Region of northwestern Sonora also indicates greater winter precipitation and cooler summers. The paleovegetation was a simple *Larrea divaricata*-dominated desertscrub with *Carnegiea gigantea, Encelia farinosa,* and rare *Juniperus californica* (Van Devender et al., 1990). At Cataviña and San Fernando, Baja California, preliminary midden analyses record pinyon-juniper woodland/chaparral in the late Wisconsin and juniper chaparral in the early Holocene instead of the modern *F. columnaris* desertscrub (Lanner and Van Devender, 1998; Peñalba and Van Devender, 1998).

Most of the records of changing distributions of plants in the late Wisconsin in southwestern Arizona and southeastern California were species living at lower elevation and latitude, reflecting the glacial winter rainfall climates (Van Devender, 1990b). Packrat middens did record late Wisconsin expansions of some desert succulents in the Lower Colorado River Valley between Baja California and Arizona, including *Agave deserti* (desert agave), *Ferocactus cylindraceus, Opuntia acanthocarpa,* and *Yucca whipplei* (our lord's candle; Van Devender, 1990b); all of these plants have moderate cold tolerances. I think that northward migrations of the warm-xeric adapted plants (especially those that live in summer rainfall regimes) in Baja California during Pleistocene glacials were unlikely.

Finally, the most likely—and most difficult to study—explanation of the Baja California-Sonora disjunct plant distribution is the long-distance transport of seeds by birds or wind across the Gulf. For some, the midriff islands in the Gulf of California between Bahía Los Angeles, Baja California, and Bahía de Kino, Sonora, may have served as stepping stones.

Summary

The transition from temperate biomes to the New World tropics in northwestern Mexico occurs between 28° and 31° north latitude in Sonora. The Sonoran Desert is the most tropical North American Desert, although most adaptations to extreme heat, light, and aridity in modern desert plants evolved in early Tertiary tropical deciduous forests beginning in the Eocene (35–54 mya). The floristic affinities of the mainland Sonoran Desert are strongest with thornscrub in Sonora and Baja California and are much less so with the tropical deciduous forest of southern Sonora. Thornscrub is the climatic, physical, and evolutionary connection between Sonoran Desert and older tropical deciduous forests.

The closest relatives of many columnar cacti in the Sonoran Desert Region are in tropical deciduous forest in the Balsas Basin, Sierra Madre Sur, and/or Valle de Tehuacán south and east of the Sierra Madre Occidental. Massive volcanic ash deposits in the late Oligocene–middle Miocene (15–30 mya) were elevated to great heights in the Sierra Madre Occidental, severing continuous distributions, and isolating the proto-Sonoran Desert to the northwest. Tropical deciduous forests were restricted to the coastal lowlands of western Mexico as environmentally limited biotic communities (grassland, conifer forest, tundra) formed along elevational gradients. In the early-middle Miocene (15–25 mya), thornscrub likely was the regional vegetation in Sonora.

The Sonoran Desert apparently formed in response to a drying trend in the late Miocene (8–15 mya), although most of the desert species probably evolved earlier in thornscrub or tropical deciduous forest. The separation from the mainland and northward movement in isolation of Baja California was an important evolutionary event in the late Miocene (ca. 5–12 mya). From the latest Miocene to the early Pleistocene, marine sediments in southern California and fossils of tropical animals in northwestern Sonora reflected reversions to more tropical climates.

During the last Pleistocene glacial (the Wisconsin), plant remains in ancient packrat middens document woodlands with *Pinus monophylla, Juniperus* spp., *Quercus turbinella,* and *Yucca brevifolia* in the present Arizona Upland subdivision of the Sonoran Desert before 11,000 years ago. Ice age climates with greater winter rainfall from the Pacific Ocean but reduced summer monsoonal rainfall from the tropical oceans were typical of 10–20% of the past 2.4 million years. Desertscrub communities dominated by *Larrea divaricata* were restricted to below 300 m elevation in the Lower Colorado River Valley of Arizona, California, Baja California, and Sonora.

Although *Carnegiea gigantea* and *Encelia farinosa* returned to Arizona from Sonora soon after the beginning of the present interglacial (the Holocene) about 11,000 years ago, Sonoran desertscrub did not form until about 9,000 years ago with the disappearance of the last woodland plants. At times in the middle Holocene (4,040–7,970 yr B.P., eleven samples from four sites), *C. gigantea* was more common than it is today in southwestern Arizona and northwestern Sonora. Relatively modern desertscrub compositions were not achieved until about 4,500 years ago, with the arrivals of *Cercidium microphyllum, Olneya tesota,* and *Stenocereus thurberi.* Relatively modern desertscrub and distributions of Sonoran Desert columnar cacti likely occurred in each of 15–20 interglacials, representing only about 5–10% of the Pleistocene.

The many species of desert plants found in both Baja California and Sonora reflect long histories of dispersals. Common distributional patterns in plants are (1) tropical species whose ranges extend southward into summer-rainfall thornscrub, (2) species found at low elevations on both shores in the Central Gulf Coast subdivision, and (3) winter-rainfall species whose ranges are mostly from central Baja California northward with limited disjunct populations in Sonora. Summer-rainfall species may have dispersed around the head of the Gulf of California during warm, tropical climates in the Pliocene or Pleistocene interglacials. Seeds of winter-rainfall species were likely dispersed across the Gulf by birds or storm winds.

Resumen

La transición desde los biomas templados hasta los trópicos del nuevo mundo, en el noroeste de México ocurre en el estado de Sonora, entre los 31° y los 28° de latitud norte. El Desierto Sonorense, es el desierto norteamericano más tropical, aunque la mayoría de las adaptaciones al calor extremo, luz y aridez por las plantas modernas del desierto evolucionaron en los bosques tropicales deciduos (selva baja caducifolia) del Terciario temprano, empezando en el Eoceno (35–54 mya [millones de años]). Las afinidades florísticas del Desierto Sonorense continental son más fuertes con el matorral espinoso en Sonora y Baja California pero mucho menos con el bosque tropical deciduo del sur de Sonora. El matorral espinoso es la conexión climática, física y evolutiva, entre el Desierto Sonorense y los bosques tropicales decíduos más antiguos.

Los parientes más cercanos de muchos cactus columnares en la región del Desierto Sonorense están en los bosques tropicales decíduos de la Cuenca del Balsas, en la Sierra Madre del Sur, y/o Valle de Tehuacán al sur y este de la Sierra Madre Occidental. Los depósitos macizos de cenizas volcánicas en el Oligoceno tardío-Mioceno medio (15–30 mya), fueron elevados a gran altura en la Sierra Madre Occidental, separando así las distribuciones contínuas, y aislando el proto-Desierto Sonorense al noroeste. Los Bosques tropicales decíduos fueron restringidos a las tierras bajas de la costa del oeste de México, como comunidades bióticas limitadas ambientalmente (pastizales, bosques de coníferas, y tundra) a lo largo de gradientes de altitud. En el Mioceno temprano-medio (15–25 mya), el matorral espinoso probablemente fue la vegetación regional en Sonora.

El Desierto Sonorense aparentemente se formó en respuesta a una tendencia de sequía en el Mioceno tardío (15 a 8 mya), sin embargo, la mayoría de las especies del desierto probablemente evolucionaron más temprano en el matorral espinoso ó el bosque tropical deciduo. La separación del continente y el movimiento hacia el norte, en el aislamiento de Baja California fue un evento evolutivo importante en el Mioceno tardío (ca. 5–12 mya). Desde el Mioceno tardío al Pleistoceno temprano sedimentos marinos al sur de California y fósiles de animales tropicales en el noroeste de Sonora reflejan inversión a climas más tropicales.

Durante el último glacial del Pleistoceno (el Wisconsiniano), restos de plantas en antiguos depósitos de *Neotoma* spp. documentan bosques con *Pinus monophylla, Juniperus* spp., *Quercus turbinella,* y *Yucca brevifolia,* en la actual subdivisión Arizona Upland del Desierto Sonorense anterior a 11,000 años. Los climas de la edad de hielo, con la mayor lluvia de invierno del Océano Pacífico pero con una reducida lluvia de los monsones del verano, que son provenientes de los océanos tropicales, fueron típicos de 10–20% de los últimos 2.4 millones de años. Las comunidades de matorral del desierto dominadas, por *Larrea divaricata* fueron restringidas, a menos de 300 m de elevación en el Lower Colorado River Valley de Arizona, California, Baja California y Sonora.

Aunque *Carnegiea gigantea* y *Encelia farinosa* regresaron a Arizona de Sonora, poco después del inicio del presente interglaciar (el Holoceno) hace cerca de 11,000 años, el matorral del desierto sonorense no se formó hasta hace cerca de 9,000 años con la desaparición de los bosques. En tiempos del Holoceno medio (4,040–7,970 años B.P., 11 muestras de cuatro lugares), *C. gigantea* fue más común, deloque es hoy al suroeste de Arizona y noroeste de Sonora. Composiciones de matorral del desierto relativamente modernas, no fueron logradas hasta hace cerca de 4,500 años, con la llegada de *Cercidium microphyllum, Olneya tesota,* y *Stenocereus thurberi.* El matorral del desierto relativamente moderno y las distribuciones de cactus columnares del Desierto Sonorense probablemente ocurrieron en cada uno de los 15–20 interglaciales, representando solamente cerca de 5–10% del Pleistoceno.

El gran número de especies de plantas del desierto encontradas en Baja California y Sonora, reflejan la larga historia de dispersiones. Los patrones comúnes de distribución en plantas son (1) especies tropicales cuyos rangos se extienden hacia el sur dentro del matorral espinoso con lluvia de verano; (2) especies encontradas en elevaciones bajas en ambas costas de la subdivisión de la costa central del golfo; y (3) especies de lluvia de invierno cuyos rangos son esencialmente del centro de Baja California hacia el norte, con limitadas poblaciones disyuntas en Sonora. Las especies de lluvias de verano pueden haber sido dispersadas alrededor del alto Golfo de California durante los climas cálidos tropicales en el Plioceno ó un interglacial del Pleistoceno. Las semillas de especies de las lluvias del invierno fueron probablemente dispersadas al otro lado del Golfo, por las aves ó los vientos de las tormentas.

ACKNOWLEDGMENTS

Paul S. Martin sparked my interest in the "secrets of the past," ancient packrat middens, and collecting plants. Everett Lindsay provided the literature on the fascinating early Tertiary vertebrate fossils from high latitudes. Charles H. Lowe shared with me his understanding and appreciation for the natural history of the Southwest, the importance of environmental gradients, and stimulated my interests in evolution and historical biogeography. Ana Lilia Reina G. translated the Summary into Spanish.

REFERENCES

Anderson, R. S., and T. R. Van Devender. 1991. Comparison of pollen and macrofossils in packrat (*Neotoma*) middens: a chronological sequence from the Waterman Mountains of southern Arizona, U.S.A. *Review of Palaeobotany and Palynology* 68:1–28.

———. 1995. Vegetation history and paleoclimates of the coastal lowlands of Sonora, Mexico—pollen records from packrat middens. *Journal of Arid Environments* 30:295–306.

Axelrod, D. I. 1979. Age and origin of the Sonoran Desert. *California Academy of Sciences Occasional Paper* 132:1–74.

Axelrod, D. I., and H. P. Bailey. 1969. Paleotemperature analysis of Tertiary floras. *Palaeogeography, Palaeoclimatology, Palaeoecology* 6:163–95.

Betancourt, J. L., T. R. Van Devender, and P. S. Martin, eds. 1990. *Packrat Middens. The Last 40,000 Years of Biotic Change.* Tucson: University of Arizona Press.

Bloom, A. L. 1983. Sea level and coastal changes. In *Late Quaternary of the United States: the Holocene,* ed. H. E. Wright, Jr., 42–51. Minneapolis: University of Minnesota Press.

Búrquez, M. A., Y. A. Martínez, R. S. Felger, and D. Yetman. 1999. Vegetation and habitat diversity at the southern edge of the Sonoran Desert. In *Ecology and Conservation of Sonoran Desert Plants: a Tribute to the Desert Laboratory,* ed. R. H. Robichaux, 36–67. Tucson: University of Arizona Press.

Cochemé, J. J., and A. Demant. 1991. Geology of the Yécora area, north Sierra Madre, northern Sierra Madre Occidental, Mexico. *Geological Society of America Special Paper* 254:81–94.

Cole, K. L. 1986. The Lower Colorado Valley: a Pleistocene desert. *Quaternary Research* 25: 392–400.

Cornejo, D. O. 1994. Morphological evolution and biogeography of Mexican columnar cacti, Tribe Pachycereeae, Cactaceae. Ph.D. diss., University of Texas, Austin.

Davis, G. P., Jr. 1982. *Man and Wildlife in Arizona: the American Exploration Period 1824–1865.* Phoenix: Arizona Game and Fish Department.

Dawson, M. R., R. W. West, W. Langston, and J. H. Hutchinson. 1976. Paleogene terrestrial vertebrates: northernmost occurrence, Ellesmere Island, Canada. *Science* 192:781–82.

Delcourt, P. A., and H. R. Delcourt. 1993. Paleoclimates, paleovegetation, and paleofloras during the late Quaternary. In *Flora of North America North of Mexico,* ed. Flora of North America Editorial Committee, 71–94. New York: Oxford University Press.

Dilcher, D. L. 1973. A paleoclimatic interpretation of the Eocene floras of southeastern North America. In *Vegetation and Vegetation History of Northern Latin America,* ed. A. Graham, 39–59. Amsterdam: Elsevier.

Estes, R., and J. H. Hutchison. 1980. Eocene lower vertebrates from Ellesmere Island, Canadian Arctic Archipelago. *Palaeogeography, Palaeoclimatology, Palaeoecology* 30:325–47.

Gibson, A. C., and K. E. Horak. 1978. Systematic anatomy and phylogeny of Mexican columnar cacti. *Annals of the Missouri Botanical Garden* 65:999–1057.

Graham, A. 1993. History of the vegetation: Cretaceous (Maastrichian)-Tertiary. In *Flora of North America North of Mexico,* ed. Flora of North America Editorial Committee, 57–70. New York: Oxford University Press.

Graham, A., and D. Dilcher. 1995. The Cenozoic record of tropical dry forests in northern Latin America and the southern United States. In *Seasonally Dry Tropical Forests,* eds. S. H. Bullock, H. A. Mooney, and E. Medina, 124–45. New York: Cambridge University Press.

Gregory, K. M. 1994. Palaeoclimate and palaeoelevation of the 35 ma Florissant flora, Front Range, Colorado. *Palaeoclimates* 1:23–57.

Grismer, L. L. 1994. The origin and evolution of the peninsular herpetofauna of Baja California. *Herpetological Review* 2:51–106.

Henry, C. D., and J. J. Aranda. 1992. The real southern Basin and Range: mid- to late Cenozoic extension in Mexico. *Geology* 20:701–4.

Hickey, L. J., R. M. West, M. R. Dawson, and D. K. Choi. 1983. Arctic terrestrial biota: paleomagnetic evidence of age disparity with mid-northern latitudes during the late Cretaceous and early Tertiary. *Science* 221:1153–56.

Imbrie, J., and K. P. Imbrie. 1979. *Ice Ages, Solving the Mystery.* Hillside, N.J.: Enslow.

Lanner, R. M., and T. R. Van Devender. 1998. The recent history of pines in the American Southwest. In *The Ecology and Biogeography of Pinus,* ed. D. M. Richardson, 171–82. Cambridge: Cambridge University Press.

Leopold, E. B., and H. D. MacGinitie. 1972. Development and affinities of Tertiary floras in the Rocky Mountains. In *Floristics and Paleofloristics of Asia and Eastern North America,* ed. A. Graham, 147–200. Amsterdam: Elsevier.

Leyden, B. W. 1984. Guatemalan forest synthesis after Pleistocene aridity. *Proceedings of the National Academy of Sciences U.S.A.* 81:4856–59.

Leyden, B. W., M. Brenner, D. A. Hodell, and J. H. Curtis. 1994. Orbital and internal forcing of climate on the Yucatán Peninsula for the past 36 ka. *Palaeogeography, Palaeoclimatology, Palaeoecology* 109:193–210.

Lindsay, E. H. 1984. Late Cenozoic mammals from northwestern Mexico. *Journal of Vertebrate Paleontology* 4:208–15.

McAuliffe, J. R., and T. R. Van Devender. 1998. A 22,000-year record of vegetation in north-central Sonoran Desert. *Palaeogeography, Palaeoclimatology, Palaeoecology* 141:253–75.

MacGinitie, H. D. 1953. *Fossil plants of the Florissant Beds, Colorado.* Carnegie Institution of Washington Publication No. 599.

Martin, P. S. 1984. Pleistocene overkill: the global model. In *Pleistocene Extinctions,* eds. P. S. Martin and R. Klein, 354–403. Tucson: University of Arizona Press.

Menges, C. M., and P. A. Pearthree. 1989. Late Cenozoic tectonism in central Sonora and its impact on regional landscape evolution. *Arizona Geological Society Digest* 17:649–80.

Morafka, D. J., G. A. Adest, L. M. Reyes, L. G. Aguirre, and S. S. Lieberman. 1992. Differentiation of North American Deserts: a phylogenetic evaluation of a vicariance model. *Tulane Studies in Zoology and Botany, Supplementary Publication* 1:195–226.

Neilson, R. P. 1986. High-resolution climatic analysis and southwest biogeography. *Science* 232:27–34.

Norell, M. A. 1989. Late Cenozoic lizards of the Anza Borrego Desert, California. *Natural History Museum of Los Angeles County Contributions in Science* 414:1–31.

Peñalba, M. C., and T. R. Van Devender. 1998. Cambios de vegetación y clima en Baja California, México, durante los ultimos 20,000 años. *Geología del Noroeste* 2:21–23.

Porter, S. C. 1989. Some geological implications of average Quaternary glacial conditions. *Quaternary Research* 32:245–61.

Roldán, J., and K. F. Clark. 1992. An overview of the geology and mineral deposits of the northern Sierra Madre Occidental and adjacent areas. In *Geology and Mineral Resources of the Northern Sierra Madre Occidental, Mexico,* eds. K. F. Clark, J. Roldán, and R. H. Schmidt, 39–65. El Paso: El Paso Geological Society.

Rondeau, R. J., T. R. Van Devender, C. D. Bertelsen, P. Jenkins, R. K. Wilson, and M. A. Dimmitt. 1996. Annotated flora and vegetation of the Tucson Mountains, Pima County, Arizona. *Desert Plants* 12:3–46.

Shaw, C. A., and H. G. McDonald. 1987. First record of giant anteater (Xenartha, Myrmecophagidae) in North America. *Science* 26:186–88.

Shreve, F. 1964. Vegetation of the Sonoran Desert. In *Vegetation and Flora of the Sonoran Desert,* eds. F. Shreve and I. L. Wiggins, 9–186. Stanford: Stanford University Press.

Spencer, J. E., and P. J. Patchett. 1997. Sr isotope evidence for a lacustrine origin for the upper Miocene to Pliocene Bouse Formation, lower Colorado River trough, and implications uplift. *Geological Society of America Bulletin* 109:767–78.

Swanson, E., and D. Wark. 1988. Mid-Tertiary sialic volcanism in Chihuahua, Mexico. In *Stratigraphy, Tectonics and Resources of Parts of the Sierra Madre Occidental Province, Mexico,* eds. K. F. Clark, P. C. Goodell, and J. M. Hoffer, 229–39. El Paso: El Paso Geological Society.

Thompson, R. S., C. Whitlock, P. J. Bartlein, S. P. Harrison, and W. G. Spaulding. 1993. In *Global Climates since the Last Glacial Maximum,* eds. H. E. Wright, Jr., J. E. Kutzbach, T. Webb III, W. F. Ruddiman, F. A. Street-Perrott, and P. J. Bartlein, 468–513. Minneapolis: University of Minnesota Press.

Tiffany, B. H. 1985. Geological factors and the evolution of plants. In *Geological Factors and the Evolution of Plants,* ed. B. H. Tiffany, 1–10. New Haven: Yale University Press.

Turner, R. M., J. E. Bowers, and T. L. Burgess. 1995. *Sonoran Desert Plants. An Ecological Atlas.* Tucson: University of Arizona Press.

Turner, R. M., and D. E. Brown. 1982. Sonoran desertscrub. *Desert Plants* 4:121–81.

Van Devender, T. R. 1990a. Late Quaternary vegetation and climate of the Chihuahuan Desert, United States and Mexico. In *Packrat middens. The Last 40,000 Years of Biotic Change,* eds. J. L. Betancourt, T. R. Van Devender, and P. S. Martin, 104–33. Tucson: University of Arizona Press.

———. 1990b. Late Quaternary vegetation and climate of the Sonoran Desert, United States and Mexico. In *Packrat middens. The Last 40,000 Years of Biotic Change,* eds. J. L. Betancourt, T. R. Van Devender, and P. S. Martin, 134–65. Tucson: University of Arizona Press.

———. 1995. Desert grassland history: changing climates, evolution, biogeography, and community dynamics. In *The Desert Grassland,* eds. M. P. McClaran and T. R. Van Devender, 68–99. Tucson: University of Arizona Press.

Van Devender, T.R., and J. F. Wiens, J. F. 1993. Holocene changes in the flora of Ragged Top, south-central Arizona. *Madroño* 40:246–64.

Van Devender, T. R., P. S. Martin, R. S. Thompson, K. L. Cole, A.J.T. Jull, A. Long, L. J. Toolin, and D. J. Donahue. 1985a. Fossil packrat middens and the tandem accelerator mass spectrometer. *Nature* 317:610–13.

Van Devender, T. R., A. M. Rea, and M. L. Smith. 1985b. The Sangamon interglacial vertebrate fauna from Rancho la Brisca, Sonora. *Transactions of the San Diego Society of Natural History* 21:23–55.

Van Devender, T. R., R. S. Thompson, and J. L. Betancourt. 1987. Vegetation history of the deserts of southwestern North America; the nature and timing of the late Wisconsin-Holocene transition. In *North America and Adjacent Oceans during the Last Deglaciation*, eds. W. F. Ruddiman and H. E. Wright, Jr., 323–52. Boulder: Geological Society of America.

Van Devender, T. R., T. L. Burgess, R. S. Felger, and R. M. Turner. 1990. Holocene vegetation of the Hornaday Mountains of northwestern Sonora, Mexico. *Proceedings of the San Diego Society of Natural History* 2:1–19.

Van Devender, T. R., J. I. Mead, and A. M. Rea. 1991. Late Quaternary plants from Picacho Peak, Arizona, with special emphasis on *Scaphiopus hammondi* (western spadefoot). *Southwestern Naturalist* 36:302–14.

Van Devender, T. R., C. H. Lowe, H. K. McCrystal, and H. E. Lawler. 1992. Viewpoint: Reconsider arrangements for some North American amphibians and reptiles. *Herpetological Review* 23:10–14.

Van Devender, T. R., T. L. Burgess, J. C. Piper, and R. M. Turner. 1994a. Paleoclimatic implications of Holocene plant remains from the Sierra Bacha, Sonora, Mexico. *Quaternary Research* 41:99–108.

Van Devender, T. R., C. H. Lowe, and H. E. Lawler. 1994b. Factors influencing the distribution of the neotropical vine snake (*Oxybelis aeneus*) in Arizona and Sonora, México. *Herpetological Natural History* 2:25–42.

Van Devender, T. R., A. C. Sanders, R. K. Wilson, and S. A. Meyer. 2000. Flora and vegetation of the Río Cuchujaqui, a tropical deciduous forest near Alamos, Sonora, México. In *The Tropical Deciduous Forest of the Alamos: Biodiversity of a Threatened Ecosystem*, eds. R. H. Robichaux and D. Yetman, 36–101. Tucson: University of Arizona Press.

Wells, P. V., and J. H. Hunziker. 1976. Origin of the creosote bush (*Larrea*) deserts of southwestern North America. *Annals of the Missouri Botanical Garden* 63:833–61.

Winograd, I. J., J. M. Landwehr, K. R. Ludwig, T. B. Coplen, and A. C. Riggs. 1997. Duration and structure of the past four interglaciations. *Quaternary Research* 48:141–54.

Wolfe, J. A. 1971. Tertiary climatic fluctuations and methods of analysis of Tertiary floras. *Palaeogeography, Palaeontology, and Palaeoecology* 9:27–57.

———. 1977. Paleogene floras from the Gulf of Alaska Region. *Geological Survey Professional Paper No. 997.*

———. 1993. A method of obtaining climatic parameters from leaf assemblages. *United States Geological Survey Professional Paper No. 1964.*

Wolfe, J. A., and D. Hopkins, D. 1967. Climatic changes recorded by Tertiary land floras in northwestern North America. In *Tertiary Correlations and Climatic Changes in the Pacific, Symposium*, ed. K. Hatai, 67–76. 11th Pacific Scientific Congress, Tokyo.

CHAPTER 2

Phytogeography of the Columnar Cacti (Tribe Pachycereeae) in Mexico: A Cladistic Approach

Patricia Dávila-Aranda
Salvador Arias-Montes
Rafael Lira-Saade
José Luis Villaseñor
Alfonso Valiente-Banuet

Introduction

The family Cactaceae is distributed mainly in the arid and semiarid zones of North and South America. Some genera, such as *Opuntia* are widely distributed in the Northern and Southern Hemispheres, whereas others, such as those in the tribes Trichocereeae and Notocacteae are restricted to South America; tribes Cacteae and Pachycereeae are well developed in North America (Gibson and Nobel, 1986; Barthlott and Hunt, 1993). In particular, the Pachycereeae, one of the nine tribes of the subfamily Cactoideae, includes the columnar species that are mainly distributed in North America. Its northern limit is the southwestern United States, and it extends southwards to the West Indies and northern South America. Although the Pachycereeae has a large distributional range, most of its species are located in Mexico with a few species in Guatemala and the southwestern United States. The floristic richness of this tribe is well represented by such genera as *Neobuxbaumia* (nine species), *Pachycereus* (seven species), and *Stenocereus* (22 species). In addition, this tribe has considerable biological and economic importance as discussed by Bravo-Hollis (1978), Gibson and Horak (1978), and Gibson and Nobel (1986). Gibson and Nobel (1986) postulated that the Pachycereeae radiated in Mexico and that the species with more primitive (plesiomorphic) characters are located in southern Mexico, especially in the Tehuacán-Cuicatlán Valley, whereas the most derived taxa seem to have radiated in two opposite directions: northwards into northwestern Mexico and the southwestern United States and southwards into Mesoamerica and the West Indies. Buxbaum (1961) made the first attempt to study the phylogeny and biogeography of the Cactaceae, although the tribe Pachycereeae was not then defined. At present, there is no morphological evidence of a derived character (synapomorphy) that can support the monophyly of the tribe (but see chapter 3). Although Gibson et al. (1986) stated that the woody skeleton composed of a ring of parallel rods is a derived feature that occurs within the entire tribe, this trait is also found in the South American genus *Neoraimondia* in the tribe Browningieae. The columnar habit (which includes arborescent, tree-like, and columnar plants) that characterizes most species of Pachycereeae is also present in other tribes, including Browningieae, Cereeae, Notocacteae, and Trichocereeae. Despite the lack of strong morphological support for the monophyly of Pachycereeae, it still can be considered a model group that can help answer questions related to its phytogeography, phylogenetic relationships, and predator and/or pollinator interactions. In Mexico, some of these topics are being examined by Turner et al. (1995), Arias-Montes et al. (1997), Valiente-Banuet et al. (1997), and Teresa Terrazas (pers. com., 1998), among others.

In this chapter, we will summarize the phytogeographic knowledge of the Pachycereeae in Mexico by considering distribution patterns, floristic affinities, and species richness, as well as levels of endemism. Additionally, we present a phytogeographic analysis to correlate the phylogenies that have been proposed and the distribution patterns of the Pachycereeae. The distributional information included in this work was obtained from the review of 374 herbarium specimens deposited in the National Herbarium of Mexico, the U.S. National Herbarium, and the Missouri Botanical Garden Herbarium. In addition, diverse floristic and monographic references were reviewed (Bravo-Hollis, 1978; Gibson and Horak, 1978; Wiggins, 1980; Martínez-Alvarado, 1985; Hunt, 1992; Turner et al., 1995; Arias et al., 1997; Cornejo and Simpson, 1997; Bravo-Hollis and Arias, forthcoming). Introduced species, as well as herbarium samples obtained from cultivation or those without relevant locality data were excluded from our analysis.

Floristic Richness

The Cactaceae contains approximately 100–110 genera and probably more than 1,500 species (Arias-Montes et al., 1997). Mexico contains at least 51 genera and 850 species, representing 49% of the genera and 56% of the species of the family. The Pachycereeae includes 58 species in 13 genera (Table 2.1; Fig. 2.1), of which five are monotypic (*Backebergia, Bergerocactus, Carnegiea, Escontria,* and *Mitrocereus*); *Stenocereus* is the most diverse genus with 22 species, followed by *Neobuxbaumia* with nine and *Pachycereus* with seven species (Table 2.2).

General Distribution Patterns

Knowledge of distribution patterns enables one to understand the origin, areas of diversification, and evolutionary trends of any group. In the following paragraphs, we summarize the generic and specific distribution patterns of the Pachycereeae.

TABLE 2.1

List of Species of Pachycereeae and Their Geographic Distribution

Species	*Country (States)*
Anisocereus gaumeri	MX (Chis, Ver, Yuc)
A. lepidanthus	GT (Pro)
Backebergia militaris	MX (Gro, Mich, Jal)
Bergerocactus emoryi	US (Ca), MX (BC)
Carnegiea gigantea	US (Az), MX (Son)
Cephalocereus apicicephalium	MX (Chis, Oax)
C. columna-trajani	MX (Oax, Pue)
C. nizandensis	MX (Oax)
C. senilis	MX (Hgo, Ver)
C. totolapensis	MX (Oax)
Escontria chiotilla	MX (Gro, Mich, Oax, Pue)
Lophocereus gatesii	MX (BCS)
L. schottii	US (Az), MX (BC, BCS, Sin, Son)
Mitrocereus fulviceps	MX (Oax, Pue)
Myrtillocactus cochal	MX (BC, BCS)
M. eichlamii	GT (Pro)
M. geometrizans	MX (Ags, Dgo, Gro, Gto, Hgo, Jal, Mex, Mich, Mor, NL, Oax, Pue, Qro, SLP, Tamps, Ver, Zac)
M. schenckii	MX (Oax, Pue)
Neobuxbaumia euphorbioides	MX (SLP, Tamps, Ver)
N. macrocephala	MX (Pue)
N. mezcalaensis	MX (Gro, Jal, Mich, Mor, Oax, Pue)
N. multiareolata	MX (Gro)
N. polylopha	MX (Gto, Hgo, Qro, SLP)
N. sanchez-mejoradae	MX (Oax)
N. scoparia	MX (Oax, Ver)
N. squamulosa	MX (Col, Gro, Mich)
N. tetetzo	MX (Oax, Pue)
Pachycereus grandis	MX (Mex, Mich, Mor, Pue, Oax)
P. hollianus	MX (Oax, Pue)
P. marginatus	MX (Gro, Gto, Hgo, Jal, Méx, Mich, Mor, Oax, Pue, Qro, SLP, Zac)
P. pecten-aboriginum	MX (BCS, Chih, Chis, Jal, Mich, Gro, Nay, Oax, Sin, Son)
P. pringlei	MX (BC, BCS, Son)
P. tepamo	MX (Mich)
P. weberi	MX (Gro, Mor, Oax, Pue)
Polaskia chende	MX (Oax, Pue)
P. chichipe	MX (Oax, Pue)
Stenocereus alamosensis	MX (Sin, Son)
S. aragonii	CR (Gua, Pun)
S. beneckei	MX (Gro, Mex, Mor, Pue)
S. chacalapensis	MX (Oax)
S. chrysocarpus	MX (Mich)

(continued)

TABLE 2.1 *(continued)*

Species	*Country (States)*
S. dumortieri	MX (Gto. Gro, Hgo, Jal, Mex, Mich, Mor, Oax, Pue, Qro, Ver)
S. eichlamii	MX (Chis, Yuc), Gt (Pro), HN (Yo), ES (Son), NC (Est)
S. eruca	MX (BCS)
S. fimbriatus	CU, HT, DO, JM, PR
S. fricci	MX (Col, Jal, Mich)
S. griseus	MX (Gro, Oax, Pue, Tamps, Ver, Yuc), TT, LA, CO, VE
S. gummosus	MX (BC, BCS, Son)
S. kerberi	MX (Col, Jal, Sin)
S. martinezii	MX (Sin)
S. montanus	MX (Chih, Sin, Son)
S. pruinosus	MX (Chis, Gro, Oax, Pue, Ver, Yuc), GT (Hue, Pro, Za)
S. queretaroensis	MX (Ags, Col, Gto, Jal, Mich, Qro)
S. quevedonis	MX (Gro, Jal, Mich, Nay, Sin)
S. standleyi	MX (Col, Gro, Mich, Nay, Sin)
S. stellatus	MX (Mor, Oax, Pue)
S. thurberi	US (Az), MX (BC, BCS, Sin, Son)
S. treleasei	MX (Oax, Pue)

Abbreviations: United States (US): Arizona (AZ), California (Ca). *Mexico* (MX): Aguascalientes (Ags), Baja California (BC), Baja California Sur (BCS), Chihuahua (Chih), Chiapas (Chis), Colima (Col), Durango (Dgo), Guanajuato (Gto), Guerrero (Gro), Hidalgo (Hgo), Jalisco (Jal), Mexico (Mex), Michoacán (Mich), Morelos (Mor), Nayarit (Nay), Nuevo Leon (NL), Oaxaca (Oax), Puebla (Pue), Queretaro (Qro), San Luis Potosi (SLP), Sinaloa (Sin), Sonora (Son), Tamaulipas (Tamps), Veracruz (Ver), Yucatán (Yuc), Zacatecas (Zac). *Guatemala* (GT): Huehuetenango (Hue), El Progreso (Pro), Zacapa (Za). *Honduras* (HN): Yoro (Yo). *El Salvador* (ES): Sonsonate (Son). *Nicaragua* (NC): Esteli (Est). *Costa Rica* (CR): Guanacaste (Gua), Puntarenas (Pun). *Cuba* (CU). *Haiti* (HT). *Republica Dominicana* (DO). *Jamaica* (JM). *Puerto Rico* (PR). *Trinidad and Tobago* (TT). *Lesser Antilles* (LA). *Colombia* (CO). *Venezuela* (VE).

Except for *Anisocereus lepidanthus* and *Myrtillocactus eichlamii* in Guatemala, *Stenocereus aragonii* of Costa Rica, and *S. fimbriatus* in the West Indies, the remaining 54 species of the tribe are located in Mexico. The tribe is distributed all along the Pacific coast, in central and southern Mexico, and in central and northern South America (Fig. 2.1). The tribe seems to be less well represented in terms of species richness in the Altiplano province, where other tribes of Cactoideae (Cacteae, Echinocereeae) are more important. The general distribution pattern of Pachycereeae indicates that the highest species richness is in southern Mexico, between 17° and 19° north latitude and 96° and 99° west longitude (Fig. 2.2). Another region of high species richness occurs in the central and southern parts of Baja California, as well as in the central part of the state of Sonora.

To better understand these distribution patterns, we analyzed them using the 17 floristic provinces of Mexico recognized by Rzedowski (1978) as a framework (Fig. 2.3). This floristic classification is, in general terms, similar to the regional climatic framework, although there are some floristic differences that cannot be climatically explained. Based on this floristic classification, the distribution patterns of all the taxa

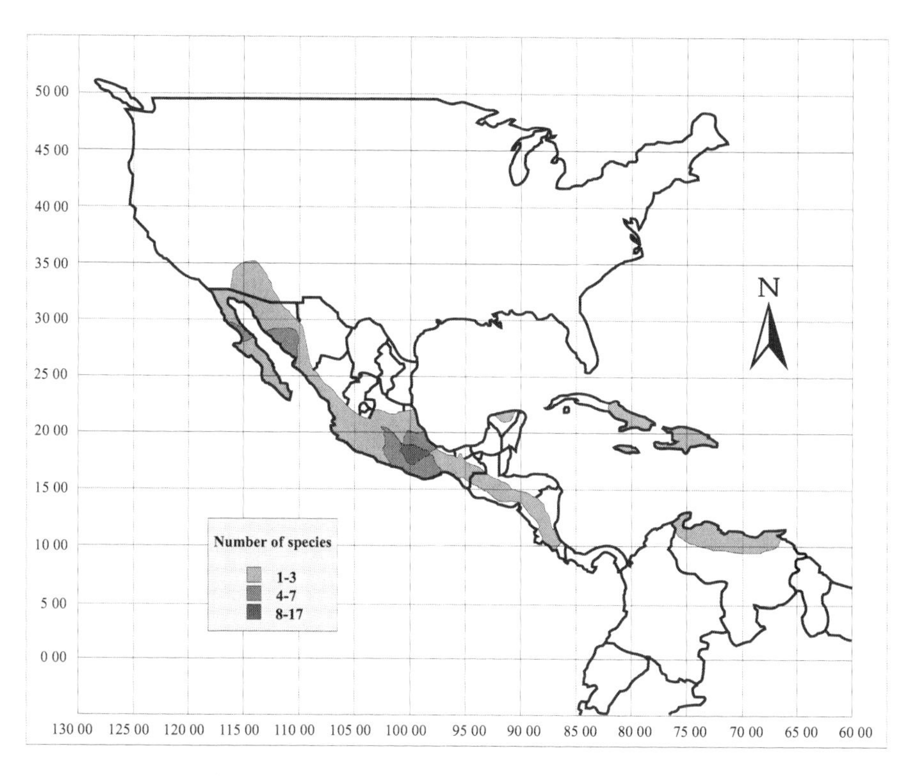

FIGURE 2.1. Geographic distribution of the tribe Pachycereeae.

of Pachycereeae located in Megamexico 3 (sensu Rzedowski, 1991) were recorded (Table 2.3). Members of Pachycereeae are widely distributed in most of the Mexican floristic provinces; however, the tribe is absent in the following provinces: Isla Guadalupe, Sierra Madre Oriental, Sierra Madre Occidental, Serranías Transistmicas, Islas Revillagigedo, and Soconusco. A few genera are restricted to only one province: *Backebergia,* a monotypic genus restricted to Depresión del Balsas province in central Mexico, and *Polaskia,* whose two species are restricted to the Valle de Tehuacán-Cuicatlán province.

Five genera of the Pachycereeae occur in only two floristic provinces. These include *Carnegiea,* present in the Planicie Costera del Noroeste province, as well as in a small area in the northwestern corner of the Altiplano province; *Lophocereus,* located in the Baja California and Planicie Costera del Noroeste provinces; *Mitrocereus,* restricted to the Valle de Tehuacán-Cuicatlán and the Serranías Meridionales provinces; *Bergerocactus,* present in the California and Baja California provinces; and *Anisocereus,* found in the Peninsula de Yucatán and Costa Pacífica provinces, as well as in Central America. Genera occurring in three different floristic provinces include *Cephalocereus,* present in the Altiplano, Valle de Tehuacán-Cuicatlán, and Costa

TABLE 2.2

Diversity and Endemism of the Genera of Pachycereeae in Mexico and MegaMexico 3[a]

Genera	*Number of Species*	*Number of Species in Mexico*	*Number of Species Endemic to Mexico*	*Number of Species Endemic to Megamexico 3*
Anisocereus	2	1	1	2
Backebergia	1	1	1	1
Bergerocactus	1	1	—	1
Carnegiea	1	1	—	1
Cephalocereus	5	5	5	5
Escontria	1	1	1	1
Lophocereus	2	2	1	2
Mitrocereus	1	1	1	1
Myrtillocactus	4	3	3	4
Neobuxbaumia	9	9	9	9
Pachycereus	7	7	7	7
Polaskia	2	2	2	2
Stenocereus	22	20	16	20
Total (percent)	58	54 (93)	47 (81)	55 (95)

[a] Sensu Rzedowski (1991); — = not available.

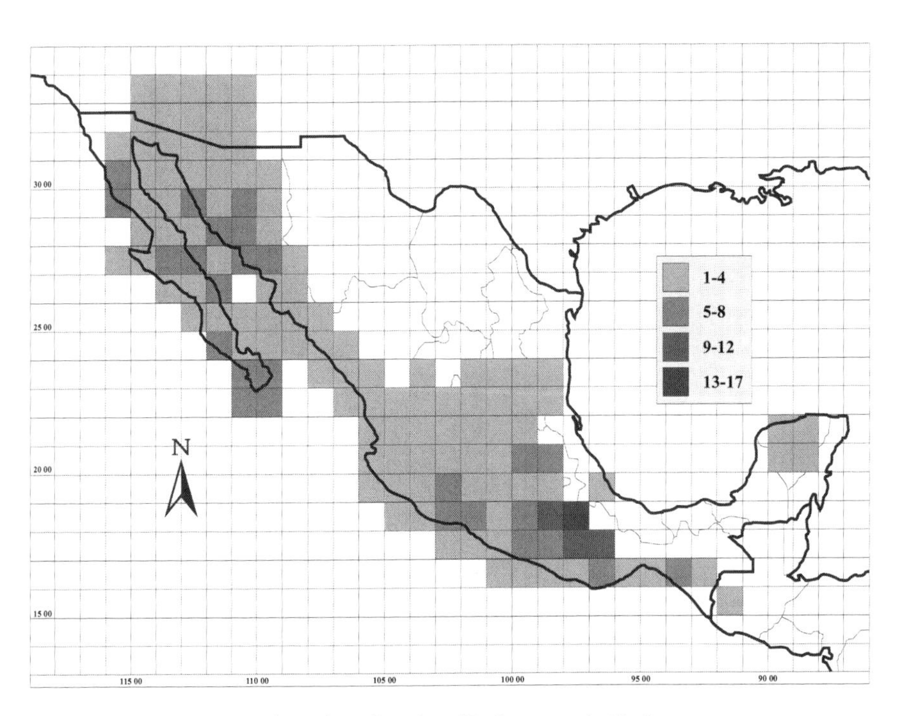

FIGURE 2.2. Distribution and number of species of Pachycereeae in Mexico.

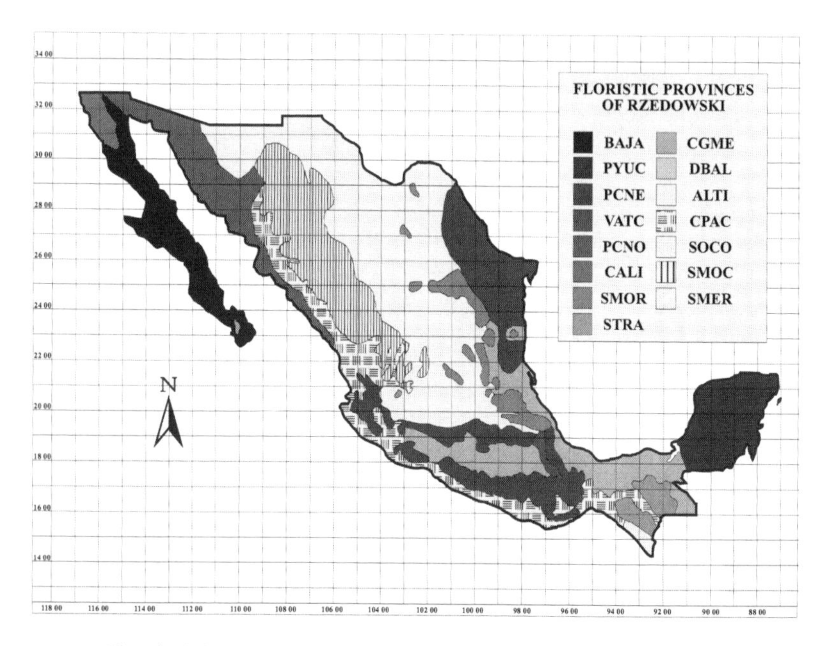

FIGURE 2.3. The Floristic Provinces recognized by Rzedowski (1978). ALTI (Altiplano Province); BAJA (Baja California Province); CALI (California Province); CGME (Costa del Golfo de México Province); CPAC (Costa Pacífica Province); DBAL (Depresión del Balsas Province); PCNE (Planicie Costera del Noreste Province); PCNO (Planicie Costera del Noroeste Province); PYUC (Península de Yucatán Province); SMER (Serranías Meridionales Province); SMOC (Sierra Madre Occidental Province); SMOR (Sierra Madre Oriental Province); STRA (Serranías Transistmicas Province); VATC (Valle de Tehuacán Province).

Pacífica provinces; and the monotypic *Escontria,* which is found in the Depresión del Balsas, Costa Pacífica, and the Valle de Tehuacán-Cuicatlán provinces.

The remaining genera (*Myrtillocactus, Neobuxbaumia, Pachycereus,* and *Stenocereus*) occur in six or more floristic provinces. Thus *Myrtillocactus* occurs in California, Serranías Meridionales, Baja California, Altiplano, Valle de Tehuacán-Cuicatlán, Depresión del Balsas, and Costa Pacífica provinces, as well as in Central America; *Neobuxbaumia* is distributed in Planicie Costera del Noreste, Valle de Tehuacán-Cuicatlán, Costa del Golfo de México, Depresión del Balsas, Costa Pacífica, and Serranías Meridionales provinces; *Pachycereus* is found in Baja California, Planicie Costera del Noroeste, Altiplano, Valle de Tehuacán-Cuicatlán, Costa Pacífica, Depresión del Balsas, and Serranías Meridionales provinces. Finally, *Stenocereus* has the widest distribution within the Pachycereeae. It is present in ten floristic provinces (Serranías Meridionales, Planicie Costera del Noroeste, Baja California, Altiplano, Planicie Costera del Noreste, Valle de Tehuacán-Cuicatlán, Costa Pacífica, Depresión del Balsas, Costa del Golfo de México, and Península de Yucatán), as well as in Central America (including Guatemala, Costa Rica, and the West Indies). In summary, the Valle de Tehuacán-Cuicatlán, the Depresión del Balsas, and the Costa Pacífica

TABLE 2.3

Distribution of Pachycereeae Species within Rzedowski's (1978) Floristic Provinces of Mexico

Species	*Province*[a]											
	1	*2*	*3*	*4*	*5*	*6*	*7*	*8*	*9*	*10*	*11*	*12*
Anisocereus gaumeri								X		X	X	
A. lepidanthus												X
Backebergia militaris									X			
Bergerocactus emoryi	X											
Carnegiea gigantea				X								
Cephalocereus apicicephalium								X				
C. columna-trajani							X					
C. nizandensis								X				
C. senilis					X							
C. totolapensis								X				
Escontria chiotilla							X	X	X			
Lophocereus gatesii			X									
L. schottii			X	X								
Mitrocereus fulviceps		X					X					
Myrtillocac cochal	X		X									
M. eichlamii											X	
M. geometrizans		X			X		X		X			
M. schenckii							X	X				
Neobuxbaumia euphorbioides						X				X		
N. macrocephala							X					
N. mezcalaensis							X	X	X			
N. multiareolata									X			
N. polylopha					X							
N. sanchez-mejoradae		X										
N. scoparia								X		X		
N. squamulosa								X				
N. tetetzo							X	X				
Pachycereus grandis							X		X			
P. hollianus							X					
P. marginatus		X			X		X	X	X			
P. pecten-aboriginum			X	X				X	X			
P. pringlei			X	X								
P. tepamo									X			
P. weberi							X		X			
Polaskia chende							X					
P. chichipe							X					
Stenocereus alamosensis				X								
S. aragonii												X
S. beneckei									X			
S. chacalapensis								X				
S. chrysocarpus									X			
S. dumortieri		X			X		X		X			
S. eichlamii												X
S. eruca			X									
S. fimbriatus												X
S. fricci								X	X			
S. griseus						X			X	X	X	
S. gummosus			X	X								
S. kerberi								X				
S. martinezii								X				
S. montanus								X				
S. pruinosus							X	X	X	X	X	X

(continued)

TABLE 2.3 *(continued)*

	Province[a]											
Species	*1*	*2*	*3*	*4*	*5*	*6*	*7*	*8*	*9*	*10*	*11*	*12*
S. queretaroensis		X							X			
S. quevedonis					X			X	X			
S. standleyi								X	X			
S. stellatus		X					X		X			
S. thurberi			X	X	X							
S. treleasei		X										

[a] Floristic provinces: 1 = California; 2 = Serranías Meridionales; 3 = Baja California; 4 = Planicie Costera del Noroeste; 5 = Altiplano; 6 = Planicie Costera del Noreste; 7 = Valle de Tehuacán-Cuicatlán; 8 = Costa Pacífica; 9 = Depresión del Balsas; 10 = Costa del Golfo de México; 11 = Península de Yucatán; 12 = Central America (Guatemala, Costa Rica, West Indies).

provinces are, floristically speaking, the best-represented, with eight of the 13 genera of the tribe occurring in each of them.

Endemism

A high degree of endemism is a remarkable feature of the Pachycereeae. Seven genera (54%), including three (*Backebergia, Escontria,* and *Mitrocereus)* of the five monotypic ones, and 47 species (81%) of the tribe are endemic to Mexico (Tables 2.1–2.4). However, if the area of the analysis is expanded in terms of the so-called Megamexico 3 (sensu Rzedowski, 1991), in which arid regions of the southwestern United States in the Sonoran and Chihuahuan Deserts and that part of Central America from Guatemala to northern Nicaragua are included, 12 of the genera (92%) and 55 species (95%) of the tribe are endemic to the area. The only genus whose distribution extends beyond the boundaries of Megamexico 3 is *Stenocereus,* which includes three species that occur from northern Nicaragua to South America. Although there are endemic species in many floristic provinces, the highest percentage of endemism is found in the Serranías Meridionales, Depresión del Balsas, and Valle de Tehuacán-Cuicatlán provinces. In contrast, levels of endemism are lower in the California, Altiplano, Planicie Costera del Noreste, Planicie Costera del Noroeste, and most of the Mesoamerican provinces.

Phytogeographic Analysis

The distribution patterns of the Pachycereeae can be correlated with phylogenetic hypotheses proposed by Cornejo and Simpson (1997) and Cota and Wallace (1997). Although these two phylogenies show many similarities, they also differ in some respects.

TABLE 2.4

Genera of Pachycereeae Endemic to Mexico and Megamexico 3[a]

Location	*Genera*
Mexico (7 genera)	*Backebergia, Cephalocereus, Escontria, Mitrocereus, Neobuxbaumia, Pachycereus,* and *Polaskia*
Megamexico 3 (12 genera)	all above plus *Anisocereus, Bergerocactus, Carnegiea, Lophocereus,* and *Myrtillocactus*

[a] Sensu Rzedowski (1991).

Using morphological and anatomical data, Cornejo and Simpson (1997) showed that the subtribes Pachycereinae and Stenocereinae are monophyletic. In addition, the cladogram shows that *Cephalocereus, Isolatocereus,* and *Stenocereus* (excluding *Anisocereus*) are well defined genera. In contrast, *Myrtillocactus* includes *Escontria* and *Polaskia,* whereas *Carnegiea* includes *Backebergia, Mitrocereus,* and *Neobuxbaumia.* Finally, they supported the inclusion of *Lophocereus* in *Pachycereus.*

The phylogenetic hypothesis of Cota and Wallace (1997), based on DNA analyses, also showed that Pachycereinae and Stenocereinae are monophyletic. The Pachycereinae includes two groups. One group forms a polytomous clade made up of *Bergerocactus, Lemairocereus,* and *Neobuxbaumia,* which is not supported by an apomorphy. The second group is a monophyletic clade formed by *Carnegiea* as a single terminal taxon in one lineage, and *Lophocereus* and *Pachycereus* in another lineage. These results do not unequivocally support the inclusion of *Neobuxbaumia* in *Carnegiea,* but do support the inclusion of *Lophocereus* in *Pachycereus.* However, the Stenocereinae includes two monophyletic groups and an isolated lineage that is not supported by an apomorphy. The first clade is a monophyletic group formed by *Myrtillocactus* as a single terminal taxon in one lineage and *Escontria* and *Polaskia* as a second lineage. The second clade is also monophyletic and includes all five species of *Stenocereus* examined by these authors. Finally, an isolated and phylogenetically nonsupported lineage is represented by *Isolatocereus.* The data of this analysis thus agree with that of Cornejo and Simpson in concluding that *Escontria, Myrtillocatus,* and *Polaskia* might represent a single genus, whereas *Stenocereus* seems well defined and separated from *Isolatocereus.*

By using Rzedowski's floristic provinces as the phytogeographic framework, and the Parsimony Analysis of Endemicity proposed by Morrone (1994), we performed two analyses. For this purpose we constructed two data matrices including the floristic provinces of Rzedowski as rows and the genera as columns. For these analyses we used Hennig 86, version 1.5 for PC-IBM (Farris, 1988). The cladograms or trees were calculated using the option "ie" of the program to ensure an exhaustive search of the minimum length trees (Farris, 1988). The Nelson consensus cladogram was also obtained by using the "nelson" option of the same program.

The first analysis included all floristic provinces and the following genera (sensu S. Arias, unpublished data): *Anisocereus, Backebergia, Bergerocactus, Carnegiea, Ceph-*

TABLE 2.5
Data Matrix Used in First Analysis[a]

Floristic Provinces[b]	*Genera*[c]												
	Ani	*Bac*	*Ber*	*Car*	*Cep*	*Esc*	*Lop*	*Mit*	*Myr*	*Neo*	*Pac*	*Pol*	*Ste*
1	0	0	1	0	0	0	0	0	1	0	0	0	0
2	0	0	0	0	0	0	0	1	1	1	1	0	1
3	0	0	1	0	0	0	1	0	1	0	1	0	1
4	0	0	0	1	0	0	1	0	0	0	1	0	1
5	0	0	0	1	1	0	0	0	1	0	1	0	1
6	0	0	0	0	0	0	0	0	0	1	0	0	1
7	0	0	0	0	1	1	0	1	1	1	1	1	1
8	1	0	0	0	1	1	0	0	1	1	1	0	1
9	0	1	0	0	0	1	0	0	1	1	1	0	1
10	1	0	0	0	0	0	0	0	0	1	0	0	1
11	1	0	0	0	0	0	0	0	0	0	0	0	1
12	1	0	0	0	0	0	0	0	1	0	0	0	1

[a] Genera according to S. Arias (unpublished data). Codes: 0 = absence; 1 = presence.

[b] 1 = California; 2 = Serranías Meridionales; 3 = Baja California; 4 = Planicie Costera del Noroeste; 5 = Altiplano; 6 = Planicie Costera del Noreste; 7 = Valle de Tehuacán-Cuicatlán; 8 = Costa Pacífica; 9 = Depresión del Balsas; 10 = Costa del Golfo de México; 11 Península de Yucatán; 12 = Central America (includes Guatemala, Costa Rica, and West Indies).

[c] Ani = *Anisocereus;* Bac = *Backebergia;* Ber = *Bergerocactus;* Car = *Carnegiea;* Cep = *Cephalocereus;* Esc = *Escontria;* Lop = *Lophocereus;* Mit = *Mitrocereus;* Myr = *Myrtillocactus;* Neo = *Neobuxbaumia;* Pac = *Pachycereus;* Pol = *Polaskia;* Ste = *Stenocereus.*

alocereus, Escontria, Lophocereus, Mitrocereus, Myrtillocactus, Neobuxbaumia, Pachycereus, Polaskia, and *Stenocereus* (Table 2.5). In this analysis, a total of four trees with a length of 22, a consistency index of 59, and a retention index of 65 was obtained. The consensus tree (Fig. 2.4a) was selected for analysis. Although this cladogram shows some unresolved lineages (polytomies), it is possible to define some "monophyletic clades." The first and largest clade is supported by *Stenocereus,* which represents the "synapomorphic taxon." This clade includes most of the provinces where the Pachycereeae occurs except the California province, and all the genera. Nested in this major clade are two well-defined clades. The first one is not supported by an apomorphy and includes Central America as well as the Península de Yucatán, Costa del Golfo, and Planicie Costera del Noreste provinces. The other clade is supported by *Pachycereus* as the "synapomorphic taxon." In this clade are included the provinces Altiplano, Planicie Costera del Noroeste, Baja California, Cadenas Montañosas Meridionales, Depresión del Balsas, Costa Pacífica, and Valle de Tehuacán-Cuicatlán, the last three forming a monophyletic clade supported by *Escontria,* a member of the subtribe Stenocereinae.

The second analysis involved the same floristic provinces as in the first, but only included the eight genera considered in the two existing phylogenies (Cornejo and Simpson, 1997; Cota and Wallace, 1997). These genera are *Anisocereus, Bergerocactus, Carnegiea* (including *Backebergia, Mitrocereus,* and *Neobuxbaumia*), *Cephalocereus, Isolatocereus, Myrtillocatus* (including *Esontria* and *Polaskia*), *Pachycereus* (including

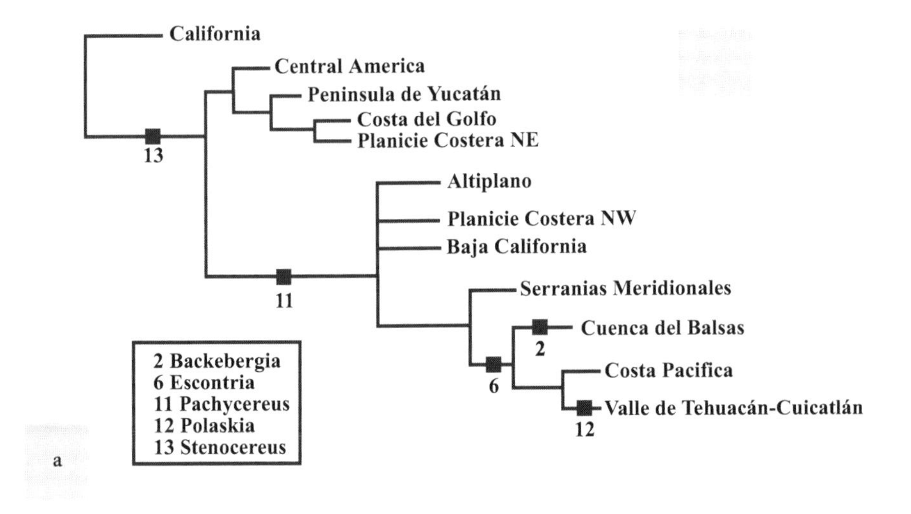

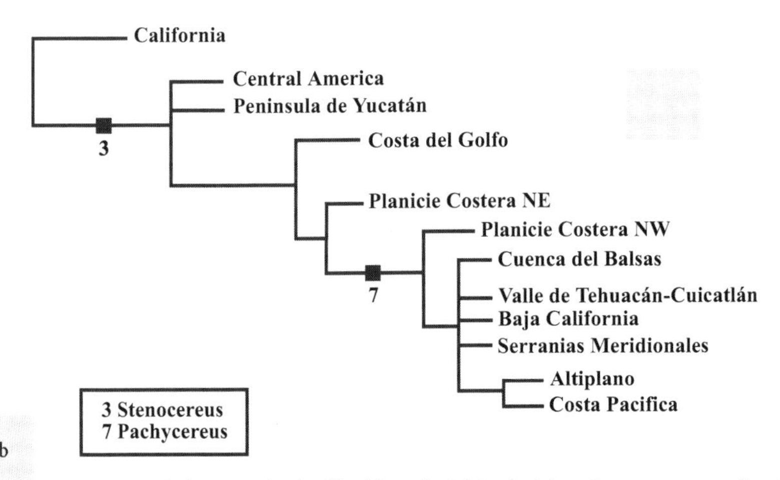

FIGURE 2.4. Area cladograms obtained in this study. (a) Analysis based on genera according to S. Arias (unpublished data); (b) Analysis based on genera according to Cornejo and Simpson (1997) and Cota and Wallace (1997).

Lophocereus), and *Stenocereus* (Table 2.6). A total of five trees was obtained in this analysis, with a length of 14, a consistency index of 57, and a retention index of 76 (Fig. 2.4b). As in the first analysis, the consensus tree was also selected for analysis. This cladogram shows some similarities with the one previously described. The California province is also an isolated branch, and the largest and monophyletic clade includes all the remaining floristic provinces and is supported by *Stenocereus* as the "synapomorphic taxon." Nested in this clade are several polytomies and a monophyletic clade supported by *Pachycereus* as the "synapomorphic taxon." This clade in-

TABLE 2.6
Data Matrix Used in Second Analysis[a]

Floristic Provinces[b]	*Genera*[c]							
	Ani	*Myr*	*Ste*	*Iso*	*Cep*	*Car*	*Pac*	*Ber*
1	0	1	0	0	0	0	0	1
2	0	1	1	1	0	1	1	0
3	0	1	1	0	0	0	1	1
4	0	0	1	0	0	1	1	0
5	0	1	1	1	1	1	1	0
6	0	0	1	0	0	1	0	0
7	0	1	1	0	1	1	1	0
8	1	1	1	1	1	1	1	0
9	0	1	1	1	0	1	1	0
10	1	0	1	0	0	1	0	0
11	1	0	1	0	0	0	0	0
12	1	1	1	0	0	0	0	0

[a] Genera according to Cornejo and Simpson (1997) and Cota and Wallace (1997). Codes: 0 = absence; 1 = presence.

[b] 1 = California; 2 = Serranías Meridionales; 3 = Baja California; 4 = Planicie Costera del Noroeste; 5 = Altiplano; 6 = Planicie Costera del Noreste; 7 = Valle de Tehuacán-Cuicatlán; 8 = Costa Pacífica; 9 = Depresión del Balsas; 10 = Costa del Golfo de México; 11 Península de Yucatán; 12 = Central America (includes Guatemala, Costa Rica, and West Indies).

[c] Ani = *Anisocereus;* Myr = *Myrtillocactus* (*Myrtillocactus, Polaskia* and *Escontria*); Ste = *Stenocereus;* Iso = *Isolatocereus dumortieri;* Cep = *Cephalocereus;* Car = *Carnegiea* (*Backebergia, Mitrocereus, Carnegiea,* and *Neobuxbaumia*); Pac = *Pachycereus* (*Lophocereus and Pachycereus*); Ber = *Bergerocactus.*

cludes a polytomy formed by the floristic provinces Planicie Costera del Noroeste, Deperesión del Balsas, Valle de Tehuacán-Cuicatlán, Baja California, and Serranías Meridionales, as well as a non-monophyletic clade formed by Costa Pacífica and Altiplano provinces.

Evolutionary and Biogeographic Implications

The topology of these two cladograms strongly suggests that the evolution of the Pachycereeae has probably taken place in a large area that includes most of the Mexican provinces and Central America. In addition, based on the groups defined in the cladograms, it seems that the tribe evolved from an ancestral stock of *Stenocereus,* which represents the most diverse genus with the widest distribution pattern. The cladograms also show that *Stenocereus* is the basal taxon of the group. Therefore, if it is assumed that the older elements of the genus are in southern Mexico, as suggested by Gibson and Horak (1978), it is possible to assume that the southern species of *Stenocereus* might represent the closest relatives of the oldest genetic stock of the tribe. If this is true, then the fact that *Stenocereus* is the only genus of the tribe that extends beyond the boundaries of Megamexico 3 (sensu Rzedowski, 1991) towards Central America, the West Indies, and South America supports Gibson and Nobel's (1986) hypothesis regarding the bi-directional evolutionary radiation of this tribe.

In particular, the arrangement of taxa in the area cladogram derived from the first analysis (Fig. 2.4a) allows us to hypothesize that the subtribe Pachycereinae is more recent than the Stenocerinae and that its evolution has probably taken place by vicariant events occurring in the wide ancestral region. This hypothesis contrasts with the conclusions about the origin of the subtribes Pachycerinae and Stenocerinae in Gibson and Horak (1978), Cornejo and Simpson (1997), and Cota and Wallace (1997), who proposed a contemporaneous origin for both lineages. The absence of members of Pachycereinae in the lowlands of southern Mexico supports our hypothesis.

The third clade of the area cladogram derived from the first analysis (Fig. 2.4a), in which *Escontria* is the "synapomorphic taxon" may represent a recently derived lineage within the Stenocereinae, although this is difficult to explain. The provinces included in this clade are Costa Pacífica, Depresión del Balsas, and Valle de Tehuacán-Cuicatlán. From these provinces, the Depresión del Balsas is supported by *Backebergia* and *Polaskia* is the "autapomorphic taxon" of the Valle de Tehuacán-Cuicatlán. Interestingly, *Backebergia* belongs to the subtribe Pachycereinae, whereas *Polaskia* is included in the Stenocereinae. The high representation of Pachycereeae and the presence of endemic genera in the Depresión del Balsas and Valle de Tehuacán-Cuicatlán provinces suggests that these areas represent an important diversification center of the group.

In summary, our area cladograms support the recently proposed evolutionary hypotheses of the tribe Pachycereeae. In addition, it appears that both subtribes evolved in almost all of the Mexican provinces, although the members of the subtribe Pachycereinae are likely to be of a more recent origin. The available information suggests that *Stenocereus* genetic stock represents the basal group and that there has been a bidirectional radiation of the tribe. Finally, the differences between the two area cladograms are probably due to different concepts of the genera in the Pachycereeae held by different authors. Consequently the taxonomic limits of *Escontria, Myrtillocactus,* and *Polaskia,* as well as the those of *Carnegiea, Backebergia, Mitrocereus, Neobuxbaumia,* and *Pachycereus* (including *Lophocereus*) need to be clarified using all available data to better understand their relationships. More comprehensive studies in the tribe using molecular, anatomical, morphological, and phytochemical evidence might help redefine the taxonomic limits of the genera, as well as clarify their phylogenetic and phytogeographic relationships.

Summary

The Pachycereeae is one of the nine tribes of the subfamily Cactoideae. It includes two subtribes (Pachycereinae and Stenocereinae) of columnar species, whose northern limit is the southwestern United States, and extends southwards to the West Indies and northern South America. Although the Pachycereeae has a large distribution range, most of its species are located in Mexico. In this chapter we summarize the phytogeographic knowledge of this group by considering distribution patterns, floristic affinities and species richness, as well as levels of endemism. Additionally, by using Rzedowski's floristic provinces as the phytogeographic framework, and the Parsimony Analysis of Endemicity, we present a phytogeographic analysis to correlate the phylogenies that have been proposed and the distribution patterns of the Pachycereeae. Our results indicate that the tribe includes 58 species in 13 genera. Five of the genera are monotypic, and *Stenocereus* is the most diverse genus with 22 species, followed by *Neobuxbaumia* with nine and *Pachycereus* with seven species. Seven genera and 47 species (81%) of the tribe are endemic to Mexico, most of which thrive in the Mexican floristic provinces of Serranías Meridionales, Depresión del Balsas, and Valle de Tehuacán-Cuicatlán. The two area cladograms obtained in the phytogeographic analysis support recently proposed evolutionary hypotheses of the tribe Pachycereeae. In addition, it appears that both subtribes evolved in most of the Mexican provinces, although the species of the subtribe Pachycereinae are likely to be of a more recent origin. The available information also suggests that a *Stenocereus* genetic stock represents the basal group and that there has been a bidirectional radiation of the tribe.

Resumen

La tribu Pachycereeae es una de las nueve que conforman la subfamilia Cactoideae. Incluye a dos subtribus (Pachycereinae y Stenocereinae) de especies columnares, cuyo límite norte de distribución es al suroeste de los Estados Unidos, extendiéndose hacia el sur hasta de las Antillas y al norte de América del Sur. Aunque las Pachycereeae tienen una amplia distribución la mayoría de sus especies crecen en México. En este capítulo presentamos un resumen del conocimiento fitogeográfico de este grupo, para lo cual se han considerado los patrones de distribución, sus las afinidades florísticas y la riqueza de especies, así como los niveles de endemismo. Adicionalmente, con el objeto de correlacionar los patrones de distribución y las dos filogenias, hasta ahora propuestas para el grupo, se llevó a cabo un análisis fitogeográfico, tomando como marco de referencia fitogeográfico a las provincias florísticas de Rzedowski, y haciendo uso del Análisis de Parsimonia de la Endemicidad. Los resultados obtenidos indicaron qué la tribu incluye 58 especies en 13 géneros. Cinco de los géneros son monotípicos, mientras *Stenocereus* es el más diverso con 22 especies, seguido por *Neobuxbaumia* con nueve especies y *Pachycereus* con siete especies. Siete géneros y 47 especies (81%) de la tribu son endémicos en México, la mayoría de los cuales prosperan en las provincias florísticas mexicanas Serranías Meridionales, Depresión del Balsas y Valle de Tehuacán-Cuicatlán. Los dos cladogramas de área que se obtuvieron en el análisis fitogeográfico apoyan las hipótesis evolutivas hasta ahora propuestas para la tribu Pachycereeae. Adicionalmente, parece ser que ambas subtribus evolucionaron en casi todas las provincias florísticas mexicanas. Aunque los elementos de la subtribu Pachycereinae probablemente tienen un origen más reciente. La información disponible también nos sugiere qué el grupo basal pudiera ser uno relacionado con *Stenocereus* y que ha habido una radiación bi-direccional en la tribu.

ACKNOWLEDGMENTS

We thank the curators of the National Herbarium of Mexico, the Missouri Botanical Garden Herbarium, and the U.S. National Herbarium for the loan of specimens, and Mayra Hernández (Facultad de Estudios Superiores-Iztacala, Universidad Nacional Autónoma de México [UNAM]), Jorge Saldivar and Enrique Ortiz (Instituto de Biología, UNAM) for their assistance in the preparation of the figures.

REFERENCES

Arias-Montes, S., S. Gama-López, and L. U. Guzmán C. 1997. Cactaceae. In *Flora del Valle de Tehuacán-Cuicatlán, Fascicle 14,* eds. P. Dávila, J. L. Villaseñor, R. Medina, and O. Téllez. Mexico: Instituto de Biología, Universidad Nacional Autónoma de México.

Barthlott, W., and D. Hunt. 1993. Cactaceae. In *The Families and Genera of Vascular Plants,* ed. K. Kubitzki, 161–97. Vol. 2. Berlin: Springer-Verlag.

Bravo-Hollis, H. 1978. *Las Cactáceas de México.* Mexico: Universidad Nacional Autónoma de México.

Bravo-Hollis, H., and S. Arias. Forthcoming. Cactaceae. In *Flora Mesoamericana,* eds. G. Davidse, M. Sousa, and S. Knapp. Vol. 4. Mexico: Universidad Nacional Autónoma de México.

Buxbaum, F. 1961. Die Entwicklungslinien der tribus Pachycereae. *F. Buxbaum Botanical Studies* 12:1–107.

Cornejo, D., and B. Simpson. 1997. Analysis of form and function in North American columnar cacti (Tribe Pachycereeae). *American Journal of Botany* 84:1482–1501.

Cota, H., and R. S. Wallace. 1997. Chloroplast DNA evidence for divergence in *Ferocactus* and its relationships to North American columnar cacti (Cactaceae: Cactoideae). *Systematic Botany* 22:529–42.

Farris, J. S. 1988. Hennig 86 Reference. Port Jefferson Station, New York.

Gibson, A. C., and K. E. Horak. 1978. Systematic anatomy and phylogeny of Mexican columnar cacti. *Annals of the Missouri Botanical Garden* 65:999–1057.

Gibson, A. C., and P. S. Nobel. 1986. *The Cactus Primer.* Cambridge: Harvard University Press.

Gibson, A. C., K. C. Spencer, R. Bajaj, and J. L. McLaughlin. 1986. The ever-changing landscape of cactus systematics. *Annals of the Missouri Botanical Garden* 73:532–55.

Hunt, D., compiler. 1992. *CITES Cactaceae Checklist.* Kew, Surrey: Royal Botanic Gardens.

Martínez-Alvarado, D. 1985. Las cactáceas del estado de Morelos. B.S. thesis. Universidad Autónoma del Estado de Morelos, Cuernavaca.

Morrone, J. J. 1994. On the identification of areas of endemism. *Systematic Biology* 43:438–41.

Rzedowski , J. 1978. *Vegetación de México.* Mexico: Limusa.

———. 1991. Diversidad y origen de la flora fanerogámica de México. *Acta Botanica Mexicana* 14:3–21.

Turner, R. M., J. E. Bowers, and T. L. Burgess. 1995. *Sonoran Desert Plants: an Ecological Atlas.* Tucson: University of Arizona Press.

Valiente-Banuet, A., A. Rojas-Martínez, A. Casas, M. C. Arizmendi, and P. Dávila. 1997. Pollination biology of two winter-blooming giant columnar cacti in the Tehuacán Valley, central Mexico. *Journal of Arid Environments* 37:331–41.

Wiggins, I. L. 1980. *Flora of Baja California.* Stanford: Stanford University Press.

CHAPTER 3

The Phylogeny and Systematics of Columnar Cacti: An Overview

Robert S. Wallace

Introduction

The evolution of cacti has engendered in many a respect for Nature's ability to create a wonderful array of biological diversity and to maintain life in normally inhospitable places as a result of the forces of adaptation and selection. Systematists and other biologists attempting to elucidate the complex interrelationships among the various cactus lineages have been repeatedly confounded by numerous instances of parallel evolution, neoteny, and phenotypic plasticity that occur with unnerving frequency throughout the family. The columnar cacti are a significant subset of this biological variability, and the tasks involved with investigating their evolutionary relationships are daunting. Nevertheless, the following overview presents a general summary of columnar cactus phylogeny as determined in my laboratory through the use of molecular characters, specifically those obtained using variation in the chloroplast DNA molecule (*cp*DNA). Phylogenetic data obtained in these investigations are derived from analysis of structural variation (inversions, insertions, deletions), restriction site variation, and DNA sequence variation based on nucleotide substitutions.

In attempting to review the phylogenetic relationships for the cacti that may be defined as "columnar" (i.e., those having a cereoid habit), it may be easier to define the limits of discussion by stating what groups will not be included. Given a broad definition of "columnar," those cactus lineages that are excluded are the subfamilies Pereskioideae and Opuntioideae, and the tribes Hylocereeae, Rhipsalideae, and Cacteae of subfamily Cactoideae. Furthermore, it is beyond the scope of this review to address the phylogeny and systematics of interspecific relationships of individual genera, and thus only tribal and inferred intergeneric relationships will be considered. Future molecular systematic research on specific genera or tribes will enable more detailed analyses of lower-rank relationships; however, such studies must be conducted within an appropriate phylogenetic context (e.g., selection of appropriate outgroups may be more confidently made when an overall phylogeny is known; Watrous and Wheeler, 1981).

Several authors have recently reviewed the history of cactus classification (Gibson and Nobel, 1986; Gibson et al., 1986; Barthlott, 1988). These accounts provide an overview of the transition from the classificatory philosophies based on artificial or phenetic taxonomic systems (in which dissimilarities are accentuated, presumably to assist in identification) to systems based on a phylogenetic interpretation. The latter philosophy uses shared, specialized features as the primary criteria for systematic evaluation and follows explicit analytical techniques to construct hypotheses of evolutionary relationships, expressed as phylogenetic "trees" (cladograms). Notable among previous cactologists was the Austrian botanist Franz Buxbaum, whose studies of vegetative and floral morphologies, as well as seed structures and seedling development produced the first phylogenetic classification of the Cactaceae. Buxbaum's (1958) classification of the subfamily Cactoideae (then considered Cereoideae) established the basis for virtually all recent classifications of that subfamily by creating a number of tribes whose members were presumed to be phylogenetically related; that is, sharing a common ancestor following a divergence from the previous common ancestor. This was a significant departure from the highly artificial system of Britton and Rose (1919–1923), whose intergeneric groupings were based on overall similarity of both vegetative and floral morphologies. Modern interpretation of evolutionary processes that give rise to biological variation recognize the phenomenon of convergent or parallel evolution, which has been shown to be rampant within the Cactaceae. The recurrence of evolutionary parallelisms in the cacti (particularly with respect to vegetative and floral characters) has contributed significantly to the considerable taxonomic confusion that has plagued the group since its original description as a distinct lineage by Linnaeus in 1753. Buxbaum's classification has been very influential in placement of groups of genera into tribes within subfamily Cactoideae in most recent classifications of the Cactaceae (Gibson and Nobel, 1986; Hunt and Taylor, 1986, 1990; Barthlott, 1988; Barthlott and Hunt, 1993)

In an attempt to circumvent problems associated with the complexities of recognized morphological parallelisms in the Cactaceae and to assist in the systematic evaluation of its evolution, Barthlott and Voit (1979) began comparative studies of seed morphology. Characters of the seed are hypothesized to be considerably more conservative in their evolution than are characters related to vegetative structures or pollination syndromes and floral morphology. These seed studies, together with morphological data and the previous classification of Buxbaum (1958), enabled Barthlott (1988) to propose a revised classification, in which hypotheses of evolutionary polarity between genera were included. The hypothesized relationships were expressed in a two-dimensional "bubble" diagram, in which the position of a genus relative to others in a tribal or subfamilial group indicated its relative level of evolutionary advancement. These new data, in addition to other data discussed by a group of cactologists associated with the International Organization for Succulent Plant Study (IOS), produced several revised classifications of the cacti (Hunt and Taylor, 1986, 1990; Hunt, 1992). Continued discussions of the IOS Cactaceae Working Party (later the Cactaceae Consensus Group, now named the International Cactaceae Systematics Group) enabled Barthlott and Hunt (1993) to propose the most recent familial classification. This is the classification that will be followed throughout the remainder of this chapter except in those instances where more recent data from my molecular studies have necessitated changes in generic or tribal placement of taxa.

The goal of my research has been to provide a phylogenetic interpretation of the systematics and evolution within the various cactus lineages based on morphologically independent (i.e., molecular) data. Furthermore, the evolutionary hypotheses of relationships (cladograms) that comparative molecular studies provide may be particularly useful in attempting to interpret morphological, ecological, and biogeographic changes through time.

This chapter will specifically examine the columnar cactus lineages and summarize the accumulated morphological and molecular systematic data to provide an overview of their inferred phylogeny and intergeneric relationships, from which other biological data may be more confidently interpreted.

Plant Materials and Methods Used

The general methods used for the study of DNA variation in my laboratory are outlined in Wallace (1995). In this section I summarize the sources of plant material and lab procedures that produced the data for this study.

Plant Material

Living plant material was obtained from the field, from botanical gardens, and from other researchers (see Acknowledgments). Voucher material is deposited in relevant herbaria, or living reference specimens are maintained at the Huntington Botanical Garden, the Desert Botanical Garden, the Royal Botanic Gardens, Kew, or the Städtische Sukkulentensammlung Zürich. Outgroup taxa (Watrous and Wheeler, 1981) were selected for their appropriateness to the ingroup being examined (e.g., the Portulacaceae for Cactaceae-wide investigations, and *Pereskia* and *Maihuenia* for studies of subfamily Cactoideae). From previous investigations (Wallace 1995; Wallace, unpublished data), it was determined that *Calymmanthium* was basal to all other lineages within the Cactoideae. It was used as an outgroup for studies of most tribes within the subfamily. A list of plant samples and their respective accession numbers used for the *cp*DNA and phylogenetic analyses discussed in this chapter is available from the author.

Laboratory Methods

Methods used to isolate DNA and perform polymerase chain reactions (PCR) and subsequent sequencing were accomplished following the procedures outlined in Wallace (1995) and Wallace and Cota (1996). For the majority of the DNA sequences investigated, the intron of the chloroplast-encoded gene *rpl*16 (ca. 1,100–1,300 base pairs [bp]) was used due to its rapid substitution rate relative to other plastid sequences (Jordan et al., 1996; Kelchner and Wendel, 1996). In addition to the *rpl*16 intron, plastid sequences for a coding region (*rbc*L; e.g., Chase et al., 1993; ca. 1,400 bp) and for another noncoding region (the *trn*L-*trn*F intergenic spacer; e.g., Taberlet et al., 1991; ca. 650 bp) were also used to evaluate the phylogenetic structure within the Cactaceae. (Note that the plastid gene abbreviations are those used by Shinozaki et al., 1986.) Primer sequences and experimental protocols are available from the author.

Phylogenetic Analysis

Initial alignment of sequences was carried out using the program CLUSTAL-W; manual alignment was then performed to optimize overall alignment and to identify indels (insertion/deletion events). Parsimony-based cladistic analyses were performed on all aligned sequences using the software packages PAUP (Version 3.1.1) or PAUP* (Phylogenetic Analysis Using Parsimony; Swofford, 1993). Indels were not used for nucleotide substitution-based analyses; they were scored as discrete events and added to the data matrix as one-state characters. Consensus trees were calculated when appropriate, and estimates of clade stability assessed using both Bremer (decay) analysis (Bremer, 1988) and Bootstrap analysis (Felsenstein, 1985).

Molecular Characters and Columnar Cactus Evolution

All cacti definable as "columnar" are members of subfamily Cactoideae; they share not only the cereoid habit but also have several molecular synapomorphies in their plastid genomes. All cacti examined have a 6-kb inversion in the large single-copy region of their *cp*DNA (Wallace, 1995). Members of subfamily Cactoideae have lost the intron to the plastid gene *rpo*C1, a deletion of approximately 740 base pairs (Wallace and Cota, 1996). Recent sequencing studies of the plastid gene *trn*L (Applequist and Wallace, unpublished data) also disclosed a roughly 367-bp deletion within its intron shared by all members of subfamily Cactoideae. The majority of detectable molecular characters used for these systematic studies of cactus phylogeny (other than major structural rearrangements as elucidated by Southern hybridizations to specific *cp*DNA probes; Wallace, 1995) have been obtained by direct sequencing of (PCR) targeted DNA regions of the plastid genome. The following phylogenetic summaries are based on these sequencing studies.

Columnar Cactus Monophyly and Tribal Affinities

Following sequence alignment and parsimony-based cladistic analyses, the phylogeny determined using sequence data for the *rbc*L gene is shown as a strict consensus tree in Fig. 3.1. Based on this conservative analysis of over 22,200 equally parsimonious trees, the overall topology is consistent with most modern classifications in terms of defining subfamilial and tribal lineages. The assessment of monophyly for the columnar lineages within subfamily Cactoideae is clear—the "Pachycereeae-Leptocereeae" clade was found to be monophyletic and sister to *Corryocactus.* The Browningieae, Cereeae, and Trichocereeae clade (the "BCT clade") also formed a monophyletic group (except for *Armatocereus,* see below), although resolution within and between these tribal groups was not sufficient to clearly discriminate between them. Interestingly, the Hylocereeae were found to be the basal lineage in this *rbc*L analysis, albeit with a weakly supported node below the polychotomy for the remaining Cactoideae. In other *cp*DNA analyses of *rpl*16 intron and *trn*L-*trn*F intergenic spacer sequences, *Calymmanthium* was basal in both cases, with Hylocereeae in a more derived position allied with the Pachycereeae-Leptocereeae clade. Studies by Nyffeler (pers. comm.) of plastid *mat*K sequences also ally the Hylocereeae to the Leptocereeae-Pachycereeae clade. This anomaly in the position of the Hylocereeae in the *rbc*L analysis is under investigation, but further discussion of this epiphytic lineage is beyond the scope of this chapter.

Analyses of the columnar cactus lineages based on the more rapidly changing *rpl*16 intron can be seen in Fig. 3.2. Although the strict consensus tree provides an overall topology similar to that seen in the *rbc*L tree, the position of several of the taxa is different. One of the problems associated with using this intron region, which results from its relatively rapid substitution rate, is that changes observed in some taxa are so extensive as to make sequence alignment impossible. Additionally, in three tribes of columnar cacti (Browningieae, in part; Cereeae; and Trichocereeae), a deletion of approximately 300 bp in the most phylogenetically informative portion of the intron (domain IV) has been lost, reducing the efficacy of this *cp*DNA region for providing phylogenetic resolution in these tribes.

The following discussions of the inferred phylogenetic relationships within and between columnar cactus lineages are based on a synthesis of the available molecular data, as well as information from other systematic studies. Future studies will undoubtedly modify several of these conclusions; however, it is hoped that this overview will provide the basic framework for further investigations of these cactus groups.

South American Lineages

With the exception of *Calymmanthium* and *Eulychnia* (Tribe Notocacteae), the three predominantly South American columnar tribes (the BCT clade) appear to be a monophyletic assemblage, with noted exceptions (see discussion of Browningieae below). Members of these tribes (with exceptions noted) all share a circa 300-bp coterminal synapomorphic deletion in domain IV of the *rpl*16 intron. Moreover, their vegetative anatomy has been shown to be similar (Mauseth, 1996) and distinct from that of other cactus lineages. Additionally, similarities in vegetative and floral morphologies of members of the tribes Cereeae and Trichocereae lend support for their close

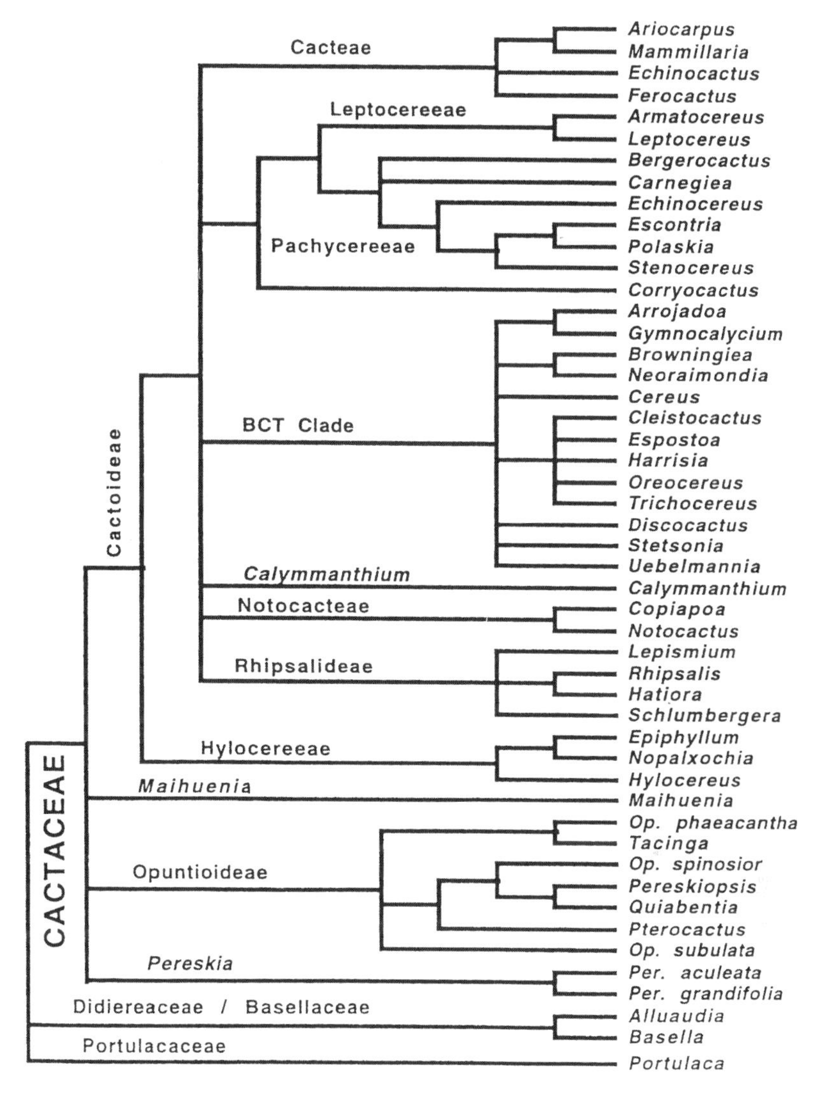

FIGURE 3.1. Strict consensus tree of 22,400 equally parsimonious trees obtained from analysis of the plastid-encoded *rbc*L gene for the family Cactaceae. A total of 1,434 bp of sequence was used for comparisons. The "BCT Clade" is composed of members of Tribes Browningieae (in part), Cereeae and Trichocereeae. Tree length = 763 steps; CI = 0.779; RI = 0.823.

relationship, although certain apomorphic distinctions have occurred (e.g., axillary hairs in the floral scales in Trichocereeae).

***Calymmanthium* Ritter.** The uniqueness of *Calymmanthium,* a monotypic genus (*C. substerile*) endemic to northern Peru, has been noted previously (Barthlott and Hunt, 1993). The plant is a large shrub or small tree with stems having relatively few

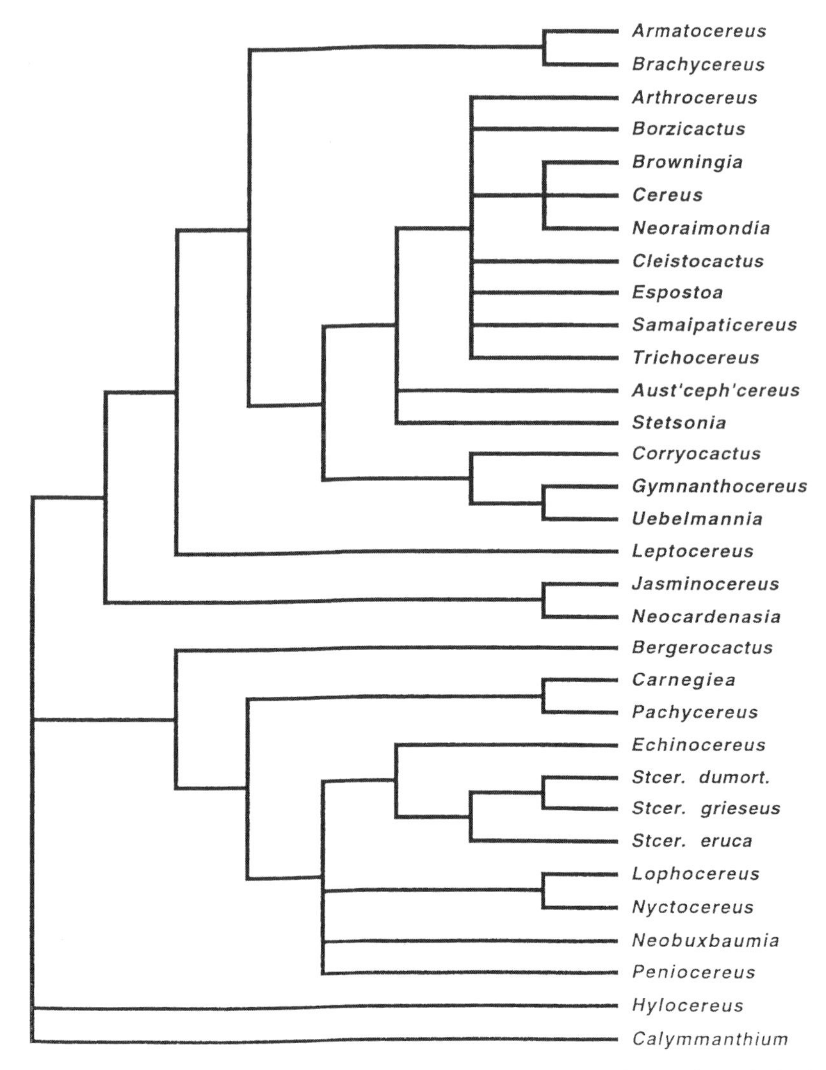

FIGURE 3.2. Strict consensus tree of 32,700 equally parsimonious trees from analysis of *rpl*16 intron sequences in columnar cactus lineages; rooted with *Calymmanthium* as outgroup. Tree length = 298 steps, CI = 0.755; RI = 0.781.

ribs (three or four); its flowers have a complex structure, unique in the Cactaceae, with an inner floral perianth partially shielded by an outer floral tube. The outer perianth has felted areoles with bristly spines. Systematic studies of sequence variation of the gene *rpl*16 intron and the *trn*L-*trn*F intergenic spacer (Wallace and Dickie, unpublished data) both place *Calymmanthium* as the sister group to all other members of subfamily Cactoideae. Furthermore, *Calymmanthium* lacks the 300-bp *rpl*16 intron deletion found in most taxa of the other three tribes of South American columnar cacti; this argues for its exclusion from tribe Browningieae as proposed by Barthlott and Hunt (1993). The phylogenetically basal position of *Calymmanthium* may be indicative of what characters the early Cactoideae ancestor may have possessed. These "primitive" characters are not only relevant for the consideration of the early evolution of columnar cacti per se, but also with respect to the ancestry of *all* members of subfamily Cactoideae. This evidence supports the hypothesis that the early cactoid ancestor was likely a shrub or small tree; it may have had weak or possibly scandent stems with few ribs, branching basitonically to mesotonically, and flowers with scaly bracts. This suite of characters appears in several lineages within the subfamily Cactoideae. The molecular data, its geographic distribution, and the uniqueness of its floral development indicate that *C. substerile* is a relict lineage within the Cactoideae, and one that is critical in its evolutionary interpretation. Unfortunately, this species has not been adequately studied, particularly with respect to its stem anatomy. When it is, I suspect that its wood and other anatomical characters will be found to be more plesiomorphic than those of *Leptocereus quadricostatus*, a relatively primitive species investigated by Mauseth and Ross (1988).

Tribe Browningieae. Members of tribe Browningieae are a relatively small group of shrubby or large, arborescent columnar cacti that typically have imbricate, scaly floral bracts. They are distributed primarily in the central Andes region (northern Chile, Bolivia, and Peru) and form a small but morphologically cohesive group of columnar cacti restricted to this geographic region. Buxbaum (1966) noted the relatively primitive floral structures of the Browningieae and recognized this group as distinct; he had previously included several of these taxa in tribe Cereeae (Buxbaum, 1958). Included in this tribe are the genera *Browningia, Armatocereus, Jasminocereus* (a monotypic Galápagos Islands endemic genus), *Neoraimondia,* and *Stetsonia* (perhaps referable to Cereeae?). Unique within the Cactaceae is the unusual development of two types of areoles in *Neoraimondia* (including *Neocardenasia*), nonflowering areoles that are similar to typical areoles of other cacti, and flowering areoles that continue growth indeterminately as lateral reproductive axes (Ostalaza, 1982; Mauseth 1997). Few systematic studies have been conducted on the Browningieae, although Gibson (1992) did review the relationship between *Browningia* and *Gymnanthocereus* and their possible extended phylogenetic ties to other columnar cactus lineages. From the representative taxa of the tribe studied for DNA variation, it may be concluded that *Calymmanthium* should be excluded from the tribe (and perhaps recognized as its own distinct tribe) and that the Browningieae are presently united as a tribe based on plesiomorphies, rather than on derived characteristics.

Subsequent studies of sequence variation in the *trn*L intron (Applequist and Wallace, unpublished data) and the *psb*A-*trn*H intergenic spacer (Wallace, unpublished data) have shown tribe Browningieae to be paraphyletic, as currently circumscribed (Fig. 3.3). Members of the tribe fall into two distinctly different clades, marked by *cp*DNA deletions or synapomorphic nucleotide substitutions. The first clade, corre-

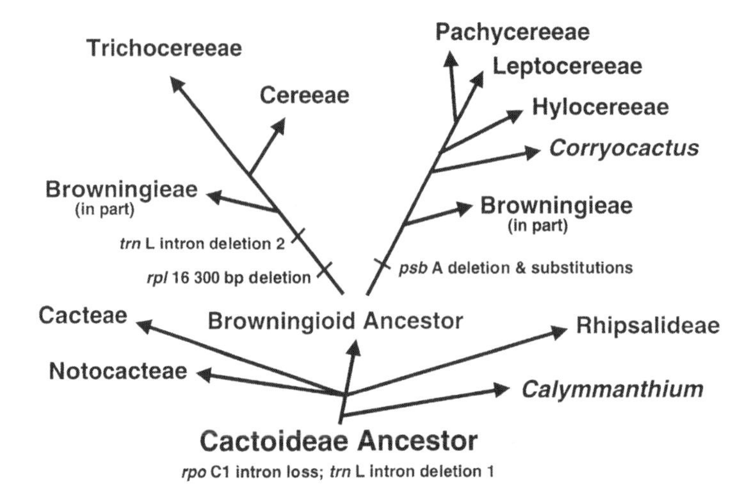

FIGURE 3.3. Inferrd evolutionary relationships among the columnar cactus tribes within subfamily Cactoideae based on molecular evidence. The tribe Browningieae is paraphyletic; two distinct clades are defined, each with affinities to different columnar cactus groups.

sponding to the BCT clade mentioned previously, includes *Browningiea, Gymnanthocereus, Gymnocereus, Neoraimondia,* and *Stetsonia,* which have a circa 268-bp deletion in the *trn*L intron that is found no where else in the Cactaceae. They also share the circa 300-bp deletion in the *rpl*16 intron noted above. The second clade includes *Armatocereus, Castellanosia,* and *Jasminocereus,* which lack both of these deletions, and share synapomorphic substitutions and a small deletion in the *psb*A-*trn*H intergenic spacer with the Pachycereeae, Leptocereeae, and *Corryocactus.* This explains the anomalous position of *Armatocereus* as the sister group to *Leptocereus* in the *rbc*L tree (Fig. 3.1), although its position in the *rpl*16 tree (Fig. 3.2) is enigmatic.

Given the phylogenetically dichotomous nature of the Browningieae as demonstrated by these molecular systematic studies, the question remains as to its taxonomic treatment. It is clear that the two subgroups each ally to distinct (and presumably more derived) columnar cactus lineages, and thus retaining the Browningieae as a tribe would perpetuate its paraphyly. Alternatively, the two clades may be divided into distinct groups, and if shown to be monophyletic, would stand as independent tribes, each sister to the more derived tribes. Further systematic studies of this evolutionarily transitional group are being undertaken.

Tribe Cereeae. Tribe Cereeae has its maximum species diversity centered in eastern South America, which is in contrast to the other groups that are primarily western-centered. This group contains cacti that range from large, robust trees to thin-stemmed shrubs, and provides an example of extreme stem reduction to a globular form, as seen in *Melocactus.* Members of tribe Cereeae are distinct from those in the Trichocereeae in that the floral tube scales are without hairs. A number of anatomical (Mauseth, 1996) and systematic studies (Braun, 1988; Taylor and Zappi, 1989; Taylor, 1991; Zappi 1994) have been conducted that permit examination of the tribe in a phylogenetic framework. Beginning with the most plesiomorphic genus *Cereus,* both Braun (1988) and Taylor and Zappi (1989) put forth evolutionary scenarios based on cladistic analyses of morphological and other characters and hypothesized transitions through the morphotypes of the extant genera that culminate in the globular genus *Melocactus.* Sequencing studies of the *rpl*16 intron indicate that all members of the Cereeae examined have lost a 300-bp segment in domain IV of this intron, as in the Trichocereeae and part of the Browningieae. In general the Cereeae are predominantly adapted for bat (and insect) pollination; however, in *Arrojadoa* and *Micranthocereus,* the floral syndrome has shifted to hummingbird pollination, which recurs again in the genus *Melocactus* (Taylor, 1991). Although it is now placed in the tribe Browningieae, some molecular evidence suggests that the genus *Stetsonia* may have closer affinities to members of tribe Cereeae. With an ongoing reevaluation of phylogeny in tribe Browningieae and the systematic questions surrounding this tribe, the final taxonomic position of *Stetsonia* may be within Cereeae.

Another genus that may also be referable to tribe Cereeae is *Uebelmannia,* a Brazilian endemic genus from Minas Gerais. Nyffeler (1997) presented anatomical evidence that this genus does not appear to be related to the Notocacteae; data from *cp*DNA studies also support its removal from the Notocacteae with likely placement in the Cereeae (see discussion below under Notocacteae).

Tribe Trichocereeae. Among all of the various groups within the Cactaceae, the tribe Trichocereeae stands out as one of the most systematically confused. It is one of the largest tribes of cacti in subfamily Cactoideae, and it runs the range of morphological extremes seen in the family: from large trees to lax, trailing, or clump-forming shrubs to globular forms that can be quite robust or extremely diminutive. As a tribe, it manifests virtually every pollination syndrome known in the Cactaceae, and given its widespread geographic distribution, plasticity of stem morphology, and wide ecological range, its classificatory and phylogenetic problems are among the most challenging to resolve. Considerable discussion has ensued regarding generic limits in this tribe (actually, this is one of the main issues in the classification of Cactaceae, regardless of group being considered).

As originally described by Buxbaum (1958), the Trichocereeae share a character of having hairs in the scales of the floral tube, which is in contrast to the "naked" floral scale axils seen in members of tribe Cereeae. Buxbaum further divided the tribe into three subtribes: subtribe Trichocereinae, in which the robust, thick-stemmed treelike genera are placed; subtribe Rebutinae, composed of mostly diminutive, clump-forming cacti with small, variously colored funnelform-to-radiate flowers; and subtribe Borzicactinae, which contains columnar, short-columnar, and globular forms, often with zygomophic or partially zygomorphic flowers adapted for hummingbird pollination. Members of the Trichocereeae have an extensive geographic range from Argentina and Chile northward to Ecuador and Brazil; with the inclusion *Harrisia* in the tribe, its range continues to the Caribbean and Florida.

The classification of the Trichocereeae is still in a state of flux, due in no small part to the paucity of comprehensive systematic studies of the tribe. One of the few recent studies to address phylogenetic and systematic questions in the Trichocereeae was that of Bregman (1992), who examined seed morphology in an attempt to resolve the phylogenetic relationships in subtribe Borzicactinae. Many systematic questions still persist and the need for comparative, character-based systematic investigations is obvious.

Short of new research on this tribe, the works of Backeberg (1958–1962, 1977), Buxbaum (1958, 1969), and Ritter (1979–1981) form the basis for present systematic considerations of this tribe. Many of the systematic questions discussed by the IOS Cactaceae Working Party (Hunt and Taylor, 1986, 1990) deal specifically with trichocereoid taxa.

Space limitations here preclude discussion of various problems yet to be resolved in the Trichocereeae. Molecular data have confirmed several prior hypotheses of relationships between certain genera (e.g., *Trichocereus, Echinopsis* s. str.), and have also provided support for the placement of certain genera within the Trichocereeae that had been assigned elsewhere in the Cactoideae. For example, *Harrisia* (including *Eriocereus* and *Roseocereus*) has affinities here and with no other genera of the Hylocereeae (Buxbaum, 1958; Gibson and Nobel, 1986) or the Echinocereeae (sensu Barthlott and Hunt, 1993). Wallace (1997) also used molecular criteria to exclude *Mediocactus hahnianus* from *Harrisia* and ally it to other *Echinopsis.* Similarly, *Samaipaticereus* also has affinities within the Trichocereeae (Barthlott and Hunt, 1993) and not with the Leptocereeae (Buxbaum, 1958; Gibson and Nobel, 1986); *Gymnocalycium* is also referable to the Trichocereeae and not to the Notocacteae. A substantial amount of research has yet to be done to resolve the many phylogenetic questions, particularly with regard to generic delimitation and definition.

Tribe Notocacteae. The Notocacteae constitutes a group of predominantly globular cacti; however, several of its members may certainly be considered columnar or cereoid, notably the genus *Eulychnia.* This genus of about four to six species native to coastal Chile and Peru forms shrubs to large trees with apical, diurnal, scaly flowers often with extensive areolar wool and/or spiny bristles on the hypanthium. Recent anatomical studies of *Eulychnia* (Nyffeler et al., 1997) disclosed a type of idioblastic sclerid that is unique to this genus. Further morphological studies will undoubtedly add systematically useful data, as will inclusion of samples of *Eulychnia* in additional molecular studies of the Notocacteae. Nyffeler (pers. comm.) has investigated the position of *Eulychnia* using plastid *mat*K sequences, and found it to be most closely related to the North American tribes Pachycereeae and Leptocereeae. Additional sequencing of *Eulychnia* is required to ascertain the phylogenetic position of this genus, as well as to investigate its disjunct distribution in Chile and Peru, relative to the geographic distribution of its inferred closest relatives.

In contrast to the shrub/tree genus *Eulychnia* is the Patagonian genus *Austrocactus,* also classified at present within the tribe Notocacteae (Buxbaum, 1958; Gibson and Nobel, 1986; Barthlott and Hunt, 1993). These plants are slender-stemmed, low-growing to sprawling in habit, with flowers vested in woolly hairs and/or bristles. Although not shrubby or arborescent cacti, they are, nonetheless, columnar in structure and warrant inclusion in discussions about evolution of columnar cacti. This genus (together with several sympatric opuntioid taxa) has one of the most southerly distributions of any of the cacti in South America, and is capable of tolerating mesic and cold conditions. Future systematic studies of *Austrocactus* in the context of a robust phylogeny for the subfamily Cactoideae will enable a more confident tribal placement for this austral lineage.

Another columnar genus (albeit with only short columnar stems) that has been referred to the Notocacteae (Gibson and Nobel, 1986; Barthlott and Hunt, 1993) is *Uebelmannia,* a group of about five species endemic to the mountains of Minas Gerais, Brazil. This disjunct distribution with respect to the majority of the globular Notocacteae found much farther south and west in Chile, Bolivia, and Argentina is telling, in that *cp*DNA variation studies of *rbc*L and the *rpl*16 intron show that *Uebelmannia* is more closely related to the BCT clade, although at present its primary affinities are not completely resolved. It also shares the 300-bp deletion in domain IV of the *rpl*16 intron, which is not found in any members of tribe Notocacteae sampled to date. Anatomical studies by Nyffeler (1997) also show that *Uebelmannia* is not related to other members of tribe Notocacteae. In their studies of intergeneric relationships in tribe Cereeae, Taylor and Zappi (1989) included *Uebelmannia* in their cladistic analysis; it was found to be derived within tribe Cereeae and related to *Melocactus* as a sister lineage. This may represent its true phylogenetic position, although molecular data are lacking to support this conclusion. Because domain IV of the *rpl*16 intron has been lost in the BCT clade, it is difficult to provide finer phylogenetic resolution in this group with this *cp*DNA marker, as the greatest number of nucleotide substitutions occur in this region. Further studies of DNA variation in tribe Cereeae will hopefully resolve the placement of *Uebelmannia.*

The genus *Corryocactus* (including *Erdisia*), of Chile, Bolivia, and Peru has been placed in tribe Notocacteae by previous authors (Gibson and Nobel, 1986; Barthlott and Hunt, 1993); however, my molecular studies show it to be related to members of tribes Leptocereeae and Pachycereeae (Fig. 3.1), which are primarily North American. Interestingly, Buxbaum placed it within tribe Leptocereeae, a group of "several genera of very primitive character, especially with very primitive flowers" (Buxbaum, 1958, p. 179). His prescient placement of this genus allied to *Leptocereus* foretold its realignment from the tribe Notocacteae of recent classifications to a position that is transitional/basal with respect to the North American columnar cacti, and that may warrant recognition of the genus as a distinct new tribe of the Cactoideae. Comparative anatomical and molecular systematic studies of *Corryocactus* are being performed by R. Nyffeler (pers. comm.), and changes in its taxonomic status await further corroborative data.

North American Lineages

The lineages of columnar cacti in North America are undoubtedly derived from South American ancestors, and the extent to which they have radiated is testament to their ability to migrate, presumably with the aid of avian or bat dispersers, and to adapt to local environmental and edaphic conditions. Noncolumnar lineages in the North American cactus flora include a group of cacti having predominantly barrel or globular form (the tribe Cacteae) and a group of lianas or holoepiphytes (the tribe Hylocereeae). All other members of subfamily Cactoideae in North America may be considered columnar, as they are ultimately derived from columnar lineages, even if they do not, at present, manifest shrubby or arborescent habits. The molecular data provide evidence for a single major migration northward from southern ancestors, particularly with respect to *Corryocactus* discussed previously. Upon arrival in suitable North American habitats, subsequent diversification occurred, which split the ancestral group into two or three primary lineages: tribes Leptocereeae and Pachycereeae and perhaps tribe Hylocereeae (Fig. 3.4).

Tribe Leptocereeae. As originally described by Buxbaum (1958), tribe Leptocereeae contained a variety of presumably primitive genera that in recent classifications have been assigned to a number of different tribes. Gibson and Nobel (1986) modified Buxbaum's classification and reassigned *Corryocactus* and *Eulychnia* to tribe Notocacteae, and *Leptocereus* and *Zehntnerella* to tribe Trichocereee. Barthlott and Hunt (1993) considered both *Leptocereus* and *Acanthocereus* to be relatively plesiomorphic members of the tribe Echinocereeae. DNA studies have essentially eliminated the Echinocereeae sensu Barthlott and Hunt (1993), realigning *Harrisia* (including *Eriocereus* and *Roseocereus*) to tribe Trichocereeae as discussed above, *Peniocereus/ Nyctocereus* and *Echinocereus* to tribe Pachycereeae (see below), and suggesting that the remaining genera—*Leptocereus* and *Acanthocereus* (including *Dendrocereus*)—be recognized with a newly defined tribe Leptocereeae proposed here.

Leptocereus, a genus of approximately eight to 12 or more species, evolved within the Caribbean Islands (Cuba, Hispaniola, and Puerto Rico) as shrubs or small trees, typically with three to eight ribs. Their flowers are diurnal or nocturnal, in a cephalium-like structure in *L. quadricostatus* (Mauseth and Ross, 1988), with a relatively short floral tube and a spreading perianth. Anatomical studies by Mauseth and Ross (1988) confirm *Leptocereus* to be a relatively plesiomorphic lineage as put forth by Buxbaum (1958); however, its status as the "most primitive" member of subfamily Cactoideae is not supported by molecular data. As discussed above for *Calymmanthium,* this genus is supported as the basal lineage in the subfamily. *Leptocereus* oc-

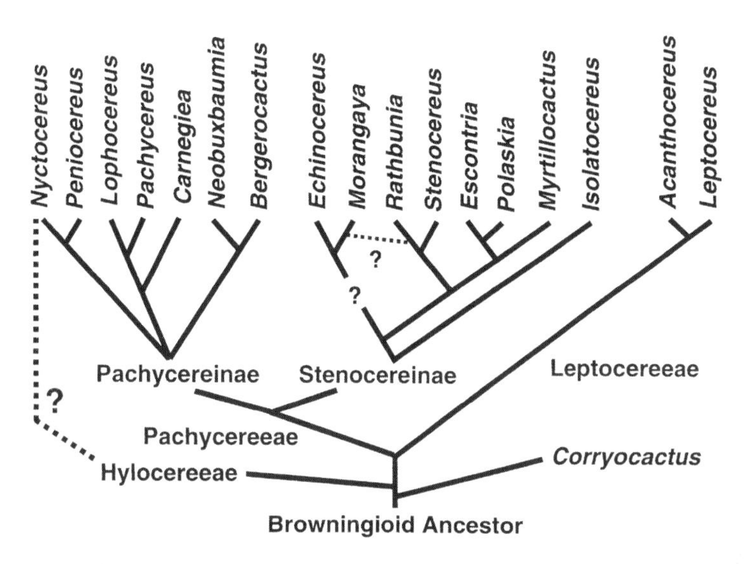

FIGURE 3.4. Evolutionary relationships among North American columnar cacti. All lineages are derived from a browningioid ancestor; the genus *Corryocactus* is a memeber of this clade, with no direct relationship to tribe Notocacteae. Subtribes Pachycereinae and Stenocereinae of tribe Pacycereeae are supported, which are sister to the Leptocereeae lineage. Dashed lines represent possible, but yet unresolved relationships.

cupies a more phylogenetically derived position, but retains a number of plesiomorphic features. Systematic studies by A. Areces-Mallea (pers. comm.) on the genus *Leptocereus* are in progress at present and will provide a synoptic review of its classification and phylogeny.

Members of the genus *Acanthocereus* are found throughout tropical and subtropical regions of the Caribbean, Mexico, Central America, and Florida (Benson 1982; as *Cereus*) as sprawling, occasionally trailing shrubs with lax branches or (in the Cuban endemic species *Dendrocereus nudiflorus* now subsumed within *Acanthocereus*) small, thick-stemmed trees; it has few ribs (three to five) and ferocious spines. The large, funnelform, nocturnal flowers have felted and short-spined areoles, but are without areolar hairs. Previously allied to the Hylocereeae (Buxbaum, 1958; Gibson and Nobel, 1986) and to the Echinocereeae (Barthlott and Hunt, 1993), DNA studies show it to be a sister group to *Leptocereus,* with no other discernible affinities to previously associated taxa.

As defined here, the tribe Leptocereeae is composed of two genera: *Leptocereus* and *Acanthocereus* (includes *Dendrocereus*). This clade was found to be the sister group to the tribe Pachycereeae based on molecular evidence, and represents the eastern branch of evolution in the North American columnar cacti. With limited habitat area (although likely possessing the same dispersal potential), the Leptocereeae has not been as successful as the Pachycereeae in speciating into a diverse number of lineages.

Tribe Pachycereeae. Within subfamily Cactoideae, one of the more thoroughly investigated tribes is the Pachycereeae. A variety of anatomical and phytochemical studies of this tribe conducted by A. C. Gibson and collaborators (Gibson and Horak, 1978; Gibson, 1982; Gibson and Nobel, 1986; Gibson et al., 1986; Gibson 1988a–d, 1991) provides the basis for the establishment of several robust hypotheses regarding the phylogenetic divergences and systematic relationships within the tribe. Despite this extensive study of a very well defined lineage of cacti, there are still a number of evolutionary questions to be addressed, especially those relating to the relationships of the Pachycereeae to other tribes of the Cactoideae, as well as aspects of intergeneric and interspecific phylogeny within the tribe.

Studies of *cp*DNA restriction site variation by Cota and Wallace (1997) compared the putatively close relationship of *Ferocactus* (tribe Cacteae) to the Pachycereeae (in particular, subtribe Stenocereinae), which Gibson hypothesized as being closely related due to the shared feature of scaly/papery floral bracts seen in *Escontria* and in several species of *Ferocactus*, along with specific microcharacters of the stem anatomy. Results of the *cp*DNA study showed that these shared characters are another example of morphological parallelism in the Cactaceae, and no direct relationship between the Pachycereeae and *Ferocactus* was supported. Perhaps more important for this discussion is that the delimitation of the two subtribes Pachycereinae and Stenocereinae was strongly supported by the restriction site data.

Additional studies of sequence variation in the *rbc*L and *rpl*16 genes provide additional support for the two subtribes and support for other systematic conclusions (Figs. 3.1–3.4). As an example, the genus *Bergerocactus* was found to have phylogenetic affinities within subtribe Pachycereinae, rather than in the tribes Hylocereeae (Gibson and Nobel, 1986) or Echinocereeae (Buxbaum, 1958). The genera *Pachycereus, Carnegiea, Neobuxbaumia,* and *Cephalocereus* form the core of the subtribe as defined by Gibson and Horak (1978). Interestingly, recent preliminary data (Wallace and Stansberry, unpublished data) have shown that parts of the *Peniocereus/Nyctocereus* lineage are allied to subtribe Pachycereinae and may represent a geophytic side branch in the evolution of a subtribe whose members are normally shrubs to large trees. These genera also share nocturnal, large white flowers with the rest of the subtribe Pachycereinae; their previous association with the geophytic genus *Wilcoxia* (now a section of *Echinocereus;* Taylor 1985; Wallace and Forquer, 1995) is an example of parallel evolution. The previous associations of *Peniocereus* and *Nyctocereus* to tribe Hylocereeae (Buxbaum, 1958; Gibson and Nobel, 1986) may well be confirmed by ongoing investigations of these lineages; the association of tribe Hylocereeae to the *Peniocereus/Nyctocereus* clade has been supported by some molecular data (Fig. 3.4). Ongoing species-level studies of the Pachycereeae and related lineages in North America will better resolve the phylogenetic relationships, from which more detailed studies of morphological and anatomical changes may be made.

The subtribe Stenocereinae is the other primary lineage within the Pachycereeae and is distinctly definable based on a suite of anatomical and phytochemical characters, as discussed by Gibson (1988a). Among the significant results from the molecular investigations of this tribe is that recognition of the monotypic genus *Isolatocereus* (*I. dumortieri*) is supported (Cota and Wallace, 1997), which corroborates Gibson's (1991) conclusions about this taxon as being the basal extant lineage of the subtribe Stenocereinae. The relationship between *Escontria, Polaskia,* and *Myrtillocactus* as outlined in Gibson and Horak (1978) was also well supported by molecular data (Cota and Wallace, 1997). The distinctive papery floral bracts of *Escontria* (Gibson, 1988d) are certainly a retained (or a secondary reversion to) plesiomorphic character state not found in *Polaskia* (Gibson, 1988b) or in *Myrtillocactus* (Gibson, 1988c), despite the relatively derived position of these genera within the subtribe. The relationship of *Rathbunia* to *Stenocereus* was also confirmed by molecular data; however, the most interesting result relative to the Stenocereinae is the inclusion of the genus *Echinocereus* within this subtribe. This genus of short-stemmed columnar plants occurs at the northern limits of the distribution of the Pachycereeae, and they may be viewed as environmentally dwarfed pachycereoid cacti. The restriction-site study of *Morangaya* and *Echinocereus* conducted by Wallace and Forquer (1995) included only one member of the Pachycereeae (*Bergerocactus*) that is allied to subtribe Pachycereinae: Had they used a member of the Stenocereinae (in particular, *Rathbunia*), it is likely that a closer relationship between *Echinocereus* and this subtribe would have been demonstrated. Additional molecular studies of the Stenocereinae will likely resolve the relationship between *Stenocereus, Rathbunia,* and *Echinocereus,* which, interestingly (with the exception of *Bergerocactus*), comprised the tribe Echinocereeae of Buxbaum (1958).

Biogeography and Evolution of Columnar Cactus

Center-of-origin hypotheses for the cacti have been briefly reviewed by Gibson and Nobel (1986), Leuenberger (1986), and Mauseth (1990). The biogeographic appraisal of South American cacti by Buxbaum (1969) adds some commentary to an under-investigated aspect of cactus biology. Without a phylogenetic context—within which biogeographic hypotheses may be framed—investigation of the biogeographic relationships for the cacti as a whole is difficult at best. Based on insights from the study of molecular variation, it is possible to reevaluate several previous hypotheses regarding the geographic origins of this family and put forth some preliminary conclusions about the origin, diversification, and geographic dispersal of major lineages within the Cactaceae.

I propose the following biogeographic scenario for the evolution of columnar cactus lineages (and cactus lineages in general). This scenario is based on concordance of topologies in area cladograms from individual molecular systematic studies of the subfamily Pereskioideae (summarized in Wallace, 1995), the subfamily Opuntioideae (Dickie, 1996; Dickie and Wallace, in review), and on several tribal and generic studies within subfamily Cactoideae.

The molecular data from several independent studies on phylogenetically distinct lineages all show appreciable concordance with respect to their basal extant lineages and their present distribution in the central Andes. The middle Andean region of northern Chile, Bolivia, and Peru seems likely as the center of origin for the Cactaceae, because a number of plesiomorphic cactus groups are endemic to this region (e.g., *Austrocylindropuntia, Browningia, Calymmanthium, Copiapoa, Miqueliopuntia, Pereskia diaz-romeroana, P. humboltii, P. weberiana*). Following phylogenetic radiation into four major lineages from the original cactus ancestor (Wallace, 1995), subsequent migration along three primary routes allowed for further radiation and specialization in morphology and habit, perhaps becoming more xeromorphically adapted in the process (Fig. 3.5). A northern and southern migration route (presumably following the xeric corridor created by uplift of the Andes) permitted migration in these directions, particularly if ornithochorous (and perhaps chiropterochorous) dispersal followed present-day migration paths. A transcontinental, west-to-east dispersal also occurred, as evidenced by the extensive radiations observed in a variety of cactus lineages: tribe Cereeae, tribe Rhipsalideae, flat-stemmed *Opuntia* lineages of the *O. inamoena/palmadora* complex (including the genus *Tacinga*), and the Trichocereeae lineage represented by *Harrisia* (including *Eriocereus*).

Along the northern migration path, ancestral lineages from South America eventually were dispersed into the xeric regions of Mexico, which provided ample habitat for radiation into more derived cactus lineages such as those seen in tribe Pachycereeae, tribe Leptocereeae (to the east), *Opuntia* (prickly pears), *Cylindropuntia* (chollas), and in the remarkable diversity of tribe Cacteae (e.g., *Coryphantha, Ferocactus, Mammillaria*). The interesting west-east relationship between tribes Pachycereeae and Leptocereeae (as defined here), respectively, is paralleled with a similar biogeographic pattern in the flat-stemmed opuntioid cacti—that is, between the prickly-pear opuntias (on the mainland) and the genus *Consolea* (which is endemic to the Caribbean region, as is *Leptocereus*).

With respect to basal groups within the columnar cactus lineages, tribe Browningieae apparently has not dispersed far from its ancestral habitats, and remains in Peru, Bolivia, and northern Chile. Migration to the south and east by members of tribe Trichocereeae on both eastern and western slopes of the Andes produced the greatest morphological and species diversity seen in South American columnar cacti. The Notocacteae have also stayed in southern South America, spreading southward into the Atacama region of Chile and eastward into Argentina, southern Brazil, and Uruguay. Tribe Cereeae, with maximal diversity in eastern Brazil, is a product of transcontinental, west-east migration, with subsequent northward migration into the Caribbean along the east-coast corridor (presumably by bird dispersal), as evidenced by the genera *Cereus, Pilosocereus* (Zappi, 1994), and *Melocactus* (Taylor, 1991).

As noted above, the Pachycereeae and Leptocereeae diverged when following a northward migration route, perhaps at the time when North America rejoined north-

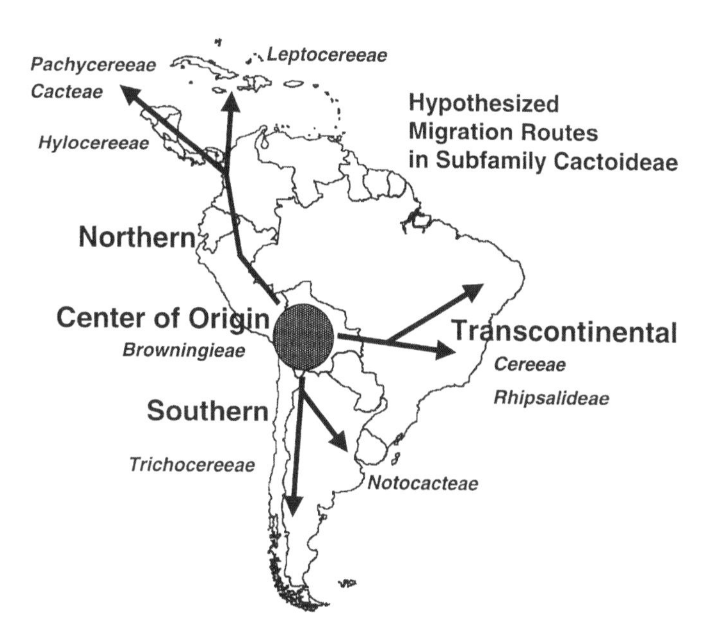

FIGURE 3.5. Hypothetical migration routes for the subfamily Cactoideae from a presumed mid-Andean center of origin for the Cactaceae. Within subfamily Cactoideae, tribes Trichocereeae and Notocacteae predominate an evolutionary radiation along the southern routes. Tribes Cereeae and Rhipsalideae diversified in the west-to-east transcontinental migration, and tribes Pachycereeae, Leptocereeae, Hylocereeae, and Cacteae resulted from northward migration into North America and the Caribbean. Members of tribe Browningieae and *Calymmanthium* remain restricted to the central Andes region.

ern South America, or as part of the previous landbridge between these continents in the early Cretaceous. This divergence must have occurred subsequent to their having shared a common ancestor with the extant genus *Corryocactus,* which is found in Peru, Bolivia, and Chile. The timing of any of these migrational events must be considered with respect to the timing of the North American and South American plates coming together once again in the late Eocene (Gibson and Nobel, 1986). Alternatively, a more recent dispersal for certain cactus groups involving the eastward movement of the Caribbean Islands may also be considered. These islands have certain endemic genera (e.g., *Consolea, Leptocereus*), which have been shown to be sister to other related groups on the North American mainland. Further phylogenetic studies are needed to infer migratory routes in the context of paleoclimate and geographical proximity.

The basic hypothesis of proposing three migration paths from an Andean cactus ancestor is consistent with the hypothesis of Raven and Axelrod (1974), who state that the cactus family evolved subsequent to the Andean orogeny, and that North American taxa are derived from South American ancestors. Still remaining are questions about the age of the Cactaceae; estimates have ranged from 100 million years to 20 million years or less (Mauseth, 1990). Further studies will be required to investigate the possibility of alternate migration paths and subsequent timing of the dispersal events relative to hypothesized paleoclimate and other biotic factors.

Morphological Trends in Columnar Cacti

From observations made in habitat and with cultivated cactus specimens, a number of trends in morphological change within the columnar lineages can be identified. These morphological modifications seem to be consistent in each lineage and follow similar polarity-of-character-state transitions.

In his *Morphology of Cacti,* Buxbaum (1950, p. 76) identified differences in branching pattern as a possible distinguishing character in various cactus groups. Based on observations of a number of representative specimens from all columnar cactus lineages, it appears that basitonic branching (that is, new growth predominantly from the base of an existing stem) is the plesiomorphic or ancestral condition. Examples of this branching type occur in *Neoraimondia, Haageocereus,* and certain *Trichocereus/Echinopsis* (e.g., *T. chilensis*). Transition to mesotonic branching (i.e., from the mid-stem region, but always subapical) is more derived. Examples of mesotonic branching are *Carnegiea, Pachycereus, Escontria, Cereus, Browningia candelaris,* and *Armatocereus.* Acrotonic branching (from the apex of the stem) is not found within most columnar lineages, but examples include members of *Rhipsalis* or many *Opuntia* and *Cylindropuntia* species. Branching pattern may be loosely used as a general indicator of a relative degree of morphological advancement within a particular lineage, although admittedly this correlation has not been empirically tested.

Another trend observed in columnar cacti is the tendency to become globular in form in taxa possessing more advanced specializations within the columnar lineages. This can be seen in the genus *Melocactus* in tribe Cereeae, in *Neoporteria/Eriosyce, Parodia,* and *Frailea* in the Notocacteae, and in *Gymnocalycium, Rebutia, Oroya,* and *Matucana* in tribe Trichocereeae. This tendency may represent a change in surface area:volume ratio in response to local floristic context, water stress, or as an adaptive pollination or antiherbivory strategy. Further study is needed to correlate various biotic factors with prevalence of the evolution of the globular habit from columnar forms.

Similarly, the independent development of cephalia in derived groups within each columnar lineage appears to be a successful adaptive strategy. Separating reproductive primordia from nonreproductive primordia, development of accessory protective structures (e.g., bristles or hairs), and enhancing floral presentation with multiple, closely spaced flowers (as opposed to a more separated floral distribution) may have distinct advantages that are adaptive to a broad pollinator spectrum. Buxbaum (1959) noted several adaptive characters in South American cacti with hairy cephalia and the cephalia (or pseudocephalia) seen in the Pachycereeae (e.g., in *Backebergia/*

Pachycereus militaris) must be evolutionarily distinct from that seen in *Melocactus* or *Discocactus*, yet their adaptive benefits are likely quite similar.

The columnar cacti have also evidenced several shifts in their pollination syndromes. If it is assumed that the Cactaceae evolved from an entomophilous ancestor, subsequent specialization to bat or hummingbird pollination would represent a derived condition. Terminal taxa in lineages that were plesiomorphically bat pollinated but are now insect pollinated would represent a reversion, or a secondarily derived syndrome. Phylogenies are necessary to elucidate such relationships, which will be examined in future investigations.

Finally, it is strikingly apparent when all columnar lineages are compared that there are several examples of amphitropical parallels occurring in these cacti. Examples include the multiple development of hummingbird pollination syndromes in tribes Trichocereeae (e.g., *Cleistocactus, Matucana, Oreocereus*), Cereeae (e.g., *Arrojadoa, Melocactus*), and Pachycereeae (e.g., *Echinocereus, Rathbunia*), and the unusual creeping habit observed in populations of *Stenocereus eruca* in Baja California, which has a morphological parallel in the Peruvian coastal endemic *Haageocereus tenuis.* Other examples of similar parallelisms undoubtedly served to confuse previous cactologists, and with a phylogenetic interpretation of these lineages, it is possible to examine these phenomena in greater detail with more significant biological relevance.

Systematic Implications of Molecular Data and the Cactaceae

A summary of the hypothesized relationships between major lineages of columnar cacti is presented in Fig. 3.3. This summary incorporates data from molecular variation, morphological affinities, and biogeographic considerations. Subsequent studies will evaluate the basic topology of this hypothetical cladogram, and will add branches to provide further resolution at lower taxonomic ranks. Discussions and investigations of select genera and individual species that have been taxonomically problematic in the past (e.g., Hunt and Taylor, 1991, 1992) raise basic phylogenetic questions to be resolved using molecular methods that have direct taxonomic relevance.

From the above discussion, it should be apparent that the phylogenetic information provided by molecular studies can do much to assist in the systematic (and ultimately taxonomic) reevaluation of the Cactaceae. Although these studies by themselves can establish basic hypotheses of relationships, additional comprehensive comparative studies—such as those by Gibson (1973), Mauseth (1993), Mauseth and Plemons-Rodriguez (1998)—and other phytochemical studies as outlined in Wallace (1986) are needed to provide corroborative evidence and to enable evaluation of evolutionary trends in morphological development. Ultimately, the synthesis of this information will lead to even more robust hypotheses regarding cactus phylogeny and the evolutionary history of a marvelous assemblage of plants.

Summary

Lineages of the cactus subfamily Cactoideae, which has a columnar or cereoid habit, are reviewed in the context of molecular systematic studies of *cp*DNA variation. The phylogenetic relationships elucidated by molecular studies provide data from which taxonomic realignments may be made and provide a new systematic context for future studies of morphological evolution and biogeography of these cacti. Two primary clades for the columnar cacti are identified: a clade consisting of members of tribes Browningieae (in part), Cereeae and Trichocereeae, and another composed of tribe Browningieae (in part), tribe Pachycereeae, and a redefined tribe Leptocereeae. The evolutionary positions of phylogenetically critical genera (such as *Calymmanthium* and *Corryocactus*) are discussed with respect to the divergences of columnar groups within subfamily Cactoideae. The columnar cacti, like all cacti, have a South American ancestry. A biogeographic scenario of columnar cactus evolution is presented.

Resumen

Los linajes de la subfamilia Cactoideae de la familia Cactaceae que tienen un hábito columnar ó cereoideo se repasan en el contexto de estudios sistemáticos moleculares de la variación del DNA en el cloroplasto. Las relaciones filogenéticas aclaradas por los estudios moleculares proporcionan datos de los cuales las realineaciones taxonómicas que pueden ser hechas, y proporcionan un nuevo contexto sistemático para los estudios futuros en la evolución morfológica y de la biogeografía de estos cactos. Los dos clados primarios para cactos acolumnares se identifican como: uno que consiste en la tribu Browningieae (en parte), la tribu Cereeae y la Trichocereeae, y otro integrado, por la Browningieae (en parte). La tribu Pachycereeae y una tribu redefinida Leptocereeae. Las posiciones evolutivas de los géneros filogenéticamente críticos tales como: *Calymmanthium* y *Corryocactus* se discuten con respecto a las divergencias de grupos acolumnares dentro de la Subfamilia Cactoideae. Los cactos acolumnares tienen un ancestro ascendencia sudamericano. Un esquema biogeográfico de la evolución de los cactos acolumnares presentados.

ACKNOWLEDGMENTS

The majority of the plant material used in phylogenetic studies of the columnar cacti discussed here is from the Huntington Botanical Garden (San Marino, California), the Desert Botanical Garden (Phoenix, Arizona), the Royal Botanic Gardens, Kew, and the Städtische Sukkulentensammlung Zürich. I thank the staff at these institutions for their ongoing support of my research. Colleagues Edward F. Anderson, J. Hugo Cota, Arthur C. Gibson, Fred Kattermann, Louise Lippold, Carlos Ostalaza N., Bruce D. Parfitt, Robert Ross, and Nigel P. Taylor were also instrumental in obtaining plant material. Discussions with members of the IOS Cactaceae Consensus Working Group of the International Organization for Succulent Plant Study are very much appreciated. The manuscript was improved significantly by the senior editor and an anonymous reviewer. Funding support for this project was provided by grants from the National Science Foundation (DEB 92-07767 and DEB 95-27884), the Cactus and Succulent Society of America Research Fund, the Wallace Genetics Foundation, and the National Geographic Society (Grant Number 5473-95), which supported field research in Mexico, Chile, and Peru as well as molecular systematic studies of the columnar cacti.

REFERENCES

Backeberg, C. 1958–1962. *Die Cactaceae.* 6 Vols. Jena: Gustav Fischer Verlag.

———. 1977. *The Cactus Lexicon.* Poole, England: Blandford Press. [English edition.]

Barthlott, W. 1988. Über die systematische Gliederungen der Cactaceae. *Beiträge zur Biologie der Pflanzen* 63:17–40.

Barthlott, W., and G. Voit. 1979. Mikromorphologie der Samenschalen und Taxonomie der Cactaceae: Ein raster-elektronenmikroskopischer Überblick. *Plant Systematics and Evolution* 132:205–29.

Barthlott, W., and D. R. Hunt. 1993. Cactaceae. In *The Families and Genera of Vascular Plants,* eds. K. Kubitzki, J. G. Rohwer, and V. Bittrich. Vol. 2. 161–97. Berlin: Springer-Verlag.

Benson, L. 1982. *The Cacti of the United States and Canada.* Stanford: Stanford University Press.

Braun, P. 1988. On the taxonomy of the Brazilian Cereeae (Cactaceae). *Bradleya* 6:85–99.

Bregman, R. 1992. Seed studies in the subtribe Borzicactinae Buxbaum (Cactaceae); morphology, taxonomy, phylogeny and biogeography. *Botanischer Jahrbuch für Systematik* 114:201–50.

Bremer, K. 1988. The limits of amino acid sequence data in angiosperm phylogenetic reconstruction. *Evolution* 42:795–803.

Britton, N.L., and J. N. Rose. 1919–1923. *The Cactaceae.* Vols. 2 and 3. New York: Dover Publications. [1963 reprint.]

Buxbaum, F. 1950. *The Morphology of Cacti. Section 1. Roots and Stems.* Pasadena, Calif.: Abbey Garden Press.

———. 1958. The phylogenetic division of the subfamily Cereoideae, Cactaceae. *Madroño* 14:177–216.

———. 1959. Die behaartblütigen Cephalientrager Südamerikas. *Österreichen Botanischer Zeitschrift* 106:138–158.

———. 1966. The origin of Tribe Browningieae. *Cactus and Succulent Journal (U.S.)* 38:43–46.

———. 1969. Die Entwicklungswege der Kakteen in Südamerika. In *Biogeography and Ecology in South America,* eds. E. J. Fittkau, J. Illes, H. Klinge, G. H. Schwabe, and H. Sioli, 583–623. The Hague: Dr. W. Junk.

Chase, M. W., D. E. Soltis, R. G. Olmstead, D. Morgan, D. H. Les, B. D. Mishler, et al. 1993. Phylogenetics of seed plants: An analysis of nucleotide sequences from the plastid gene *rbc*L. *Annals of the Missouri Botanical Garden* 80:528–80.

Cota, J. H., and R. S. Wallace. 1997. Chloroplast DNA evidence for divergence in *Ferocactus* and its relationships to North American columnar cacti (Cactaceae: Cactoideae). *Systematic Botany* 22:529–42.

Dickie, S. L. 1996. Molecular systematic study of subfamily Opuntioideae (Cactaceae). Master's thesis, Iowa State University, Ames, Iowa.

Dickie, S. L., and R. S. Wallace. 2001. Phylogeny and evolution in the Subfamily Opuntioideae (Cactaceae): Insights from *rpl*16 intron sequence variation. *Systematic Botany* (in review).

Felsenstein, J. 1985. Confidence limits on phylogenies: an approach using the bootstrap. *Evolution* 39:783–91.

Gibson, A. C. 1973. Comparative anatomy of secondary xylem in Cactoideae (Cactaceae). *Biotropica* 5:29–65.

———. 1982. Phylogenetic relationships of Pachycereeae. In *Ecological Genetics and Evolution. The Cactus-Yeast-Drosophila Model System,* eds. J.S.F. Barker and W. T. Starmer, 3–16. Sydney: Academic Press.

———. 1988a. The systematics and evolution of subtribe Stenocereinae. 1. Composition and definition of the subtribe. *Cactus and Succulent Journal (U.S.)* 60:11–16.

———. 1988b. The systematics and evolution of subtribe Stenocereinae. 2. *Polaskia. Cactus and Succulent Journal (U.S.)* 60:55–62.

———. 1988c. The systematics and evolution of subtribe Stenocereinae. 3. *Myrtillocactus. Cactus and Succulent Journal (U.S.)* 60:109–16.

———. 1988d. The systematics and evolution of subtribe Stenocereinae. 4. *Escontria. Cactus and Succulent Journal (U.S.)* 60:161–67.

———. 1991. The systematics and evolution of subtribe Stenocereinae. 11. *Stenocereus dumortieri* versus *Isolatocereus dumortieri. Cactus and Succulent Journal (U.S.)* 63:184–90.

———. 1992. The Peruvian Browningias and *Gymnanthocereus. Cactus and Succulent Journal (U.S.)* 64:62–68.

Gibson, A. C., and K. E. Horak. 1978. Systematic anatomy and phylogeny of Mexican columnar cacti. *Annals of the Missouri Botanical Garden* 65:999–1057.

Gibson, A. C., and P. S. Nobel. 1986. *The Cactus Primer.* Cambridge, Mass.: Harvard University Press.

Gibson, A. C., K. C. Spencer, R. Bajaj, and J. L. McLaughlin. 1986. The ever-changing landscape of cactus systematics. *Annals of the Missouri Botanical Garden* 73:532–55.

Hunt, D. 1992. *CITES Cactaceae Checklist.* Kent, England: Royal Botanic Gardens, Kew.

Hunt, D., and N. Taylor. 1986. The genera of the Cactaceae: towards a new consensus. *Bradleya* 4:65–78.

———. 1990. The genera of the Cactaceae: progress towards consensus. *Bradleya* 8:85–107.

———. 1991. Notes on miscellaneous genera of Cactaceae. *Bradleya* 9:81–92.

———. 1992. Notes on miscellaneous genera of Cactaceae (2). *Bradleya* 10:17–32.

Jordan, W. C., M. W. Courtney, and J. E. Neigel. 1996. Low levels of intraspecific genetic variation at a rapidly evolving chloroplast DNA locus in North American duckweeds (Lemnaceae). *American Journal of Botany* 83:430–39.

Kelchner, S. A, and J. F. Wendel. 1996. Hairpins create minute inversions in non-coding regions of chloroplast DNA. *Current Genetics* 30:259–62.

Leuenberger, B. E. 1986. *Pereskia* (Cactaceae). *Memoirs of the New York Botanical Garden* 41:1–140.

Mauseth, J. D. 1990. Continental drift, climate, and the evolution of cacti. *Cactus and Succulent Journal (U.S.)* 62:302–8.

———. 1993. Medullary bundles and the evolution of cacti. *American Journal of Botany* 80:928–32.

———. 1996. Comparative anatomy of subfamilies Cereeae and Browningieae (Cactaceae). *Bradleya* 14:66–81.

———. 1997. Comparative anatomy of *Neoraimondia roseiflora* and *Neocardenasia herzogiana* (Cactaceae). *Haseltonia* 5:37–50.

Mauseth, J. D., and B. Plemons-Rodriguez. 1998. Evolution of extreme xeromorphic characters in wood: a study of nine evolutionary lines in Cactaceae. *American Journal of Botany* 85:209–18.

Mauseth, J. D., and R. G. Ross. 1988. Systematic anatomy of the primitive cereoid cactus *Leptocereus quadricostatus. Bradleya* 6:49–64.

Nyffeler, R. 1997. Stem anatomy of *Uebelmannia* (Cactaceae)—with special reference to *Uebelmannia gummifera. Botanica Acta* 110:489–95.

Nyffeler, R., U. Eggli, and B. E. Leuenberger. 1997. Noteworthy idioblastic sclerids in the stems of *Eulychnia* (Cactaceae). *American Journal of Botany* 84:1192–97.

Ostalaza, N. C. 1982. *Neoraimondia roseiflora* (Werd.) Backbg. *Boletin de Lima* 19:89–93.

Raven, P. H., and D. L. Axelrod. 1974. Angiosperm biogeography and past continental movements. *Annals of the Missouri Botanical Garden* 61:539–673.

Ritter, F. 1979–1981. *Kakteen in Sudamerika.* 4 vols. Spangenberg, Germany: Selbstverlag.

Shinozaki, K., M. Ohme, T. Tanaka, T. Wakasugi, N. Hayashida, et al. 1986. The complete nucleotide sequence of the tobacco chloroplast genome: its gene organization and expression. *EMBO Journal* 5:2043–49.

Swofford, D. L. 1993. *PAUP: Phylogenetic Analysis Using Parsimony,* version 3.1.1. Washington, D.C.: Smithsonian Institution.

Taberlet, P., L. Geilly, G. Pautau, and J. Bouvet. 1991. Universal primers for amplification of three non-coding regions of chloroplast DNA. *Plant Molecular Biology* 17:1105–9.

Taylor, N. P. 1985. *The Genus Echinocereus.* Kent, England: Royal Botanic Gardens, Kew.

———. 1991. The genus *Melocactus* in Central and South America. *Bradleya* 9:1–80.

Taylor, N. P., and D. C. Zappi. 1989. An alternative view of generic delimitation and relationships in tribe Cereeae (Cactaceae). *Bradleya* 7:13–40.

Wallace, R. S. 1986. Biochemical taxonomy and the Cactaceae: an introduction and review. *Cactus and Succulent Journal (U.S.)* 58:35–38.

———. 1995. Molecular systematic study of the Cactaceae: Using chloroplast DNA to elucidate cactus phylogeny. *Bradleya* 13:1–12.

———. 1997. The phylogenetic position of *Mediocactus hahnianus. Cactaceae Consensus Initiatives* 4:11–12.

Wallace, R. S., and J. H. Cota. 1996. An intron loss in the chloroplast gene *rpo*C1 supports a monophyletic origin for the subfamily Cactoideae of the Cactaceae. *Current Genetics* 29:275–81.

Wallace, R. S., and E. D. Forquer. 1995. Molecular evidence for the systematic placement of *Echinocereus pensilis* (K. Brandegee) J. Purpus (Cactaceae: Cactoideae: Echinocereeae) *Haseltonia* 3:71–76.

Watrous, L. E., and Q. D. Wheeler. 1981. The outgroup comparison method of character analysis. *Systematic Zoology* 30:1–11.

Zappi, D. C. 1994. *Pilosocereus* (Cactaceae): the genus in Brazil. *Succulent Plant Research.* Vol. 3. Sherborne, England: David Hunt Publishers.

CHAPTER 4

Phylogenetic Relationships of Pachycereeae: A Cladistic Analysis Based on Anatomical-Morphological Data

Teresa Terrazas
Sofía Loza-Cornejo

Introduction

Within the subfamily Cactoideae, the tribe Pachycereeae has undergone a substantial radiation of columnar and shrubby species in Mexico. Pachycereeae sensu Barthlott and Hunt (1993) has ten genera and nearly 60 species. Although most taxa are endemic to Mexico, a few species are distributed in Central America, northern South America, and the West Indies. In this chapter, we first synthesize the taxonomic history of the tribe, as well as the anatomical, morphological, and chemical characters that have been used to suggest phylogenetic relationships within it (Buxbaum, 1961; Gibson and Horak, 1978; Gibson, 1982; Gibson et al., 1986). We then present the results of a cladistic analysis for some members of the tribe based on anatomical and morphological data in order to identify apomorphic characters and to propose a working hypothesis for future studies.

Taxonomic History

Buxbaum (1958) proposed the tribe Pachycereeae containing six genera in his phylogenetic classification of the subfamily. Three years later, he offered a modified interpretation of the tribe by recognizing 13 genera and grouping them into five subtribes with tribe Leptocereeae as the putative ancestor (Table 4.1). His classification was based on external characters of flowers, fruits, seeds, and seedlings. He was also the first to recognize the importance of seed structure, funicular pigment cells in the flower at anthesis, and the presence of triterpenes for aligning some species of this tribe.

In their well-known studies, Gibson and Horak (1978) and Gibson et al. (1986) suggested several modifications to Buxbaum's classification, based on the discovery of silica bodies in epidermal and hypodermal cells of species of *Stenocereus.* Thus, they emended the limits of the genus *Stenocereus,* based mainly on silica bodies in dermal tissue, to include *Machaerocereus* and *Rathbunia.* They also proposed two subtribes: the Stenocereinae, which groups *Escontria, Myrtillocactus, Polaskia,* and *Stenocereus* together (based on the presence of triterpenes; dull, rough seeds; and funicular pigment cells); and the Pachycereinae (based on the presence of alkaloids; smooth, glossy seeds; and a tendency to possess crystals in the dermal system). Unfortunately, Gibson and collaborators never presented the data matrix or the character state coding for most of the features they discussed.

In 1990, the Cactaceae working group (established by the International Organization for the Study of Succulent Plants [IOS]) published the list of accepted taxa and placed the tribe Pachycereeae in group VI (Hunt and Taylor, 1990), recognizing only seven genera (Table 4.1). In this proposal, *Bergerocactus* was transferred from the tribe Echinocereeae of Buxbaum (1958) to Pachycereeae; *Rathbunia* was reestablished based on a proposal to conserve the genus name; *Backebergia, Lemaireocereus, Lophocereus, Mitrocereus,* and *Pterocereus* were recognized as part of *Pachycereus; Carnegiea* was defined in a broader sense to include *Anisocereus, Neobuxbaumia, Neodawsonia,* and *Pseudomitrocereus;* and *Escontria* and *Polaskia* were treated as part of *Myrtillocactus.* In the most recent classification of the Cactaceae (Barthlott and Hunt, 1993), ten genera were included in Pachycereeae (Table 4.1). However, we have been unable to find a reference where the key characters of the emended genera are presented.

More recently, a cladogram was presented by Cornejo and Simpson (1997) supporting subtribes Pachycereinae and Stenocereinae. These authors recognized most of the genera that have been submerged in *Carnegiea, Myrtillocactus, Pachycereus,* and *Stenocereus.* Notably, most of the recognized species of Stenocereinae were included; but *Cephalocereus,* as defined by Barthlott and Hunt (1993), was not included in their study. This topology was probably based on morphological characters; however, as in previous studies, the authors did not indicate which characters were used, but mentioned that the topology was adapted from phylogenetic information provided by Gibson and coworkers. Cota and Wallace (1997), investigating the relationships between *Ferocactus* and columnar cacti of subtribe Stenocereinae (especially *Escontria*), studied eight species of this subtribe and six of the Pachycereinae. Their tree topology, based on *cp*DNA restriction-site variation, supports the monophyly of Pachycereeae and the recognition of subtribes Pachycereinae and Stenocereinae. Their results also suggest that *Pachycereus,* as defined by Barthlott and Hunt (1993), is paraphyletic. To date, no cladistic analysis has tested the anatomical and morphological characters that have been proposed as synapomorphies for the two subtribes. One of the objectives of this chapter is to present the results of a cladistic analysis of several members of the Pachycereeae.

TABLE 4.1

Generic Classification of Pachycereeae Proposed by Different Authors[a]

Buxbaum (1958)	*Buxbaum (1961)*	*Buxbaum (1975)*	*Gibson and Horak (1978)*	*IOS (1989)*	*Barthlott and Hunt (1993)*
Carnegiea	Pterocereinae	Pterocereinae	Pachycereinae	*Bergerocactus*	*Bergerocactus*
Cephalocereus	*Escontria*	*Escontria*	*Backebergia*	*Carnegiea*	*Carnegiea*
Lemaireocereus	*Pterocereus*	*Pterocereus*	*Carnegiea*	*Cephalocereus*	*Cephalocereus*
Mitrocereus	Pachycereinae	Pachycereinae	*Cephalocereus*	*Myrtillocactus*	*Escontria*
Neobuxbaumia	*Heliobravoa*	*Heliobravoa*	*Lophocereus*	*Pachycereus*	*Myrtillocactus*
Pachycereus	*Pachycereus*	*Pachycereus*	*Mitrocereus*	*Rathbunia*	*Neobuxbaumia*
	Pseudomitrocereus	*Pseudomitrocereus*	*Neobuxbaumia*	*Stenocereus*	*Pachycereus*
	Stenocereinae	Stenocereinae	*Pachycereus*		*Polaskia*
	Carnegiea	*Carnegiea*	Stenocereinae		*Rathbunia*
	Lophocereus	*Cephalocereus*	*Escontria*		*Stenocereus*
	Stenocereus	*Lophocereus*	*Myrtillocactus*		
	Cephalocereinae	*Machaerocereus*	*Polaskia*		
	Cephalocereus	*Mitrocereus*	*Stenocereus*		
	Mitrocereus	*Neobuxbaumia*			
	Neobuxbaumia	*Rathbunia*	*Anisocereus?*		
	Myrtillocactinae	*Stenocereus*	*Pterocereus?*		
	Myrtillocactus	Myrtillocactoinae	*Lemainocereus?*		
	Polaskia	*Myrtillocactus*			
		Polaskia			

[a] Modified from Gibson (1988a).

Characters

Few anatomical, morphological, and chemical characters and no character coding have been explicitly presented in previous studies of the Pachycereeae. A character synthesis for the genera recognized by Barthlott and Hunt (1993) will be given below, based on our own observations and on published data (Buxbaum, 1961; Gibson, 1973, 1988a–e, 1989a,b, 1990a,b, 1991; Bravo-Hollis, 1978; Gibson and Horak, 1978; Gibson and Nobel, 1986; Gibson et al., 1986; Bravo-Hollis and Sánchez-Mejorada, 1991; Arias-Montes et al., 1997).

Anatomical Characters

Anatomical characters have been considered by Gibson and collaborators to be the key to grouping the genera of Pachycereeae into subtribes. The number of species studied anatomically is shown in Table 4.2. This summary indicates that nearly 50% of the species recognized by Hunt (1992) have been studied based on stem anatomy, whereas only 28% have been studied based on wood. Here we summarize anatomical features from pith to dermal tissue of stems of Pachycereeae.

Pith. In columnar cacti, pith size is related to species habit. Pith is typically composed of parenchyma cells and has a poorly developed vascular system. Pith mucilage cells are absent in species of *Cephalocereus* (two species), *Neobuxbaumia* (one species), *Pachycereus* (three species), and *Polaskia* (two species). Mucilage cells are few, small,

TABLE 4.2
Number of Species Studied Anatomically by Gibson (1973, 1988a–e, 1989a,b, 1990a,b) and Gibson and Horak (1978) in Genera of the Tribe Pachycereeae

Genus[a]	*Anatomical Characteristic*	
	Stem	*Wood*
Bergerocactus (1)	1	1
Carnegiea (1)	1	1
Cephalocereus (5)	1	1
Escontria (1)	1	1
Myrtillocactus (4)	3	2
Neobuxbaumia (8)	2	1
Pachycereus (13)	7	5
Polaskia (2)	2	0
Rathbunia (2)	2	1
Stenocereus (21)	13	5

[a] Species recognized by Hunt (1992).

and scattered in *Escontria* and *Myrtillocactus* (three species), *Pachycereus* (five species), *Stenocereus dumortieri, S. martinezii,* and *S. pruinosus,* but are more numerous in other species of *Stenocereus.* According to Gibson (1982), highly derived columnar cacti have numerous mucilage cells in the pith and cortex. Notably, mucilage cells are present in all studied members of tribe Echinocereeae (Mauseth et al., 1998), but their occurrence is variable in members of Browningieae and Cereeae (Mauseth, 1996). Medullary vascular bundles are commonly smaller and fewer in number than cortical bundles in Pachycereeae, as is true in other Cactoideae (Mauseth, 1996; Mauseth et al., 1998). Pith in *Pereskia* lacks medullary bundles and is considered a relict (Mauseth and Landrum, 1997). Medullary vascular bundles are mostly collateral with little secondary growth and without primary phloem fibers. *Rathbunia alamosensis* and *R. kerberi* show amphivasal medullary vascular bundles that are related to secondary growth. This is probably a synapomorphy for both taxa. Apart from mucilage cells, crystals have been reported in *Pachycereus fulviceps, P. weberi,* and *Neobuxbaumia mezcalaensis.* Occurrence of mucilage cells and crystals in this region was coded as two-state characters for the cladistic analysis (Appendix 4.1).

Stele. The stele consists of a ring of various collateral bundles surrounding the pith. Vascular bundles are separated by parenchyma, and different patterns of fusion occur when vascular cambium establishes and pith enlargement stops. A delay in fiber development has been observed after vascular cambium is established in species of *Neobuxbaumia* (T. Terrazas, unpubl. data). Our study has confirmed the occurrence of extraphloematic fibers in each vascular bundle in members of Pachycereeae. However, some species develop these fibers during the first stages of primary vascular tissue differentiation near the apex, as observed in *Stenocereus* (three species) and

Pachycereus weberi. Species of other genera show a relatively later differentiation, as seen in *Bergerocactus, Carnegiea, Cephalocereus* (one species), *Escontria, Myrtillocactus* (three species), *Polaskia* (two species), the remaining species of *Pachycereus* (six species), and *Stenocereus* (seven species). Because lack of extraphloematic fibers is considered relictual in Cactaceae (Mauseth and Landrum, 1997), their early development is interpreted here as relictual in Pachycereeae. The character, extraphloematic fibers, was coded as a two-state character (Appendix 4.1).

Wood. In the Pachycereeae, wood is diffuse-porous and vessels are mostly solitary but with some groups of two to five vessels. Tangential vessel diameter varies from 49 µm in *Pachycereus schottii* to 110 µm in *Pachycereus pringlei.* Lengths of vessel elements range from 178 µm in *Rathbunia alamosensis* to 420 µm in *Neobuxbaumia mezcalaensis.* Perforation plates are simple and mostly transverse, with intervascular pitting ranging from alternate to scalariform. Deposits are absent; however, small crystals are deposited on external cell walls in *Pachycereus hollianus.* Libriform fibers may be exclusively nonseptate, septate, or mixed, depending on the species. Axial parenchyma ranges from scanty to vasicentric in strands of two cells, and deposits are mostly absent, except in *Stenocereus chrysocarpus* and *S. queretaroensis,* which have prismatic crystals in the parenchyma strands. Primary and secondary rays are distinctive in most taxa. Primary rays are longer, varying from 4 mm in *Stenocereus stellatus* to 8 mm in *Carnegiea gigantea,* whereas secondary rays are shorter, ranging from 1 mm in *Myrtillocactus geometrizans* to 3 mm in *Pachycereus hollianus.* Perforated ray cells have been observed in seven of the ten genera (Terrazas, 2000). Starch grains and crystals occur in ray cells, whereas small silica grains have only been seen in *Pachycereus weberi* (Terrazas, unpubl. data). Qualitative wood features within Pachycereeae appear to be remarkably similar to those reported for *Leptocereus* and other fibrous species of Echinocereeae (Mauseth et al., 1998), as well as those for tribes Browningieae and Cereeae (Mauseth, 1996). Gibson (1973) mentioned that there is a strong allometric relationship between tracheary elements and stem size. Based on this statement, vessel element length was the character used in the cladistic analysis presented here. Most qualitative characters just described are autapomorphies for few species and were not included in the phylogenetic analysis.

Cortex. The diversity of tactile texture in the cortical tissue corresponds, at the anatomical level, to differences between species in number, size, and distribution of parenchyma cells, idioblastic mucilage cells, and crystal-bearing cells. In all species of Pachycereeae that have been examined, two regions are well defined: the palisade cortex (just below the hypodermis), and the inner cortex (located in the central rib and between the bases of the ribs and the stele). Within these regions, Gibson and Horak (1978) recognized three basic patterns of mucilage distribution: (1) medium-to-large mucilage cells that are generally evenly spaced in the main stem and ribs, but are not abundant near the hypodermis, as in *Carnegiea, Escontria, Myrtillocactus* (three species), *Neobuxbaumia* (three species), and *Pachycereus* (six species); (2) mucilage cells that are closely packed throughout the main stem and rib cortex and are generally found near the hypodermis, as in species of *Bergerocactus emoryi, Cephalocereus* (two species), *Pachycereus hollianus,* and *Stenocereus* (six species); and (3) large mucilage cells that are located beneath the hypodermis, as in *Stenocereus thurberi* and *S. martinezii.* Occurrence of mucilage cells in the cortex was coded as a binary character (Appendix 4.1).

According to Gibson and Horak (1978) and Gibson (1982), mucilage cells in the cortex are absent in species of *Pachycereus* (four species) and *Polaskia* (two species), but are present in more specialized cacti. Nevertheless, we observed mucilage cells in *Pachycereus hollianus.* Because *Pereskia* possesses numerous mucilage cells in its cortex, this is considered to be the ancestral condition (Mauseth and Landrum, 1997). Gibson and Horak (1978) indicated that usually only highly derived species have mucilage cells in the pith and cortex. Loss or gain of stem mucilage cells in certain species has likely occurred several times in different lineages. As in Pachycereeae, mucilage cells occurrence is variable in members of tribes Cereeae and Browningieae, and in *Leptocereus* and *Neoabbottia* of the tribe Echinocereeae (Taylor and Zappi, 1989; Mauseth, 1996; Mauseth et al., 1998). Gibson (1982) hypothesized an irreversible trend from zero to many mucilage structures, assuming that mucilage has a function in increasing matrix capacity for water retention. However, other authors have stated that the functional significance of mucilage is not yet fully understood (Nyffeler and Eggli, 1997).

Gibson and Horak (1978) mentioned that numerous crystal-bearing cells are present in several species of *Pachycereus,* as well as in *Neobuxbaumia* (two species) and *Cephalocereus* (two species), and are unusually abundant in *Pachycereus fulviceps.* They considered this feature a probable synapomorphy for the subtribe Pachycereinae. These authors also indicated that many species of *Stenocereus* have large aggregates of cuboidal crystals near the stele and scattered aggregates of crystals in the center of the rib. However, they have not been observed in six species of *Stenocereus.* Presently, we do not know whether the same type of crystals occurs in Pachycereinae and Stenocereinae. A thorough review of the types of crystals and their anatomical distribution is needed to properly code these features for phylogenetic analysis. Crystals have been reported in genera of tribes Browningieae and Cereeae, except in *Armatocereus* (in Browningieae) and *Cereus* (in Cereeae) (Mauseth, 1996).

A vast system of cortical vascular bundles exists in members of the Pachycereeae. Cortical bundles show secondary growth without xylary fibers, a feature shared with tribes Browningieae, Cereeae, and some members of Echinocereeae (Mauseth, 1996; Mauseth et al., 1998). With the exception of *Bergerocactus emoryi,* extraphloem fibers in cortical bundles are absent; however, few secondary phloematic sclereids occur in *Stenocereus fricii* and *S. stellatus.*

Dermal system. Most members of Pachycereeae have a unistratose epidermis with a cuticle that varies in thickness in different species. Multistratose epidermis has formed, however, in several species of this tribe (Gibson and Horak, 1978; Loza-Cornejo and Terrazas, unpubl. data). The divisions that produce the multistratose epidermis start in very young tissues at the stem apex. However, these divisions may be very regular and mostly periclinal, so that the mature epidermis consists of pairs of cells, as in *Pachycereus pecten-aboriginum, P. tepamo,* and *Stenocereus dumortieri.* Multistratose epidermis occurs in the remaining species, but the divisions are irregular, as in *Neobuxbaumia tetetzo, Polaskia chende, P. chichipe, Stenocereus stellatus,* and *S. treleasei,* where oblique, anticlinal, and periclinal divisions that occur in small patches produce an irregular epidermis called "bullate" by Gibson and Horak (1978). In species of *Carnegiea, Cephalocereus* (two species), *Escontria, Myrtillocactus* (three species), *Pachycereus* (two species), and *Stenocereus* (two species), the epidermis proliferates by irregular divisions of some epidermal cells after the cuticle has been deposited and as a consequence, the epidermal surface cannot expand. This character state of the epidermis is termed "internal divisions." Multistratose epidermis of both irregular patterns has also been reported for several species of *Armatocereus, Browningia, Dendrocereus, Monvillea, Neoabbottia,* and *Pilosocereus* (Mauseth, 1996; Mauseth et al., 1998); it also occurs sporadically in *Pereskia* (Mauseth and Landrum, 1997).

With the available data, including our own observations, we decided to identify the multistratose epidermis character state only when divisions are very regular and start very early in development, as in *Pachycereus pecten-aboriginum* and *P. tepamo.* Uni- or multistratose epidermis was coded as a binary character, whereas presence or absence of secondary divisions in the epidermal cells was coded as an unordered multistate character. Other dermal features such as stoma type, position of stomatal apparatus relative to the epidermal surface, and undulation of cell wall should be studied in more representatives of Pachycereeae, as different patterns of variation have been reported in *Bergerocactus emoryi, Escontria chiotilla, Myrtillocactus cochal, Pachycereus* (two species), and *Stenocereus* (two species) (Gibson and Horak, 1978; Eggli, 1984).

In most Pachycereeae, the cuticle is smooth and varies in thickness from nearly 1 µm in some species of *Stenocereus* to 30 µm in species of *Myrtillocactus.* Those taxa with thicker cuticle over the outer tangential wall may show wavelike ridges, as in *Carnegiea* or form hornlike protrusions, as in *Escontria, Myrtillocactus* (three species), *Pachycereus marginatus,* and *P. hollianus.* Cuticle accumulation pattern was coded as a three-state character for the cladistic analysis (Appendix 4.1).

Silica grains in epidermal cells occur in all species of *Rathbunia* (two species) and *Stenocereus* (14 species) that have been sampled; this seems to be a true synapomorphy for both taxa. With the exception of *Echinocereus pensilis,* silica grains in the epidermal cells have not been found in any other member of the Cactoideae (Loza-Cornejo and Terrazas, unpubl. data). The presence of prismatic crystals may have evolved independently in different lineages of this subfamily (Mauseth et al., 1998).

In the Pachycereeae, they occur in *Bergerocactus emoryi* (rarely), *Cephalocereus* (two species), *Neobuxbaumia* (two species), and *Pachycereus fulviceps*. Epidermal inclusions were coded as a multistate character (Appendix 4.1).

The hypodermis is collenchymatous in all Pachycereeae; however, the number of cellular strata and total width vary among taxa. Although thickness of the hypodermis needs to be studied in more samples per species, we coded it in three categories, which represent the variation presently known (Appendix 4.1). Silica grains occurred in species of *Rathbunia* (two species) and *Stenocereus* (14 species), whereas prismatic crystals have been observed in *Cephalocereus* (two species), *Neobuxbaumia* (two species), and *Pachycereus fulviceps*. Hypodermal crystals have been reported in species of *Armatocereus* and *Jasminocereus* of the tribe Browningieae; in species of *Melocactus, Monvillea,* and *Pilosocereus* of the tribe Cereeae; and in species of *Dendrocereus* and *Neoabbottia* of the tribe Echinocereeae (Mauseth, 1996; Mauseth et al., 1998). Hypodermal inclusions were coded as an unordered multistate character (Appendix 4.1).

Morphological Characters

Vegetative and reproductive features that may be useful for phylogenetic analysis and their state of knowledge are summarized here. Three authors have synthesized morphological characters for most species of Pachycereeae and Stenocereinae (Buxbaum, 1961; Bravo-Hollis, 1978; Gibson, 1988b–e, 1989a,b, 1990a,b, 1991).

Vegetative morphology. The Pachycereeae is recognized by its tall, arborescent members with unbranched or highly branched stems. In such genera as *Cephalocereus, Pachycereus,* and *Stenocereus,* habit may vary from short basitomic shrubs to tall trees with numerous basitomic or acrotomic secondary branches or unbranched stems. Thus, the establishment of character states for habit is difficult: Branch abundance and stem definitions appear to be related to environmental factors (Cornejo and Simpson, 1997). Gibson (1973) suggested that habit and tracheary elements are highly allometrically related. Based on this assertion, vessel element length, which is a more conservative measure, was used in this study as an indirect way to code habit characteristics (Appendix 4.1). Moreover, other parameters of tree architecture (e.g., the angles between the stem and branches and between two branches) should be evaluated within the Pachycereeae. Buxbaum (1950) suggested that the angle of branch formation is characteristic for each species; however, this feature has been overlooked in the Pachycereeae as well as in other columnar cacti.

Number of ribs varies widely within Pachycereeae. *Escontria, Myrtillocactus,* and *Rathbunia* have the fewest ribs, whereas *Carnegiea, Cephalocereus,* and *Neobuxbaumia* have the most. Species of *Pachycereus* and *Stenocereus* exhibit wide variation in rib number. This variation appears to be unrelated to habit: Within *Stenocereus, S. gummosus* and *S. montanus* have different growth forms but similar numbers of ribs (seven or eight). Rib number was coded as a binary character for the cladistic analysis (Appendix 4.1).

Interareolar morphology and areole shape need to be analyzed carefully to understand their possible value in determining relationships. However, areole color appears to be a highly discrete character. Red-colored areoles are found in seven species of *Stenocereus* (Gibson, 1990a,b; Arreola-Nava, 2000). Arreola-Nava (2000) showed that the red color is caused by a dark-brown deposit that occludes the head-cell lumina of the glandular trichomes. Occurrence of glandular trichomes was included in the analysis. The presence of a horizontal notch midway between areoles is a distinctive feature in *Stenocereus beneckei, S. martinezii, S. quevedonis,* and *S. thurberi,* but is absent in the remaining species of *Stenocereus* and other genera. Thus occurrence of a horizontal notch was also included in the analysis.

Spine number varies widely within the largest genera of Pachycereeae. This feature was not included in the analysis because of its variability. Species of Mexican columnar cacti show a differentiation into central and radial spines with the exception of *Bergerocactus emoryi, Pachycereus gaumeri, P. lepidanthus, P. marginatus, P. pringlei, Stenocereus beneckei, S. chacalapensis,* and *S. eichlamii.* Further analysis of spine morphology is needed as a way to find characters that perhaps will be useful for delimiting genera (Schill et al., 1973).

Reproductive morphology. Flower-bearing areoles in the Pachycereeae vary from being distributed in a cephalium to the complete absence of differentiation of both flowering region and its areoles. Cephalium occurs in all species of *Cephalocereus* and two species of *Pachycereus, P. fulviceps* and *P. militaris.* In these species, the cephalium may be apical, lateral, or annular; it may show such modifications as loss of rib structure and reduction of distance between areoles, as well as the presence of numerous long trichomes and bristles and lack of spines in areoles. In *Carnegiea, Neobuxbaumia* (three species), and *Pachycereus* (five species), in contrast, the modifications are restricted to areoles in those branches specialized for flowering. According to Buxbaum (1964a,b), the cephalium is a character without phylogenetic information because it has evolved several times within the Cactoideae. However, in recent studies in the Cereeae (Taylor and Zappi, 1989) and in *Pilosocereus* (Zappi, 1990), the presence of specialized reproductive branches has been considered a derived feature. Reproductive branch specialization was coded as a multistate character in the analysis included here (Appendix 4.1). Developmental studies are needed to verify that transitional flowering regions are an intermediate character state between the highly specialized cephalium and the absence of this specialization.

Buxbaum (1961) also recognized a cephaloid zone within the areole in members of the Cactoideae. Cephaloid zones have been observed in *Escontria, Myrtillocactus,* and some species of *Stenocereus* (Buxbaum, 1964a,b; Bravo-Hollis, 1978; Gibson, 1990a). The cephaloid zones may produce one or more flowers annually for years. However, the duration of their activity has not been quantified, and it is not known if they are homologous to the specialized branches of *Carnegiea, Neobuxbaumia,* and *Pachycereus* mentioned above. Gibson and Nobel (1986) also reported that species of

Neobuxbaumia (three species) and *Pachycereus* (three species) possess areoles that produce flowers for many years. Studies attempting to clarify reproductive branch morphology are currently in progress.

Although floral morphology provides key characters useful in the taxonomy of most dicot species, in Pachycereeae, several floral features have evolved in parallel due to their close relationship with pollinators (Gibson and Nobel, 1986). In addition, lack of good flower collections and of uniform information in species descriptions makes it difficult to code most floral features for a phylogenetic analysis. Thus only six floral features were included in our analysis. However, we also discuss other important characters to point out the need for future studies. Areoles in Pachycereeae commonly produce only one flower, with the exception of *Myrtillocactus* (four species), *Pachycereus gatesii, P. marginatus,* and *P. schottii,* which may bear two or more flowers per areole. Number of flowers per areole was included in the analysis.

Flowers of Pachycereeae are mostly medium-sized, but small flowers occur in *Escontria, Myrtillocactus* (three species), *Pachycereus* (two species), and *Polaskia* (one species). With the exception of *Rathbunia,* all flowers have radial symmetry. The distinctive pericarpel is variously scaly: Chartaceous scales distinguish *Escontria,* and scales with bristle tips occur in *Polaskia chende.* However, scales in the pericarpel should be studied further to identify which characters to include in phylogenetic analyses. The two types of scales mentioned here are autapomorphies of *Escontria* and *Polaskia chende.* The podaria of the pericarpel in *Neobuxbaumia euphorbioides* and *N. scoparia* are specialized as extrafloral nectaries. This feature has not been reported in other species of *Neobuxbaumia* or other taxa of Pachycereeae. Areoles of the pericarpel usually possess trichomes, spines, or bristles; rarely are they naked, as in *Escontria, Polaskia chichipe,* and *Pachycereus* (three species). However, in some descriptions, differences among species with respect to the type of structures present in the areoles of the pericarpel have been described. The receptacle tube is commonly naked, but it may also show the same structures as the pericarpel (e.g., short spines occur in the areoles of several species of *Stenocereus*). Receptacle tube size was coded as a binary character in the analysis presented here (Appendix 4.1). Tepal shape and color also vary among genera but were not included in the analysis due to lack of data. Tepal position was included in the analysis because it is constant among genera.

Species of Pachycereeae possess a nectar chamber, but its shape and closure vary among genera. Buxbaum (1953) indicated that there are several types of nectar chamber closure in Mexican columnar cacti. Unfortunately, he did not mention the species that belong to each type. In the open nectar chamber, the primary stamens are inserted at the same height, and the chamber therefore has an upper limit; it is open because the filaments are only slightly inflected. In the half-open nectar chamber, the primary stamens are thickened at their base and connected by an axial protrusion, as occurs in *Cephalocereus columna-trajani.* The next two types are characterized by the presence of a diaphragm. The *Neobuxbaumia* type is distinguished by several rows of primary and secondary stamens on the upper surface of the diaphragm, whereas in the *Mitrocereus* type, the stamen bases are fused without any axial tissue and are inserted at the inner border of the diaphragm (as in *Pachycereus militaris*). Detailed observations are needed to place each species in one of the types proposed by Buxbaum. We also need developmental studies to determine whether these types of nectar chamber closure are homologous. With the information presented in published descriptions, we were unable to assign species to any of the above types. Two binary characters—size of the nectar chamber and condition of being open or closed without regard to the pattern of closure—were included in our analysis. Other features of stamens and stigma need to be recorded in more species. For example, the number of lobules in the stigma has been reported in a few species of *Cephalocereus* (two species), *Neobuxbaumia* (four species), *Pachycereus* (five species), and *Stenocereus* (eight species). The character, funicular pigment cells, proposed by Gibson and Horak (1978) as a synapomorphy of the Stenocereinae, was included in the analysis as a binary character (Appendix 4.1).

Pollen grain morphology needs to be studied in more species of *Stenocereus* and *Cephalocereus* to complete the initial survey of Cactaceae done by Leuenberger (1976), who studied 20 species of nine genera of Pachycereeae sensu Barthlott and Hunt (1993). Gama-López et al., (1997) reported pollen shape differences among seven species of *Pachycereus.* Their study suggests that pollen features may be useful characters in future cladistic analyses.

All Pachycereeae produce a berry that varies in color and shape within genera. In the present study, only the presence or lack of fruit areoles with trichomes, bristles, or spines was included in the analysis as well as the persistent of these areoles at maturity (Appendix 4.1). Fruit morphology has been inconsistently described in the literature, making it difficult to code all of its features. The same is true for seed morphology; however, coding presented by Gibson et al. (1986) was used for seeds. Characters related to seed coat brightness and appearance of seed surface were included in the analysis presented here. Most seeds of Pachycereeae are black and bright, and these features are plesiomorphic for this tribe. A prominent raphe (Gibson and Horak, 1978) distinguishes the seeds of ten species of *Pachycereus;* information for the remaining species of this genus is lacking. This seed character was also included in the analysis. Position of hilum, raphe, and cuticular striation seem to be good characters that should be studied in all members of the tribe in such a way as to code them following the recommendations by Barthlott and Voit (1979).

Chemical Characters

The studies of Gibson and collaborators have sampled nearly 44% of the species of the Pachycereeae for alkaloids and triterpenes (Table 4.3); most members of subtribe Stenocereinae have been studied (Gibson, 1982; Gibson et al., 1986). The ability to synthesize either alkaloids or glycosidic triterpenes has been used by these authors as characters for grouping taxa in the two subtribes of Pachycereeae. Taxa in the Pachycereinae synthesize mostly tetrahydroisoquinoline alkaloids; however,

TABLE 4.3
Number of Species in Genera of the Tribe Pachycereeae sensu Barthlott and Hunt (1993) That Can Synthesize Alkaloids and Triterpenes[a]

Genus[b]	*Alkaloids*	*Glycosidic Triterpenes*	*Nonglycosidic Triterpenes*	*None*
Bergerocactus (1)			1	
Carnegiea (1)	1			
Cephalocereus (5)	1		1	2
Escontria (1)	1	1		
Myrtillocactus (4)		4		
Neobuxbaumia (8)	1			3
Pachycereus (13)	10		1	2
Polaskia (2)	1	2		
Rathbunia (2)		2		
Stenocereus (21)	5	17	2	

[a] Data are from Gibson and Horak (1978); Gibson (1982); and Gibson et al. (1986).
[b] Species recognized by Hunt (1992).

several species of *Cephalocereus, Neobuxbaumia* and *Pachycereus* lack alkaloids. Traces of nonglycosidic triterpenes have been detected in *Cephalocereus senilis* and *Pachycereus fulviceps* (Gibson and Horak, 1978). Additional studies are needed in species of *Cephalocereus* and *Neobuxbaumia* to confirm their inability to synthesize alkaloids or other compounds. Most members of the subtribe Stenocereinae have the ability to synthesize glycosidic triterpenes; however alkaloids have also been identified in *Escontria, Polaskia* (one species), and in several species of *Stenocereus: S. beneckei, S. dumortieri, S. eruca, S. stellatus,* and *S. treleasei.* The synthesis of alkaloids and triterpenes follows different chemical pathways, thus they were coded as two distinct characters (Appendix 4.1). However, only the ability to synthesize glycosidic triterpenes or tetrahydroisoquinoline alkaloids were included in the analysis to avoid inferences about biochemical phylogeny within each group of compounds. When the compounds have been only reported as unidentified alkaloids, their occurrence was not included in our analysis (e.g., *Escontria, Stenocereus dumortieri, S. stellatus*).

Phylogenetic Analysis

Our main purpose here is to test the hypothesis that genera as defined by Barthlott and Hunt (1993) are monophyletic and to identify their apomorphic characters by means of a cladistic analysis. Twenty-five taxa were used in this analysis. Character scores for most of the 29 characters (Appendix 4.2) are based on revisionary and monographic studies (Gibson, 1973, 1988b–e, 1989a,b, 1990a,b, 1991; Bravo-Hollis, 1978; Gibson and Horak, 1978; Gibson et al., 1986; Bravo-Hollis and Sánchez-Mejorada, 1991; Arias-Montes et al., 1997) and supplemented by our observations. Because

Buxbaum (1961) suggested that species of *Leptocereus* may be the ancestor of Pachycereeae, we used *Leptocereus* as the outgroup of Pachycereeae for our analysis (but see chapter 3). The analysis was performed using PAUP version 3.1.1 (phylogenetic analysis using parsimony; Swofford, 1993). Heuristic search with 50 replicates was conducted with the COLLAPSE, MULPARS, STEPEST DESCENT, ACCTRAN, and TBR branch-swapping options. Internal support for relationships was measured by decay analysis (Bremer, 1988). The analysis resulted in 12 equally parsimonious trees of 89 steps. The topologies represented in these trees, however, are all quite similar, and differ mainly in the relationships among hypothesized terminal taxa within large clades. A representative cladogram (Fig. 4.1) and the strict consensus tree (Fig. 4.2) are presented here.

The phylogenetic analysis does not completely support the monophyly of clades previously suggested by Gibson (1982), Cornejo and Simpson (1997), and Cota and Wallace (1997). The strict consensus tree, however, shows three strongly supported clades (Fig. 4.2). *Stenocereus* is hypothesized to be monophyletic on the basis of its distinct silica bodies in the dermal tissue (characters 4 and 6) (Fig. 4.1). This result is congruent with other phylogenetic studies (Gibson et al., 1986; Cornejo and Simpson, 1997; Cota and Wallace, 1997). Although there are only three representatives of the genus *Stenocereus* with red-colored areoles in our analysis, they form a monophyletic group. *S. dumortieri* appears in a basal position within the *Stenocereus* lineage. The recognition of this species as the monotypic genus *Isolatocereus* should await further evidence. Furthermore, the inclusion of more species of *Stenocereus* in a phylogenetic analysis may support its divergence from the *Stenocereus* clade.

Our analysis further suggests that *Escontria, Polaskia,* and *Myrtillocactus* represent a monophyletic clade, supporting the close alliance among these three genera as suggested previously (Gibson and Horak, 1978; Gibson et al., 1986; Cornejo and Simpson, 1997; Cota and Wallace, 1997). The results also provide evidence to maintain the three genera as distinct taxa, as recognized in the most recent classification (Barthlott and Hunt, 1993). These genera are characterized by short, non-funnel-shaped flowers with a small, narrow nectar chamber, which are commonly pollinated by bees; by fruits that lack areoles; and by very small, dull seeds, as well as lack of mucilage cells in the pith.

The proposed monophyly of subtribe Stenocereinae (Gibson et al., 1986; Cota and Wallace, 1997) is not supported by our analysis. Thus, the synapomorphies suggested by Gibson (1982) and Gibson et al. (1986) of funicular pigment cells and ability to synthesize triterpenes appear to have evolved in the ancestor of the group, whereas rough, dull seeds have evolved independently in more than two lineages. Gibson (1982) and Gibson et al. (1986) have previously suggested that tribe Pachycereinae possesses only plesiomorphic characters and that no synapomorphy defines it. However, they grouped taxa in this subtribe based on the presence of alkaloids; smooth, glossy seeds; and crystals in the dermal system. In our phylogeny,

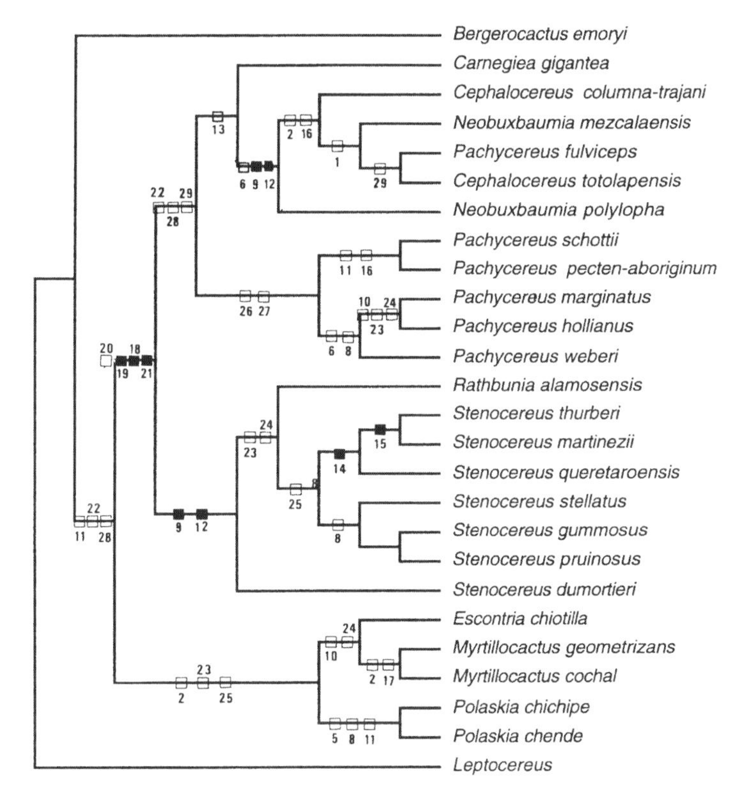

FIGURE 4.1. Representative most parsimonious cladogram. Tree length = 89, CI (consistency index) = 0.43, RI (retention index) = 0.72. Character changes indicated by squares (solid = unique; empty = homoplasious, gains or reversals). Numbers indicate characters (see Appendix 4.1). *Leptocereus* is the outgroup.

this clade is weakly supported by the lack of funicular pigment cells and the ability to synthesize alkaloids proposed by Gibson et al. (1986) as the synapomorphies for the subtribe. The other two characters suggested by Gibson et al. (1986) do not define the clade. Glossy seed is a plesiomorphic character for the Pachycereeae, but smooth seeds and a prominent raphe in the seed group five species of *Pachycereus* in a clade. The analysis also suggests that crystals in the dermal system define the *Cephalocereus-Neobuxbaumia-Pachycereus fulviceps* clade, which forms a polytomy in the strict consensus tree (Fig. 4.2). Members of this clade are characterized by a loss of the ability to synthesize alkaloids. A more comprehensive study of these chemical compounds needs to be done, especially for *Pachycereus, Cephalocereus,* and *Neobuxbaumia* as defined by Barthlott and Hunt (1993). Based on the number of ribs, *Carnegiea* was basal to the *Cephalocereus-Neobuxbaumia-Pachycereus fulviceps* clade. The inclusion of other members of these genera in future analyses will

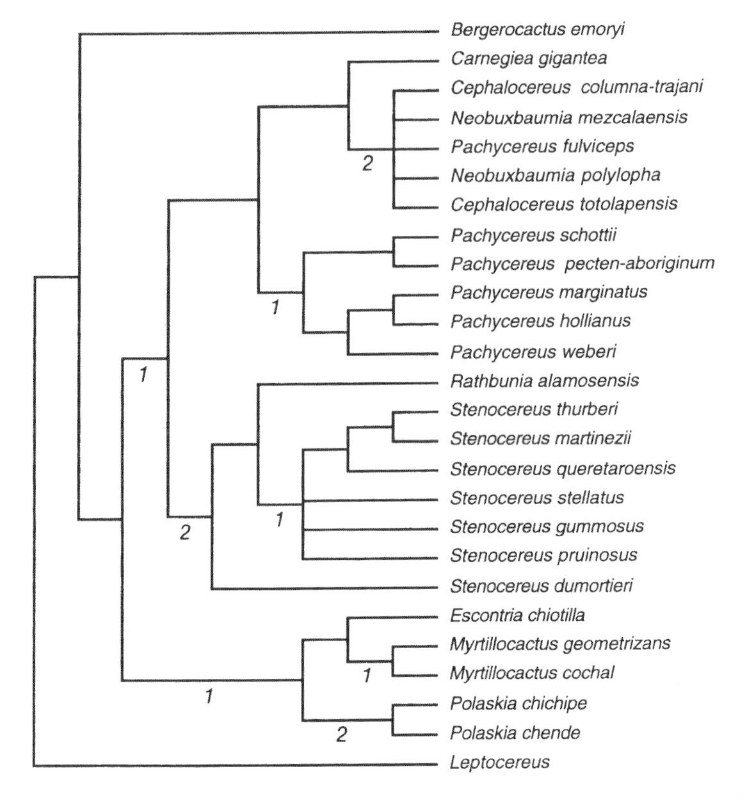

FIGURE 4.2. Strict consensus of the 12 most parsimonious trees. Decay index is shown below branches.

help to understand their limits or might support the IOS proposal that they form a single genus; the position of *Pachycereus fulviceps* (*Mitrocereus fulviceps*) especially needs to be reevaluated.

The monophyly of *Pachycereus,* as defined by Barthlott and Hunt (1993), is not supported by the data presented here (Figs. 4.1, 4.2). The inclusion of the remaining species of this genus will help to understand its limits. *Bergerocactus emoryi* has an isolated basal position within the tribe. This evidence suggests that an analysis including taxa from other tribes would help elucidate *Bergerocactus* tribal relationships. *Bergerocactus* may belong within another tribe, as has been suggested by other authors (Buxbaum, 1961; Gibson and Horak, 1978).

Finally, did we simply fail to include characters that might support the sister taxa relationships within the genera of Pachycereeae? There are unverified apomorphies for the genera (e.g., floral, fruit, and seed features) that need to be incorporated into future analyses. Our results also show that certain anatomical characters (1, 3, 5, 6, 9, and 11) are highly homoplasious, and their coding should be reevaluated.

Conclusions

To hypothesize sister relationships and better understand generic limits, more species from each genus must be included in future analyses. Such well supported lineages as *Stenocereus* and *Escontria-Polaskia-Myrtillocactus* should be analyzed separately to improve the resolution of internal topology. More comprehensive morphological studies are needed to be able to code and include other characters in future cladistic analyses, especially for reproductive features. We must complete a search for the newly discovered anatomical features (e.g., silica grains in wood, amphivasal medullary bundles) in other species, so that these features may be included in future cladistic analyses. Finally, we need to combine anatomical-morphological data with molecular data in phylogenetic analyses to gain a better understanding of the evolution of Mexican columnar cacti.

Summary

The main purposes of this chapter are to summarize the anatomical, morphological, and chemical data available for tribe Pachycereeae; to perform a cladistic analysis to identify apomorphic characters; and to propose a working hypothesis for future studies. Since Buxbaum's study (1958), the number of genera recognized as members of Pachycereeae has been unstable. The most recent classification suggests that this tribe contains ten genera (Barthlott and Hunt, 1993). The stem anatomy of nearly 50% of the species recognized by Hunt (1992) has been studied; the wood of 28% has been studied. Some vegetative and reproductive morphological characters have been inconsistently reported in the literature, making it difficult to incorporate these characters into a cladistic analysis, whereas nearly 44% of the species of Pachycereeae have been sampled for alkaloids and triterpenes. The results of our phylogenetic analysis do not completely support the monophyly of the clades that has been previously suggested. *Stenocereus* forms a monophyletic clade, defined by the distinctive silica bodies in the dermal tissue. A second clade supports the close relationship between *Escontria, Polaskia,* and *Myrtillocactus,* as has been suggested by other authors. *Pachycereus* is paraphyletic as defined by Barthlott and Hunt (1993). A *Cephalocereus-Neobuxbaumia-Pachycereus fulviceps* clade and *Carnegiea* alliance should be reevaluated with the inclusion of more species of these genera. To elucidate sister relationships among species of Pachycereeae, we need to focus new research on developmental studies of reproductive branches and floral features and to generate anatomical and chemical (alkaloids and triterpenes) data for currently unstudied species. An effort should be made to combine morphological and anatomical characters with molecular data in a way that promotes understanding phylogenetic relationships among members of Pachycereeae.

Resumen

Los principales objetivos de este capítulo son sintetizar la información anatómica, morfológica y química, disponible en la literatura especializada para la tribu Pachycereeae; realizar un análisis cladístico con la finalidad de identificar los caracteres apomórficos y proponer una hipótesis de relaciones de parentesco para futuros estudios. Desde la publicación de Buxbaum en 1958, el número de géneros que conforman la tribu Pachycereeae ha sido inestable. Sin embargo, una de los clasificaciones más recientes reconoce diez géneros (Barthlott y Hunt, 1993). Se ha estudiado la anatomía del tallo para cerca del 50% de las especies aceptadas por Hunt (1992) y sólo para un 28% su madera. La información existente en la literatura es inconsistente para algunos caracteres morfológicos vegetativos y reproductivos, dificultando su incorporación a los análisis cladísticos. La capacidad de sintetizar alcaloides y triterpenos se ha estudiado en cerca de un 44% de las especies. Los resultados del análisis cladístico sugieren que no todos los clados propuestos previamente son monofiléticos. *Stenocereus* forma un clado monofilético, definido por la presencia de cuerpos de sílice en el tejido dérmico. Un segundo clado apoya la estrecha relación de *Escontria, Polaskia,* y *Myrtillocactus,* también hipotetizada por otros autores. *Pachycereus* es un género parafilético de acuerdo a la definición de Barthlott y Hunt (1993). El grupo formado por *Cephalocereus-Neobuxbaumia-Pachycereus fulviceps* y *Carnegiea* deberá ser re-evaluado con la inclusión de otras especies de estos géneros. Con el fin de tener un mejor entendimiento de las relaciones de hermandad entre las especies de Pachycereeae, es necesario dirigir futuras investigaciones a estudios de ontogenia para las ramas reproductivas y algunos caracteres florales; así como producir la información anatómica y de compuestos químicos (alcaloides y triterpenos) para las especies a la fecha no estudiadas. Se sugiere combinar caracteres morfológicos y anatómicos con los datos moleculares con objeto de tener un mejor entendimiento de las relaciones de parentesco entre los miembros de Pachycereeae.

ACKNOWLEDGMENT

Research for this chapter was supported by CONABIO (L074).

APPENDIX 4.1

Characters Used in the Cladistic Analysis

Character Number	*Character*
1	pith inclusions lacking (0); crystals (1)
2	mucilage cells in pith lacking (0), present (1)
3	extra-phloematic fibers early differentiation (0); late differentiation (1)
4	vessel element length > 250 μ (0); < 250 μ (1)
5	mucilage cells in cortex present (0); lacking (1)
6	cortex inclusion crystals present (0); lacking (1)
7	epidermis one layer (0); multiple layers (1)
8	epidermal secondary divisions lacking (0); internal (1); bullate (2) [unordered]
9	epidermis inclusions lacking (0); crystals (1); silica grains (2) [unordered]
10	cuticle smooth (0); striate (1); horned (2) [unordered]
11	hypodermis thickness < 150 μ (0); 150–300 μ (1); > 300 μ [unordered]
12	hypodermis inclusions lacking (0); crystals (1); silica grains (2) [unordered]
13	stems with ≤ 10 ribs (1); > 11 ribs (2)
14	glandular trichomes lacking (0); present (1)
15	horizontal notch between areoles lacking (0); present (1)
16	reproductive branch differentiation lacking (0); poorly developed (1); well developed (2) [unordered]
17	flowers per areole > two (0); one (1); two (2) [unordered]
18	receptacle tube short < 2 cm (0); large > 2 cm (1)
19	tepal position open (0); closed (1)
20	nectar chamber lacking (0); short (1); long (2) [unordered]
21	nectar chamber open (0); closed (1)
22	funicular pigment cells lacking (0); present (1)
23	fruit lacking areoles (0); with areoles (1)
24	mature fruit naked (0); with caducous areole (1), with persistent areole (2) [unordered]
25	seed surface glossy (0); dull (1)
26	seed surface appearance smooth (0); rough (1)
27	prominent raphe in the seed lacking (0); present (1)
28	triterpenes lacking (0); glycosidic (1)
29	alkaloids lacking (0); present (1)

APPENDIX 4.2

Character Values for Taxa Used in the Cladistic Analysis[a]

Species	Character																												
	1	*2*	*3*	*4*	*5*	*6*	*7*	*8*	*9*	*10*	*11*	*12*	*13*	*14*	*15*	*16*	*17*	*18*	*19*	*20*	*21*	*22*	*23*	*24*	*25*	*26*	*27*	*28*	*29*
Bergerocactus emoryi	0	1	1	1	0	1	0	0	0	0	0	0	1	0	0	0	1	0	0	0	0	0	1	2	0	1	0	0	0
Carnegiea gigantea	?	1	1	0	0	1	0	1	0	1	2	0	1	0	0	1	1	1	1	2	1	0	1	2	0	1	0	0	1
Cephalocereus columnatrajani	0	0	1	0	0	0	0	1	1	0	2	1	1	0	0	2	1	1	1	2	1	0	1	2	0	1	0	0	1
Escontria chiotilla	0	0	1	0	0	1	0	0	0	2	2	0	0	0	0	0	1	0	0	1	0	1	0	0	1	1	0	1	0
Polaskia chichipe	0	0	1	0	1	1	0	2	0	0	1	0	0	0	0	0	1	0	0	1	0	1	0	1	1	1	0	1	0
Myrtillocactus geometrizans	0	1	1	0	0	1	0	1	0	2	?	0	0	0	0	0	0	0	0	1	0	1	0	0	1	1	0	1	0
Neobuxbaumia mezcalaensis	1	0	1	0	0	0	0	1	1	0	1	1	1	0	0	0	1	1	1	2	1	0	1	2	0	1	0	0	1
Rathbunia alamosensis	0	1	1	1	0	1	0	1	2	0	1	2	0	0	0	0	1	1	1	2	1	1	0	1	0	1	0	1	0
Pachycereus schottii	0	1	1	0	0	1	0	1	0	0	0	0	0	0	0	1	2	1	1	1	1	0	0	0	0	0	1	0	1
Pachycereus marginatus	0	1	1	0	1	1	0	0	0	2	2	0	0	0	0	0	2	1	1	2	1	0	0	1	0	0	1	0	1
Pachycereus pectenaborigenum	0	0	1	0	0	1	1	—	0	0	1	0	0	0	0	1	1	1	1	2	1	0	1	2	0	0	1	0	1
Pachycereus weberi	1	1	0	0	0	0	0	0	0	0	2	0	0	0	0	0	1	1	1	2	1	0	1	2	0	0	1	0	1
Stenocereus thurberi	0	1	1	0	0	1	0	1	2	0	2	2	1	1	1	0	1	1	1	2	1	1	0	1	0	1	0	1	0
Stenocereus stellatus	0	1	0	0	0	1	0	2	2	0	2	2	0	0	0	0	1	1	1	2	1	1	0	1	1	1	0	1	0
Stenocereus gummosus	0	1	1	0	0	1	0	0	2	0	1	2	0	0	0	0	1	1	1	2	1	1	0	1	1	1	0	1	0
Stenocereus pruinosus	0	1	1	0	0	1	0	0	2	0	2	2	0	0	0	0	1	1	1	2	1	1	0	1	?	1	0	1	0
Pachycereus fulviceps	1	0	1	0	0	0	0	0	1	1	2	1	1	0	0	2	1	1	1	2	1	0	1	2	0	0	1	0	0
Polaskia chende	0	0	1	0	1	1	0	2	0	0	1	0	0	0	0	0	1	0	0	1	0	1	0	2	1	1	0	1	0
Myrtillocactus cochal	0	1	1	1	0	1	0	1	0	2	?	0	0	0	0	0	0	0	0	1	0	1	0	0	1	1	0	1	0
Stenocereus dumortieri	?	?	1	0	0	1	1	—	2	0	2	2	0	0	0	0	1	1	1	2	1	1	1	2	?	1	0	1	0
Neobuxbaumia polylopha	0	1	1	0	?	0	0	0	1	0	1	1	1	0	0	0	1	1	1	2	1	0	1	2	0	1	0	0	0
Stenocereus martinezii	0	1	1	0	0	1	0	1	2	0	1	2	0	1	1	0	1	1	1	2	1	1	0	1	1	1	0	1	0
Pachycereus hollianus	0	1	1	0	0	0	0	0	0	2	0	0	0	0	0	0	1	1	1	2	1	0	0	1	0	0	1	0	1
Stenocereus queretaroensis	0	1	1	0	0	1	0	1	2	0	2	2	0	1	0	0	1	1	1	2	1	1	0	1	1	1	0	1	0
Cephalocereus totolapensis	?	0	1	0	0	0	0	1	1	0	2	1	1	0	0	2	1	1	1	2	1	0	1	2	0	1	0	0	0
Leptocereus sp.	0	1	0	0	0/1	0	0	1	1	0	0	0	0	0	0	?	1	0	0	1	?	0	0/1	0	0	1	0	?	?

[a]For character abbreviations, see Appendix 4.1. ? = condition unknown, — = not applied.

REFERENCES

Arias-Montes, S., S. Gama-López, and L. U. Guzmán C. 1997. Cactaceae. In *Flora del Valle de Tehuacán-Cuicatlán,* eds. P. Dávila, J. L. Villaseñor, R. Medina, and O. Téllez. Mexico: Instituto de Biología, Universidad Nacional Autónoma de México.

Arreola-Nava, H. J. 2000. Sistemática de las especies de *Stenocereus* (A. Berger) Riccob. con aréolas morenas (Cactoideae-Cactaceae). Master's thesis, Colegio de Postgraduados, Montecillo, Estado de México, Mexico.

Barthlott, W., and D. R. Hunt. 1993. Cactaceae. In *The Families and Genera of Vascular Plants,* ed. K. Kubitzki, 161–97. Berlin: Springer-Verlag.

Barthlott, W., and G. Voit. 1979. Mikromorphologie der Samenschalen und Taxonomie der Cactaceae: ein raster-elektronenmikroskopischer Überblick. *Plant Systematics and Evolution* 132:205–29.

Bravo-Hollis, H. 1978. *Las Cactáceas de México.* Vol. 1. Mexico: Universidad Nacional Autónoma de México.

Bravo-Hollis, H., and H. Sánchez-Mejorada. 1991. *Las Cactáceas de México.* Vol. 2. Mexico: Universidad Nacional Autónoma de México.

Bremer, K. 1988. The limits of amino acid sequence data in angiosperm phylogenetic reconstruction. *Evolution* 42:795–803.

Buxbaum, F. 1950. *Morphology of Cacti. Section I. Roots and Stems.* Pasadena Calif.: Abbey Garden Press.

———. 1953. *Morphology of Cacti. Section II. Flowers.* Pasadena Calif.: Abbey Garden Press.

———. 1958. The phylogenetic division of the subfamily Cereoideae, Cactaceae. *Madroño* 14:177–206.

———. 1961. Die Entwicklungslinien der Tribus Pachycereeae F. Buxb. (Cactaceae-Cereoideae). In *Botanische Studien.* Vol. 12. Jena: Veb Gustav Fischer Verlag.

———. 1964a. Was ist ein Cephalium? *Kakteen und Sukkulenten* 15:28–31.

———. 1964b. Was ist ein Cephalium? *Kakteen und Sukkulenten* 15:43–48.

Cornejo, D. O., and B. B. Simpson. 1997. Analysis of form and function in North American columnar cacti (tribe Pachycereeae). *American Journal of Botany* 84:1482–1501.

Cota, J. H., and R. S. Wallace. 1997. Chloroplast DNA evidence for divergence in *Ferocactus* and its relationships to North American columnar cacti (Cactaceae: Cactoideae). *Systematic Botany* 22:529–42.

Eggli, U. 1984. Stomatal types of Cactaceae. *Plant Systematics and Evolution* 146:197–214.

Gama-López, S., S. Arias-Montes, and J. L. Alvarado. 1997. Morfología de los granos de polen de las especies del género *Pachycereus* (Cactaceae, Pachycereeae). In *Resúmenes I Congreso Nacional sobre Cactáceas,* 65. Montecillo, Mexico: Direccion de Publicaciones y Materiales Educativòs, IPN.

Gibson, A. C. 1973. Comparative anatomy of secondary xylem in Cactoideae (Cactaceae). *Biotropica* 5:29–65.

———. 1982. Phylogenetic relationships of Pachycereeae. In *Ecological Genetics and Evolution. The Cactus-Yeast-Drosophylla Model System,* eds. J.S.F. Baker and W. T. Starmer, 3–16. Sydney: Academic Press.

———. 1988a. The systematics and evolution of subtribe Stenocereinae. 1. Composition and definition of the subtribe. *Cactus and Succulent Journal* 60:11–16.

———. 1988b. The systematics and evolution of subtribe Stenocereinae. 2. *Polaskia. Cactus and Succulent Journal* 60:55–62.

———. 1988c. The systematics and evolution of subtribe Stenocereinae. 3. *Myrtillocactus. Cactus and Succulent Journal* 60:109–16.

———. 1988d. The systematics and evolution of subtribe Stenocereinae. 4. *Escontria. Cactus and Succulent Journal* 60:161–67.

———. 1988e. The systematics and evolution of subtribe Stenocereinae. 5. *Cina* and its relatives. *Cactus and Succulent Journal* 60:283–88.

———. 1989a. The systematics and evolution of subtribe Stenocereinae. 6. *Stenocereus stellatus* and *Stenocereus treleasei. Cactus and Succulent Journal* 61:26–32.

———. 1989b. The systematics and evolution of subtribe Stenocereinae. 7. The Macherocerei Members of *Stenocereus. Cactus and Succulent Journal* 61:104–12.

———. 1990a. The systematics and evolution of subtribe Stenocereinae. 8. Organ pipe cactus and its closest relatives. *Cactus and Succulent Journal* 62:13–24.

———. 1990b. The systematics and evolution of subtribe Stenocereinae. 9. *Stenocereus queretaroensis* and its closest relatives. *Cactus and Succulent Journal* 62:170–76.

———. 1991. The systematics and evolution of subtribe Stenocereinae. 11. *Stenocereus dumortieri* versus *Isolatocereus dumortieri. Cactus and Succulent Journal* 63:184–90.

Gibson, A. C., and K. E. Horak. 1978. Systematic anatomy and phylogeny of Mexican columnar cacti. *Annals of the Missouri Botanical Garden* 65:999–1057.

Gibson, A. C., and P. S. Nobel. 1986. *The Cactus Primer.* Cambridge, Mass.: Harvard University Press.

Gibson, A. C., K. C. Spencer, R. Bajaj, and J. L. McLaughlin. 1986. The ever-changing landscape of cactus systematics. *Annals of the Missouri Botanical Garden* 73:532–55.

Hunt, D. R. 1992. *CITES Cactaceae Checklist.* Kent, England: Royal Botanic Gardens, Kew.

Hunt, D. R., and N. Taylor. 1990. The genera of Cactaceae: progress towards a consensus. *Bradleya* 8:85–107.

Leuenberger, B. E. 1976. Die Pollenmorphologie der Cactaceae und ihre Bedeutung für die Systematik. *Dissertationes Botanicae* 31:1–321.

Mauseth, J. D. 1996. Comparative anatomy of tribes Cereeae and Browningieae (Cactaceae). *Bradleya* 14:66–81.

Mauseth, J. D., and J. V. Landrum. 1997. Relictual vegetative anatomical characters in Cactaceae: the genus *Pereskia. International Journal of Plant Science* 110:55–64.

Mauseth, J. D., T. Terrazas, and S. Loza-Cornejo. 1998. Anatomy of relictual members of subfamily Cactoideae, IOS 1a (Cactaceae). *Bradleya* 16:41–43.

Nyffeler, R., and U. Eggli. 1997. Comparative stem anatomy and systematics of *Eriosyce* sensu lato (Cactaceae). *Annals of Botany* 80:767–86.

Schill, R., W. Barthlott, and N. Ehler. 1973. Mikromorphologie der Cactaceen-Dornen. *Akademie der Wissenschaften und der Literatur* 6:263–79.

Swofford, D. L. (1993). PAUP: Phylogenetic analysis using parsimony. Version 3.1.1. Washington, D.C.: Smithsonian Institution.

Taylor, N. P., and D. C. Zappi. 1989. An alternative view of generic delimitation and relationships in tribe Cereeae (Cactaceae). *Bradleya* 7:13–40.

Terrazas, T. 2000. Occurrence of perforated ray cells in genera of Pachycereeae. *IAWA Journal* 21:457–62.

Zappi, D. C. 1990. *Pilosocereus* (Cactaceae). The genus in Brazil. *Succulent Plant Research* 3:1–159.

CHAPTER 5

Phylogeny and Convergence in Cactophilic Bats

NANCY B. SIMMONS
ANDREA L. WETTERER

Introduction

Bats are important pollinators of columnar cacti (Alcorn et al., 1959, 1961, 1962; McGregor et al., 1962; Howell, 1974a; Nassar, 1991; Petit, 1995, 1997; Sahley, 1995; Fleming et al., 1996; Valiente-Banuet et al., 1996, 1997b; Ceballos et al., 1997; Nassar et al., 1997; Horner et al., 1998), and recent research has shown that bats are also dispersal vectors for cactus seeds (Sosa and Soriano, 1996; Valiente-Banuet et al., 1996). With only one exception, bats that visit columnar cacti belong to a single family, Phyllostomidae (New World leaf-nosed bats). Members of several phyllostomid genera visit cacti regularly, and some species rely heavily on cactus pollen, nectar, and fruits as nutrient sources during some or all parts of the year. Until recently, disagreements concerning phylogeny of phyllostomids have made it difficult to investigate the evolution of cactus-visiting behavior and associated morphologies in bats. The number of times this behavior may have evolved has remained uncertain, as has the degree and significance of any morphological convergence in these bats. A new phylogeny of Phyllostomidae (Wetterer et al., 2000) now provides a comparative context for investigating these and other evolutionary questions.

Cactophilic Bats

Plants that depend on bats for pollination and/or seed dispersal are sometimes termed "chiropterophilous" or "chiropterophilic" (van der Pijl, 1956, 1961; Baker, 1961; Howell, 1974a; Valiente-Banuet et al., 1996; Nassar et al., 1997). To our knowledge, there is no equivalent term for bats that depend on cacti for some or all of their nutrients during the flowering and fruiting seasons. We propose the term "cactophilic" (Greek, "prickly-plant loving") for bat species that regularly visit cacti to feed on cactus products such as nectar, pollen, and fruit. Maintenance of bat populations in some arid regions may be possible only because of the availability of cactus products (Sosa and Soriano, 1996; Petit, 1997), and migratory patterns of some bat species may be structured to follow a "nectar corridor" of blooming columnar cacti in the spring (Fleming et al., 1993). However, the degree to which different bat species and populations depend on cacti varies considerably with taxon, season, and locality, and no bat species depends entirely on cactus products to meet all of its nutritional requirements (Fleming, 1982; Arita, 1991; Fleming et al., 1993; Ceballos et al., 1997; Petit, 1997; Ruiz et al., 1997). For the purposes of discussion, we draw a distinction between "obligate cactophiles" (species or populations that depend exclusively or predominantly on cactus products for nutrients during at least some seasons of the year) and "opportunistic cactophiles" (species or populations that use cactus products when available but do not appear to rely on them for survival). These categories represent theoretical ends of a continuum, with most taxa and populations of cactus-visiting bats falling somewhere in between. Few if any bat species are truly obligate cactophiles in the sense that the entire species depends on cacti for survival, but many populations may require cactus products in order to remain viable.

Eighteen bat species are known to visit cacti at least occasionally (Table 5.1), but data for many species are limited. *Antrozous pallidus,* the only nonphyllostomid, appears to be an opportunistic cactophile. Limited to arid areas of North America, it feeds principally on arthropods and small vertebrates gleaned from surfaces (Hatt, 1923; Huey, 1936; Borrell, 1942; Orr, 1954; O'Shea and Vaughan, 1977; Bell, 1982; Hermanson and O'Shea, 1983). Cactus products are probably used by this species only as a supplementary and perhaps accidental food source (Barbour and Davis, 1969; Howell, 1980; Fleming, 1991; Herrera et al., 1993). Howell (1980) proposed that ingestion of cactus fruit by these bats might be an incidental result of attempting to capture juice-feeding noctuid moths in fruit cavities, and Barbour and Davis (1969) and Herrera et al. (1993) suggested that pallid bats may visit cactus flowers to eat insects and only inadvertently consume pollen and/or nectar. However, we consider it equally likely that these bats are simply opportunists who intentionally utilize whatever food sources are available, including cactus nectar, pollen, and fruit. Howell (1980) found that the feces of *Antrozous pallidus* in her study area contained 25% cactus fruit pulp and seeds, a quantity that suggests to us that these bats deliberately feed on cactus products. In this respect, they appear to be similar to many phyllostomid bats, which are well known for their eclectic diets (Gardner, 1977; Ferrarezzi and Gimenez, 1996). Although Herrera et al. (1993, p. 604) suggested that nectar in cardon cactus flowers "would be difficult to extract by a bat lacking the morphological characteristics of nectar-feeding bats (e.g., small ears, long muzzle and tongue)," we note that lack of such specializations does not necessarily preclude obtaining food from cactus flowers, as demonstrated by records of other nonspecialized species visiting cactus flowers (see accounts below for *Phyllostomus discolor,*

Artibeus jamaicensis, and *A. intermedius*). These taxa have relatively short muzzles and short tongues, but nevertheless appear to feed on cactus flowers.

Phyllostomus discolor is an opportunistic cactophile that occasionally visits flowering columnar cacti in the Venezuelan Andes (Sosa and Soriano, 1996). *P. discolor* ranges from southern Mexico to northern Argentina (Koopman, 1994) and is a common species in rainforest habitats (Voss and Emmons, 1996). It has a diverse diet that includes insects, fruit, pollen, nectar, and flower parts in varying proportions in different habitats and seasons (Heithaus et al., 1975; Gardner 1977; Sazima and Sazima, 1977; Humphrey et al., 1983). There is only one report of this species visiting cacti (Sosa and Soriano, 1996), and it seems unlikely that *P. discolor* depends to any large extent on cactus products.

Artibeus jamaicensis similarly appears to be an opportunistic cactophile, although some populations appear to make greater use of cactus products than the species described above. Broadly distributed from central Mexico to southern Brazil, this species is common in lowland forest habitats (Marques-Aguiar, 1994; Voss and Emmons, 1996). It is primarily frugivorous, although it also feeds on flower products, leaves, and insects (Heithaus et al., 1975; Gardner, 1977; Handley and Leigh, 1991; Handley et al., 1991; Ramírez-Pulido et al., 1993). Population densities of *A. jamaicensis* may be high in lowland rainforests, where these bats appear to be fig specialists (Gardner et al., 1991; Leigh and Handley, 1991; Handley and Leigh, 1991; Handley et al., 1991). Observations of this species using cactus products in arid habitats in Mexico (Valiente-Banuet et al., 1996, 1997a) suggest that *A. jamaicensis* is behaviorally flexible and may opportunistically use whatever plant resources are available in its local habitat. Presence of cactus resources may be critical for local populations (e.g., in arid habitats in Mexico); it is not clear whether *A. jamaicensis* could maintain viable populations in such regions without these resources.

Artibeus intermedius occurs sympatrically with *A. jamaicensis* throughout much of Mexico, Central America, and parts of Colombia (Davis, 1984; Koopman, 1994; Reid, 1997). This taxon, sometimes considered a subspecies of *A. lituratus* (Hall, 1981; Marques-Aguiar, 1994), is common in habitats ranging from tropical thorn scrub to lowland rainforest (Davis, 1984). Dietary habits of *A. intermedius* appear to be broadly similar to those of *A. jamaicensis* (Villa-R., 1967; Fleming et al., 1972; Heithaus et al., 1975; Gardner, 1977; Ramírez-Pulido et al., 1993). In arid regions, *A. intermedius* may be an opportunistic cactophile. It has been captured at flowering columnar cacti in the Tehuacán Valley of Mexico, but in far fewer numbers than its sympatric congener, *A. jamaicensis* (Valiente-Banuet et al., 1997a).

Another opportunistic cactophile is *Sturnira lilium*, a small stenodermatine whose range extends from Sonora to Uruguay (Gannon et al., 1989; Koopman, 1993, 1994). This species inhabits a wide range of habitats, including dry tropical forests, humid subtropical forests, and lowland rainforests (Handley, 1976; Gannon et al., 1989; Voss and Emmons, 1996). *S. lilium* is primarily frugivorous, although it also

TABLE 5.1

Summary of Cactophilic Bat Species and Cactus Products Utilized

Bat Taxon	*Cactus Taxon*	*Products Utilized*	*References*
Family Phyllostomidae			
Subfamily Glossophaginae			
Tribe Glossophagini			
Anoura geoffroyi	*Myrtillocactus*[a] indeterminate (columnar)	nectar, pollen	Alvarez and González (1970)
	Pilosocereus lanuginosus (columnar)	nectar, pollen	J. Nassar (pers. comm.)
Choeronycteris mexicana	*Neobuxbaumia tetetzo* (columnar)	nectar, pollen, fruit	Valiente-Banuet et al. (1996)
	N. mezcalaensis (columnar)	nectar, pollen	Valiente-Banuet et al. (1997b)
	N. macrocephala (columnar)	nectar, pollen	Valiente-Banuet et al. (1997b)
	Pachycereus weberi (columnar)	nectar, pollen	Valiente-Banuet et al. (1997a)
	P. pecten-arboriginum (columnar)	nectar, pollen	T. Fleming (pers. comm.)
	Pilosocereus chrysacanthus (columnar)	nectar, pollen	Valiente-Banuet et al. (1997a)
	Cephalocereus hoppenstedtii (columnar)	nectar, pollen	Valiente-Banuet et al. (1996)
	Myrtillocactus[a] indeterminate (columnar)	nectar, pollen	Alvarez and González (1970)
	Hylocereus undatus (epiphytic)	nectar, pollen	Valiente-Banuet et al. (1996)
	Platyopuntia indeterminate (bushlike)	nectar, pollen	Alvarez and González (1970)
Glossophaga longirostris	Cactaceae indeterminate	nectar, pollen, fruit	Soriano et al. (1991); Sosa and Soriano (1996); Santos et al. (1996); Petit (1997); Ruiz et al. (1997)
	Stenocereus griseus (columnar)	nectar, pollen, fruit	Nassar (1991); Soriano et al. (1991); Sosa and Soriano (1996); Nassar et al. (1997); Petit (1997); Ruiz et al. (1997)
	Subpilosus horrispinus (columnar)	nectar, pollen	Nassar et al. (1997)
	S. repandus (columnar)	nectar, pollen, fruit	Soriano et al. (1991); Petit (1995, 1997); Sosa and Soriano (1996); Nassar et al. (1997)
	Pilosocereus lanuginosus (columnar)	fruit	Petit (1997)
	P. moritzianus (columnar)	nectar, pollen	Nassar et al. (1997)
	P. tillianus (columnar)	nectar, pollen, fruit	Soriano et al. (1991); Sosa and Soriano (1996)
	P. sp. (columnar)	fruit	Ruiz et al. (1997)

Glossophaga soricina	*Cereus atroviridis* (columnar)	nectar, pollen	Lemke (1984)
	Stenocereus griseus (columnar)	nectar, pollen	J. Nassar (pers. comm.)
	Pachycereus weberi (columnar)	nectar, pollen	Valiente-Banuet et al. (1997a)
	Myrtillocactus[a]	nectar, pollen	Alvarez and Gonzalez (1970)
	Platyopuntia indeterminate (bushlike)	nectar, pollen	Alvarez and González (1970)
	Echinocactus indeterminate (spherical)	nectar, pollen	Alvarez and González (1970)
Leptonycteris curasoae	Cactaceae indeterminate	nectar, pollen	Sosa and Soriano (1996); Petit (1997)
	Stenocereus griseus (columnar)	nectar, pollen, fruit	Nassar (1991); Petit (1995, 1997); Sosa and Soriano (1996); Nassar et al. (1997)
	Subpilocereus horrispinus (columnar)	nectar, pollen	Nassar et al. (1997)
	S. repandus (columnar)	nectar, pollen, fruit	Petit (1995, 1997); Sosa and Soriano (1996); Nassar et al. (1997)
	Pilosocereus lanuginosus (columnar)	fruit	Petit (1997)
	P. moritzianus (columnar)	nectar, pollen	Nassar et al. (1997)
Leptonycteris yerbabuenae[b]	*Carnegiea gigantea* (columnar)	nectar, pollen, fruit	Hayward and Cockrum (1971); Howell (1974); Fleming and Sosa (1994); Horner et al. (1998)
	Cephalocereus hopenstedtii (columnar)	nectar, pollen	Valiente-Banuet et al. (1996)
	Neobuxbaumia tetezo (columnar)	nectar, pollen, fruit	Valiente-Banuet et al. (1996)
	N. mezcalaensis (columnar)	nectar, pollen	Valiente-Banuet et al. (1997b)
	N. macrocephala (columnar)	nectar, pollen	Valiente-Banuet et al. (1997b)
	Pachycereus pringlei (columnar)	nectar, pollen, fruit	Fleming and Sosa (1994); Fleming et al. (1996); Horner et al. (1998)
	P. pecten-aboriginum (columnar)	nectar, pollen	T. Fleming (pers. comm.)
	P. weberi (columnar)	nectar, pollen, fruit	Valiente-Banuet et al. (1997b)
	Pilosocereus chrysacanthus (columnar)	nectar, pollen	Valiente-Banuet et al. (1997a)
	Stenocereus thurberi (columnar)	nectar, pollen, fruit	Howell (1980); Fleming and Sosa (1994); Fleming et al. (1996); Horner et al. (1998)
	S. stellatus (columnar)	fruit	Valiente-Banuet et al. (1996)
	Myrtillocactus[a] indeterminate (columnar)	nectar, pollen	Alvarez and González (1970)
	Hylocereus undatus (epiphyte)	nectar, pollen	Valiente-Banuet et al. (1996)

(continued)

TABLE 5.1 *(continued)*

Bat Taxon	*Cactus Taxon*	*Products Utilized*	*References*
	Platyopuntia indeterminate (bushlike)	nectar, pollen	Alvarez and González (1970)
Leptonycteris nivalis	*Neobuxbaumia mezcalaensis* (columnar)	nectar, pollen	Valiente-Banuet et al. (1997b)
	N. macrocephala (columnar)	nectar, pollen	Valiente-Banuet et al. (1997b)
	Pachycereus weberi (columnar)	nectar, pollen	Valiente-Banuet et al. (1997a)
	Pilosocereus chrysacanthus (columnar)	nectar, pollen	Valiente-Banuet et al. (1997a)
	Stenocereus thurberi (columnar)	nectar, pollen, fruit	Dalquest (1953)
	Myrtillocactus[a] indeterminate (columnar)	nectar, pollen	Alvarez and González (1970)
Monophyllus redmani	*Leptocereus* indeterminate (columnar)	nectar, pollen	Silva Taboada (1979)
	Acanthocereus nudiflorus (columnar)	nectar, pollen	Silva Taboada (1979)
	Harrisia indeterminate (bushlike)	nectar, pollen	Silva Taboada (1979)
	Selenicereus indeterminate (epiphyte)	nectar, pollen	Silva Taboada (1979)
Tribe Lonchophyllini			
Platalina genovensium	*Weberbauerocereus weberbaueri* (columnar)	nectar, pollen	Sahley (1995); Sahley and Baraybar (1996)
Subfamily Phyllonycterinae			
Phyllonycteris poeyi	*Leptocereus* indeterminate (columnar)	nectar, pollen	Silva Taboada (1979)
	Acanthocereus nudiflorus (columnar)	nectar, pollen	Silva Taboada (1979)
	Harrisia indeterminate (bushlike)	nectar, pollen	Silva Taboada (1979)
	Selenicereus indeterminate (epiphyte)	nectar, pollen	Silva Taboada (1979)
Subfamily Brachyphyllinae			
Brachyphylla nana	Cactaceae indeterminate	nectar, pollen	Silva Taboada (1979)
Subfamily Carolliinae			
Carollia perspicillata	Cactaceae indeterminate	fruit	Santos et al. (1996)
Subfamily Stenodermatinae			
Artibeus jamaicensis	*Neobuxbaumia tetetzo* (columnar)	nectar, pollen, fruit	Valiente-Banuet et al. (1996)
	Pachycereus weberi (columnar)	nectar, pollen	Valiente-Banuet et al. (1997a)

	Stenocereus stellatus (columnar)	fruit	A. Rojas-Martinez and A. Valiente-Banuet (pers. comm.)
Artibeus intermedius	*Pachycereus weberi* (columnar)	nectar, pollen[c]	Valiente-Banuet et al. (1997a)
Chiroderma salvini	*P. weberi* (columnar)	nectar, pollen[c]	Valiente-Banuet et al. (1997a)
Sturnira lilium	Cactaceae indeterminate	fruit	Santos et al. (1996)
	Neobuxbaumia tetetzo (columnar)	fruit	A. Rojas-Martinez and A. Valiente-Banuet (pers. comm.)
	Pachycereus fulviceps (columnar)	fruit	A. Rojas-Martinez and A. Valiente-Banuet (pers. comm.)
	P. weberi (columnar)	nectar, pollen	Valiente-Banuet et al. (1997a)
	Pilosocereus chrysacanthus (columnar)	nectar, pollen	Valiente-Banuet et al. (1997a)
Subfamily Phyllostominae			
Phyllostomus discolor	*Stenocereus griseus* (columnar)	nectar, pollen	Sosa and Soriano (1996)
	Subpilocereus repandus (columnar)	nectar, pollen	Sosa and Soriano (1996)
	Pilosocereus tillianus (columnar)	nectar, pollen	Sosa and Soriano (1996)
Family Antrozoidae			
Antrozous pallidus	Cactaceae indeterminate	nectar, pollen[d]	Barbour and Davis (1969); Fleming (1991); Herrera et al. (1993)
	Pachycereus pringeli (columnar)	nectar, pollen[d]	T. Fleming (pers. comm.)
	Stenocereus thurberi (columnar)	fruit	Howell (1980)

[a] Includes *Lemaireocereus*.

[b] Although this taxon has been considered a subspecies of *Leptonycteris curasoae* by many authors, including Arita and Humphrey (1988), Koopman (1993), and Wilkinson and Fleming (1996), we follow Hensley and Wilkins (1988) and Koopman (1994) in recognizing *L. curasoae* and *L. yerbabuenae* as distinct species. These taxa have nonoverlapping geographic ranges and apparently exhibit consistent minor differences in dental and external morphology (Koopman, 1981, 1994; Hensley and Wilkins, 1988). These data alone might not be considered conclusive, but recent phylogenetic analyses of nucleotide sequence data (295 bp from the control region of mitochondrial DNA) indicate that individuals of *L. curasoae* and *L. yerbabuenae* represent separate although closely related clades that diverged approximately 540,000 years ago (Wilkinson and Fleming, 1996). In our view, it is most reasonable to conclude that *L. curasoae* and *L. yerbabuenae* are distinct species in the absence of any compelling evidence to the contrary. *Leptonycteris sanborni*, a name commonly used in the literature prior to the middle 1980s, is a junior synonym of *L. yerbabuenae* (Arita and Humphrey, 1988; Koopman, 1993).

[c] A single bat was captured in association with flowering cacti, but no pollen was found on the body; see text for discussion.

[d] Intentional ingestion of cactus flower products not confirmed; see text for discussion.

eats nectar, pollen, and insects (Heithaus et al., 1975; Gardner, 1977; Gannon et al., 1989). Although not a frequent cactus visitor, analyses of fecal samples and pollen wipes from *S. lilium* have shown that this bat feeds on cactus fruits in both Colombia and Mexico, and may also visit cactus flowers (Table 5.1). It seems likely that *S. lilium* utilizes cactus products on an opportunistic basis in all arid regions where its range overlaps that of cactus species.

The same may also be true of *Chiroderma salvini,* a medium-sized stenodermatine that ranges from Chihuahua to northern Bolivia (Koopman, 1993, 1994). Little is known about the diet of this species, but it is presumed to be primarily furgivorous (Goodwin, 1946; Jones et al., 1972; Gardner, 1977). There is one record of this species visiting flowering columnar cacti in Mexico (Valiente-Banuet et al., 1997a).

Another apparently opportunistic species is *Carollia perspicillata,* a small, widely distributed frugivore. Ranging from Oaxaca and Veracruz to southern Brazil, *C. perspicillata* is found in diverse habitats including thorn scrublands, dry deciduous forests, and lowland rainforests (Pine, 1972; Fleming, 1988; Cloutier and Thomas, 1992; Koopman, 1993, 1994; Voss and Emmons, 1996). *C. perspicillata* feeds principally on small understory fruits, supplementing its diet with variable quantities of nectar, pollen, and insects in different regions and seasons (Fleming et al., 1972; Heithaus et al., 1975; Sazima, 1976; Gardner, 1977; Bonaccorso, 1979; Fleming and Heithaus, 1986; Fleming, 1988; Cloutier and Thomas, 1992). *C. perspicillata* is apparently a dietary generalist, including in its diet fruits from more than 24 plant families and flower products from at least six families (Fleming, 1988). In this context, it is not surprising that *C. perspicillata* may occasionally feed on cactus fruits in some regions (e.g., arid zones in Colombia; Santos et al., 1995). However, it seems unlikely that this species relies on cacti to any great extent.

The bat fauna of the Greater Antilles includes several endemic taxa, some of which are cactophilic. *Brachyphylla nana,* restricted to Cuba, Hispaniola, the Cayman Islands, and the southern Bahamas (Swanepoel and Genoways, 1978, 1983b; Koopman, 1993, 1994), is one such species. Apparently common in most habitats on the islands, *B. nana* has a diverse diet that includes pollen, nectar, fruit, and insects (Silva Taboada and Pine, 1969; Gardner, 1977; Swanepoel and Genoways, 1978, 1983b; Silva Taboada, 1979). In a study of *B. nana* in Cuba, Silva Taboada (1979) reported finding pollen from six plant families—including Cactaceae—in the stomachs of these bats. However, *B. nana* is far from being a nectar and pollen specialist: 32% of the 85 sample stomachs contained only insects, and another 20% contained a mixture of pollen and insect parts. It therefore seems likely that *B. nana* is an opportunistic cactophile that includes cactus nectar and pollen in its diet when they are available in the local environment.

Phyllonycteris poeyi, another Antillean endemic, is confined to the islands of Cuba and Hispaniola (Koopman, 1993, 1994). Like *Brachyphylla nana, P. poeyi* has an eclectic diet including nectar, pollen, fruit, and insects (Silva Taboada and Pine, 1969; Gardner, 1977; Silva Taboada, 1979). Analysis of stomach contents of 205 Cuban individuals indicates that most of the diet of *P. poeyi* consists of nectar and pollen, although fruits are also regularly consumed. Flowers of at least seven plant families are visited by these bats, and pollen of several species of cactus have been found in their stomachs (Table 5.1; Silva Taboada, 1979). The diversity of plants visited by *P. poeyi* suggests that they are opportunistic cactophiles.

The remaining ten species of cactophilic bats—including those that appear to be most heavily dependant on cacti—all belong to the phyllostomid subfamily Glossophaginae (New World nectar-feeding bats). Cactophilic taxa include species of *Anoura, Monophyllus, Glossophaga, Leptonycteris, Choeronycteris,* and *Platalina.*

There are only two records of *Anoura geoffroyi* visiting cactus flowers (Table 5.2), and it is unlikely that this species uses cactus products on a regular basis. Widely distributed in the Neotropics (Koopman, 1994), *A. geoffroyi* is most often found in forest habitats, where it feeds on insects as well as fruit and flower products of a variety of plants (Howell and Burch, 1974; Gardner, 1977). It seems most likely that *A. geoffroyi* is an opportunistic cactophile that uses cactus products only when they are abundant in the local environment. This seems to have been the case in Venezuela, where J. Nassar (pers. comm.) captured a single pollen-covered individual of *A. geoffroyi* in a patch of thorn forest containing a large number of *Pilosocereus lanuginosus* that were in full bloom.

TABLE 5.2
Selected Measurements of Glossophagine Bats[a]

Taxon	*Forearm Length (mm)*	*Condylobasal Length (mm)*	*Weight (g)*	*References*
Tribe Lonchophyllini				
*Platalina genovensium**	46–50	29–31	16–22.5	Solari[b]
Lionycteris spurrelli	32–38	17–19	6–11	Koopman (1994)
Lonchophylla hesperia	36–41	24–27	10.5–12	S. Solari[b]; B. Patterson[c]
L. bokermanni	38–42	23–25	not available	
L. handleyi	44–48	25–29	16–17	S. Solari[b]; Reid (1997)
L. robusta	39–46	24–26	14–19.5	Reid (1997); Emmons (1997); Albuja (1982)
L. mordax	32–37	20–23	7–9	Reid (1997); Emmons (1997)
L. dekeyseri	34–38	20–22	not available	
L. thomasi	31–34	18–21	5–10	Reid (1997); Emmons (1997); Albuja (1982); Simmons and Voss (1998)
Tribe Glossophagini				
*Monophyllus redmani**	35–43	19–23	8–13	Klingener et al. (1978)
M. plethodon	38–46	19–23	12.5–17.5	Koopman (1994)
*Glossophaga longirostris**	35–42	21–23	10–16	Webster and Handley (1986)
*G. soricina**	32–39	19–22	7–12	Reid (1997); Albuja (1982); Simmons and Voss (1998)

(continued)

TABLE 5.2 *(continued)*

Taxon	*Forearm Length (mm)*	*Condylobasal Length (mm)*	*Weight (g)*	*References*
G. morenoi	32–37	19–22	7–9	Reid (1997)
G. leachii	34–39	17–20	9–11	Reid (1997)
G. commissarisi	31–37	17–20	6–11	Koopman (1994)
*Leptonycteris nivalis**	50–60	26–29	25–35	Reid (1997)
*L. yerbabuenae**	50–56	25–29	20–27	Reid (1997)
*L. curasoae**	50–56	26–28	20–27	Reid (1997)
Anoura luismanueli	33–37	19–21	7.5–10	Molinari (1994)
A. latidens	40–46	23–24	13.5–17	S. Solari[b]; C. Handley[d]
A. cultrata	38–44	22–26	12–23	Reid (1997); Emmons (1997)
A. caudifera	34–39	21–24	8.5–13	Emmons (1997); Simmons and Voss (1998); Molinari (1994)
*A. geoffroyi**	39–47	24–26	12.5–20	Reid (1997); Emmons (1997); Albuja (1982)
Scleronycteris ega	34–35	21–22	not available	
Lichonycteris obscura	30–36	16–19	6–10.5	Reid (1997); Emmons (1997); Simmons and Voss (1998)
Hylonycteris underwoodi	31–37	19–22	6–12	Reid (1997); Emmons (1997)
Choeroniscus periosus	40–42	29–30	not available	
C. minor[e]	33–38	20–25	7–12	Simmons and Voss (1998)
C. godmani	32–36	18–21	5–13	Reid (1997)
*Choeronycteris mexicana**	42–47	28–30	14–19	Reid (1997)
Musonycteris harrisoni	40–43	30–34	6–12.6	C. Handley[d]; H. Arita (pers. comm.); M. Tschapka (pers. comm.)

[a]Taxa known to be cactophilic are marked with an asterisk. Information on body weights should be interpreted with caution; some of these estimates may include data from subadults and/or pregnant females.

[b]Data from specimens in the Museo de Historia Natural, Universidad Nacional Mayor de San Marcos, Lima, Peru (Sergio Solari, pers. comm.). Jiménez and Pefaur (1982) reported a body weight of 47.0 g for *Platalina genovensium,* but this appears to be in error; this value is more likely a forearm measurement.

[c]Data from specimens in the Field Museum of Natural History (FMNH; B. Patterson, pers. comm.).

[d]Data from specimens in the National Museum of Natural History (C. Handley, pers. comm.).

[e]*C. minor* is a senior synonym of *C. intermedius;* see Simmons and Voss (1998).

Monophyllus redmani is endemic to the Greater Antilles, where it occurs on Cuba, Hispaniola, the southern Bahamas, Jamaica, and Puerto Rico (Homan and Jones, 1975a; Koopman, 1981, 1993, 1994). The diet of this bat includes nectar, pollen, and insects (Gardner, 1977; Silva Taboada, 1979; Rodriguez-Duran and Lewis, 1987; Rodriguez-Duran et al., 1993). Although flower products may be the principal food source of this bat, insects are also very important, occurring in 72% of 65 stomach samples analyzed by Silva Taboada (1979) and over 60% of fecal samples studied by Rodriguez-Duran and Lewis (1987) and Rodriguez-Duran et al.

(1993). Pollen of four cactus species (Table 5.1) has been found in stomach samples from *M. redmani* (Silva Taboada, 1979), but this represents only a portion of the plant diversity exploited by this bat. Flowers of at least ten plant families are apparently visited by *M. redmani* (Silva Taboada, 1979). It seems likely that *M. redmani* is an opportunistic cactophile, visiting cactus flowers when available but also utilizing many other food sources.

Another taxon with broad dietary habits is *Glossophaga soricina,* a species common in most neotropical habitats from tropical Mexico to northern Argentina (Koopman, 1981, 1994; Webster, 1993; Voss and Emmons, 1996). Its diet consists of fruit, nectar, pollen, flower parts, and insects, with relative proportions varying with season and locality (Howell and Burch, 1974; Heithaus et al., 1975; Gardner, 1977; Alvarez et al., 1991). Although some populations in arid areas use cactus products regularly, these bats visit other plants in the same habitat more frequently (e.g., *Agave, Crescentia;* Lemke, 1984). Investigations of diet using carbon stable isotope analysis suggest that *G. soricina* in southern Mexico feeds mostly on C3 plants (e.g., Bignoniaceae, Bombacaceae, Convolvulaceae, Leguminosae), even when CAM plants (Cactaceae, Agavaceae) are available (Fleming et al., 1993). We therefore consider *G. soricina* to be an opportunistic cactophile.

The same does not seem to be true of *Glossophaga longirostris,* a species confined to northern South America and the southern Lesser Antilles (Webster, 1993; Webster and Handley, 1986; Webster et al., 1998). This species is most frequently found in dry habitats (Handley, 1976) and is closely associated with cacti in many areas, including arid zones in Colombia, Venezuela, and the Lesser Antilles (Alvarez and González, 1970; Nassar, 1991; Soriano et al., 1991; Petit, 1995, 1997; Sosa and Soriano, 1996; Valiente-Banuet et al., 1996; Nassar et al., 1997; Ruiz et al., 1997). In at least some of these regions (e.g., Curaçao and arid areas in the Venezuelan Andes), *G. longirostris* apparently depends on cactus products for survival, and patterns of reproduction reflect cactus flower and fruit availability (Sosa and Soriano, 1996; Petit, 1997). These populations appear to be obligate cactophiles. However, not all populations of *G. longirostris* live in arid areas (Handley, 1976), and forest populations may rely entirely on products of other plants.

Leptonycteris is a small genus of glossophagines that includes three species: *L. curasoae, L. yerbabuenae* (= *L. sanborni*), and *L. nivalis* (see footnotes to Table 5.1 for discussion of taxonomy). All of these taxa are known to utilize cactus products (Table 5.1), and many populations may be obligate cactophiles. *L. curasoae* is confined to arid areas of northwestern Colombia and northern Venezuela as well as nearby islands (Koopman, 1981, 1994). *L. curasoae* relies on at least two species of cacti throughout its range (e.g., *Stenocereus griseus, Subpilocereus repandus*). This bat has been shown to be a critical pollinator of these and other species of columnar cacti (Nassar, 1991; Petit, 1995, 1997; Nassar et al., 1997). Populations of *L. curasoae* in at least some areas (e.g., Curaçao) are clearly obligate cactophiles

(Petit, 1997), and it is possible that all populations depend on cactus products for survival.

Leptonycteris yerbabuenae, perhaps the most studied of the cactophilic bats, has a broad geographic range that extends from El Salvador north to Arizona (Koopman, 1981, 1994). Recent work has suggested that some populations migrate north from central Mexico along a "nectar corridor" of blooming columnar cacti in the spring and move south in the fall, following a similar pattern of blooming paniculate agaves (Fleming et al., 1993). Although other populations of *L. yerbabuenae* do not appear to rely heavily on cactus products (e.g., in Jalisco, Mexico; Ceballos et al., 1997), the migratory population in Mexico appears to be an obligate cactophilic group.

Little is known about dietary habits of the third species, *Leptonycteris nivalis,* which ranges from southern Texas to Guatemala (Hensley and Wilkins, 1988; Koopman, 1994). Use of cactus products has been documented in *L. nivalis* (Table 5.1), and some preliminary carbon stable isotope data suggest that this species may feed extensively on CAM plants in parts of its range (Fleming et al., 1993; Moreno, 2000). However, the degree to which *L. nivalis* depends on cacti as opposed to agaves has not yet been determined.

At least some species of columnar cacti (e.g., *Neobuxbaumia macrocephala, N. mezcalaensis, N. tetetzo*) apparently depend on *Leptonycteris* species and another glossophagine bat, *Choeronycteris mexicana,* for pollination (Valiente-Banuet et al., 1996). *C. mexicana* ranges from the southwestern United States to Honduras (Koopman, 1981, 1994) and inhabits a wide range of habitats, including tropical deciduous forests (Arroyo-Cabrales et al., 1987). Their diet includes fruits, pollen, nectar, and probably insects (Gardner, 1977; Arroyo-Cabrales et al., 1987). These bats are frequent visitors to flowering and fruiting columnar cacti in Mexico (Valiente-Banuet et al., 1996) and they probably rely heavily on cactus products throughout much of the year in arid habitats. This appears to be confirmed by preliminary carbon stable isotope data, which suggest that *C. mexicana* may feed extensively on CAM plants in Mexico (Fleming et al., 1993). We expect that these populations of *C. mexicana* are obligate cactophiles, although populations in other areas (e.g., forest habitats) probably are not. In southeastern Arizona, this species feeds heavily at flowers of *Agave palmeri* (K. Hinman, pers. comm.).

Platalina is a monotypic genus confined to mid- to high-elevation arid regions of western Peru (Koopman, 1981, 1994). *P. genovensium* is a regular visitor to columnar cacti and appears to depend on cactus resources for survival throughout most of its range (Sahley, 1995; Sahley and Baraybar, 1994, 1996). These bats may therefore be classified as obligate cactophiles. However, little is known about the natural history of *P. genovensium,* and it is possible that some populations are not obligate cactophiles. Insect remains are common in fecal samples from *P. genovensium,* and carbon stable isotope analyses of muscle tissue samples from these bats have revealed ^{12}C:^{13}C ratios indicative of a diet including some resources other than CAM plants (Sahley and Baraybar, 1996).

Bat Phylogeny and the Origins of Cactophily

Most cactophilic bats belong to a single family, Phyllostomidae. The opportunistic cactophile *Antrozous pallidus* is usually included in the family Vespertilionidae, but was recently removed to its own family (Antrozoidae), based on results of phylogenetic analyses (Simmons, 1998; Simmons and Geisler, 1998). Antrozoidae is not closely related to Phyllostomidae (Simmons, 1998; Simmons and Geisler, 1998), and use of cactus products clearly evolved independently in these two groups. Within Phyllostomidae, cactophily is found in members of six subfamilies: Brachyphyllinae, Carolliinae, Glossophaginae, Phyllostominae, Stenodermatinae, and Phyllonycterinae (Table 5.1). All of the phyllostomines (*Phyllostomus discolor*), brachyphyllines (*Brachyphylla nana*), phyllonycterines (*Phyllonycteris poeyi*), carolliines (*Carollia perspicillata*), and stenodermatines (*Artibeus jamaicensis, A. intermedius, Sturnira lilium, Chiroderma salvini*) on this list are opportunistic cactophiles; all taxa that include obligate cactophilic populations belong to the subfamily Glossophaginae.

Phyllostomidae is one of the most diverse families of bats. It currently includes more than 50 genera and 140 species that variously feed on insects, small vertebrates, fruit, nectar, pollen, flower parts, buds, leaves, and even blood (Gardner, 1977; Hill and Smith, 1984; Koopman, 1993; Wetterer et al., 2000). As a result of this diversity, phyllostomids have been the subject of numerous phylogenetic studies (for a review, see Wetterer et al., 2000). Although many data sets have been used to investigate relationships at different taxonomic levels, there has been little agreement concerning interrelationships of genera and subfamilies until recently. Relationships of glossophagines have been particularly controversial, with different data sets and analytical methods suggesting very different phylogenetic conclusions (Griffiths, 1982, 1983; Haiduk and Baker, 1982; Warner, 1983; Smith and Hood, 1984; Honeycutt and Sarich, 1987; Baker et al., 1989; Van Den Bussche, 1991, 1992; Gimenez et al., 1996). However, a recently completed analysis of relationships of phyllostomids (Wetterer et al., 2000) has resolved many outstanding questions.

Wetterer et al. (2000) used a data set of diverse morphological characters (including all those previously cited by other authors), rDNA restriction sites, and features of the sex chromosomes to investigate intergeneric relationships of phyllostomids. In a series of parsimony analyses, they found support for monophyly of all traditionally recognized phyllostomid subfamilies (Brachyphyllinae, Carolliinae, Desmodontinae, Glossophaginae, Phyllonycterinae, Phyllostominae, and Stenodermatinae) and many lower-level clades. Although Griffiths (1982) recognized Lonchophyllinae as a taxon distinct from Glossophaginae and suggested that they may not be closely related, Wetterer et al. (2000) found strong support for a sister-group relationship between these clades. To reflect this relationship, Wetterer et al. (2000) suggested a revised classification recognizing two tribes (Glossophagini and Lonchophyllini) within the subfamily Glossophaginae. Both tribes contain cactophilic taxa: Lonchophyllini includes *Platalina,* and Glossophagini includes *Anoura, Choeronycteris, Glossophaga, Leptonycteris,* and *Monophyllus.* The other phyllostomid genera that include cactophilic species or populations (*Artibeus, Brachyphylla, Carollia, Phyllonycteris, Phyllostomus,* and *Sturnira*) belong to other subfamilies. *Phyllonycteris* belongs to a clade (Phyllonycterinae) that is the sister-taxon of Glossophaginae. Together, these two subfamilies (Phyllonycterinae + Glossophaginae) comprise the clade Hirsutoglossa ("hairy-tongued"; Wetterer et al., 2000). *Brachyphylla,* sole member of the subfamily Brachyphyllinae, occupies an unresolved position near the base of the phyllostomid tree (Figure 5.1; Wetterer et al., 2000). The remaining genera (*Artibeus, Carollia, Chiroderma, Phyllostomus,* and *Sturnira*) are not closely related to Glossophaginae or to each other.

A posteriori mapping of characters or ecological traits onto phylogenies offers a powerful method for generating and testing evolutionary inferences (Brooks and McLennan, 1991; Swofford and Maddison, 1992). Mapping cactophily onto the Wetterer et al. tree indicates that this habit evolved multiple times in phyllostomids: once each in Brachyphyllinae, Carolliinae, and Phyllostominae; at least three times in Stenodermatinae; and at least once (possibly several times) in Hirsutoglossa (Fig. 5.1). Fig. 5.1 suggests that cactophily is primitive for Hirsutoglossa (and therefore also for Glossophaginae), but this pattern appears to be a spurious result of the method employed, which overlooked within-genus variation in dietary habits of glossophagines. To map cactophily on the tree in Fig. 5.1, we had to adopt a very broad definition of cactophily (i.e., each genus scored as having "no known cactophilic members" or "cactophily in some or all species"). Better resolution can be obtained by scoring each species separately.

A more detailed view of Hirsutoglossa is shown in Fig. 5.2. Topology of this tree was derived from the Wetterer et al. tree by expanding each genus to show all of its constituent species. Relationships within most genera were left unresolved to reflect our lack of understanding of interspecific relationships of these groups. One exception is *Leptonycteris,* for which species relationships were resolved following the results of Wilkinson and Fleming's (1996) analysis of mitochondrial gene sequences. The structure of the tree in Fig. 5.2 indicates multiple convergent origins of cactophily in Hirsutoglossa. In most cases, evolutionary diet shifts have apparently been limited to acquisition of opportunistic cactophily; obligate cactophily has apparently only evolved a few times (e.g., in *Glossophaga, Leptonycteris, Platalina,* and perhaps *Choeronycteris*). It seems reasonable to assume that obligate cactophily evolved from opportunistic cactophily in most or all cases, but this cannot be tested given data currently available.

Among nonglossophagine phyllostomids, the only genus known to include more than one cactophilic species is *Artibeus.* Recent phylogenies of *Artibeus* based on morphology (Marques-Aguiar, 1994) and nucleotide sequence data (Van Den Bussche et al., 1998) agree that *A. jamaicensis* and *A. intermedius* are not particularly closely related within the genus. Mapping cactophily on either the Marques-Aguiar (1994) tree

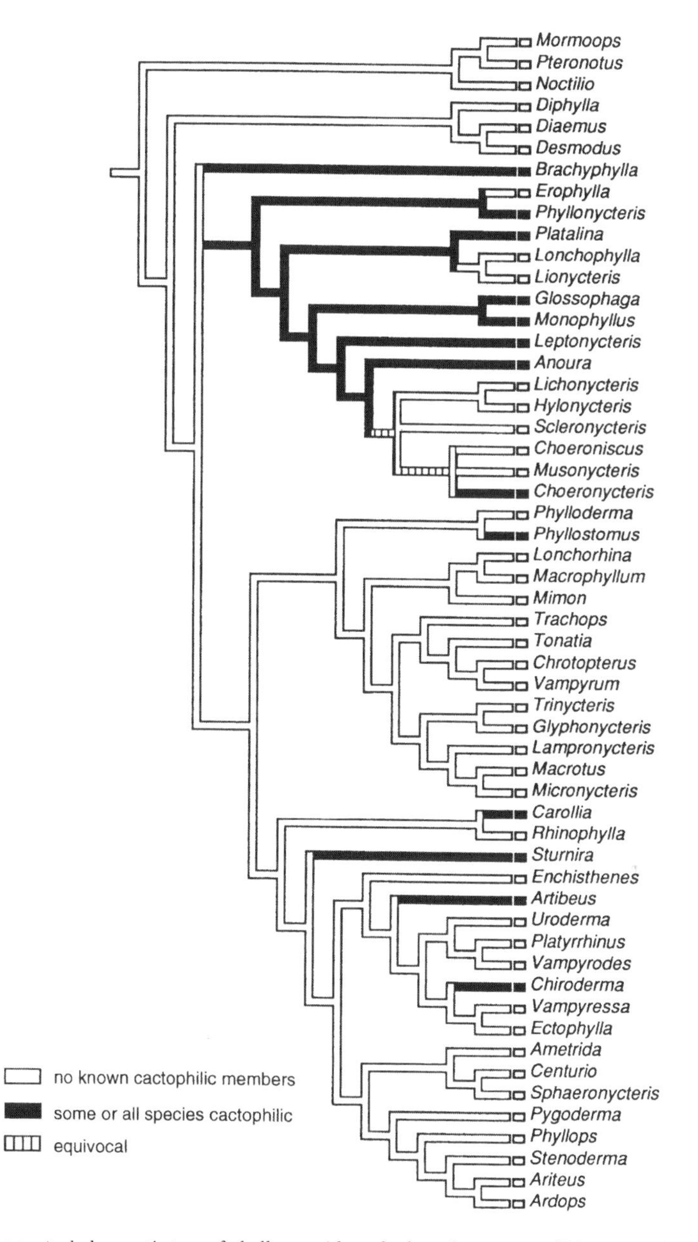

FIGURE 5.1. A phylogenetic tree of phyllostomids and selected outgroups (*Mormoops, Pteronotus,* and *Noctilio*) showing the distribution of cactophily among genera. Tree topology and generic nomenclature follow Wetterer et al. (2000). This hypothesis suggests that cactophily has evolved at least six times in Phyllostomidae.

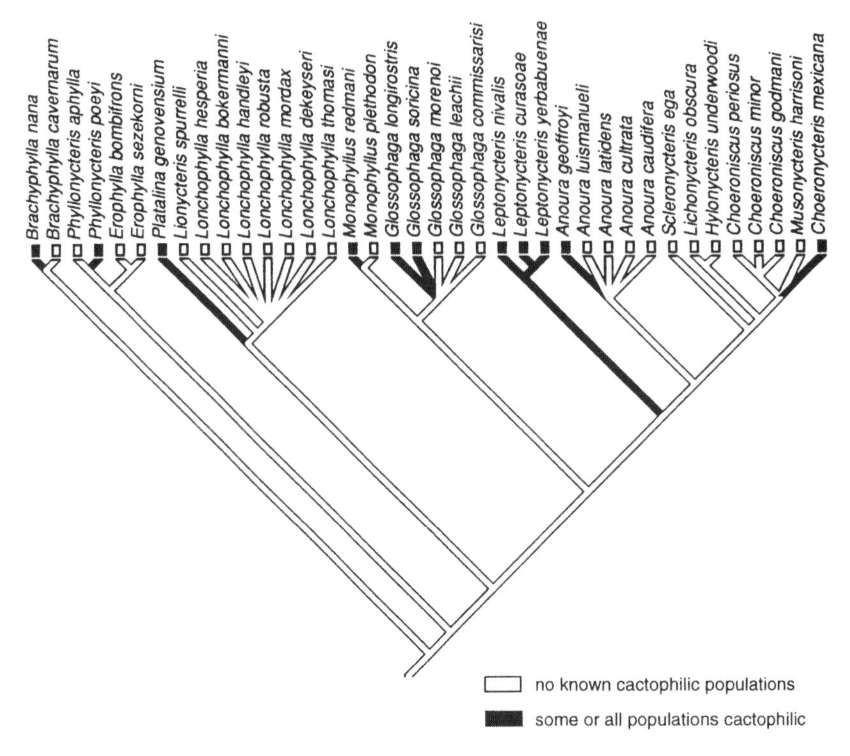

FIGURE 5.2. A phylogenetic tree of glossophagine phyllostomids showing the distribution of cactophily among species. This tree was derived from that shown in Fig. 5.1 by expanding each glossophagine genus to show all of its constituent species; relationships among species of *Leptonycteris.*

or the Van Den Bussche et al. (1998) tree (not shown in Fig. 5.2) suggests that cactophily is not primitive for the genus *Artibeus,* but instead evolved independently in *A. jamaicensis* and *A. intermedius.*

A notable feature of the evolutionary patterns described above is that most cactophilic bats belong to clades otherwise characterized by noncactophily. The only exceptions are *Glossophaga* (containing two cactophilic species and three noncactophiles) and *Leptonycteris* (all species cactophilic). This suggests that speciation has only rarely occurred within cactophilic lineages. Adoption of cactophilic habits most often seems to be a population-level phenomenon that occurs within species, not a species-level trait that is inherited by sister-taxa from a common ancestor. In this context, the evolutionary histories of *Glossophaga* and *Leptonycteris* are of particular interest because they may represent cases in which speciation *has* occurred within a cactophilic group. Nothing is known about phylogenetic relationships of species of *Glossophaga,* but Wilkinson and Fleming's (1996) study of *Leptonycteris* provides a context for investigating this phenomenon in at least one lineage.

Wilkinson and Fleming (1996) calculated divergence times for *Leptonycteris* lineages based on corrected genetic distances among groups. They hypothesized that *L. curasoae* and *L. yerbabuenae* diverged approximately 540,000 years ago, and that the lineage leading to these species diverged from *L. nivalis* approximately one million years ago. If the common ancestor of all three species was cactophilic (which is the most parsimonious interpretation of the available data; see Fig. 5.2), this suggests that cactophily evolved in *Leptonycteris* over one million years ago. Wilkinson and Fleming (1996, pp. 335–36), who treated *L. yerbabuenae* as a subspecies of *L. curasoae,* concluded that

> If these dates accurately reflect evolutionary history, they suggest that these taxa are the products of climatic events occurring during the late Pliocene and Pleistocene. Prior to the uplift of the Mexican plateau and its flanking Sierra Madres in the late Pliocene about 2 million years ago . . . a single arid-adapted species of *Leptonycteris* is likely to have occurred in Mexico. After the uplift, this species must have split into the current two species—an upland *L. nivalis* and a lowland *L. curasoae*. . . . The presence of *L. curasoae* in Mexico and northern South America suggests that an arid or semi-arid corridor connected these regions during at least one Pleistocene glacial advance. . . . During the mid-Pleistocene this corridor may have broken, producing the two subspecies of *L. curasoae* [*L. c. curasoae* and *L. c. yerbabuenae*].

If correct, this hypothesis suggests that the extant species of *Leptonycteris* evolved as a result of two vicariance events that split widespread cactophilic species into separate lineages with restricted geographic distributions. It is possible that reliance on cactus products may have contributed to lineage splitting in each case. Rather than switching to other food sources and maintaining populations in less-arid habitats between cactus/agave zones, these bats apparently retained their ancestral association with arid habits, cacti, and agaves. Gene flow between bat populations was interrupted, and this ultimately led to speciation. Although there were doubtless other factors involved, the distribution of columnar cacti may have influenced the course of evolution in *Leptonycteris.*

An obvious requirement for the evolution of cactophily in any given group is overlap of the geographic range of the bats with those of potentially chiropterophilic cacti, most of which are found in relatively arid environments. Cactophily clearly cannot evolve in bat taxa that are restricted to lowland rainforests, cloud forests, or other habitats that do not support cactus populations. In cases where the range of a bat species covers multiple habitats including regions rich in chiropterophilic columnar cacti, evolution of opportunistic cactophily may be a relatively common phenomenon, at least among bats already specialized for a nectarivorous, frugivorous, or omnivorous diet (e.g., most phyllostomids). It is probably no accident that most of the opportunistic cactophilic species identified above have broad geographic ranges and are found in multiple habitats. Evolution of obligate cactophily may require a stronger association between a bat lineage (or population) and cactus-rich environments.

Although this resembles the classic "chicken or egg" problem, close correspondence of the geographic range of a bat species or population with the range of chiropterophilic cacti may be the most important correlate of obligate cactophily.

Morphology of Cactophilic Bats: Phylogenetic Context

It is difficult to identify any morphological or ecological features common to all cactophilic bats if one considers the complete taxonomic list of cactus-visiting bats (Table 5.1). All are microchiropteran (echolocating) bats, and all have geographic ranges that overlap with those of columnar cacti, but these bats have little else in common other than a taste for cactus products. However, it is only within Glossophaginae that obligate cactophily has evolved, and it is here that we can most profitably look for patterns. Various authors have noted that cactophilic bats of this clade are characterized by (1) a long, filiform tongue with brush-like papillae at the tip; (2) a long, narrow rostrum and palate; (3) a reduced dentition; (4) small ears; and (5) hovering flight (Howell, 1974b, 1979; Fleming, 1982; Tuttle, 1991; Herrera et al., 1993; Nassar et al., 1997). However, these features characterize most or all glossophagines and therefore appear to have evolved well before any of the presumed origins of cactophily in any glossophagine lineage.

Presence of a long, filiform tongue with brushlike papillae at the tip characterizes Hirsutoglossa (Glossophaginae + Phyllonycterinae; Wetterer et al., 2000). This morphology is widely regarded as an adaptation for nectar feeding, allowing access to nectar reserves hidden deep within flowers (e.g., Park and Hall, 1951; Wille, 1954; Phillips, 1971; Winkelmann, 1971; Greenbaum and Phillips, 1974; Griffiths, 1978, 1982; Hill and Smith, 1984). A relatively long, narrow rostrum and palate and a reduced dentition also characterize Hirsutoglossa (Phillips, 1971; Freeman, 1995; Wetterer et al., 2000). The most extreme expressions of these traits are found among glossophagines (e.g., in the clade comprising *Choeroniscus, Choeronycteris, Hylonycteris, Lichonycteris, Musonycteris,* and *Scleronycteris*), but many of these bats are not known to be cactophiles. Dental reduction is thought to result from a decreased need for chewing when aqueous nectar and tiny pollen grains form a significant part of the diet (Phillips, 1971; Hill and Smith, 1984; Freeman, 1995). Reduction of the incisor dentition apparently facilitates tongue extrusion (Hill and Smith, 1984; Freeman, 1995) and may reduce the chance of the bat accidentally cutting its tongue on its incisors (Fenton, 1992). A long, narrow rostrum and palate are correlated with increased tongue length and also effectively increase the reach of the muzzle into narrow spaces (e.g., into the corolla and nectar chamber of flowers). All glossophagines have relatively small ears and a short noseleaf (Wetterer et al., 2000), features that facilitate unencumbered insertion of the head into flowers.

Like the cranium and dentition, the flight mechanism of glossophagines (and perhaps all hirsutoglossans) also differs from that seen in other bat clades. All glossophagine species studied thus far share a common morphology that is characterized by relatively high wing loading, below average aspect ratio, and large, rounded wingtips (Smith and Starrett, 1979; Norberg and Rayner, 1987). This combination of factors makes these bats particularly well suited for hovering flight (Norberg and Rayner, 1987). Many glossophagines are known to hover when feeding from flowers, and this trait is probably primitive for the subfamily (Goodwin and Greenhall, 1961; Struhsaker, 1961; Strickler, 1978; Howell, 1979; Heithaus, 1982; Lemke, 1984; von Helversen and Reyer, 1984). The evolution of hovering therefore seems to predate the evolution of cactophilic behavior.

The suite of features described above is best explained as a complex that evolved in concert with a dietary shift to increased reliance on flower products. The "nectarivory" craniodental complex and the ability to hover while feeding may represent preadaptations for effective cactophily. Glossophagines are particularly well equipped to extract nectar from flowers, including those of columnar cacti. Although many phyllostomids are frugivorous, obligate cactophily may only be possible in species that can also efficiently obtain nectar; glossophagines are better suited for this task than any other clade of New World bats.

This raises an interesting question: Are there any morphological traits that characterize cactophilic bats but not noncactophiles? Limiting our discussion to glossophagines once more, examination of measurement data suggests one interesting pattern. Comparisons of forearm length (a highly constrained dimension that is indicative of both wing and body size), condylobasal length (a measure of skull length), and body weight among the 32 currently recognized glossophagine species (Table 5.2) indicate that cactophilic bats often have longer forearms, longer skulls, and greater body weights than their closest relatives. For example, *Platalina genovensium* is the largest member of Lonchophyllini, and all three species of *Leptonycteris* have longer forearms and higher body weights than any other members of Glossophagini. Although there is overlap in measurements among species of *Glossophaga, G. longirostris* is typically larger than its congeners in all three measures. *Choeronycteris* is characterized by a longer forearm and greater body weight than most other members of its clade. Only seven glossophagine species are characterized by forearm length ≥40 mm, condylobasal length ≥25 mm, and body weight >14 g, and five are known cactophiles (i.e., *Choeronycteris mexicana, Leptonycteris curasoae, L. nivalis, L. yerbabuenae,* and *Platalina genovensium*). The remaining two species in this size class are *Lonchophylla handleyi* and *Choeroniscus periosus.* Although little is known about the natural history of these species, there is no reason to believe that they are cactophilic. *L. handleyi* occurs east of the Andes in Ecuador, Peru, and southern Colombia (Hill, 1980; Albuja, 1982; Koopman, 1993, 1994). Locations of collection localities suggest that it is limited to lowland and foothill forest habitats in western Amazonia. Similarly, *C. periosus* (including *C. ponsi;* Pirlot, 1967) is known only from forested regions in western Colombia, western Ecuador, and northwestern Venezuela (Handley, 1966; Pirlot, 1967; Koopman, 1993, 1994). Given the data currently available, it seems unlikely that either *L. handleyi* or *C. periosus* is cactophilic, because there is no evidence that these bats ever inhabit arid habitats. However, consideration of habitat associations drew our attention to yet another glossophagine species, *Musonycteris harrisoni*, which has a large condlyobasal measure and long forearm similar to many cactophilic taxa (Table 5.2).

Other Potentially Cactophilic Bats

Musonycteris harrisoni is a remarkable but poorly known bat found only in southwestern Mexico (Jalisco southeastward to Guerrero; Webster et al., 1982; Koopman, 1993, 1994). The longest nosed of all glossophagines, it has an extremely long, narrow rostrum that is more than one-half the length of the skull (Schaldach and McLaughlin, 1960). Condlyobasal length in *M. harrisoni* (30.0–34.0 mm) far exceeds that of many cactophiles, and forearm length in this species (40.0–43.0 mm) is also long for a glossophagine (Table 5.2). Only the relatively small body weight of *M. harrisoni* (6.0–11.5 g) is somewhat anomalous for a cactophile. However, at least two glossophagines in this weight class are cactophilic (*Glossophaga soricina* and *Monophyllus redmani*), indicating that small body size does not necessarily preclude cactophily.

Virtually nothing is known about the natural history of *Musonycteris harrisoni*. All specimens reported in the literature (a total of fewer than 20) were captured in arid regions either dominated by or near scrub forest (Schaldach and McLaughlin, 1960; Webster et al., 1982). The first known specimens were collected in a banana grove, where the bats were apparently feeding from banana flowers (Schaldach and McLaughlin, 1960). It is presumed that *Musonycteris harrisoni* feeds on nectar, pollen, and some insects, as do other long-nosed glossophagines (Gardner, 1977), but no dietary data are available.

The geographic range of *Musonycteris harrisoni* in southwestern Mexico overlaps that of at least nine other glossophagine species, including several cactophiles (i.e., *Anoura geoffroyi, Choeronycteris mexicana, Glossophaga soricina, Leptonycteris nivalis, L. yerbabuenae;* Hall, 1981; Koopman, 1981; Valiente-Banuet et al., 1996). This region is characterized by high cactus diversity, with various habitats supporting at least six and sometimes more than 20 species of columnar cacti (Valiente-Banuet et al., 1996). Columnar cacti are in bloom at least six months of the year in this region (Valiente-Banuet et al., 1996). Almost 70% of these cacti are night-blooming forms, and 60% have floral characteristics that suggest that they are chiropterophilic (Valiente-Banuet et al., 1996). Given the geographic range of *M. harrisoni*, its morphology (e.g., forearm ≥40 mm, condylobasal length ≥30 mm), and the abundance of bat-accessible cactus products in this region, we hypothesize that this species is cactophilic. The extremely long rostrum of this taxon is approached in only two other glossophagine species (*Platalina genovensium* and *C. mexicana*), both of which may be obligate cactophiles. It is hard to imagine that *M. harrisoni* does not utilize cactus products (particularly nectar and pollen) if they are available in the local environment. We predict that future studies of cactus pollinators—particularly field projects using mistnets deployed around flowering columnar cacti in Jalisco, Colima, Michoacán, and Guerreo—will ultimately demonstrate that *M. harrisoni* frequently visits cacti in these areas. The restricted geographic range and low body weight of *M. harrisoni* suggest that it is not a migratory species (see below for additional discussion of body weight and migration).

In view of the above conclusions, it is interesting to note that the closest relative (sister taxon) of *Musonycteris harrisoni* may be *Choeronycteris mexicana.* Schaldach and McLaughlin (1960) noted the similarity between these taxa in their original description of *M. harrisoni.* Handley (1966) compared skull morphology of *Musonycteris, Choeronycteris,* and all species of *Choeroniscus,* and concluded that *M. harrisoni* was best considered an exceptionally long-muzzled species of *Choeronycteris.* Phillips (1971) reviewed the problem and chose to regard *Musonycteris* and *Choeronycteris* as distinct genera, based on basicranial and dental differences. Webster et al. (1982) similarly noted karyotypic differences between these taxa; however, neither study refuted a close relationship between *Musonycteris* and *Choeronycteris.* Most recently, Wetterer et al. (2000) found that *Musonycteris, Choeronycteris,* and *Choeroniscus* form a clade, but they could not resolve relationships within this group (Figure 5.1). Should *Musonycteris* ultimately be shown to be the sister-taxon of *Choeronycteris,* it is possible that cactophily evolved only once in this group, in the most recent common ancestor of these taxa. If so, these bats may represent another instance of speciation within a cactophilic lineage.

Consideration of body size, phylogenetic relationships, and geographic ranges also suggest that at least two other phyllostomid species may be cactophiles: *Monophyllus plethodon* and *Brachyphylla cavernarum.* Both species are found throughout the Lesser Antilles from Anguilla to St. Lucia, a region from which there are no known cactophilic bats. Complete absence of cactophilic bats from the northern Lesser Antilles seems unlikely, as many columnar cacti that produce chiropterophilic-type flowers (large flowers that are white or pale-colored, exhibit nocturnal anthesis, and have large quantities of nectar and pollen) occur on these islands (e.g., species of *Cereus, Harrisia, Leptocereus, Pilosocereus*). We suggest that the probable pollinators of these cacti may include *M. plethodon* and *B. cavernarum,* both of which are sister species of known cactophilic taxa.

Monophyllus plethodon is a glossophagine bat endemic to the Lesser Antilles from Anguilla to St. Vincent and Barbados, although it is also known from fossils from Puerto Rico (Homan and Jones, 1975b; Koopman, 1981, 1993, 1994). *M. plethodon* is the larger of the two species of *Monophyllus* (Table 5.2), a feature that initially drew our attention to this species. We noted earlier that cactophilic glossophagines tend to have longer forearms, longer skulls, and greater body weights than their closest relatives. In *Monophyllus,* however, the known cactophilic species (*M. redmani*) is notably smaller than its sister species (Table 5.2). This apparent anomaly would be eliminated if both species of *Monophyllus* were shown to be cactophilic. Unfortunately, little is known about the natural history of *M. plethodon.* It has been captured in habitats ranging from xeric woodlands to dense rainforest (Homan and Jones, 1975b), but nothing is know about its diet. *M. plethodon* is presumed to have a diet similar to that of *M. redmani,* which feeds on nectar, pollen, and insects. If this is the case, it is likely that *M. plethodon* may also be an opportunistic cactophile, feeding from cactus flowers when available.

Brachyphylla cavernarum is a brachyphylline bat confined to Puerto Rico, the Virgin Islands, and the Lesser Antilles as far south as St. Vincent and Barbados (Swanepoel and Genoways, 1978, 1983a; Koopman, 1993, 1994). This species has been collected in habitats ranging from humid forest to dry scrub forest (Swanepoel and Genoways, 1983a). The diet of *B. cavernarum* is apparently eclectic, including pollen, nectar, fruit, and insects (Bond and Seaman, 1958; Nellis, 1971; Gardner, 1977; Nellis and Ehle, 1977; Swanepoel and Genoways, 1978, 1983a). Flowers and fruits of many plant families provide food for this bat, which appears to feed opportunistically (Nellis, 1971; Nellis and Ehle, 1977). In this respect is it quite similar to its sister species, *B. nana*. Given that *B. nana* feeds on cactus products when available, it seems likely that *B. cavernarum* does the same.

Why Is Size Important?

It is well known that body size and proportions have numerous physiological, mechanical, and ecological implications (Peters, 1983; Schmidt-Nielson, 1984; Norberg and Rayner, 1987). Our observation that cactophilic glossophagines typically are characterized by greater forearm length, condylobasal length, and body weight than their noncactophilic relatives raises questions about just why this is so. Formally addressing this complex problem is beyond the scope of this study, but we can offer some hypotheses. First, increased skull length in glossophagines—particularly cactophilic genera—is not associated with a similar increase in skull breadth (Schaldach and McLaughlin, 1960; Handley, 1966; Swanepoel and Genoways, 1979; Webster, 1993; Freeman, 1995). This means that cactophilic taxa tend to have a longer rostrum than their relatives. Hill and Smith (1984) hypothesized that rostral length in glossophagines may closely parallel the length of the flowers on which a particular species feeds. It is probably not a coincidence that long-muzzled glossophagines (e.g., *Choeronycteris, Musonycteris, Platalina*) may all be cactophiles.

The flowers of columnar cactus are large, relatively long, and are typically funnel-shaped or tubular in form. For example, the mean length of individual flowers of saguaro (*Carnegiea gigantea*) and cardon (*Pachycereus pringlei*) cacti is over 100 mm, and mean width of the perianth is approximately 25 mm (Fleming et al., 1996). Organ pipe cactus (*Stenocereus thurberi*) flowers are slightly smaller, with a mean length of approximately 80 mm and a mean perianth width of just under 20 mm (Fleming et al., 1996). Mexican species of *Neobuxbaumia* have flowers with a mean length of approximately 50 mm and a mean width of 16–23 mm (Valiente-Banuet et al., 1997b). Columnar cactus species in South America and nearby islands also have large flowers: *Subpilocereus repandus* has flowers with a mean length of approximately 80 mm and a mean perianth width of 21 mm, and *Stenocereus griseus* has a mean length of 50 mm and a mean perianth width of 21 mm (Nassar et al., 1997). *Subpilocereus horrispinus* has a mean length of approximately 60 mm and a mean perianth width of 21 mm, and *Pilosocereus mortizianus* and *P. lanuginosus* have flowers with a mean length of just over 40 mm and a mean perianth width of 23–25 mm (Nassar et al., 1997). None of these flowers has an internal depth of less than 35 mm (Nassar et al., 1997), and all are large compared with the bats that visit them (see Table 5.2).

As is clear from Tuttle's (1991) photographs of feeding *Leptonycteris,* a bat must insert most or all of its head into a cactus flower to reach the nectar at the bottom. To effectively feed while hovering, the bat must simultaneously get its tongue into the nectar and keep its wings clear of interference. A long rostrum may increase the reach of a bat and thus increase feeding efficiency, and relatively large body size may facilitate hovering by increasing the distance between the head and shoulders (thus placing the wings farther from the edges of flowers). Longer forearms, presumably associated with larger wings, are probably less subject to interference caused by contact with flower petals while the bat is hovering. In the larger cactophilic glossophagines, larger wings may also be necessary to support greater body weight while maintaining the wing loading and aspect ratio appropriate for hovering (see discussions of scaling in Norberg and Rayner, 1987). As a result of cactus flower morphology, larger bats with longer forearms and rostra may be at a selective advantage when attempting to obtain nectar from flowers of columnar cacti. The size and shape of these flowers may have helped to drive the evolution of increased rostral length, forearm length, and body size in cactophilic glossophagines. Conversely, taxa already characterized by some or all of these morphological traits may have been essentially preadapted for effective cactophily. Note, however, that the relative values of increased rostral length, forearm length, and body size to bats visiting cactus flowers have yet to be tested experimentally, so we have no way of knowing if any aspects of these hypotheses are correct.

Other factors that may be important for understanding the morphology of cactophilic bats include features of the habitats in which these bats live (e.g., arid, relatively open, often subject to major shifts in temperature) and the distances that bats must travel on a daily basis (e.g., between roost and foraging sites, among plants while foraging) and during migrations. Increased body size may be advantageous for bats living in arid environments for several reasons. Relative surface area decreases with increasing size, so maintenance of appropriate water balance may be easier in larger-bodied forms (Peters, 1983; Schmidt-Nielson, 1984). It has been calculated that all else being equal, larger birds take longer to starve than smaller birds, and that smaller birds are more adversely affected by low temperatures than larger birds (Peters, 1983).

It seems likely that the same is true of bats. Increased body size may therefore help insulate cactophilic bats living in arid environments from the effects of both food shortages and temperature variation. However, several smaller-sized glossophagine species successfully inhabit arid areas in Mexico (e.g., *Glossophaga soricina, Musonycteris harrisoni*) and Venezuela (e.g., *Choeroniscus godmani*), so large body size is clearly not a prerequisite for glossophagine life in arid regions. But it is interesting to note that only the larger glossophagines (*Choeronycteris mexicana, Leptonycteris nivalis, and L. yerbabuenae*) have geographic ranges that extend into the subtropical/temperate zone of northern Mexico and the southwestern United States (Hall, 1981). Higher body weight may be more important in these regions due to increased seasonal and daily temperature variation and potentially less reliable food sources.

Increased body size also has advantages in terms of energy efficiency. The relative cost of transport (horizontal flight) decreases with increasing size and increasing wing length, factors that may be important for small bats that must commute substantial distances to feed (Norberg and Rayner, 1987; Sahley et al., 1993). Due to the often patchy nature of resource distribution in arid environments, cactophilic bats in many areas may have to cover relatively long distances—with considerable energy expenditure—during their nightly foraging bouts (Howell, 1979; Sahley et al., 1993; Horner et al., 1998). For example, one population of *Leptonycteris yerbabuenae* regularly commutes 20–30 km each way between roost sites on Isla Tiburon in the Gulf of California and foraging areas on the mainland, where they feed from flowers of three species of columnar cacti (Sahley et al., 1993). Bats in this population fly for approximately five hours each night and cover a total distance of about 100 km per night (Horner et al., 1998). Foraging flights (including commuting) account for approximately 21% of the bat's daily time budget but about 44% of its daily energy budget (Horner et al., 1998). Increased body size and longer wings may facilitate cactophily in these bats by making relatively long-distance foraging flights energetically affordable.

Long-distance migration may also be energetically cheaper for larger bats, due to the reduced cost of transport (Norberg and Rayner, 1987). Some populations of *Leptonycteris yerbabuenae* in Mexico are thought to migrate hundreds of kilometers yearly, with many individuals traveling over 1,500 km each way between central Mexico and the Sonoran Desert (Fleming et al., 1993; Wilkinson and Fleming, 1996; Ceballos et al., 1997). Mexican populations of *L. nivalis* and *Choeronycteris mexicana* are also thought to be migratory, moving north to the Chihuahuan or Sonoran deserts in the spring and south again in the fall (Barbour and Davis, 1969; Cockrum, 1991; Fleming et al., 1993; Moreno, 2000). It is possible that these bats evolved larger body sizes in part to lower the energy cost of migration; however, it is equally likely that larger body sizes evolved for entirely different reasons, and that any reduction in the energy cost of migration may have been of secondary importance. It is also possible that large body size has acted as a preadaptation for evolution of migratory habits in these bats. Regardless of the reasons for increased size, it seems clear that cactophilic bats may accrue many advantages by having relatively large body size, particularly in geographic regions subject to large seasonal variations in temperature and food availability.

Given the observations discussed above, we must ask yet another question: Why aren't cactophilic glossophagines larger than they are? The answer to this question probably lies in their diet and hovering habits. All nectarivorous bats apparently eat substantial quantities of fruit (Gardner, 1977), but frugivory does not appear associated with any intrinsic limits on body size. For example, frugivorous phyllostomids in several genera (e.g., *Artibeus, Platyrrhinus, Sturnira*) commonly achieve weights of more than 50 grams, and some weigh as much as 90 g (e.g., *Phyllostomus hastatus;* Emmons, 1997). Frugivorous megachiropterans often weigh more than 200 g and may reach weights of one kg or more (*Pteropus vampyrus;* Norberg and Rayner, 1987; *P. neohibernicus;* Flannery, 1995). In contrast, nectar-feeding bats on all continents (including macroglossine megachiropterans that are only distantly related to glossophagines) appear to be size-limited (Norberg and Rayner, 1987; Freeman, 1995). With the exception of some Australian flying foxes of the genus *Pteropus* (Ratcliffe, 1932; Richards, 1983), most bats that feed regularly from flowers are small. Few bat species thought to rely extensively on nectar and pollen for food have a mean weight of more than 45 g, and most are much smaller (Norberg and Rayner, 1987; Emmons, 1997; Table 5.2). Flower-visitors that hover while feeding (rather than landing on the flower) are apparently restricted in size for aerodynamic and mechanical reasons (Norberg and Rayner, 1987). However, the demands of hovering flight are not the only factor that may constrain body size. Flowers generally produce a small quantity of nectar per day, and the amount that can be obtained by a bat in a reasonable amount of time may be limited (Howell and Hartl, 1980). Although nectar availability is clearly not a limiting factor in some situations (e.g., *Leptonycteris yerbabuenae* foraging in the spring at cactus flowers in Bahia Kino, Sonora; Horner et al., 1998), this may not be true at other localities and times of year (e.g., during seasons of low flower production). Nectar may therefore be a limiting resource that effectively constrains body size to that which can be supported by the quantity of nectar that the bat can obtain on a regular basis. The fit between size of the flower and size of the bat is also important in cases where flowers are bell-shaped, as they are in columnar cacti. Baker (1961) noted that only smaller bats can visit bell-shaped flowers, whereas bats of all sizes can effectively obtain nectar from ball- or brush-shaped flowers. In the case of columnar cacti, the size and shape of the flowers likely constrains body size of the bats that feed from them.

Cacti in Rainforest Habitats

The preceding discussion focused on columnar cactus–bat interactions in arid and semiarid habitats, and our arguments concerning body size are based on this ecological situation. However, not all cacti grow in arid regions. A number of cactus species are rainforest epiphytes, and these may also be pollinated by glossophagine bats (M. Tschapka, pers. comm.). Continued studies of pollination biology of rainforest cacti may significantly expand the list of bat species known to be opportunistic cactophiles. Many of the glossophagines implicated in pollination of rainforest epiphytic cacti (e.g., *Glossophaga comissarisi, Hylonycteris underwoodi, Lichonycteris obscura;* M. Tschapka, pers. comm.) are relatively small-bodied species (Table 5.2). The association of large body size with cactophily may subsequently be shown to apply only to bats associated with columnar cacti in arid and semiarid habitats.

Summary

Many columnar cactus species apparently depend on bats for pollination and seed dispersal. In return, cactophilic bats obtain nectar, pollen, and fruit that may form a critical part of their diet. At least 18 bat species are known to visit columnar cacti in different regions of North, Central, and South America. A review of the natural history, phylogenetic relationships, and morphology of these bats suggests that cactophily has evolved many times: once in Antrozoidae, and a minimum of 13 times in Phyllostomidae. Many cactophilic bats are opportunistic cactus feeders that do not appear to depend on cactus products for survival (e.g., *Anoura geoffroyi, Antrozous pallidus, Artibeus intermedius, A. jamaicensis, Brachyphylla nana, Carollia perspicillata, Chiroderma salvini, Glossophaga soricina, Monophyllus redmani, Phyllonycteris poeyi, Phyllostomus discolor, Sturnira lilium*). However, obligate cactophily characterizes at least some populations of other species (e.g., *Glossophaga longirostris, Leptonycteris curasoae, L. yerbabuenae, Platalina genovensium,* and probably *Choeronycteris mexicana* and *L. nivalis*). Consideration of the distribution of cactophily both within and among species suggests that speciation has rarely occurred within cactophilic lineages. Adoption of cactophilic habits most often seems to be a population-level phenomenon that occurs within species, not a species-level trait that is inherited by sister-taxa from a common ancestor. However, some exceptions exist, most notably in the genus *Leptonycteris.*

Morphological comparisons indicate that cactophilic glossophagines (especially obligate cactophiles) tend to have longer skulls, longer forearms, and greater body weights than their noncactophilic relatives. All of these features apparently play important roles in facilitating cactophily in the arid and semiarid habitats where columnar cacti most commonly occur. Increased rostral length may aid cactophilic bats in reaching the nectar in the large, tubular flowers typical of columnar cacti. Increased body size may facilitate life in arid habitats through favorable effects on maintenance of water balance, temperature regulation, and the costs of flight. Based on its morphology and geographic range, we predict that the little-known glossophagine *Musonycteris harrisoni,* a southwestern Mexican endemic, is a cactophilic bat. Two species endemic to the Lesser Antilles—*Monophyllus plethodon* and *Brachyphylla cavernarum*—may also be cactophilic and could prove to be important pollinators of columnar cacti in this region.

Resumen

Muchas especies de cactos columnares dependen de los murciélagos para su polinización y dispersión de las semillas. En cambio, los murciélagos cactófilos obtienen el néctar, polen y frutos, que pueden ser una fracción crítica de su dieta. Se conocen por lo menos dieciocho especies de murciélagos que visitan cactos columnares en diferentes regiones de Norte, Centro y Suramérica. La revisión de la historia natural, relaciones filogenéticas y morfología de estos murciélagos nos sugiere que la cactofilia ha evolucionado muchas veces: una en Antrozoidae y un mínimo de trece veces en Phyllostomidae. Muchos murciélagos cactófilos se alimentan de cactos, de manera oportunista y no parecen tener que depender de productos de los cactos, para su supervivencia (p.e., *Anoura geoffroyi, Antrozous pallidus, Artibeus intermedius, A. jamaicensis, Brachyphylla nana, Carollia perspicillata, Chiroderma salvini, Glossophaga soricina, Monophyllus redmani, Phyllonycteris poeyi, Phyllostomus discolor, Sturnira lilium*). Sin embargo, la cactofilia obligatoria se caracteriza por lo menos en algunas poblaciones de otras especies (p.e., *Glossophaga longirostris, Leptonycteris curasoae, L. yerbabuenae, Platalina genovensium,* y, probablemente, *Choeronycteris mexicana* y *L. nivalis*). Al considerar la distribucion de la cactofilia al interior y entre las especies se advierte que los linajes cactófilos rara vez se han mezclado. La adopción de hábitos cactófilos a menudo parece ser más un fenómeno a nivel de poblaciones que ocurre en el interior de las especies. No es un rasgo a nivel de especies que dos especies hermanas heredaron a partir de un ancestro común. Sin embargo, algunas excepciones existen; más notoriamente en el género *Leptonycteris.*

Las comparaciones morfológicas indican que los glosofáginos cactófilos (especialmente los cactófilos obligatorios), tienden a tener cráneos más largos, antebrazos más largos y mayor peso corporal que sus parientes no cactófilos. Todas estas características aparentemente juegan papeles importantes en la facilitación de la cactofilia. En los hábitats áridos y semi-áridos donde los cactos columnares se encuentran comúnmente, el aumento en la longitud rostral podría ayudar a los murciélagos cactófilos a alcanzar el néctar en las flores grandes y tubulares típicas de los cactos columnares. El incremento en el tamaño corporal podría facilitar la vida en los hábitats áridos por medio de efectos favorables en el mantenimiento del equilibrio hídrico, la termorregulación y costos del vuelo. Basándonos en su morfología y distribución geográfica, predecimos que el glosofágino muy poco conocido *Musonycteris harrisoni,* un endémico del suroeste mexicano, es un murciélago cactófilo. También existen dos especies endémicas en las Antillas Menores—*Monophyllus plethodon* y *Brachyphylla cavernarum*—por tanto podrían ser cactófilas e importantes polinizadores de cactos columnares en esta región.

ACKNOWLEDGMENTS

This chapter would not have been possible without the help of many individuals, first and foremost T. Fleming and A. Valiente-Banuet. We thank them for the invitation to participate, and also for enduring numerous requests for information, advice, and references. H. Arita, T. Fleming, C. Handley, J. Nassar, B. Patterson, S. Solari, and M. Tschapka kindly shared unpublished data with us, for which we are grateful. We thank T. Fleming, J. Nassar, M. Tschapka, A. Valiente-Banuet, and R. Wallace for reviewing earlier versions of this manuscript and making many helpful suggestions. Special thanks also to T. Conway for help with the bibliography and figures, L. Davalos for translation of the resumen, and to P. Brunauer, who always managed to find what we needed in the library. This study was supported in part by NSF Research Grants DEB-9106868 and DEB-9873663 (NBS), and an American Museum of Natural History Graduate Fellowship (ALW).

REFERENCES

Albuja, L. 1982. *Murcielagos del Ecuador.* Quito, Ecuador: Escuela Politécnica Nacional, Departmento de Ciéncieas Biológicas.

Alcorn, S. M., S. E. McGregor, G. D. Butler, and E. B. Kurtz. 1959. Pollination requirements of the saguaro (*Carnegiea gigantea*). *Cactus and Succulent Journal* 31:39–41.

Alcorn, S. M., S. E. McGregor, and G. Olin. 1961. Pollination of saguaro cactus by doves, nectar-feeding bats, and honey bees. *Science* 133:1594–95.

———. 1962. Pollination requirements of the organpipe cactus. *Cactus and Succulent Journal* 34:134–38.

Alvarez, J., M. R. Willig, J. K. Jones, and W. D. Webster. 1991. *Glossophaga soricina. Mammalian Species* 379:1–7.

Alvarez, T., and L. González Q. 1970. Análisis polínicio del contenido gástrico de murciélagos Glossophaginae de México. *Anales de la Escuela Nacional de Ciencias Biológicas, Mexico* 18:137–65.

Arita, H. T. 1991. Spatial segregation in long-nosed bats, *Leptonycteris nivalis* and *Leptonycteris curasoae,* in Mexico. *Journal of Mammalogy* 72:706–14.

Arita, H. T., and S. R. Humphrey. 1988. Revision taxonomica de los murcielagos magueyeros del genero *Leptonycteris* (Chiroptera: Phyllosomidae). *Acta Zoologica de Mexico* 29:1–60.

Arroyo-Cabrales, J., R. R. Hollander, and J. K. Jones. 1987. *Choeronycteris mexicana. Mammalian Species* 291:1–5.

Baker, H. G. 1961. The adaptation of flowering plants to nocturnal and crepuscular pollinators. *Quarterly Review of Biology* 36:64–73.

Baker, R. J., C. S. Hood, and R. L. Honeycutt. 1989. Phylogenetic relationships and classification of the higher categories of the New World bat family Phyllostomidae. *Systematic Zoology* 38:228–38.

Barbour, R. W., and W. H. Davis. 1969. *Bats of America.* Lexington: University of Kentucky Press.

Bell, G. P. 1982. Behavioral and ecological aspects of gleaning by a desert insectivorous bat, *Antrozous pallidus* (Chiroptera: Vespertilionidae). *Behavioral Ecology and Sociobiology* 10:217–23.

Bonaccorso, F. G. 1979. Foraging and reproductive ecology in a Panamanian bat community. *Bulletin of the Florida State Museum of Biological Sciences* 24:359–408.

Bond, R. M., and G. A. Seaman. 1958. Notes on a colony of *Brachyphylla cavernarum. Journal of Mammalogy* 39:150–51.

Borell, A. E. 1942. Feeding habits of the pallid bat. *Journal of Mammalogy* 23:337.

Brooks, D. R., and D. A. McLennan. 1991. *Phylogeny, Ecology, and Behavior.* Chicago: University of Chicago Press.

Ceballos, G., T. H. Fleming, C. Chávez, and J. Nassar. 1997. Population dynamics of *Leptonycteris curasoae* (Chiroptera: Phyllostomidae) in Jalisco, Mexico. *Journal of Mammalogy* 78:1220–30.

Cloutier, D., and D. W. Thomas. 1992. *Carollia perspicillata. Mammalian Species* 417:1–9.

Cockrum, E. L. 1991. Seasonal distribution of northwestern populations of the long-nosed bats, *Leptonycteris sanborni* Family Phyllostomidae. *Anales del Instituto Biología Universidad Nactional Autonoma de México, Serie Zoologica* 62:181–202.

Dalquest, W. W. 1953. Mammals of the Mexican state of San Luis Potosi. *Louisiana State University Studies, Biological Sciences Series* 1:1–229.

Davis, W. B. 1984. Review of the large fruit-eating bats of the *Artibeus "lituratus"* complex (Chiroptera: Phyllostomidae) in Middle America. *Occasional Papers, The Museum, Texas Tech University* 93:1–16.

Emmons, L. H. 1997. *Neotropical Rainforst Mammals: a Field Guide.* 2nd ed. Chicago: University of Chicago Press.

Fenton, M. B. 1992. *Bats.* New York: Facts on File.

Ferrarezzi, H., and E. A. Gimenez. 1996. Systematic patterns and the evolution of feeding habits in Chiroptera (Archonta: Mammalia). *Journal of Comparative Biology* 1:75–94.

Flannery, T. 1995. *Mammals of New Guinea: Revised and Updated Edition.* Ithaca: Cornell University Press.

Fleming, T. H. 1982. Foraging strategies of plant-visiting bats. In *Ecology of Bats,* ed. T. H. Kunz, 287–325. New York: Plenum Press.

———. 1988. *The Short-Tailed Fruit Bat: a Study in Plant–Animal Interactions.* Chicago: University of Chicago Press.

———. 1991. Following the nectar trail. *Bats* 9(4):4–7.

Fleming, T. H., and E. R. Heithaus. 1986. Seasonal foraging behavior of the frugivorous bat *Carollia perspicillata. Journal of Mammalogy* 67:660–71.

Fleming, T. H., and V. J. Sosa. 1994. Effects of nectarivorous and frugivorous mammals on reproductive success of plants. *Journal of Mammalogy* 75:845–51.

Fleming, T. H., E. T. Hooper, and D. E. Wilson. 1972. Three Central American bat communities: structure, reproductive cycles, and movement patterns. *Ecology* 53:655–70.

Fleming, T. H., R. A. Nuñez, and L.S.L. Sternberg. 1993. Seasonal changes in the diets of migrant and nonmigrant nectarivorous bats as revealed by carbon stable isotope analysis. *Oecologia* 94:72–75.

Fleming, T. H., M. D. Tuttle, and M. A. Horner. 1996. Pollination biology and the relative importance of nocturnal and diurnal pollinators in three species of Sonoran Desert columnar cacti. *Southwestern Naturalist* 41:257–69.

Fleming, T. H., C. T. Sahley, J. N. Holland, J. Nason, and J. L. Hamrich. 2001. Sonoran Desert columnar cacti and the evolution of generalized pollination systems. *Ecological Monographs* 71:511–30.

Freeman, P. W. 1995. Nectarivorous feeding mechanisms in bats. *Biological Journal of the Linnean Society* 56:439–63.

Gannon, M. R., M. R. Willig, and J. K. Jones. 1989. *Sturnira lilium. Mammalian Species* 333: 1–5.

Gardner, A. L. 1977. Feeding habits. In *Biology of Bats of the New World Family Phyllostomatidae,* Part II, eds. R. J. Baker, J. K. Jones, and D. C. Carter, *Special Publications, The Museum, Texas Tech University,* 13:293–350.

Gardner, A. L., C. O. Handley, and D. E. Wilson. 1991. Survival and relative abundance. In *Demography and Natural History of the Common Fruit Bat,* Artibeus jamaicensis, *on Barro Col-*

orado Island, Panamá, eds. C. O. Handley, D. E. Wilson, and A. L. Gardner, *Smithsonian Contributions to Zoology,* 511:53–76.

Gimenez, E. A., H. Ferrarezzi, and V. A. Taddei. 1996. Lingual morphology and cladistic analysis of the New World nectar-feeding bats (Chiroptera: Phyllostomidae). *Journal of Comparative Biology* 1:41–63.

Goodwin, G. G. 1946. Mammals of Costa Rica. *Bulletin of the American Museum of Natural History* 87:273–473.

Goodwin, G. G., and A. M. Greenhall. 1961. A review of the bats of Trinidad and Tobago. *Bulletin of the American Museum of Natural History* 112:191–301.

Greenbaum, I. F., and C. J. Phillips. 1974. Comparative anatomy and general histology of tongues of long-nosed bats (*Leptonycteris sanborni* and *L. nivalis*) with reference to infestation of oral mites. *Journal of Mammalogy* 55:489–504.

Griffiths, T. A. 1978. Muscular and vascular adaptations for nectar-feeding in the glossophagine bats *Monophyllus* and *Glossophaga. Journal of Mammalogy* 59:414–18.

———. 1982. Systematics of the New World nectar-feeding bats (Mammalia, Phyllostomidae) based on morphology of the hyoid and lingual regions. *American Museum Novitates* 2742:1–45.

———. 1983. On the phylogeny of the Glossophaginae and proper use of outgroup analysis. *Systematic Zoology* 32:283–85.

Haiduk, M. W., and J. R. Baker. 1982. Cladistical analysis of nectar feeding bats (Glossophaginae: Phyllostomidae). *Systematic Zoology* 31:252–65.

Hall, E. R. 1981. *The Mammals of North America.* 2nd ed. New York: Wiley.

Handley, C. O. 1966. Descriptions of new bats (*Choeroniscus* and *Rhinophylla*) from Colombia. *Proceedings of the Biological Society of Washington* 79:83–88.

———. 1976. Mammals of the Smithsonian Venezuelan Project. *Brigham Young University Science Bulletin, Biological Series* 20:1–89.

Handley, C. O., and E. G. Leigh. 1991. Diet and food supply. In *Demography and Natural History of the Common Fruit Bat,* Artibeus jamaicensis, *on Barro Colorado Island, Panamá,* eds. C. O. Handley, D. E. Wilson, and A. L. Gardner, *Smithsonian Contributions to Zoology* 511:147–50.

Handley, C. O., A. L. Gardner, and D. E. Wilson. 1991. Food habits. In *Demography and Natural History of the Common Fruit Bat,* Artibeus jamaicensis, *on Barro Colorado Island, Panamá,* eds. C. O. Handley, D. E. Wilson, and A. L. Gardner, *Smithsonian Contributions to Zoology* 511:141–46.

Hatt, R. T. 1923. Food habits of the Pacific pallid bat. *Journal of Mammalogy* 4:260–61.

Hayward, B. J., and E. L. Cockrum. 1971. The natural history of the western long-nosed bat *Leptonycteris sanborni. Western New Mexico University Research Science* 1:75–123.

Heithaus, E. R. 1982. Coevolution between bats and plants. In *Ecology of Bats,* ed. T. H. Kunz, 327–67. New York: Plenum Press.

Heithaus, E. R., T. H. Fleming, and P. A. Opler. 1975. Foraging patterns and resource utilization in seven species of bats in a seasonal tropical rainforest. *Ecology* 56:841–54.

Hensley, A. P., and K. T. Wilkins. 1988. *Leptonycteris nivalis. Mammalian Species* 307:1–4.

Hermanson, J. W., and T. J. O'Shea. 1983. *Antrozous pallidus. Mammalian Species* 213:1–8.

Herrera, L. G., T. H. Fleming, and J. S. Findley. 1993. Geographic variation in carbon composition of the pallid bat, *Antozous pallidus,* and its dietary implications. *Journal of Mammalogy* 74:601–6.

Hill, J. E. 1980. A note on *Lonchophylla* (Chiroptera: Phyllostomatidae) from Ecuador and Peru, with the description of a new species. *Bulletin of the British Museum (Natural History), Zoology Series* 38:233–36.

Hill, J. E., and J. D. Smith. 1984. *Bats: a Natural History.* Austin: University of Texas Press.

Homan, J. A., and J. K. Jones. 1975a. *Monophyllus redmani. Mammalian Species* 57:1–3.

———. 1975b. *Monophyllus plethodon. Mammalian Species* 58:1–2.

Honeycutt, R. L., and V. M. Sarich. 1987. Albumin evolution and subfamilial relationships among New World leaf-nosed bats (Family Phyllostomidae). *Journal of Mammalogy* 62:805–11.

Horner, M. A., T. H. Fleming, and C. T. Sahley. 1998. Foraging behaviour and energetics of a nectar-feeding bat, *Leptonycteris curasoae* (Chiroptera: Phyllostomidae). *Journal of Zoology, London* 244:575–86.

Howell, D. J. 1974a. Bats and pollen: physiological aspects of the syndrome of chiropterophily. *Comparative Biochemistry and Physiology* 48A:263–76.

———. 1974b. Feeding and acoustic behavior in glossophagine bats. *Journal of Mammalogy* 55:293–308.

———. 1979. Flock foraging in nectar-feeding bats: advantages to the bats and to the host plants. *American Naturalist* 114:23–49.

———. 1980. Adaptive variation in diets of desert bats has implications for evolution of feeding strategies. *Journal of Mammalogy* 61:730–33.

Howell, D. J., and D. Burch. 1974. Food habits of some Costa Rican bats. *Revista de Biologia Tropical* 21:281–94.

Howell, D. J., and D. L. Hartl. 1980. Optimal foraging in glossophagine bats: when to give up. *American Naturalist* 115:696–704.

Huey, L. M. 1936. Desert pallid bats caught in mouse traps. *Journal of Mammalogy* 17:285–86.

Humphrey, S. R., F. J. Bonaccorso, and T. L. Zinn. 1983. Guild structure of surface-gleaning bats in Panama. *Ecology* 64:284–94.

Jiménez, M. P., and J. Pefaur. 1982. Aspectos sistemáticos y ecológicos de *Platalina genovensium* (Chiroptera: Mammalia). In *Zoologia Neotropical: Actas del Octavo Congreso Latinamericano de Zoologia,* ed. P. Salinas, 707–18. Mérida, Venezuela: Octavo Congreso Latinamericano de Zoologia.

Jones, J. K., J. R. Choate, and A. Cadena. 1972. Mammals from the Mexican state of Sinaloa. II. Chiroptera. *Occasional Papers of the Museum of Natural History, University of Kansas* 6:1–29.

Klingener, D., H. H. Genoways, and R. J. Baker. 1978. Bats from southern Haiti. *Annals of the Carnegie Museum* 47:81–97.

Koopman, K. F. 1981. The distribution patterns of New World nectar-feeding bats. *Annals of the Missouri Botanical Garden* 68:352–69.

———. 1993. Order Chiroptera. In *Mammal Species of the World, a Taxonomic and Geographic Reference.* 2nd ed., eds. D. E. Wilson and D. M. Reeder, 137–241. Washington D.C.: Smithsonian Institution Press.

———. 1994. Chiroptera: Systematics. *Handbook of Zoology. Mammalia.* Vol. 8, pt. 60, 1–217. Berlin: Walter de Gruyter.

Leigh, E. G., and C. O. Handley. 1991. Population estimates. In *Demography and Natural History of the Common Fruit Bat,* Artibeus jamaicensis, *on Barro Colorado Island, Panamá,* eds. C. O. Handley, D. E. Wilson, and A. L. Gardner, *Smithsonian Contributions to Zoology* 511:53–76.

Lemke, T. O. 1984. Foraging ecology of the long-nosed bat, *Glossophaga soricina,* with respect to resource availability. *Ecology* 65:538–48.

Marques-Aguiar, S. A. 1994. A systematic review of the large species of *Artibeus* Leach, 1821 (Mammalia: Chiroptera), with some phylogenetic inferences. *Boletim do Museu Paraense Emílio Goeldi, Série Zoologia* 10:3–83.

McGregor, S. E., S. M. Alcorn, and G. Olin. 1962. Pollination and pollinating agents of the saguaro. *Ecology* 43:259–67.

Molinari, J. 1994. A new species of *Anoura* (Mammalia Chiroptera Phyllostomidae) from the Andes of northern South America. *Tropical Zoology* 7:73–86.

Moreno, A. 2000. Ecological studies of the Mexican long-nosed bat (*Leptonycteris nivalis*). Ph.D. Diss., Texas A&M University, College Station.

Nassar, J. M. 1991. Biologia reproductiva de cuatro cactaceas quiropterofila venezolanas (Ceereae: *Stenocereus griseus, Pilosocereus moritizianus, Subpilocereus repandus* y *S. horrispinus*), y estrategias de visita de los murcielagos asociados a estas. Bachelor's thesis, Universidad Central de Venezuela, Caracas, Venezuela.

Nassar, J. M., N. Ramírez, and O. Linares. 1997. Comparative pollination biology of Venezuelan columnar cacti and the role of nectar-feeding bats in their sexual reproduction. *American Journal of Botany* 84:918–27.

Nellis, D. W. 1971. Additions to the natural history of *Brachyphylla* (Chiroptera). *Caribbean Journal of Science* 11:91.

Nellis, D. W., and C. P. Ehle. 1977. Observations on the behavior of *Brachyphylla cavernarum* (Chiroptera) in Virgin Islands. *Mammalia* 41:403–9.

Norberg, U. M., and J.M.V. Rayner. 1987. Ecological morphology and flight in bats (Mammalia; Chiroptera): wing adaptations, flight performance, foraging strategy and echolocation. *Philosophical Transactions of the Royal Society of London, B* 316:335–427.

Orr, R. T. 1954. Natural history of the pallid bat, *Antrozous pallidus. Proceedings of the California Academy of Sciences* 28:165–264.

O'Shea, T. J., and T. A. Vaughan. 1977. Nocturnal and seasonal activities of the pallid bat, *Antrozous pallidus. Journal of Mammalogy* 58:269–84.

Park, H., and E. R. Hall. 1951. The gross anatomy of the tongues and stomachs of eight New World bats. *Transactions of the Kansas Academy of Science* 54:64–72.

Peters, R. H. 1983. *The Ecological Implications of Body Size.* Cambridge: Cambridge University Press.

Petit, S. 1995. The pollinators of two species of columnar cacti on Curaçao, Netherlands Antilles. *Biotropica* 27:538–41.

———. 1997. The diet and reproductive schedules of *Leptonycteris curasoae curasoae* and *Glossophaga longirostris elongata* (Chiroptera: Glossophaginae) on Curaçao. *Biotropica* 29:214–23.

Phillips, C. J. 1971. The dentition of glossophagine bats: development, morphological characteristics, variation, pathology, and evolution. *University of Kansas Museum of Natural History Miscellaneous Publications* 54:1–138.

Pine, R. H. 1972. Bats of the genus *Carollia. Texas Agricultural Experiment Station Technical Monograph* 8:1–125

Pirlot, P. 1967. Nouvelle récolte de chiroptères dans l'ouest du Venezuela. *Mammalia* 31: 260–74.

Ramírez-Pulido, J., M. A. Armella, and A. Castro-Campillo. 1993. Reproductive patterns of three Neotropical bats (Chiroptera: Phyllostomidae) in Guerrero, Mexico. *Southwestern Naturalist* 38:24–29.

Ratcliffe, F. 1932. Notes on the fruit bats (*Pteropus* spp.) of Australia. *Journal of Animal Ecology* 1:32–59.

Reid, F. A. 1997. *A Field Guide to the Mammals of Central America and Southeast Mexico.* Oxford: Oxford University Press.

Richards, G. C. 1983. Little Red Flying-Fox. In *The Australian Museum Complete Book of Australian Mammals,* ed. R. Strahan, 277–79. Sydney: Angus and Robertson.

Rodriguez-Duran, A., and A. R. Lewis. 1987. Patterns of population size, diet, and activity time for a multispecies assemblage of bats at a cave in Puerto Rico. *Caribbean Journal of Science* 23:352–60.

Rodriguez-Duran, A., A. R. Lewis, and Y. Montes. 1993. Skull morphology and diet of Antillean bat species. *Caribbean Journal of Science* 29:258–61.

Ruiz, A., M. Santos, P. J. Soriano, J. Cavelier, and A. Cadena. 1997. Relaciones mutualísticas entre el murciélago *Glossophaga longirostris* y las cactáceas columnares en la zona arida de la Tatacoa, Colombia. *Biotropica* 29:469–79.

Sahley, C. T. 1995. Bat and hummingbird pollination of two species of columnar cacti: effects on fruit production and pollen dispersal. Ph.D. diss., University of Miami, Coral Gables, Florida.

Sahley, C. T., and L. E. Baraybar. 1994. The natural history and population status of the nectar-feeding bat, *Platalina genovensium,* in southwestern Peru. *Bat Research News* 35:113.

———. 1996. Natural history of the long-snouted bat, *Platalina genovensium* (Phyllostomidae: Glossophaginae) in southwestern Peru. *Vida Silvestre Neotropical* 5:101–9.

Sahley, C. T., M. A. Horner, and T. H. Fleming. 1993. Flight speeds and mechanical power outputs of the nectar-feeding bat, *Leptonycteris curasoae* (Phyllostomidae: Glossophaginae). *Journal of Mammalogy* 74:594–600.

Santos, M., A. Ruiz, J. Cavelier, and P. Soriano. 1995. The phenology of cacti and its relationships with the bat community in a tropical dry forest-thorn shrubland of Colombia. *Bat Research News* 36:105–6.

Sazima, I. 1976. Observations on the feeding habits of phyllostomid bats (*Carollia, Anoura, Vampyrops*) in southeastern Brazil. *Journal of Mammalogy* 57:381–82.

Sazima, I., and M. Sazima. 1977. Solitary and group foraging: two flower-visiting patterns of the lesser spear-nosed bat *Phyllostomus discolor. Biotropica* 9:213–15.

Schaldach, W. J., and C. A. McLaughlin. 1960. A new genus and species of glossophagine bat from Colima, Mexico. *Los Angeles County Museum Contributions to Science* 37:1–8.

Schmidt-Nielson, K. 1984. *Scaling: Why Is Animal Size So Important?* Cambridge: Cambridge University Press.

Silva Taboada, G. 1979. *Los Murciélagos de Cuba.* Havana: Editora de la Academia de Ciencias de Cuba.

Silva Taboada, G., and R. H. Pine. 1969. Morphological and behavioral evidence for the relationship between the bat genus *Brachyphylla* and the Phyllonycterinae. *Biotropica* 1:10–19.

Simmons, N. B. 1998. A reappraisal of interfamilial relationships of bats. In *Bat Biology and Conservation,* eds. T. H. Kunz and P. A. Racey, 3–26. Washington, D.C.: Smithsonian Institution Press.

Simmons, N. B., and J. H. Geisler. 1998. Phylogenetic relationships of *Icaronycteris, Archaeonycteris, Hassianycteris,* and *Palaeochiropteryx* to extant bat lineages, with comments on the evolution of echolocation and foraging strategies in Microchiroptera. *Bulletin of the American Museum of Natural History* 235:1–182.

Simmons, N. B., and R. S. Voss. 1998. The mammals of Paracou, French Guiana: a neotropical lowland rainforest fauna. Part 1. Bats. *Bulletin of the American Museum of Natural History* 237:1–219.

Smith, J. D., and C. S. Hood. 1984. Genealogy of the New World nectar-feeding bats reexamined: a reply to Griffiths. *Systematic Zoology* 33:435–60.

Smith, J. D., and A. Starrett. 1979. Morphometric analysis of chiropteran wings. In *Biology of Bats of the New World Family Phyllostomatidae.* Part III, eds. R. J. Baker, J. K. Jones, and D. C. Carter. *Special Publications, The Museum, Texas Tech University* 16:229–316.

Soriano, P. J., M. Sosa, and O. Rossell. 1991. Hábitos alimentarios de *Glossophaga longirostris* Miller (Chiroptera: Phyllostomidae) en una zona árida de los Andes venezolanos. *Revista de Biología Tropical* 39:263–68.

Sosa, M., and P. J. Soriano. 1996. Resource availablity, diet and reproduction in *Glossophaga longirostris* (Mammalia: Chrioptera) in an arid zone of the Venezuelan Andes. *Journal of Tropical Ecology* 12:805–18.

Strickler, T. L. 1978. Functional osteology and myology of the shoulder in Chiroptera. *Contributions in Vertebrate Evolution* 4:1–198.

Struhsaker, T. 1961. Morphological factors regulating flight in bats. *Journal of Mammalogy* 42:152–59.

Swanepoel, P., and H. H. Genoways. 1978. Revision of the Antillean bats of the genus *Brachyphylla* (Mammalia: Phyllostomatidae). *Bulletin of the Carnegie Museum of Natural History* 12:1–53.

———. 1979. Morphometrics. In *Biology of Bats of the New World Family Phyllostomatidae.* Part III, eds. R. J. Baker, J. K. Jones, and D. C. Carter. *Special Publications, The Museum, Texas Tech University* 16:13–106.

———. 1983a. *Brachyphylla cavernarum. Mammalian Species* 205:1–6.

———. 1983b. *Brachyphylla nana. Mammalian Species* 206:1–3.

Swofford, D. L., and W. P. Maddison. 1992. Parsimony, character-state reconstructions, and evolutionary inferences. In *Systematics, Historical Ecology, and North American Freshwater Fishes,* ed. R. L. Mayden, 186–223. Stanford: Stanford University Press.

Tuttle, M. D. 1991. Bats, the cactus connection. *National Geographic* 179:130–40.

Valiente-Banuet, A., M. C. Arizmendi, A. Rojas-Martínez, and L. Domínguez-Canseco. 1996. Ecological relationships between columnar cacti and nectar-feeding bats in Mexico. *Journal of Tropical Ecology* 12:103–19.

Valiente-Banuet, A., A. Rojas-Martínez, A. Casas, M. C. Arizmendi, and P. Davila. 1997a. Pollination biology of two winter-blooming giant columnar cacti in the Tehuacán Valley, central Mexico. *Journal of Arid Envrionments* 37:331–41.

Valiente-Banuet, A., A. Rojas-Martínez, M. C. Arizmendi, and P. Davila. 1997b. Pollination biology of two columnar cacti (*Neobuxbaumia mezcalaensis* and *Neobuxbaumia macrophylla*) in the Tehuacán Valley, central Mexico. *American Journal of Botany* 84:452–55.

Van Den Bussche, R. A. 1991. Phylogenetic analysis of restriction site variation in the ribosomal DNA complex of New World leaf-nosed genera. *Systematic Zoology* 40:420–32.

———. 1992. Restriction-site variation and molecular systematics of New World leaf-nosed bats. *Journal of Mammalogy* 73:29–42.

Van Den Bussche, R. A., J. L. Hudgeons, and R. J. Baker. 1998. Phylogenetic accuracy, stability, and congruence: relationships within and among the New World bat genera *Artibeus, Dermanura,* and *Koopmania.* In *Bat Biology and Conservation,* eds. T. H. Kunz and P. A. Racey, 59–71. Washington: Smithsonian Institution Press.

van der Pijl, L. 1956. Remarks on pollination by bats in the genera *Freycinetia, Duabanga* and *Haplophragma,* and on chiropterophily in general. *Acta Botanica Neerlandica* 5:135–44.

———. 1961. Ecological aspects of flower evolution. II. Zoophilous flower classes. *Evolution* 15:44–59.

Villa-R., B. 1967. *Los murciélagos de México.* Mexico: Instituto Biología, Universidad Nacional Autónoma de México.

von Helversen, O., and H-U. Reyer. 1984. Nectar intake and energy expenditure in a flower-visiting bat. *Oecologia* 63:178–84.

Voss, R. S., and L. H. Emmons. 1996. Mammalian diversity in neotropical lowland rainforests: a preliminary assessment. *Bulletin of the American Museum of Natural History* 230:1–115.

Warner, R. M. 1983. Karyotypic megaevolution and phylogenetic analysis: New World nectar-feeding bats revisited. *Systematic Zoology* 32:279–82.

Webster, W. D. 1993. Systematics and evolution of bats of the genus *Glossophaga. Special Publications, The Museum, Texas Tech University* 36:1–184.

Webster, W. D., and C. O. Handley. 1986. Systematics of Miller's long-tongued bat, *Glossophaga longirostris,* with description of two new subspecies. *Occasional Papers, The Museum, Texas Tech University* 100:1–22.

Webster, W. D., L. W. Robbins, R. L. Robbins, and R. J. Baker. 1982. Comments on the status of *Musonycteris harrisoni* (Chiroptera: Phyllostomidae). *Occasional Papers, The Museum, Texas Tech University* 78:1–5.

Webster, W. D., C. O. Handley, and P. J. Soriano. 1998. *Glossophaga longirostris. Mammalian Species* 576:1–5.

Wetterer, A. L., M. V. Rockman, and N. B. Simmons. 2000. Phylogeny of phyllostomid bats: data from diverse morphological systems, sex chromosomes, and restriction sites. *Bulletin of the American Museum of Natural History* 248:1–200.

Wilkinson, G. S., and T. H. Fleming. 1996. Migration and evolution of lesser long-nosed bats *Leptonycteris curasoae,* inferred from mitochondrial DNA. *Molecular Ecology* 5:329–39.

Wille, A. 1954. Muscular adaptation of the nectar-eating bats (subfamily Glossophaginae). *Transactions of the Kansas Academy of Science* 57:315–25.

Winklemann, J. R. 1971. Adapations for nectar-feeding in glossophagine bats. Ph.D. diss., University of Michigan, Ann Arbor.

CHAPTER 6

Genetic Diversity in Columnar Cacti

J. L. Hamrick
John D. Nason
Theodore H. Fleming
Jafet M. Nassar

Introduction

The availability of molecular genetic markers during the past 30 years has significantly increased our ability to describe genetic diversity within and among populations of a wide variety of plant taxa. The most comprehensive and widely available procedure used to identify single-gene genetic diversity is isozyme electrophoresis, which identifies and quantifies genetic variation at enzymatic loci. To date, approximately 2,750 species of seed plants have been the subject of "allozyme" studies. This robust database has allowed generalizations to be made concerning the influence of life history and ecological characteristics of species on the amount and distribution of allozyme variation (e.g., Brown, 1979; Hamrick et al., 1979; Hamrick and Godt, 1989). Specifically, reviews of the plant allozyme literature have demonstrated that long-lived, woody species with large geographic ranges and outcrossing breeding systems have significantly more allozyme genetic diversity than do species with other combinations of traits (Hamrick and Godt, 1989, 1996). These species also have much less genetic differentiation among their populations than is true of species with more limited gene flow potential (Hamrick and Godt, 1989, 1996).

However, even though the number of plant allozyme studies will soon surpass 3,000, there are groups of plant species that are significantly underrepresented. One such group is the Cactaceae. To our knowledge, of the approximately 97 genera and 1,400 species of cacti found in the Western Hemisphere (Mabberley, 1997), only 11 genera and 13 species have been studied electrophoretically (Table 6.1). In contrast, more than 60 species of the genus *Pinus* have been analyzed for allozyme diversity.

There are several reasons why the Cactaceae are poorly represented in the plant population-genetics literature. First, the studies are strongly biased towards north temperate species. As a result, such species as cacti, which occur predominantly from Mexico to South America, are generally less thoroughly studied. Second, species of

TABLE 6.1
Listing of Allozyme Studies of Cactus Species

Species	*Tissue*	*Number of Populations*	*Number of Loci*	*Reference*
Carnegiea gigantea	flower buds	16	30	Hamrick et al. (unpubl. data)
Cereus repandus	seedlings	14	17	Nassar et al. (unpubl. data)
Echinocereus engelmannii	stem	6	7[a]	Neel et al. (1996)
Lophocereus schottii	stem	8	18	Parker and Hamrick (1992)
L. schottii	flower buds	21	31	Nason et al. (unpubl. data)
Melocactus curvispinus	seedlings	18	19	Nassar et al. (forthcoming a)
Opuntia basilaris	stem	1	1	Sternberg et al. (1977)
O. spinosissima	stem	2	13	Hamrick and Godt (1997)
Pachycereus pringlei	flower buds, seedlings	1	8	Murawski et al. (1994)
P. pringlei	flower buds	9	24	Fleming et al. (1998)
P. pringlei	flower buds	19	24	Hamrick et al. (unpubl. data)
Pereskia guamacho	leaves	17	19	Nassar et al. (forthcoming b)
Pilosocereus lanuginosus	seedlings	10	23	Nassar et al. (unpubl. data)
Stenocereus griseus	seedlings	15	18	Nassar et al. (unpubl. data)
S. thurberi	flower buds	20	31	Hamrick et al.
Weberbauerocereus weberbaueri	flower buds, seedlings	1	12	Sahley (1996)

[a] Number of enzyme systems. Loci were not identified.

arid habitats, as a group are underrepresented in the plant allozyme literature. Very few of the ecological dominants of arid lands have been the subject of genetic diversity studies. In contrast, there are more than 150 studies of tropical tree species. Finally, many potential investigators have hesitated to undertake studies of the Cactaceae because of the generally held belief that the mucilaginous cactus tissue makes enzyme extraction difficult, if not impossible.

In this chapter, our chief objective is to review what is known concerning allozyme genetic variation in the Cactaceae, particularly columnar cacti. We dispel the myth that it is difficult to electrophoretically analyze extracts from cactus tissue for allozyme loci. We compare the results of the few existing cactus allozyme studies with results obtained from reviews of the extensive plant allozyme literature to determine whether the levels and distribution of genetic diversity in cactus species are consistent with their life history traits.

Electrophoretic Procedures

We have found that cactus species require no special handling or extraction procedures to obtain well resolved isozyme bands. In fact, several cactus species have given excellent band resolution. Generally, the handling and extraction procedures that our laboratory has developed for a variety of gymnosperm and angiosperm species have also worked well for cactus species.

We have used a variety of cactus tissue with varying results (Table 6.1). Best expression and resolution is typically obtained from greenhouse-grown seedlings. We usually try to use seedlings that are at least one centimeter in height but have also used both larger and smaller seedlings with success. Flower-bud tissue produces nearly equal expression and resolution to that of seedlings. Adult stem tissue has not worked as well for some enzyme systems but provides good results for other enzyme systems and loci.

FIELD COLLECTING

When collecting cactus tissue in the field, we have used two approaches for its preservation and transportation. We have had success collecting flower buds and stem tissue, placing the tissue in a cooler with ice and transporting it directly to the lab within two or three days. Some loss of expression has occasionally been experienced, but this is a successful technique when dealing with limited samples (i.e., those that can be collected during a one- or two-day period) from a geographically restricted area. In our experience, flower buds preserve better than stem tissue, perhaps because their removal from the plant is less traumatic to the tissue.

When the sampling period is extended, as in sampling for geographic surveys of allozyme variation, we collect flower buds, place them on ice, and within 36–48 hours transfer the buds to a liquid-nitrogen container. When it is necessary to keep track of samples from individual plants, we wrap the buds (or cross-sections of buds) in aluminum foil and individually mark each sample. Several wrapped buds are then placed in net bags tied to nylon fishing line and are immersed in a container of liquid nitrogen. Once frozen, tissues should not be allowed to thaw. Air transport can be accomplished with a dry liquid-nitrogen carrier or the samples can be placed on dry ice.

ENZYME EXTRACTION

Enzyme extraction can be accomplished using protocols developed for other plant materials (see Kephart, 1990). Using flower-bud or stem tissue, we crush the material to a fine powder under liquid nitrogen with a mortar and pestle. Fine ocean sand may be added to improve grinding of the plant tissue. An extraction buffer is then added to the powder to make a slurry. We have had good results with the extraction buffer of Mitton et al. (1979), which was developed for coniferous species. The buffer of Wendel and Parks (1982) also works well for some species (e.g., *Lophocereus schottii*). Kephart (1990) and Wendel and Weeden (1989) provide thorough reviews of available extraction buffers. The slurry that results is then filtered through a patch of Miracloth and the liquid is absorbed on filter-paper wicks. Wicks can be used immediately or stored in a –70°C freezer with no loss of enzyme expression. Standard electrophoretic procedures and enzyme stains are then applied to resolve allozyme banding patterns. Kephart (1990) also discusses a variety of electrophoretic buffers and stains that work well for plants.

Measures of Genetic Diversity

Genetic diversity can be estimated at three levels: within species, within populations, and among populations. Within-species measures are estimated from pooled (over all populations sampled) values and are not confounded by partitioning of genetic diversity within and among populations. In contrast, genetic diversity within populations is a function of total genetic diversity within the species as well as the proportion of this total that occurs within populations. In this chapter, species values are subscripted by "s" and within-population values are subscripted by "p."

Several parameters are typically used to measure the levels and distribution of genetic diversity within species and populations and among populations. Each measure is informative but some are composite measures that incorporate information from other parameters. Genetic diversity measures used in this chapter include:

$P =$ the proportion of polymorphic loci (i.e., loci with more than one allele).

$AP =$ the mean number of alleles per polymorphic locus.

$A_e = 1/\Sigma p_i^2 =$ the effective number of alleles at a locus, where p_i is the frequency of the ith allele. This parameter is calculated for each locus and is usually averaged over all loci (monomorphic and polymorphic). Values of A_e are influenced by P, AP, and the evenness of allele frequencies.

$H_o =$ observed heterozygosity. This parameter is calculated directly from observed genotype frequencies at each locus and is averaged over all loci. The parameter is of limited value as a comparative statistic between species because it is affected by inbreeding and other evolutionary processes that may be unique to the species or population.

$H_e = 1 - \Sigma p_i^2 =$ the expected proportion of heterozygous loci per individual. Referred to as genetic diversity or genic diversity, this parameter is calculated for each locus and is averaged across monomorphic (i.e., $H_e = 0$) and polymorphic loci. It is a composite measure that summarizes genetic diversity at a locus. It is influenced by P, AP, and allele frequencies at each locus and is the most commonly used index of genetic diversity for allozyme data.

$H_T = 1 - \Sigma p_i^2 =$ the total genetic diversity at polymorphic loci, where p_i is the mean frequency of the ith allele pooled across all populations in a study.

$H_S =$ the mean genetic diversity within populations for polymorphic loci.

$G_{ST} = (H_T - H_S)/H_T =$ the proportion of total genetic diversity that occurs among populations. Values of G_{ST} are usually averaged over all polymorphic loci to estimate population divergence for the species.

$F_{IS} = (H_{ep} - H_o)/H_{ep} =$ the proportional deviation at each locus of observed from expected heterozygosity within populations. Values of F_{IS} can be averaged over all polymorphic loci.

A more complete discussion of these parameters and their use can be found in Berg and Hamrick (1997).

Genetic Diversity in Cactus Species

VARIATION WITHIN SPECIES

Allozyme studies with sufficient populations and loci are available for seven species of columnar cacti and for two other cactus species, *Melocactus curvispinus* and *Pereskia guamacho* (Table 6.2). *Stenocereus griseus* has the highest proportion of polymorphic loci ($P_s = 100\%$), whereas *Stenocereus thurberi* has the lowest (83.8%). The number of alleles per polymorphic locus ranges from 2.79 for *Carnegiea gigantea* to 3.82 for *M. curvispinus.* Genetic diversity (H_{es}) also varies among species, with *C. gigantea* having the lowest value (0.129) and *Pilosocereus lanuginosus* the highest (0.274). It is interesting that autotetraploid species, *Pachycereus pringlei* and *Pilosocereus lanuginosus,* maintain approximately the same levels of genetic diversity as diploid species.

The relatively low genetic diversity in *Carnegiea gigantea* is due to the presence of a common allele ($p > 0.95$) at 17 of its 28 polymorphic loci. *Melocactus curvispinus* and *Stenocereus griseus* also have low H_{es} values relative to their P_s and AP_s values for the same reason. This conclusion is supported by the lower H_T values for these three species relative to the other five species (Table 6.3).

VARIATION WITHIN POPULATIONS

Data at the within-population level are available for ten cactus species. A single population was sampled for *Weberbauerocereus weberbaueri* and as a result, estimates of genetic diversity for this species may not be representative of the species.

Mean values of P_p range from 45.3% for *Melocactus curvispinus* to 75.1% for *Pilosocereus lanuginosus* (Table 6.4). *Weberbauerocereus weberbaueri* has the highest mean number of alleles per polymorphic locus (2.88) and *M. curvispinus* has the least (2.17). This trend is also seen for H_{ep}: *W. weberbaueri* = 0.257 whereas *M. curvispinus* = 0.098. Where valid comparisons can be made (i.e., diploid species), H_o values are somewhat lower than H_{ep} values, indicating a deficit of heterozygous individuals relative to Hardy-Weinberg expectations. Mean F_{IS} values of six of the seven diploid species have a deficiency of heterozygotes (positive F_{IS} values), whereas *Lophocereus schottii* has a very small heterozygote excess (Table 6.3). The F_{IS} values for *Cereus*

TABLE 6.2

Levels of Overall Genetic Diversity for Nine Species of Cacti[a]

Species	P_s(%)	AP_s	A_{es}	H_{es}
Carnegiea gigantea	93.3	2.79	1.20	0.129
Cereus repandus	94.1	3.69	1.47	0.242
Lophocereus schottii	90.3	3.00	1.39	0.214
Melocactus curvispinus	89.5	3.82	1.21	0.145
Pachycereus pringlei	91.7	3.14	1.38	0.212
Pereskia guamacho	89.5	3.53	1.45	0.239
Pilosocereus lanuginosus	91.3	3.52	1.43	0.274
Stenocereus griseus	100.0	3.50	1.35	0.812
S. thurberi	83.8	3.42	1.33	0.201

[a] P_s = proportion of polymorphic loci; AP_s = number of alleles per polymorphic locus; A_{es} = effective number of alleles; H_{es} = expected proportion of loci heterozygous per individual = genetic diversity.

repandus, Pereskia guamacho, and *Stenocereus griseus* are significantly different from zero ($P < 0.01$).

An excess of homozygosity within a population can be due to at least two causes: inbreeding and population subdivision. For predominantly outcrossing species such as the cacti discussed here, self-fertilization is unlikely but biparental inbreeding between related individuals can occur, especially if seed dispersal is local. In addition, if there is population substructure (i.e., a Wahlund effect), there would also be an apparent deficiency of heterozygotes. To obtain adequate sample sizes for these cactus species, rather large spatial areas were often sampled. It is therefore possible that some population genetic structure exists within the sampled areas, giving rise to the apparent deficiency of heterozygotes observed for some of these species.

Three of these cacti—*Pachycereus pringlei, Pilosocereus lanuginosus,* and *Weberbauerocereus weberbaueri*—are autotetraploid species with apparent tetrasomic inheritance patterns. As a result, individuals have four copies of alleles at each locus.

TABLE 6.3

Distribution of Genetic Diversity among Populations of Nine Species of Cacti[a]

Species	H_T	H_S	G_{ST}	F_{IS}
Carnegiea gigantea	0.139	0.125	0.075	0.057
Cereus repandus	0.277	0.228	0.126	0.182
Lophocereus schottii	0.237	0.158	0.242	−0.003
Melocactus curvispinus	0.166	0.112	0.189	0.377
Pachycereus pringlei	0.231	0.213	0.076	—[b]
Pereskia guamacho	0.261	0.215	0.112	0.180
Pilosocereus lanuginosus	0.267	0.252	0.043	—[b]
Stenocereus griseus	0.177	0.152	0.092	0.145
S. thurberi	0.239	0.201	0.128	0.036

[a] H_T = total genetic diversity at polymorphic loci; H_S = mean within-population genetic diversity; G_{ST} = proportion of total genetic diversity due to differences among populations; F_{IS} = inbreeding coefficient within populations.

[b] Autotetraploid species.

This does not affect most of the population-level genetic diversity parameters. Observed heterozygosity is considerably elevated, however. For example, for a diploid species with two equally frequent alleles at a locus ($p = q = 0.5$), 50% of the individuals should be heterozygous. For a species with tetrasomic inheritance, in contrast, 94.4% of the individuals should be heterozygous (i.e., have at least two alleles at a locus).

VARIATION AMONG POPULATIONS

Interpopulation genetic variation is quite low for the majority of the nine cactus species with adequate data (Table 6.3). The exception is *Lophocereus schottii*, which has a relatively high G_{ST} value (0.242). A significant portion of the among-population variation in *L. schottii* (38.8%) is due to variation between its Baja California and Sonora populations. The remainder is predominantly due to variation among its Baja California populations (Table 6.5). There are two named subspecies of *L. schottii* in Baja California and our collections included both subspecies. Even taking these considerations into account, *L. schottii* has a somewhat higher G_{ST} value for its Sonora populations than do the other three Sonora species (Table 6.5). This result may be due to low pollen flow among geographically separated populations of *L. schottii.* This species is predominantly pollinated by the mutualistic moth, *Upiga virescens* (Fleming and Holland, 1998; Holland and Fleming, 1999), which may have limited long-distance dispersal abilities. The other three Sonora cactus species are pollinated by a combination of birds, bats, and insects and may have a greater potential for pollen flow. Genetic differentiation among populations of five Venezuelan cactus species (*Cereus repandus, Melocactus curvispinus, Pereskia guamacho, Pilosocereus lanuginosus* and *Stenocereus griseus*) is consistent with this interpretation. Although these five species were sampled from approximately the same geographic locations, the bee-pollinated *Pereskia guamacho* and the hummingbird-pollinated *M. curvi-*

TABLE 6.4
Levels of Within-Population Genetic Diversity for Ten Species of Cacti[a]

Species	P_p(%)	AP_p	A_{ep}	H_o	H_{ep}
Carnegiea gigantea	53.7	2.20	1.19	0.110	0.116
Cereus repandus	72.3	2.44	1.38	0.179	0.205
Lophocereus schottii	49.5	2.33	1.25	0.142	0.144
Melocactus curvispinus	45.3	2.17	1.16	0.067	0.098
Pachycereus pringlei	62.1	2.50	1.37	—[b]	0.200
Pereskia guamacho	63.4	2.42	1.37	0.170	0.202
Pilosocereus lanuginosus	76.1	2.69	1.41	—[b]	0.253
Stenocereus griseus	56.7	2.36	1.30	0.145	0.161
S. thurberi	62.4	2.36	1.30	0.157	0.169
Weberbauerocereus weberbaueri	66.6	2.88	1.24	—[b]	0.257

[a] P_p = proportion of polymorphic loci; AP_p = number of alleles per polymorphic locus; A_{ep} = effective number of alleles; H_o = observed proportion of loci polymorphic per individual; H_{ep} = expected proportion of loci heterozygous per individual = genetic diveristy.

[b] Autotetraploid species; thus each individual is heterozygous for most of the polymorphic loci.

TABLE 6.5
Levels of Genetic Diversity within Sonora and Baja California Populations of Four Species of Columnar Cacti[a]

Species	P_S(%)		AP_s		H_{es}		G_{ST}	
	Baja California	*Sonora*	*Baja California*	*Sonora*	*Baja California*	*Sonora*	*Baja California*	*Sonora*
Carnegiea gigantea	—	93.3	—	2.79	—	0.129	—	0.075
Lophocereus schottii	76.7	73.3	3.09	2.59	0.197	0.178	0.205	0.082
Pachycereus pringlei	91.7	70.8	3.00	2.82	0.220	0.200	0.062	0.062
Stenocereus thurberi	80.6	83.9	3.04	2.92	0.214	0.160	0.097	0.059

[a]See Tables 6.2 and 6.3 for definition of symbols.

spinus have somewhat higher G_{ST} values (0.112 and 0.189, respectively) than the strictly bat-pollinated *S. griseus* (0.092), *C. repandus* (0.126), and *Pilosocereus lanuginosus* (0.043) (see Table 6.3).

BAJA CALIFORNIA VS. SONORA POPULATIONS

The geographic distribution of three of the columnar cacti from northern Mexico and the southwestern United States allows comparisons among Sonora and Baja California populations of these species. Genetic diversity in the Baja California populations is generally higher than that seen in the Sonora populations for all of the genetic diversity parameters (P_S, AP_S and H_{es}, Table 6.5) for these four species. Much of the additional genetic diversity in the Baja California populations is due to the presence of several low-frequency alleles that are absent from the Sonora populations. Such genetic differences may indicate that the Baja California populations are ancestral to the Sonora populations, but such interpretations are difficult to verify when based only on genetic diversity measures.

There is also more genetic differentiation among the Baja California populations for *Lophocereus schottii* and *Stenocereus thurberi.* As discussed above, much of the variation among *L. schottii* populations in Baja California is due to sampling across its two subspecies. Genetic differentiation among Baja California populations of *S. thurberi* is nearly 65% greater than genetic differentiation among populations of *S. thurberi* from Sonora. Fleming et al. (2001) have shown that when *S. thurberi* occurs with *Pachycereus pringlei,* it is predominantly bird- and insect pollinated but where *P. pringlei* is absent, *S. thurberi* is more frequently visited by the bat *Leptonycteris curasoae.* In Baja California, *S. thurberi* is sympatric with *P. pringlei* throughout its range, whereas in northern Sonora, *P. pringlei* is absent. Thus the higher genetic differentiation among the Baja California populations of *S. thurberi* may be due to more extensive pollination by less vagile insects. In much of its Sonora range, *S. thurberi* is pollinated by the more widely foraging *L. curasoae,* leading to the possibility of greater gene exchange among its populations (Fleming et al., unpubl. data) and less genetic differentiation.

TABLE 6.6

Comparisons of the Mean Levels of Genetic Diversity for Cactus Species with Other Plant Species with Similar Life-History Traits[a]

Group	P_S(%)	AP_S	H_{es}	H_T	G_{ST}
All plant species[b]	51.3	2.89	0.150	0.224	0.228
Long-lived woody plants[b]	65.0	2.88	0.177	0.253	0.084
Animal-pollinated, outcrossing woody plants[b]	63.2	2.87	0.211	0.268	0.099
Cactus species	91.5	3.38	0.211	0.222	0.120

[a] See Tables 6.2 and 6.4 for definition of symbols.

[b] From Hamrick et al. (1992).

Comparisons with Other Plant Species

The nine cactus species with sufficient population samples generally have high levels of genetic diversity relative to the mean for other plant species (Table 6.6). The proportion of polymorphic loci (P_s) is much higher than that of the average plant species. The number of alleles at polymorphic loci (AP_s) and mean genetic diversity (H_{es}) are also higher than the mean for other plant species. Interestingly, total genetic diversity at polymorphic loci (H_T) is nearly equal to that of the average plant species. Cactus species also have a low proportion of their total genetic diversity among their populations (G_{ST}) relative to other types of plants.

Relative to plants that share many of their life-history traits—long-lived woody plants and woody plants with animal-pollinated, predominantly outcrossing breeding systems—the nine cactus species discussed here have higher proportions of polymorphic loci and more alleles per polymorphic locus and equivalent H_{es} values. The mean H_T value for the cacti is somewhat lower than the mean H_T for other woody plants. Apparently, a high proportion of the polymorphic loci of these cacti have a single very common allele and one or more relatively rare alleles.

The mean G_{ST} value for these cacti is somewhat higher than the means for other woody plants (predominantly wind-pollinated) and for animal-pollinated, outcrossing woody species. However, if *Lophocereus schottii* is removed from the analysis (see above), the mean G_{ST} value for the other eight species is 0.105, only slightly higher than that of other animal-pollinated woody species.

Summary

Studies of genetic diversity within cactus species are surprisingly few, considering their ecological importance in the arid lands of the Western Hemisphere. Data estimating genetic diversity within and among populations exist for only ten cactus species. The only geographical regions that have been partially studied are Sonora and Baja California in Mexico and northern Venezuela. Thus columnar cacti in central and southern Mexico, Central America, Andean South America, and the Caribbean are largely unstudied. The existing studies demonstrate that cactus species, on average, maintain high levels of genetic diversity. These cacti differ from many other long-lived woody species with animal-pollinated breeding systems in that an exceptionally high proportion of their loci are polymorphic and that these polymorphic loci have more alleles. However, overall genetic diversity at polymorphic loci (H_T) is lower than the mean for other plants, indicating that a high proportion of the polymorphic loci of these cacti have a very common allele and one or a few rare alleles.

Most of the genetic diversity within cactus species resides within their populations. There was some indication that predominantly insect-pollinated species have more genetic differentiation among their populations than predominantly bat-pollinated species. Additional studies are, however, needed before such generalizations can be stated with confidence.

Resumen

Los estudios de diversidad genética en especies de cactus son sorprendentemente pocos considerando su importancia ecológica en regiones áridas del hemisferio occidental. De acuerdo a nuestro conocimiento, solo existen estimaciones de niveles de diversidad genetica para diez especies de cactus. Las únicas regiones geográficas que han sido parcialmente estudiadas son Sonora y Baja California en México, y el norte de Venezuela. Por tanto, las especies de cactus columnar en el centro y sur de México, America Central, los Andes de Sur America y el Caribe permanecen sin ser estudiadas. Los estudios existentes demuestran que las especies de cactus, en promedio, mantienen niveles altos de diversidad genética. Estas especies de cactus parecen diferir de otras especies de plantas leñosas de larga vida polinizadas por animales, en las cuales existe una proporción excepcionalmente alta de loci polimórficos, los cuales tienen más alelos. Sin embargo, la diversidad genética total en locus polimórficos (H_T) es menor que el promedio para otras plantas indicando que una alta proporción de loci polimorficos de estos cactus se caracterizan por tener un alelo muy común y pocos alelos raros.

La mayor parte de la diversidad genética en especies de cactus reside dentro de las poblaciones. Se encontraron indicios de que especies polinizadas predominantemente por insectos tienen más diferenciación genética entre sus poblaciones qué las especies polinizadas predominantemente por murciélagos. Se necesitan más estudios adicionales antes de poder generalizar en estas conclusiones.

ACKNOWLEDGMENTS

We thank Mindy Burke for technical assistance in the lab, Mary Harris for assistance with the field collections, and Mary Jo Godt for searching the allozyme database for cactus species. Portions of this work were supported by NSF Grant DEB9420254 to J.L.H. and J.D.N and an NSF supplemental grant to T.H.F. and J.M.N.

REFERENCES

Berg, E. E., and J. L. Hamrick. 1997. Quantification of genetic diversity at allozyme loci. *Canadian Journal of Forest Research* 27:415–24.

Brown, A.H.D. 1979. Enzyme polymorphism in plant populations. *Theoretical Population Biology* 15:1–42.

Fleming, T. H., and J. N. Holland. 1998. The evolution of obligate mutualisms: the senita and senita moth. *Oecologia* 114:368–78.

Fleming, T. H., S. Maurice, and J. L. Hamrick. 1998. Geographic variation in the breeding system and the evolutionary stability of trioecy in *Pachycereus pringlei* (Cactaceae). *Evolutionary Ecology* 12:279–89.

Fleming, T. H., C. T. Sahley, J. N. Holland, J. Nason, and J. L. Hamrick. 2001. Sonoran Desert columnar cacti and the evolution of generalized pollination systems. *Ecological Monographs* 71:511–30.

Hamrick, J. L., and M.J.W. Godt. 1989. Allozyme diversity in plant species. In *Plant Population Genetics, Breeding and Genetic Resources,* ed. A.H.D. Brown, M. T. Clegg, A. L. Kahler, and B. S. Weir, 43–63. Sunderland, Mass.: Sinauer.

———. 1996. Effects of life history traits on genetic diversity in plant species. *Philosophical Transactions of the Royal Society of London: Biological Sciences* 351:1291–98.

———. 1997. Genetic diversity in *Opuntia spinosissim,* a rare and endangered Florida Keys cactus. Final Report, Florida Nature Conservancy.

Hamrick, J. L., M.J.W. Godt, and S. L. Sherman-Broyles. 1992. Factors influencing levels of genetic diversity in woody plant species. *New Forests* 6:95–124.

Hamrick, J. L., Y. B. Linhart, and J. B. Mitton. 1979. Relationships between life history characteristics and electrophoretically detectable genetic variation in plants. *Annual Review of Ecology and Systematics* 10:173–200.

Holland, N., and T. H. Fleming. 1999. Mutualistic interactions between *Upiga virescens* (Pyralidae), a pollination seed-consumer, and *Lophocereus schottii* (Cactaceae). *Ecology* 80: 2074–84.

Kephart, S. R. 1990. Starch gel electrophoresis of plant isozymes: a comparative analysis of techniques. *American Journal of Botany* 77:693–712.

Mabberley, D. J. 1997. *The Plant Book.* 2nd ed. London: Cambridge University Press.

Mitton, J. B., Y. B. Linhart, B. K. Sturgeon, and J. L. Hamrick. 1979. Allozyme polymorphism detected in mature needle tissue of ponderosa pine, *Pinus ponderosa* Laws. *Journal of Heredity* 70:86–89.

Murawski, D. A., T. H. Fleming, K. Ritland, and J. L. Hamrick. 1994. Mating system of *Pachycereus pringlei:* an autotetraploid cactus. *Heredity* 72:86–94.

Nassar, J. M., J. L. Hamrick, and T. H. Fleming. 2001. Genetic variation and population structure of the mixed-mating cactus, *Melocactus curvispinus* (Cactaceae). *Heredity* 87:69–79.

———. Forthcoming. Allozyme diversity and population genetic structure of the leafy cactus, *Pereskia guamacho* (Cactaceae). *American Journal of Botany.*

Neel, M. C., J. Clegg, and N. N. Ellstrand. 1996. Isozyme variation in *Echinocereus engelmannii* var. *munzii* (Cactaceae). *Conservation Biology* 10:622–31.

Parker, K. C., and J. L. Hamrick. 1992. Genetic diversity and clonal structure in a columnar cactus, *Lophocereus schottii. American Journal of Botany* 79:86–96.

Sahley, C. T. 1996. Bat and hummingbird pollination of an autotetraploid columnar cactus, *Weberbauerocereus weberbaueri* (Cactaceae). *American Journal of Botany* 83:1329–36.

Sternberg, L., I. P. Ting, and Z. Hanscom. 1977. Polymorphism of microbody malate dehydrogenase in *Opuntia basilaris. Plant Physiology* 59:329–30.

Wendel, J. F., and C. R. Parks. 1982. Genetic control of isozyme variation in *Camellina japonica* L. *Journal of Heredity* 73:197–204.

Wendel, J. F., and N. F. Weeden. 1989. Visualization and interpretation of plant isozymes. In *Isozymes in Plant Biology*, ed. D. E. Soltis and P. S. Soltis, 5–41. Portland, Ore.: Dioscorides Press.

PART II

Anatomy and Physiology

CHAPTER 7

Evolutionary Trends in Columnar Cacti under Domestication in South-Central Mexico

Alejandro Casas
Alfonso Valiente-Banuet
Javier Caballero

Introduction

The south Pacific drainage in south-central Mexico (Figure 7.1) has been identified as the richest area in species of columnar cacti in the world (Valiente-Banuet et al., 1996). Thornscrub and tropical deciduous forests are common natural landscapes in the area, and columnar cacti are among their principal components. Archaeological studies in the Tehuacán Valley (MacNeish, 1967) and Guilá Naquitz, Oaxaca (Flannery, 1986) suggest that the region was inhabited by humans probably from 14,000 B.P., and have found there the oldest evidence of plant domestication in the New World. From ancient times, the peoples of this area used a broad spectrum of plant and animal resources, with cacti being among the most important because of their abundance, easy accessibility, and diversity. In the Tehuacán Valley alone, Dávila et al. (1993) identified nearly 74 species of cacti, almost all of them producing edible products (Casas and Valiente-Banuet, 1995).

Smith (1967, 1986) reported remains of nine species of cacti (Table 7.1) from archaeological excavations in caves of the Tehuacán Valley and Guilá Naquitz, and Callen (1967) identified the following types of cactus remains in human coprolites: (1) "Opuntia" spp., (2) "Lemaireocereus" (*Pachycereus* spp. and *Stenocereus* spp., according to current nomenclature), and (3) "cactus tissue" (unidentified cacti). In the earliest coprolites found from the El Riego phase (6,500–5,000 B.C.), Callen (1967) identified these types of cactus remains to be part of a wild food diet along with *Setaria* spp. seeds, *Ceiba parvifolia* roots, *Agave* spp. leaves and meat. In the Coxcatlán phase (5,000–3,500 B.C.), stem tissue and fruits of *Opuntia* spp. and "Lemairocereus" were equally dominant materials. In the Abejas (3,500–2,300 B.C.), Ajalpan (1,500–900 B.C.), Santa María (900–200 B.C.), Palo Blanco (200 B.C.–A.D. 700) and

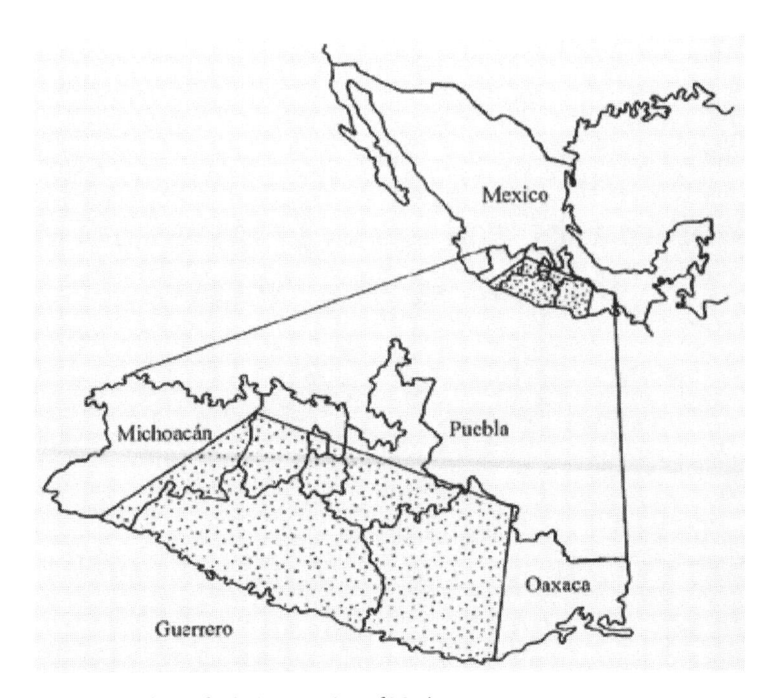

FIGURE 7.1. The South Pacific drainage region of Mexico.

Venta Salada (A.D. 700–1,540) phases, consumption of "Lemaireocereus" stem tissue, fruits, and seeds may have been more important than products of *Opuntia,* and during the Ajalpan and Santa María phases, "Lemaireocereus" may have been the principal plant constituents in human diets.

Historical information on utilization of columnar cacti can be found in the "General and Natural History of the Indies," published by Hernández de Oviedo y Valdés in 1535; the Barberini Codex from 1552 (De la Cruz and Badiano, 1964), which includes a description of "teonochtli," identified as *Stenocereus* sp. by Bravo-Hollis (1978); the Florentino Codex (Sahagún, 1970), containing information on "netzolli," which is probably *Escontria chiotilla,* and "teunochtli," which may be a species of *Stenocereus* (Casas, 1997); the books of Francisco Hernández (1959), who described *Myrtillocactus geometrizans* and a possible *Stenocereus* species also called "teonochtli"; and in the "Geographic Relations of the XVI Century," which contains a reference to the columnar cactus "teonochtli" in the Relation of Acatlán (Acuña, 1985). Sahagún (1985), as well as Del Barco (1988) describes how harvest of fruits of columnar cacti was crucial for subsistence of some prehispanic and postconquest peoples from northern Mexico, who migrated from zone to zone following the seasonal abundance of cactus fruits.

Although such sources of information indicate a long history of utilization of columnar cacti by peoples of the region, no evidence has been found of the cultivation of these plants in either prehistory or in historical times. In historical documents,

TABLE 7.1

Archaeological Remains of Cacti Found in Caves in the Tehuacán Valley[a]

Species	Phase and Radiocarbon Date (years B.P.)								
	Ajuereado 14,000–8,800	*El Riego* 8,800–7,000	*Coxcatlán* 7,000–5,400	*Abejas* 5,400–4,300	*Purrón* 4,300–3,500	*Ajalpan* 3,500–2,800	*Santa María* 2,800–2,150	*Palo Blanco* 2,150–1,300	*Venta Salada* 1,300–500
Cephalocereus hoppenstedtii[b]		X	X			X	X		
Echinocactus grandis		X						X	
Escontria chiotilla[b]			X					X	
Ferocactus flavovirens								X	X
Myrtillocactus geometrizans[b]		X	X						X
Opuntia spp.	X	X	X	X	X	X	X	X	X
Pachycereus hollianus[b]		X	X	X					
P. weberi[b]		X	X	X			X	X	
Stenocereus stellatus[b]				X					

[a] Based on MacNeish (1967) and Smith (1967).

[b] Columnar cacti.

this omission, according to Casas (1997), could be because the Spaniards did not consider fruits of columnar cacti to be important resources. For this reason, they did not describe them and many other cultivated plants. But such an omission could also be explained if cultivation of columnar cacti began after the Spanish conquest. Additional archaeological and historical evidence is needed to answer this question, but botanical studies might also provide helpful information. Comparative studies of morphological and genetic variation, reproductive biology, and other biological aspects commonly affected by human manipulation would allow the analysis of evolutionary changes of columnar cacti under domestication. This information might contribute to estimates of the antiquity of human manipulation.

Ethnobotanical research in the study area by Caballero and Mapes (1985) and Casas et al. (1994, 1996, 1997a) has revealed the existence of a wide spectrum of forms of interactions between humans and plants, including columnar cacti. Casas et al. (1996) group these forms of plant management into those occurring in situ (in the wild) and those occurring ex situ. Through management in situ, humans can take products from nature without significant perturbations, but they may also alter the structure of plant populations by increasing the quantity of target species or particular phenotypes. The main strategies of plant management in situ are:

1. Gathering, which is the taking of useful plant products directly from natural populations.
2. Tolerance, including maintenance within man-made environments of useful plants that existed before the environments were transformed by humans.
3. Enhancement, intended to increase the population density of useful plant species, including the sowing of seeds or the intentional propagation of vegetative structures in the same places occupied by wild plant populations.
4. Protection, which includes deliberate activities such as the elimination of competitors and predators, fertilization, pruning, etc., to safeguard critical wild plants.

Plant management ex situ includes interactions taking place outside natural populations in habitats created and controlled by humans, including the transplantation of entire individuals and sowing and planting of sexual or vegetative propagules. These ex situ interactions usually involve domesticated plants, but it is also possible to find them occurring with wild plants (Casas and Caballero, 1996; Casas et al., 1996, 1997a,b).

Domestication is an evolutionary process through which domesticated plants become morphophysiologically divergent from their wild ancestors (see Darwin, 1868; Harlan, 1992). Such divergence is a consequence of artificial selection and other evolutionary forces resulting from manipulation of plants by humans. Artificial selection and domestication of plants have been generally associated with cultivation (Harlan, 1992), but these processes can also act under different forms of manipulation of wild populations in situ (see Casas et al., 1997a). Gatherers usually distinguish among individual plants of the same species and select for desirable features (e.g., flavor, size, color, presence of toxic substances; Casas and Caballero, 1996; Casas et al., 1996, 1997a). The edible wild plant species and the preferred variants are tolerated, enhanced, and/or protected in situ when they are found during the clearing of forest areas, whereas those species and variants whose edible parts are not preferred by humans are eliminated.

This chapter examines cultural and biological aspects related to the use, management, and role of columnar cacti in the economy of Mexican people of the south Pacific drainage, to determine if domestication is occurring in some of these species and which factors are influencing this process. The case of *Stenocereus stellatus* is included to illustrate patterns of artificial selection and evolutionary trends resulting from domestication under different forms of management. This case study is based on comparisons of morphology and reproductive biology between wild and manipulated populations. Such comparisons are then used to determine how domestication processes might occur in other columnar cacti.

Patterns of Use and Management of Columnar Cacti

We conducted a survey among indigenous peoples of the south Pacific drainage to investigate patterns of management and mechanisms of artificial selection in different species of columnar cacti (Casas et al., 1999a). Information on factors influencing the different ways of management, such as patterns of use and spatial and temporal availability, was emphasized. This information stressed morphological features of the different species and variants preferred for different purposes, as well as the destination of products of columnar cacti and their role in the economy of households. The information was complemented with ethnobotanical data for the relevant species from the Banco de Información Etnobotánica de Plantas Mexicanas of the Jardín Botánico, Universidad Nacional Autónoma de México and from bibliographical sources, especially Bravo-Hollis (1978), Dávila et al. (1993), Valiente-Banuet et al. (1996), Casas (1997), and Casas et al. (1997a, 1999a).

Nearly 40 species of columnar cacti occur in the area, and all of them produce edible fruits that are collected and consumed by humans (Table 7.2). Nearly half of these species are giant columnar cacti (species with habit "G" in Table 7.2), some of which are about 15 m tall and are characterized by slow vegetative growth and first flowering only after several decades of growth. According to the local people, the giant cacti have not been cultivated because of these growth and flowering habits. However, 20 of the species listed in Table 7.2 (species with habit "S") are 2–8 m in height. They are relatively fast growing (first flowering occurs 6–8 years after seed germination or 2–4 years after vegetative propagation) and most of them also reproduce clonally. These species are cultivated by local people. Among these species, exceptional morphological variation apparently resulting from human management can be observed in fruits of *Escontria chiotilla, Myrtillocactus geometrizans, Pachycereus hollianus, P. marginatus, Stenocereus griseus, S. pruinosus, S. queretaroensis*, and *S. stellatus.*

TABLE 7.2

Species of Pachycereeae from the South Pacific Drainage of Mexico

Species	*Uses*[a]	*Habit*[b]	*Cultural Status*[c]	*Products*[d]	*Distribution within the Area*[e]
Backebergia militaris	1, 2	G	W	H	Gro, Mich
Cephalocereus apicicephalium	1, 2	G	W	H	Oax
C. chrysacanthus	1*, 2	G	W	H	Pue, Oax
C. collinsii	1*, 2	S	W	H	Oax
C. guerreronis	1, 2	S	W	H	Gro
C. columna-trajanis	1*, 2, 6	G	W	H	Pue
C. nizandensis	1, 2	G	W	H	Oax
C. palmeri var. *Sartorianus*	1*, 2	G	W	H	Oax
C. purpusii	1, 2	S	W	H	Mich
C. quadricentralis	1, 2	S	W	H	Oax
C. totolapensis	1, 2	G	W	H	Oax
Escontria chiotilla	1**, 2, 3, 4, 5, 7	S	W, M	H, C	Pue, Oax, Gro, Mich
Mitrocereus fulviceps	1*, 2, 6	G	W	H	Pue, Oax
Myrtillocactus geometrizans	1**, 2, 3, 7	S	W, M	H, C	Gro Oax, Mich
M. schenkii	1**, 2, 3, 7	S	W, M	H, C	Pue, Oax
Neobuxbaumia macrocephala	1, 2, 6	G	W	H	Pue
N. mezcalaensis	1**, 2, 4, 5, 6	G	W	H	Pue, Oax, Mor, Mich, Gro
N. multiareolata	1, 2	G	W	H	Gro
N. scoparia	1, 2	G	W	H	Oax
N. tetetzo	1**, 2, 4, 5, 6	G	W	H, C	Pue, Oax
Pachycereus grandis	1**, 2, 4	G	W	H	Mor, Mex, Pue
P. hollianus	1**, 2, 3, 4, 7	S	W, M, C	H	Pue
P. marginatus	1**, 2, 7	S	W, M, C	H	Pue, Oax, Mex
P. pecten-aboriginum	1**, 2, 4, 6	G	W, M	H	Mich, Gro, Oax
P. weberi	1**, 2, 3, 4, 6	G	W, M	H, C	Pue, Oax, Gro, Mich
Polaskia chende	1**, 2, 4, 7	S	W, M, C	H, C	Pue, Oax
P. chichipe	1**, 2, 4, 7	S	W, M, C	H, C	Pue, Oax
Stenocereus dumortieri	1**, 2, 3, 4, 6	G	W, M	H, C	Oax, Mor, Gro. Mich
S. griseus	1**, 2, 3, 4, 5, 7	S	C	H, C	Pue, Oax, Gro
S. pruinosus	1**, 2, 3, 4, 6	S	W, M, C	H, C	Pue, Gro Oax
S. stellatus	1**, 2, 3, 4, 5, 7	S	W, M, C	H, C	Mor, Pue, Oax, Gro
S. treleasei	1**, 2, 4, 7	S	W, M, C	H, C	Oax
S. fricii	1**, 2, 4, 7	S	W, M	H, C	Mich
S. queretaroensis	1**, 2, 7	S	W, M, C	H, C	Mich
S. quevedonis	1**, 2, 4, 7	S	W, M, C	H, C	Mich, Gro
S. chacalapensis	1**, 2	G	W	H	Oax
S. chrysocarpus	1**, 2	G	W, M	H	Mich
S. beneckei	1*, 2, 7	S	W	H	Gro, Mor, Mex,
S. standleyi	1**, 2, 4, 7	S	W, M, C	H	Mich, Gro

[a] 1 = edible fruits (*regular quality, **good quality); 2 = fodder; 3 = alcoholic beverage; 4 = edible seeds; 5 = edible stems and flowers; 6 = house construction; 7 = living fences.

[b] G = giant columnar cacti (slow growth); S = small columnar cacti (fast growth).

[c] W = wild; M = managed in situ; C = cultivated.

[d] H = consumption by household; C = commercialization.

[e] Gro = Guerrero; Mex = Mexico; Mich = Michoacán; Mor = Morelos; Oax = Oaxaca; Pue = Puebla.

Columnar cacti are used mainly for their fruits, which are consumed both fresh and dried and are used for preparing jams or as fodder for domestic animals (use "1" in Table 7.2). Although fruits of all species are sometimes consumed by humans, it is possible to distinguish (1) those species producing sweet, juicy fruits, which are considered in this study as "good quality fruits" and are commonly harvested; (2) those species whose fruits are considered of "regular quality" and are only occasionally collected because the plants are scarce, harvesting is difficult, or the fruits are not tasty; and (3) those species whose fruits do not contain juicy pulp and are consumed only during periods of food scarcity.

Branches of columnar cacti are cut and fed to domestic donkeys, cows, and goats after removal of the spines (use "2" in Table 7.2). An alcoholic drink called "colonche" or "nochoctli" may be prepared from the fruits of some species (use "3" in the table). Seeds of some species (use "4") are consumed by humans. In general, seeds obtained from fresh or dried fruits are washed, dried, and roasted for preparing traditional sauces or to be ground into an edible paste. Stems and flowers of some species (use "5") are occasionally consumed. Stems of columnar cacti seem to have been a common food in the past (Callen, 1967), but at present they are eaten only during seasons of food scarcity. In contrast, the flower buds are more commonly boiled and consumed when in season. Wood of giant columnar cacti is commonly used in construction of house roofs and fences (use "6"). *Polaskia chichipe* and *Stenocereus stellatus* are sources of fuel wood for manufacturing pottery. Individuals of 11 species are commonly grown as living fences and barriers for protection against soil erosion in terraces of cultivated slopes (use "7").

In the Tehuacán Valley, the economic value of edible cactus parts is:

- Three to five fresh fruits or six to ten dried fruits of *Escontria chiotilla, Stenocereus pruinosus,* or *S. stellatus* (see also Casas et al., 1997a) are exchanged for one liter of maize;
- Eight to ten fresh fruits or ten to fifteen dried fruits of *Neobuxbaumia tetetzo, Polaskia chende,* or *P. chichipe* for one liter of maize;
- One liter of both fresh or dried fruits of *Myrtillocactus geometrizans* for one liter of maize;
- One liter of flower buds for three liters of maize; and
- One liter of seeds for fifteen liters of maize.

These columnar cacti seem to be at present the most economically important in the region.

Peoples of the area commonly gather fruits and other useful products of columnar cacti from wild populations (cultural status "W" in Table 7.2). In general, those surveyed indicated that they gather fruits selectively, preferring larger fruits of species or variants with juicy pulp, sweeter flavor, thinner pericarp, shorter and fewer spines, and deciduous areoles. Additional forms of selection can also be observed with respect to some of these plants. For instance, peasants frequently tolerate or let stand individuals of 18 species of columnar cacti when they clear the vegetation for cultivating maize (cultural status "M" in Table 7.2). It is also common practice to plant vegetative propagules of the spared columnar cacti in cleared areas to enhance their local abundance. Because the individuals of these tolerated species compete with cultivated plants, they are selected carefully. Humans prefer to spare individuals with big fruits, sweet flavor, thin pericarp, and few spines, but tolerance of other phenotypes is also common, especially when there are no other useful species or competing phenotypes.

And finally, 11 species are cultivated (cultural status "C") mainly by planting vegetative parts in home gardens or in agricultural fields, where they serve as living fences or barriers for prevention of soil erosion. In general, humans cut vigorous branches from mature wild or cultivated individuals. They are eventually irrigated and ash is commonly deposited as fertilizer on the soil covering the main stems. Individuals cultivated in home gardens may also be derived from seedlings that sprout from seeds dispersed via bird, bat, or human feces. Because the local peoples generally do not recognize variants of columnar cactus species based on vegetative characteristics, decisions on eliminating or sparing individuals are taken after 4–10 years, when the individuals first produce fruits.

Case Study: *Stenocereus stellatus*

Stenocereus stellatus is a species endemic to central Mexico, occurring in the wild in tropical deciduous and thornscrub forests. It is also cultivated in home gardens, and some wild populations are managed in situ (Casas et al., 1997a). This species exhibits considerable morphological variation, especially in fruit characteristics, which are presumably partly under genetic control and partly influenced by environmental conditions. These conditions include altitudes ranging from 600 to 2,000 m; levels of precipitation that vary between 300 and 800 mm per year; mean annual temperatures from 17 to 24°C; and soils derived from limestone, sandstone, volcanic rocks, or alluvial deposits. The morphological variation of *S. stellatus* also seems to be the result of human manipulation. According to archaeological information obtained in caves from Tehuacán Valley, *S. stellatus* has apparently been used by humans for more than 5,000 years (MacNeish, 1967; Smith, 1967). Indigenous groups inhabiting the area currently use and manage this species mainly for its edible fruits. Casas et al. (1997a) reported that management in situ of wild populations of *S. stellatus* is carried out by sparing some desirable phenotypes and removing others during clearing of vegetation, and sometimes enhancing numbers of the desirable phenotypes by cutting and planting their branches. Cultivation is practiced mainly in home gardens, where desirable phenotypes are vegetatively propagated and new variation is incorporated through tolerance of volunteer seedlings.

Pulp color, flavor, amount of edible matter, skin thickness, and spininess of the mature fruits are the most significant characteristics used in folk classification of variants, assessment of quality of products, and selection of individuals of *Stenocereus stellatus* for preferential propagation (Casas et al., 1997a). Manipulation of this plant species by humans thus appears to involve artificial selection. This seems to be particularly intense in home gardens, where manipulation of *S. stellatus* occurs by frequent planting and replacing individuals, but it also seems to be significant in managed in situ populations (Casas et al., 1997a).

Objectives

One of the hypotheses tested in this study was that if artificial selection has been significant, both management in situ and cultivation of this species might have changed patterns of morphological variation from those occurring in unmanaged wild populations, especially in those characters that are direct targets of human selection. Thus we compared morphology among individuals from populations under different management regimes. We also compared individuals from two regions: the Tehuacán Valley and La Mixteca Baja (Fig. 7.2) to examine the extent to which patterns of morphological variation in populations can be related to environmental factors.

In addition, we examined pollination biology and breeding systems to assess the extent of modification by domestication and whether gene flow between wild and cultivated populations has been disrupted. Studies by Valiente-Banuet et al. (1996, 1997a,b) have shown that self-pollination is ineffective in all cases of columnar cacti studied in the Tehuacán Valley. Therefore, we expected wild populations of *Stenocereus stellatus* to be outcrossed. However, this pattern of reproduction may be modified in cultivated populations. In other species, mutants to self-compatibility have been favored by human selection because they can produce fruit and seed reliably in the absence of any pollinator and do not depend on sources of compatible pollen (see Proctor et al., 1996).

Temporal patterns of flower and fruit production by wild plants are commonly modified by domestication to adjust them to human convenience. Because this might have occurred in *Stenocereus stellatus,* we carried out a comparative analysis of floral phenology in wild, managed in situ, and cultivated populations. Our goal was to determine if natural or artificial selection had affected the timing of flower production and whether this had caused significant barriers to pollen exchange between populations managed in different ways and between populations of different regions.

Patterns of Variation among Individuals

We analyzed 23 morphological characters from 324 reproductive-sized individuals sampled from wild, managed in situ, and cultivated populations in Tehuacán Valley and Mixteca Baja (Table 7.3; see Casas et al., 1998, 1999b). According to ethnobotanical information (Casas et al., 1997a), a subset of the characters is the direct tar-

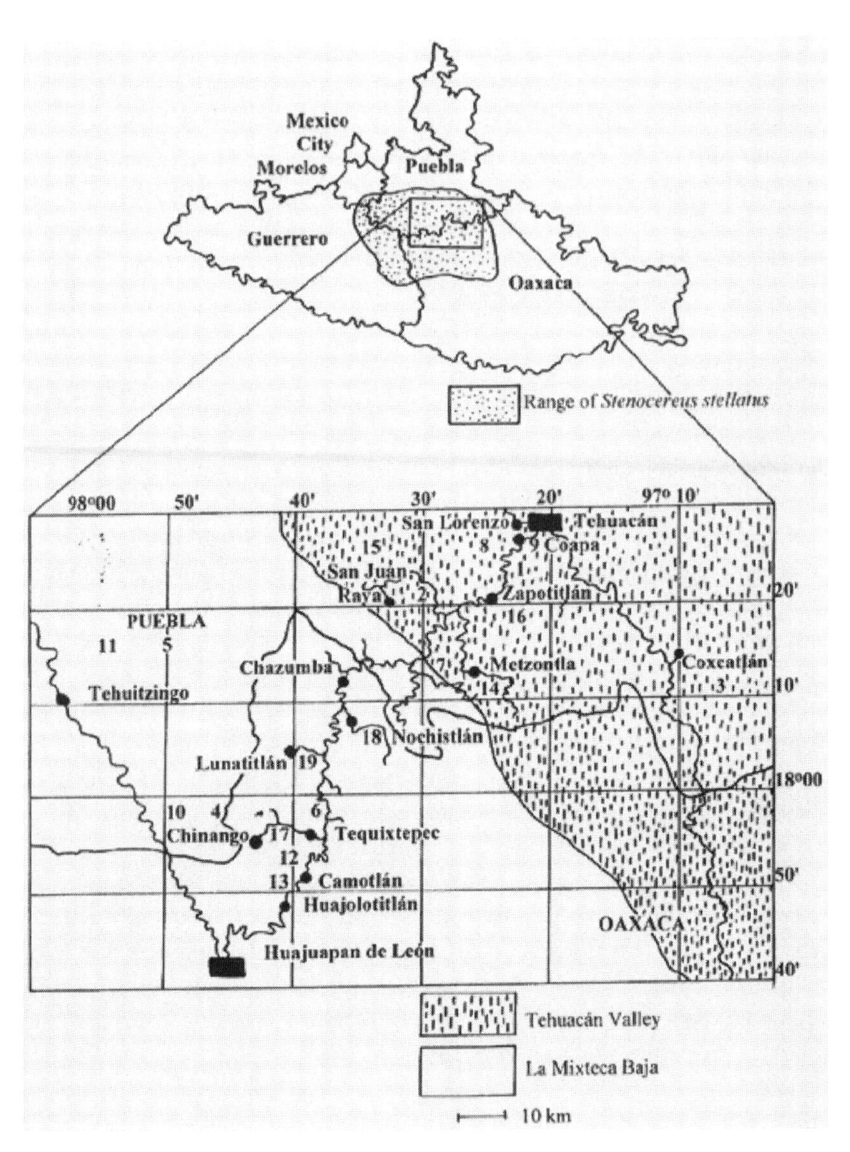

FIGURE 7.2. Range of *Stenocereus stellatus* in the Tehuacán Valley and La Mixteca Baja and our study populations. Wild populations (W): 1 = Zapotitlán-W; 2 = San Juan Raya-W; 3 = Coxcatlán-W; 4 = Chinango-W; 5 = Tepexco-W; 6 = Tequixtepec-W. Managed in situ populations (M): 7 = Metzontla-M; 8 = San Lorenzo-M; 9 = Coapa-M; 10 = Chinango-M; 11 = Tepexco-M; 12 Camotlán-M; 13 = Huajolotitlán-M. Cultivated populations (C): 14 = Metzontla-C; 15 = San Lorenzo-C; 16 = Zapotitlán-C; 17 = Chinango-C; 18 = Nochistlán-C; 19 = Lunatitlán-C.

TABLE 7.3
Morphological Characters Analyzed to Test the Hypothesis of Artificial Selection on *Stenocereus stellatus*

Characters	*Target of Selection*
Number of branches	no
Length of the highest branch	no
Diameter of the highest branch	no
Number of stem ribs	no
Stem rib width	no
Stem rib depth	no
Number of spines per areole	no
Size of the central spines	no
Distance between areoles	no
Fruit form	no
Skin color	no
Pulp color	yes
Pulp flavor	yes
Fruit size	yes
Number of areoles per fruit	yes
Number of areoles per cm^2	yes
Skin thickness	yes
Pulp as percentage of fruit weight	yes
Water as percentage of pulp weight	yes
Number of seeds per fruit	no
Mean seed weight	no
Total seed weight per fruit	no
Seed mass as percentage of pulp weight	no

get of human selection. Other characteristics were also included to determine general patterns of variation independent of human influence. We used multivariate statistical methods to analyze the patterns of variation (Casas et al., 1999b).

Morphological patterns within each region were analyzed using Principal Component Analyses (PCA), which allowed us to visualize similarities and differences between populations according to their regime of management. The resulting eigenvectors were used to identify the characteristics that best serve to define groups of individuals (Sneath and Sokal, 1973). Individuals of the two regions are distributed along continuous gradients rather than in discrete groups (Fig. 7.3). Along the first principal component axis of this figure, most of the wild individuals are grouped to the left whereas most of the cultivated ones are to the right; those from managed in situ populations occur in the middle. The most significant characters in the first principal component are fruit size and seed weight (with positive values) and density of spines on fruits (with negative values). Cultivated individuals thus have the largest and least spiny fruits and the heaviest seeds. In the second principal component, the most significant characters are amount of pulp and skin thickness. All these characters are among the main targets of artificial selection.

We used Discriminant Function Analyses (DFA) to evaluate the overlap in general morphology of the three groups (wild, managed in situ, and cultivated popula-

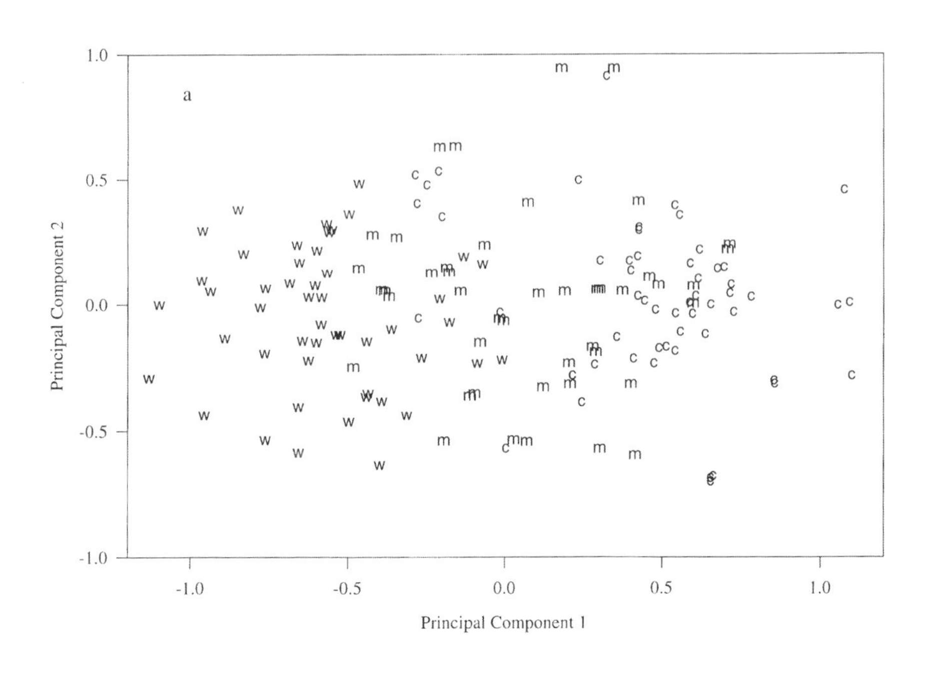

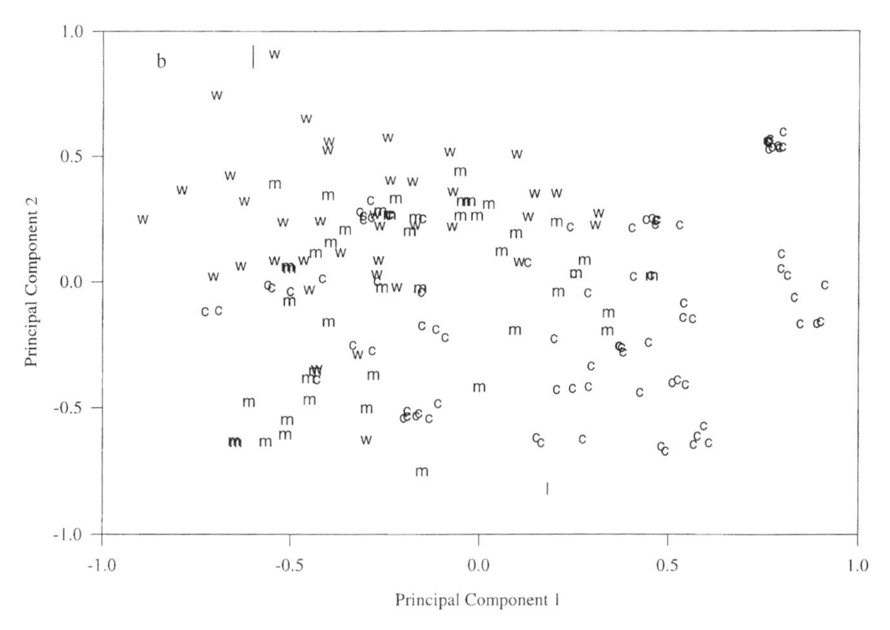

FIGURE 7.3. Ordination of individuals of *Stenocereus stellatus* in multivariate space based on the first and second principal components of a morphological analysis. (a) Populations from the Tehuacán Valley; (b) populations from La Mixteca Baja (w = wild; m = managed in situ; c = cultivated individuals).

tions) by looking at those cases where the analysis did not assign individuals to their original groups. Most of the individuals were classified correctly, indicating that in each region the three groups are distinct, although some individuals of one group are morphologically similar to individuals of the other two groups (Table 7.4). Overlap was more frequent among managed in situ and cultivated populations. However, according to multiple analyses of variance (MANOVA) there were significant differences among the three groups of populations in both regions (Casas et al. 1999b).

Patterns of Variation among Populations

We performed a cluster analysis (CA) with the unweighted pair-group method using averages (Sneath and Sokal, 1973) to examine the morphological similarity of populations in Tehuacán Valley and La Mixteca Baja. Our goal in doing this was to analyze differences related to environmental differences between the two regions, and to compare the morphologies among populations under similar methods of management across regions. CA clustered populations into two main groups (Fig. 7.4). The first group includes all populations from the Tehuacán Valley as well as three populations from La Mixteca Baja (the wild populations Tequixtepec-W and Chinango-W, and the managed in situ population Chinango-M); the second group includes the remaining populations from La Mixteca Baja. Thus populations were generally classified according to the region of provenance. The main clusters subsequently placed populations into groups and subgroups, each containing populations managed in a similar way. With the exception of the wild and managed in situ populations of Tepexco, all populations were grouped according to their method of management (Casas et al., 1999b).

Trends in Morphological Differentiation

We conducted one-way analyses of variance to study how morphological characters differed between wild, managed in situ, and cultivated populations within and across the regions studied. With these analyses, we expected to visualize trends of morpho-

TABLE 7.4

Classification of Individuals of Wild, Managed in situ and Cultivated Populations of *Stenocereus stellatus* from the Tehuacán Valley and La Mixteca Baja

Region	*Actual Group*	*Predicted Group (%)*[a] *Wild*	*Managed in situ*	*Cultivated*	*Total (%)*
Tehuacán Valley	wild	95.9	4.1	0.0	100.0
	managed in situ	2.2	88.9	8.9	100.0
	cultivated	3.6	16.4	80.0	100.0
La Mixteca Baja	wild	72.5	12.5	15.0	100.0
	managed in situ	11.1	80.0	8.9	100.0
	cultivated	6.7	7.8	85.6	100.0

[a] Predicted groups in percentage of individuals according to discriminant function analysis.

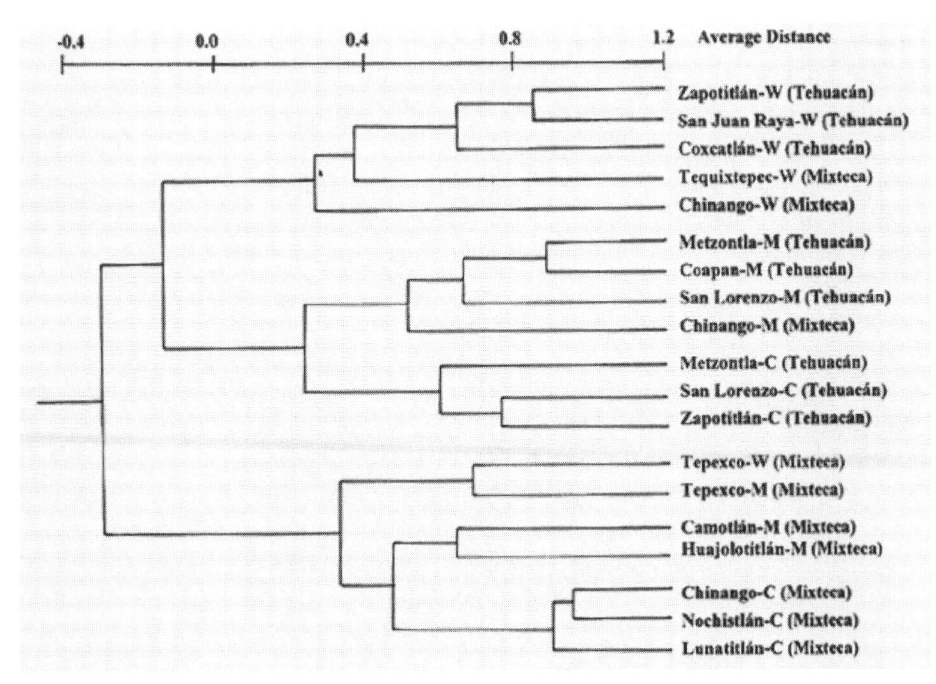

FIGURE 7.4. Classification of populations of *Stenocereus stellatus* studied based on CA of morphological data. The distance measure is based on the Euclidean Distance coefficient analyzed under the UPGMA method (Sneath and Sokal, 1973). Abbreviations: C = cultivated, M = managed in situ, W = wild.

logical differentiation influenced by artificial selection and environmental factors. Morphological characters of both vegetative parts and fruits differed significantly across the Tehuacán Valley and La Mixteca. As illustrated in Fig. 7.5, the number and dimensions of vegetative parts, fruits, and seeds were greater in La Mixteca Baja, whereas density of spines in fruits was higher in Tehuacán Valley. However, within each region, populations under different management regimes differed significantly, mainly in fruit characters. Figure 7.6 and Table 7.5 depict the character variation across the differently managed populations. Thus as the management of plants becomes more intensive, they possess larger fruits with a higher proportion of pulp, more and heavier seeds, fewer spines per cm^2, and thinner fruit skin; a higher proportion of individuals also produced fruits with green skin and sweet pulp with color other than red.

Floral Biology and Flower Visitors

Although Gibson and Horak (1978) suggested that flowers of *Stenocereus stellatus* are pollinated by hummingbirds, we investigated some aspects of floral biology of *S. stellatus* to determine the most likely pollinators and mechanisms of pollination. Time of anthesis, discharge of pollen from anthers, turgidity of stigma, and closing of flowers and amounts of nectar produced during anthesis were measured on flowers from five wild and five cultivated populations in Tehuacán Valley and La Mixteca Baja

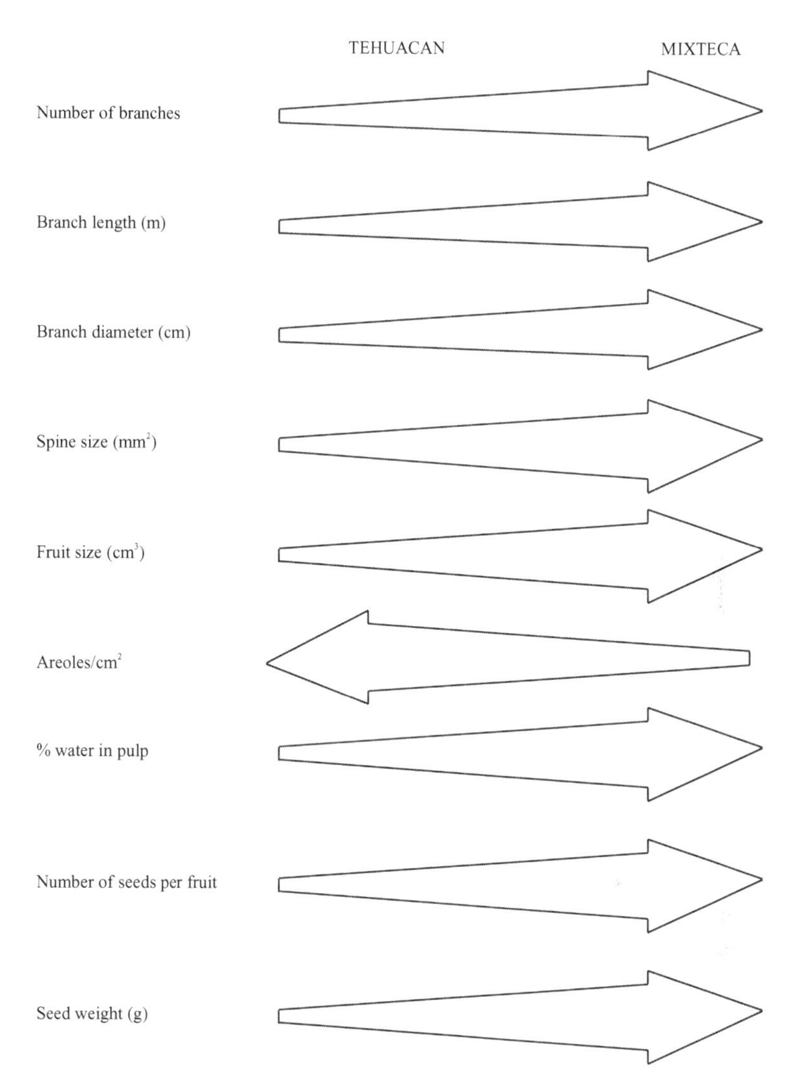

FIGURE 7.5. Trends in morphological variation between the Tehuacán Valley and La Mixteca Baja. Arrows indicate increasing or decreasing trends in each variable.

(Casas et al., 1999c). Mist nets set up in random transects were used to capture flower visitors. We set up 39 nets for 546 net-hours. Insects were captured from flowers every two hours from 2300 to 1000 on the same nights that netting was carried out. Pollen samples were collected from the bodies of the captured animals by using fuchsin jelly squares (Beattie, 1971). We mounted samples on microscope slides and scanned for pollen grains of *S. stellatus*.

Anthesis began around 2000; by 2300, the flowers were completely open. One hour later (around 2400), pollen started being released but stigmatic lobes were not

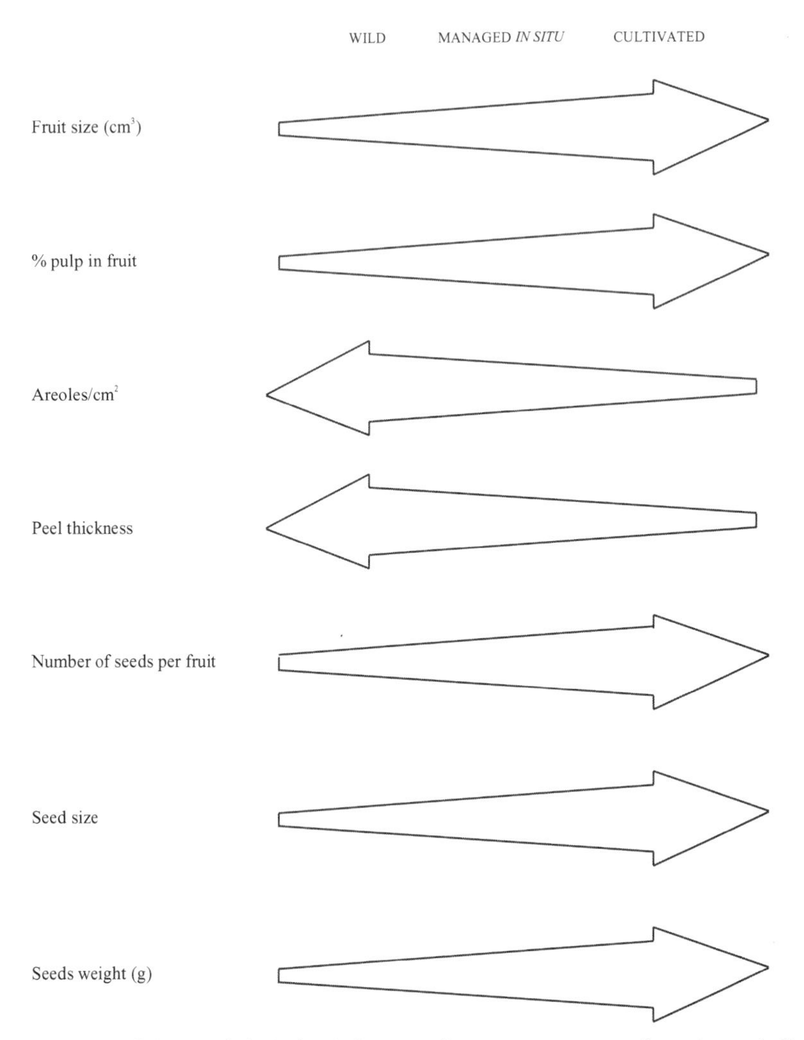

FIGURE 7.6. Trends in morphological variation according to management regimes. Arrows indicate increasing or decreasing trends in each variable.

completely open until 0200 to 0300. Flowers started to close around 0700, and around 0900 the entire flower was completely closed. Rates of nectar production were similar in all populations studied, with peak production coinciding with the time when the stigmatic lobes were completely open and turgid. These observations indicate that anthesis is predominantly nocturnal; pollen and nectar is particularly abundant at night; and nocturnal visitors can visit flowers for seven to eight hours versus only one or two hours for diurnal visitors. Pollination thus seems to occur at night, most likely from 0200 to 0500, when maximum turgidity of the stigma lobes coincides with peak nectar production. Among nocturnal visitors, hawkmoths (*Eumorpha* spp.), bats

TABLE 7.5

Character States of Qualitative Characteristics of Fruits in Individuals Sampled in Populations of *Stenocereus stellatus* from the Tehuacán Valley and La Mixteca Baja

	Tehuacán Valley (%)			*Mixteca Baja (%)*		
Character	*Wild (n = 49)*	*Managed (n = 45)*	*Cultivated (n = 45)*	*Wild (n = 40)*	*Managed (n = 45)*	*Cultivated (n = 90)*
Skin color						
Red	96.0	93.0	87.3	100.0	85.5	57.8
Green	4.0	7.0	12.7	0.0	15.5	42.2
Pulp color						
Red	100.0	100.0	89.0	95.0	86.7	56.7
Not red	0.0	0.0	11.0	5.0	13.3	43.3
Flavor						
Sour	100.0	47.0	29.0	72.5	11.0	16.7
Sweet	0.0	53.0	71.0	27.5	89.0	83.3

(*Artibeus jamaicensis, Choeronycteris mexicana, Leptonycteris curasoae yerbabuenae,* and *L. nivalis*) and beetles (*Carpophilus* spp.) were observed interacting with flowers of *S. stellatus.* Bats contacting flowers had their faces covered with pollen. They invariably touched the stigmatic surface when seeking nectar. Pollen of *S. stellatus* was identified on their bodies, and they were by far the most frequent animals captured with pollen of *S. stellatus.* Therefore bats are the most probable pollinators (Casas et al., 1999c).

Breeding Systems

Our experiments for testing breeding systems were based on methods developed by Valiente-Banuet et al. (1996). Six experimental treatments were applied to at least ten flowers per treatment in plants from four wild and four cultivated populations (Table 7.6). The treatments were (1) unmanipulated self-pollination, (2) manual self-pollination, (3) cross-pollination, (4) nocturnal pollination, (5) diurnal pollination, and (6) natural pollination. Total number of aborted flowers, mature fruits, and seeds per fruit per treatment were counted (see details in Casas et al., 1999c). We also tested for pollen incompatibility between cultivated phenotypes and wild individuals. Pollen from flowers of the "blanco," "amarillo," "morado," and "colorado" (white, yellow, purple, and red pulp, respectively) cultivated variants from Chinango and a wild individual used as a control was manually deposited on stigmas of sets of ten flowers of wild individuals per treatment.

Results of the pollination experiments (Table 7.6) indicate that, apart from the natural pollination treatments, only manual cross-pollination and nocturnal pollination treatments produced fruits; these results were similar in both wild and cultivated populations. However, pollen incompatibility between wild and domesticated types of plant species is also a possible source of reproductive isolation influencing the process of domestication, because all fruits from experimental crosses with pollen

TABLE 7.6

Numbers of Fruits and Seeds Produced Following Different Types of Pollination in Nine Populations of *Stenocereus stellatus*[a]

	Self Pollination														
	Unmanipulated		*Manual*		*Manual Cross Pollination*			*Diurnal Pollination*		*Noctural Pollination*			*Natural Pollination*		
Population[b]	*NF*	*FS (%)*	*NF*	*FS (%)*	*NF*	*FS (%)*	*NS*	*NF*	*FS (%)*	*NF*	*FS (%)*	*NS*	*NF*	*FS (%)*	*NS*
Zapotitlán-W	37	0	10	0	10	60	29,048	10	0	10	80	95,124	20	75	93,435
Coxcatlán-W	20	0	10	0	10	50	34,227	10	0	10	70	70,333	20	65	75,015
Chinango-C	35	0	20	0	10	60	68,952	15	0	20	75	1,540,125	30	73	1,472,227
Metzontla-C	30	0	10	0	—	—	—	—	—	—	—	—	—	—	—
Total	122	0	50	0	30	17 (57%)		35	0	40	30 (75%)		70	50 (71%)	

[a] NF = number of flowers in treatments; FS = percent fruit set; NS = mean number of seeds.
[b] Abbreviations: C = cultivated; W = wild.

from cultivated phenotypes aborted, whereas 60% of crosses with pollen from wild individuals successfully produced fruits (Casas et al., 1999c).

Dynamics of Flowering in Populations

We recorded the number of individuals flowering every two to four weeks between April and November 1995, in samples of the nine populations shown in Table 7.7. Following Dafni (1992), the data were processed according to the date of appearance of first flowers (flowering commencement); dates when 25%, 25–50%, and 50% or more of individuals had open flowers; and date when flowering finished (flowering termination). These data were also used to define the flowering peak (maximum number of flowers per individual and of flowering plants per population) and flowering duration of the sample in days. Number of flowering days and number of flowers at flowering peak per individual were compared by analyses of variance among populations within each region.

Table 7.7 shows that flowering occurred over a period of at least 91 days, with a few flowers blooming per individual per night. Although the beginning and termination of flowering differed by more than one month between populations, the flowering peak in all populations occurred between the first and third week of July. At peak flowering, individuals of cultivated populations had significantly more flowers in anthesis than individuals in other populations of each region. Flowering duration in individuals of cultivated populations was also significantly longer than in individuals of other populations. However, flowering periods in different populations overlapped by at least 75 days, indicating that temporal barriers for pollination between populations are unlikely.

Discussion

Ethnobotanical information indicates that at present humans make decisions about how to manipulate columnar cacti according to the quality of their products and their roles in human subsistence. Thus the species and variants cultivated or managed in situ are generally those with the most useful fruit characteristics. However, as we found in the case of *Stenocereus stellatus* (Casas et al. 1997a), cultivation is particularly intensive in areas where the commercialization of fruits or their consumption by households makes it necessary to produce more and/or better fruits. Availability of plant resources is another crucial factor influencing the way the plants are manipulated, and this seems to be a general pattern of plant management in the south Pacific drainage region. For example, Casas and Caballero (1996) and Casas et al. (1997a) found that *Leucaena esculenta* and *S. stellatus* are intensively cultivated only in those places where wild populations are scarce.

Ease of manipulation is also important. Thus, although species such as *Mitrocereus fulviceps, Neobuxbaumia mezcalaensis, N. tetetzo,* and *Pachycereus weberi* pro-

TABLE 7.7

Flowering Periods in Nine Populations of *Stenocereus stellatus*[a]

Population[b]	*Flowering Beginning*	*Percentage of Flowering Individuals with Open Flowers*							*Flowering Duration (days)*	*Mean Number of Days of Flowering ± SE per Individual*[d]	*Mean Number of Flowers ± SE per Individual at Flowering Peak*[d]
		25%	*25–50%*	*50%*	*25–50%*	*<25%*	*<10%*	*0%*			
Tehuacán											
Zapotitlán-W	Jun (1)	Jun (2)	Jun (3)	Jul (3)	Jul (4)	Aug (3)	Aug (4)	Sep (1)	91	36.75 ± 3.64	1.8 ± 0.28
S. J. Raya-W	Apr (4)	May (4)	Jun (3)	Jun (4)	Jul (3)	Jul (4)	Aug (2)	Sep (2)	119	37.57 ± 6.40	4.1 ± 1.58
Coxcatlán-W	Jun (1)	Jun (2)	Jun (3)	Jul (3)	Sep (1)	Sep (2)	Sep (3)	Sep (4)	105	42.36 ± 7.33	4.2 ± 0.79
Metzontla-M	May (4)	Jun (1)	Jun (2)	Jun (3)	Aug (1)	Aug (2)	Aug 3	Aug (4)	97	41.58 ± 4.41	7.9 ± 1.55
S. Lorenzo-M	May (2)	May (4)	Jun (3)	Jul (2)	Jul (3)	Jul (4)	Aug (4)	Sep (2)	93	39.30 ± 6.06	5.7 ± 1.57
Metzontla-C	May (3)	May (4)	Jun (2)	Jun (3)	Aug (4)	Sep (1)	Sep (2)	Sep (4)	137	66.00 ± 4.35[c]	13.7 ± 1.73[c]
Mixteca											
Chinango-W	May (1)	May (2)	May (3)	Jul (2)	Aug (3)	Aug (4)	Sep (1)	Sep (2)	99	40.21 ± 7.07	8.9 ± 2.50
Chinango-M	May (3)	Jun (3)	Jul (1)	Jul (2)	Aug (2)	Aug (3)	Aug 4	Sep (1)	113	35.72 ± 7.08	3.1 ± 1.03
Chinango-C	Apr (3)	May (3)	May (4)	Jun (1)	Aug (3)	Aug (4)	Sep (1)	Sep (3)	151	56.57 ± 2.60[b]	21.5 ± 2.60[b]

[a] Based on percentage of individuals flowering at different times of the flowering season. Numbers in parentheses correspond to the week number within each month.

[b] Abbreviations: C = cultivated; M = managed in situ; W = wild.

[c] Significantly different values among populations compared within each region ($p < 0.001$).

[d] SE = standard error.

duce good quality fruits (Table 7.2), they are not cultivated ex situ because their slow growth makes the effort of sowing seeds and taking care of seedlings unrewarding for decades. Slow growth may not be relevant for making decisions on managing wild populations of such species in situ because, under this form of management, humans simply tolerate the presence of plants during the clearing of vegetation. The effect of artificial selection favoring particular phenotypes of these species in situ is probably negligible, given the difficulties of intentional direct propagation. Furthermore, when the seeds of desirable phenotypes are sown, genetic variance makes it uncertain that the phenotypes selected are those expressed in the progeny sown. In contrast, the fixation of desirable characters in species that can be vegetatively propagated, such as *Stenocereus pruinosus, S. queretaroensis* or *S. stellatus,* is relatively easy. Although these species have self-incompatible breeding systems (Pimienta-Barrios and Nobel, 1994; Casas et al., 1999c), artificial selection is possible even on progeny from sexual reproduction, because of their relatively fast growth.

Fruits are the most commonly used parts of columnar cacti. Size, pulp color, flavor, spininess, and thickness are used to characterize fruit quality. Combinations of character states may produce a broad spectrum of varieties constituting the raw material for artificial selection at the plant community (selecting for species) or species (selecting for phenotypes) level. Artificial selection is apparently carried out by identifying and vegetatively propagating individuals with desired phenotypes from wild, managed in situ, or cultivated populations. Artificial selection is also applied when plants of desired forms are preferentially spared or protected when land is cleared or when seedlings are spared in cultivated populations until their fruits can be evaluated, after which plants are either retained or discarded.

These observations suggest that artificial selection may be applied to easily manipulated species. It has achieved significant results in such species as *Stenocereus griseus, S. pruinosus, S. queretaroensis, and S. stellatus,* as documented by Pimienta-Barrios and Nobel (1994) and Casas et al. (1997a, 1998, 1999b,c). Artificial selection may also be significant in such species as *Escontria chiotilla, Myrtillocactus geometrizans, M. shenki, Pachycereus hollianus, P. marginatus, Polaskia chichipe,* and *P. chende,* which are intensely cultivated and managed in situ and which exhibit morphological variation in characters that are targets of human preference. The case study of *S. stellatus* provides a model of in situ and ex situ artificial selection that might be helpful for analyzing patterns of domestication in this group of species. Moreover, the model of domestication under management in situ (domestication in situ) can also help identify possible processes of domestication in some giant columnar cacti, such as *Neobuxbaumia mezcalaensis, N. tetetzo, Pachycereus weberi,* and *Stenocereus dumortieri.*

In the case of *Stenocereus stellatus,* multivariate analyses generally demonstrated that there are significant morphological differences between wild, managed in situ, and cultivated populations. These results support the hypothesis that human management has significantly influenced morphological divergence of both managed in situ and cultivated populations from wild populations of this species and that, therefore, domestication may be caused not only through cultivation but also through management of wild populations. The most relevant characters for grouping individuals in PCA are those that constitute direct targets of selection by humans (Casas et al., 1997a), which reinforces the conclusion that artificial selection is the crucial factor for explaining the divergence between wild and manipulated populations. Analysis of the overlaps between populations indicates that the morphological divergence is great between wild and cultivated populations, apparently because artificial selection is particularly intense under the continuous planting and replacing of individuals cultivated in home gardens.

Although the phenotypes of managed in situ and cultivated populations originated from wild populations, some cultivated phenotypes are rare or have not been observed in the wild. This is particularly true for individuals with large fruits and pulp colors other than red. Only 2.3% of the individuals sampled in wild populations had pink or yellow pulp, and other pulp colors (purple, orange, and white) were not observed in the wild. However, nearly 42% of individuals sampled in cultivated populations of La Mixteca Baja were of these phenotypes. The question remains whether these phenotypes originated in the wild and were carried to home gardens or if they originated in home gardens and escaped to the wild. Regardless of origin, it is clear that success of such phenotypes is low in the wild and it is higher only under human protection. In other words, domestication of *S. stellatus* appears to have occurred by protecting and enhancing those individuals whose morphological characteristics are favorable to humans but that are scarce or absent in the wild.

In addition, morphological variation in *Stenocereus stellatus* appears to be influenced by environmental conditions and may also be due to genetic differentiation. For example, CA clustered populations in part according to the region from which they originated, and analyses of variance found significant differences between populations of the two regions in most of the morphological characteristics analyzed. The clearest environmental difference between the regions is annual precipitation, which is significantly higher in La Mixteca Baja (720.5–763.7 mm) than in the Tehuacán Valley (440.6–590.0 mm); this factor may have a crucial effect on the differentiation of populations.

The studies of floral biology confirmed that anthesis in *Stenocereus stellatus* is mainly nocturnal and that the cactus cannot produce seeds in the absence of nocturnal visitors, of which bats are the most likely pollinators. With bats as pollinators, significant movement of pollen between populations can be expected: Bats of the genus *Leptonycteris* have been observed by Sahley et al. (1993) commuting 30 km from their roosts to feed at cactus flowers and fruit in the Sonoran Desert. Accordingly, participation of bats as pollinators makes isolation by distance within regions unlikely between the wild, managed in situ, and cultivated populations we studied. Distances separating our wild and managed in situ populations ranged from 300 m to 4 km, and no more than 10 km separate wild and cultivated populations in the Tehuacán Valley. All populations of *S. stellatus* studied showed an outbreeding system that has been unaffected by cultivation. This is because both bat pollinators and the pattern of traditional cultivation, which usually includes several variants of *S. stellatus* in a single garden (Casas et al. 1997a), increase the chances of compatible cross-pollination and make selection for self-compatibility unnecessary. Although our experiments of cross pollination between wild and cultivated phenotypes only included cultivated phenotypes with the strongest signs of domestication, they revealed that reproductive isolation may exist between some of the variants of *S. stellatus.* This could explain, at least in part, the morphological and genetic divergence between wild, managed, and cultivated populations. Nevertheless, more detailed studies are required to determine whether barriers created by pollen incompatibility occur between or even within wild and cultivated variants. Apart from the barrier of pollen incompatibility, the absence in wild populations of phenotypes found in home gardens can also be explained by failure of seeds to germinate and establish these variants under wild conditions. This hypothesis is yet to be tested.

Our data on phenology of *Stenocereus stellatus* indicate that individuals of cultivated populations produce significantly more flowers per night than individuals of wild and managed in situ populations. Also, individuals of cultivated populations have longer flowering periods than individuals of the other populations. However, the blooming season overlaps for at least 75 days per year in all populations studied, which indicates that temporal mechanisms of reproductive isolation are not operating.

The origin of variants that are common in home gardens but rare or absent in wild populations is uncertain. Interspecific hybridization could be a source of origin of variants exclusive to home gardens. *Stenocereus pruinosus* is a species that possibly hybridizes with *S. stellatus.* Although more detailed studies are needed to confirm this interspecific hybridization, preliminary observations on the phenology of *S. pruinosus* and the information from our study indicate that these species are often sympatric and that their flowering seasons overlap between April and part of June in some populations. This suggests that there are neither temporal nor geographic barriers preventing hybridization between these species in the wild. Participation of bats in pollination of both *S. pruinosus* and *S. stellatus* may produce interspecific hybrids, but this has yet to be demonstrated. Similarly, interspecific hybridization between *S. stellatus* and *S. treleasei,* which apparently is also possible (Casas et al., 1999c), should be examined carefully to understand variation in *S. stellatus.*

In conclusion, although populations of *S. stellatus* in the two regions studied differ in morphology (probably due to environmental differences), within each region wild and manipulated populations have diverged morphologically, especially in fruit characters, presumably due to artificial selection. Although the greatest divergence from wild populations was found to occur in cultivated populations, it is also significant in wild populations managed in situ. This suggests that domestication in situ is an ongoing process in this species and that both in situ and ex situ domestication have influenced the evolution of *S. stellatus* and probably other columnar cactus species. *S. stellatus* has two general strategies of reproduction. Vegetative propagation, which is conservative in terms of variation, may be a natural strategy that permits successful phenotypes to persist in particular environments. Sexual reproduction by outcrossing, which is apparently obligatory, may be a natural strategy to reintroduce variation to permit survival in the diverse environments characteristic of the study area. Humans make use of these strategies of reproduction in *S. stellatus* for artificial selection when they propagate, mainly by vegetative means, desirable variants that have arisen by sexual reproduction. In addition, there are apparently reproductive barriers between wild and cultivated populations resulting from pollen incompatibility. The morphological divergence between these types of populations may be maintained in part by pollen incompatibility, as well as by artificial selection against wild phenotypes in home gardens, but probably also by natural selection operating against domesticated phenotypes in the wild. The general pattern of in situ and ex situ domestication may be similar in other cactus species with relatively fast vegetative growth and/or clonal propagation, and may serve as a useful model for analyzing possible processes of domestication of certain giant columnar cacti.

Summary

We identify species of columnar cacti used and managed by peoples of the south Pacific drainage of south-central Mexico. All of the 40 species of Pachycereeae occurring in the area offer useful resources, most of them exclusively gathered from wild populations, but wild populations of about 18 species are also managed in situ, and 11 species are cultivated in home gardens. We analyze the case of *Stenocereus stellatus* to examine the effects of artificial selection under different forms of management. Comparisons between wild, managed in situ, and cultivated populations reveal significant morphological differences in the plant parts selected by humans for use, suggesting that artificial selection is the main factor determining such divergence. Spatial and temporal barriers to pollen flow between wild and cultivated populations appear to be insufficient for explaining the observed morphological divergence. Failures in crosses between some cultivated variants and wild individuals suggest that this divergence could be maintained in part owing to pollen incompatibility. It probably also results from the inability of cultivated varieties to become established in the wild. We present a pattern of domestication determined by in situ and ex situ processes of artificial selection acting in conjunction, which may be useful for understanding the domestication of cactus species with coexisting wild and cultivated populations. A pattern of in situ domestication acting alone may be helpful to understand possible processes of domestication in giant columnar cacti.

Resumen

Se identifican las especies de cactáceas columnares utilizadas y manejadas por los pueblos de la vertiente del Pacífico sur de México. Las 40 especies de Pachycereae del área ofrecen productos útiles la mayor parte de los cuales son exclusivamente recolectados a partir de poblaciones silvestres. Las poblaciones silvestres de 18 especies son también manejadas in situ, y 11 especies son también cultivadas en las huertas. Se analiza el caso de *Stenocereus stellatus,* con el fin de examinar los efectos de la selección artificial, asi como los patrones de domesticación en cactáceas columnares bajo diferentes formas de manejo. Comparaciones efectuadas entre las poblaciones silvestres, manejadas in situ y cultivadas revelan que existen diferencias morfológicas significativas, especialmente en las partes de la planta seleccionadas por las personas para su uso. Esto sugiere qué la selección artificial es el factor principal que determina tal divergencia. Al parecer, las barreras espaciales ó temporales al flujo de polen entre poblaciones silvestres y cultivadas, son insuficientes para explicar la divergencia morfológica observada, no obstante, se observaron fallas al intentar cruzar individuos silvestres y de algunas variantes cultivadas, lo que sugiere que tal divergencia podría mantenerse en parte, debido a la incompatibilidad de polen. Esto probablemente resulta también de una deficiencia en la capacidad de estas variantes cultivadas para establecerse en condiciones silvestres. Se presenta un patrón de domesticación determinado por los procesos de selección artificial in situ y ex situ ocurriendo conjuntamente, lo cual resulta útil para entender la domesticación de las especies con poblaciones silvestres y cultivadas coexistiendo. Un patrón de domesticación in situ actuando solo puede contribuir a entender posibles procesos de domesticación en cactáceas columnares gigantes.

ACKNOWLEDGMENT

The authors are thankful for financial support from DGAPA-UNAM, México (research grant IN224799).

REFERENCES

Acuña, R., ed. 1985. *Relaciones Geográficas del Siglo XVI: Tlaxcala.* Vol. 2. Mexico: Universidad Nacional Autónoma de México.

Beattie, A. J. 1971. A technique for the study of insect-borne pollen. *Pan Pacific Enthomologist* 47:82.

Bravo-Hollis, H. 1978. *Las Cactáceas de México.* Vol. 1. Mexico: Universidad Nacional Autónoma de México.

Caballero, J., and C. Mapes. 1985. Gathering and subsistence patterns among the Purhepecha Indians of Mexico. *Journal of Ethnobiology* 5:31–47.

Callen, E. O. 1967. Analysis of the Tehuacán coprolites. In *Prehistory of the Tehuacán Valley.* Vol. 1. *Environment and Subsistence,* ed. D. S. Byers, 261–89. Austin: University of Texas Press.

Casas, A. 1997. Evolutionary trends of *Stenocereus stellatus* (Pfeiffer) Riccobono (Cactaceae) under domestication. Ph.D. diss., University of Reading, Reading, England.

Casas, A., and J. Caballero. 1996. Traditional management and morphological variation in *Leucaena esculenta* (Moc. et Sessé ex A.DC.) Benth. (Leguminosae: Mimosoideae) in the Mixtec region of Guerrero, Mexico. *Economic Botany* 50:167–81.

Casas, A. and A. Valiente-Banuet. 1995. Etnias, recursos genéticos y desarrollo sustentable en zonas áridas y semiáridas de México. In *IV Curso Sobre Desertificación y Desarrollo Sus-*

tentable en América Latina y el Caribe, eds. G. M. Anaya and C.S.F. Díaz, 37–66. Ciencias Agrícolas, Mexico: United Nations Environmental Program/Food and Agriculture Organization.

Casas, A., J. L. Viveros, and J. Caballero. 1994. Etnobotánica mixteca: sociedad, cultura y recursos naturales en la Montaña de Guerrero. Mexico: Instituto Nacional Indigenista–Consejo Nacional para la Cultura y las Artes.

Casas, A., M. C. Vázquez, J. L. Viveros, and J. Caballero. 1996. Plant management among the Nahua and the Mixtec from the Balsas River Basin: an ethnobotanical approach to the study of plant domestication. *Human Ecology* 24:455–78.

Casas, A., B. Pickersgill, J. Caballero, and A. Valiente-Banuet. 1997a. Ethnobotany and the process of domestication of the xoconochtli *Stenocereus stellatus* (Cactaceae) in the Tehuacán Valley and La Mixteca Baja, Mexico. *Economic Botany* 51:279–92.

Casas, A., J. Caballero, C. Mapes, and S. Zárate. 1997b. Manejo de la vegetación, domesticación de plantas y origen de la agricultura en Mesoamérica. *Boletín de la Sociedad Botánica de México* 61:31–47.

Casas, A., A. Valiente-Banuet, and J. Caballero. 1998. La domesticación de *Stenocereus stellatus* (Pfeiffer) Riccobono (Cactaceae). *Boletín de la Sociedad Botánica de México* 62:129–40.

Casas, A., J. Caballero, and A. Valiente-Banuet. 1999a. Use, management and domestication of columnar cacti in south-central Mexico: a historical perspective. *Journal of Ethnobiology* 19:71–95.

Casas, A., J. Caballero, A. Valiente-Banuet, J. A. Soriano, and P. Dávila. 1999b. Morphological variation and the process of domestication of *Stenocereus stellatus* (Cactaceae) in central Mexico. *American Journal of Botany* 86:522–33.

Casas, A., A. Valiente-Banuet, A. Rojas-Martínez, and P. Dávila. 1999c. Reproductive biology and the process of domestication of *Stenocereus stellatus* (Cactaceae) in central Mexico. *American Journal of Botany* 86:534–42.

Dafni, A. 1992. *Pollination Ecology. A Practical Approach.* The Practical Approach series. Oxford: Oxford University Press.

Darwin, C. 1868. *The Variation of Plants and Animals under Domestication.* London: John Murray.

Dávila, P., J. L. Villaseñor, R. L. Medina, A. Ramírez, A. Salinas, J. Sánchez-Ken, and P. Tenorio. 1993. *Listados Florísticos de México. X. Flora del Valle de Tehuacán-Cuicatlán.* Mexico: Universidad Nacional Autónoma de México.

De la Cruz, M., and J. Badiano. 1964. *Libellus de Medicinalibus Indorum Herbis. Codex Barberini.* Mexico: Instituto Mexicano del Seguro Social.

Del Barco, M. 1988. *Historia Natural y Crónica de la Antigua California.* Mexico: Universidad Nacional Autónoma de México.

Flannery, K. V., ed. 1986. *Guilá Naquitz.* New York: Academic Press.

Gibson, A. C., and K. E. Horak. 1978. Systematic anatomy and phylogeny of Mexican columnar cacti. *Annals of the Missouri Botanical Garden* 65:999–1057.

Harlan, J. R. 1992. Origins and processes of domestication. In *Grass Evolution and Domestication,* ed. G. P. Chapman, 159–75. Cambridge: Cambridge University Press.

Hernández de Oviedo y Valdés, G. 1535. *Historia General y Natural de las Indias.* Madrid.

Hernández, F. 1959. *Historia Natural de Nueva España.* Seven volumes. Mexico: Universidad Nacional Autónoma de México.

MacNeish, R. S. 1967. A summary of the subsistence. In *The Prehistory of the Tehuacán Valley,* ed. D. S. Byers, 290–309. Austin: University of Texas Press.

Pimienta-Barrios, E., and P. S. Nobel. 1994. Pitaya (*Stenocereus* spp., Cactaceae): an ancient and modern fruit crop of Mexico. *Economic Botany* 48:76–83.

Proctor, M., P. Yeo, and A. Lack. 1996. *Pollination Biology.* Cambridge: Cambridge University Press.

Sahagún, B. 1970. *El Manuscrito 218–20 de la Colección Palatina de la Biblioteca Medica Laurenziana. Códice Florentino.* Mexico: Gobierno de la República Mexicana.

———. 1985. *Historia General de las Cosas de Nueva España.* Mexico: Porrúa.

Sahley, C.T., M. A. Horner, and T. H. Fleming. 1993. Flight speeds and mechanical power outputs of the nectar-feeding bat *Leptonycteris curasoae* (Phyllostomidae: Glossophaginae). *Journal of Mammalogy* 74:594–600.

Smith, C. E. 1967. Plant remains. In *The Prehistory of the Tehuacán Valley,* ed. S. Byers, 220–25. Austin: University of Texas Press.

———. 1986. Preceramic plant remains from Guilá Naquitz. In *Guilá Naquitz,* ed. K. V. Flannery, 265–74. New York: Academic Press.

Sneath, P.H.A., and R. R. Sokal. 1973. *Numerical Taxonomy. The Principles and Practice of Numerical Classification.* San Francisco: Freeman.

Valiente-Banuet, A., M. C. Arizmendi, A. Rojas-Martinez, and L. Dominguez-Canseco. 1996. Ecological relationships between columnar cacti and nectar-feeding bats in Mexico. *Journal of Tropical Ecology* 12:103–19.

Valiente-Banuet, A., A. Rojas-Martínez, M. C. Arizmendi, and P. Dávila. 1997a. Pollination biology of two columnar cacti (*Neobuxbaumia mezcalaensis* and *Neobuxbaumia macrocephala*) in the Tehuacán Valley, Central Mexico. *American Journal of Botany* 84:452–55.

Valiente-Banuet, A., A. Rojas-Martínez, A. Casas, M. C. Arizmendi, and P. Dávila. 1997b. Pollination biology of two winter-blooming giant columnar cacti in the Tehuacán Valley, Mexico. *Journal of Arid Environments* 37:331–42.

CHAPTER 8

Growth Form Variations in Columnar Cacti (Cactaceae: Pachycereeae) within and between North American Habitats

Martin L. Cody

Introduction

A habitat or vegetation type may be characterized by two relatively simple indices, namely, species diversity and growth form diversity. Although both indices have been employed widely by plant ecologists, are generally considered useful and informative, and contribute to a considerable literature, neither is well understood. Indeed, explaining why one or the other diversity index has a high or low value in a particular vegetation type remains one of the outstanding challenges of the field. Our understanding of plant species diversity and its variation has advanced little since Whittaker's (1977) synthesis (but for exceptions see Tilman [1988], Rosenzweig [1995], and various chapters in Ricklefs and Schluter [1993] that emphasize historical influences). Patterns in growth form diversity (GFD), the focus of this chapter, are even less well explored. To date, the chief tool in the description of GFD has been Raunkier's (1934) system based on perennating plant parts (see e.g., Crawley, 1986), but this is at best a coarse classification of growth form (GF); the category of "phanaerophyte," for example, might refer to deciduous, evergreen, or leafless trees, shrubs, stem succulents or epiphytes of any size or shape, regardless of a wide array of adaptive morphological features. Descriptive models of tree architecture (Tomlinson and Zimmerman, 1978) are one useful refinement of the Raunkier GF classification. A comprehensive review of plant growth functions is provided by Hunt (1982), and Nicklas (1994) provides a detailed treatment of plant scaling relationships among morphological components (e.g., leaves, stems, reproductive organs) and across taxa.

Givnish (1975) employed Raunkier's scheme to show that plant GFD increases with an index of climatic diversity that measures seasonal variation in monthly mean temperatures and in the diversity, month by month, of combinations of temperature and precipitation. Arid habitats, including deserts, rank high on this index and possess the greatest variety of plant growth forms. Cody (1989, 1991) attempted to account for variations in GFD among the North American deserts using climate indices and an expanded Raunkier classification with qualified success: around 75% of GFD variation could be explained by climatic factors. Within these deserts, columnar cacti are restricted to the Sonoran Desert (they are absent from Mohave, Great Basin, and Chihuahuan deserts) where, as leafless, stem-succulent trees, they contribute notably to GFD in the hotter desert at mid- to lower elevations and latitudes. GFD variation within the stem succulents, mostly cacti, responds particularly to increasing mean annual temperature ($R^2 = 0.86$; Cody 1991).

Farther south, columnar cacti recur in the arid interior desert valleys of central and southern Mexico (Michoacán to Oaxaca), where high temperatures and low precipitation approximate Sonoran Desert climates. Beyond desert habitats, columnar cactus diversity attenuates rapidly northwestwards into the Mediterranean climate zone (e.g., northwest Baja California), but high diversity is encountered upslope on moisture gradients from Mexico south into Central America. In northwest Mexico, this vegetation gradient is described as Sonoran desertscrub–Sinaloan thornscrub–tropical deciduous forest (Van Devender and Friedman, 1998) and corresponds generally to decreasing latitude, increasing elevation, and in particular, increasing annual precipitation (75–250 mm, 250–550, and 550–750 mm are approximate ranges for the three major vegetation categories). In tropical rainforest, cactus representatives are exclusively epiphytes, but many similar GF components featuring columnar cacti are repeated across the equator in comparable desert and dry forest habitats in South America and across the Atlantic Ocean in Africa by ecologically equivalent species in different families (notably Asclepiadaceae, Crassulaceae, and Euphorbiaceae).

In this chapter, I review a simple methodology for characterization of columnar cactus growth form via branching patterns and employ it to examine intra- and inter-specific variation in GF within and across xeric habitats in North America and the Caribbean. Throughout the chapter, a community perspective is maintained, emphasizing GF similarities and differences in species coexisting within habitats.

Descriptive Tools for Branching Patterns

Branching Pattern and Ontogeny

Relative to other perennial plants of comparable biomass, columnar cacti have a branching structure of near geometric simplicity. An extreme is found in species that consist of but a single stem or trunk (e.g., several *Cephalocereus* and *Neobuxbaumia* species), but most species are branched, with branch or stem number B increasing and then falling as a relatively smooth function of height above ground H. As the leafless photosynthetic stems ramify with, to a first approximation, little change in diameter, biomass and photosynthetic area are readily estimated.

Individual growth in columnar cacti involves two components: initiation of new branches and elongation of existing branches. Fig. 8.1a shows diagrammatically growth by elongation following branch initiation within a limited height interval; more commonly, branch initiation continues (Fig. 8.1b), at increased heights above the ground, as the cactus grows. Thus within a given plant height interval, branch number is enhanced with ontogeny by both the initiation of new branches within the height interval and by elongation into the interval of branches from below.

To provide indices of species- and site-specific branching patterns, individuals of a local population are assigned to size classes based on their total branch length $\int B \cdot dH$ (or $\Sigma B \cdot \Delta H$ over all height intervals ΔH); within size classes, mean branch numbers are plotted as a function of height above ground. The result is a family of curves in which, in the simplest case, size classes are congruent and related by changes in a single parameter. This is illustrated in Fig. 8.1c, which shows a family of two-parameter parabolas in which height at initial branching a is fixed, and overall plant height b increases with size class increments.

Four examples illustrate the methodology (Fig. 8.2): *Pachycereus weberi* (Teotitlan, Oaxaca), *P. pecten-aboriginum* (Sierra de la Laguna, Baja California Sur), *Harrisia gracilis* (Mitcheltown, Jamaica), and *Pilosocereus gaumeri* (Dzibilchaltun and Chicxulub, Yucatán). In the field, individual cacti are identified and branch numbers counted at a series of heights above the ground; from these counts, the total stem or branch length of individuals is computed. Individuals of each species are then assigned to size classes based on total stem length, with size class intervals covering 2 units on a $\log_{1.6}$ (i.e., Fibonacci) scale; sample sizes are given in the figure. The branching patterns of these four species (Fig. 8.2, left-hand side) differ in several respects, including overall size of plant (maximum size class represented is 7, 6, 4, and 5, respectively), height above ground at which branches begin (range, 0.3–1.3 m) and end (range, 3.5–14.2 m), peak number of branches (7–60), and the height at which branch number reaches a maximum (1–3.75 m).

Scaling and Curve Fitting

Both axes in Fig. 8.2 are logarithmically (base e) scaled such that branch numbers >1 ($\ln B > 0$) are shown in the diagrams. The form and symmetry of the branching pattern are clearly influenced by the scaling of the axes, particularly of the abscissa. A logarithmic scaling renders the diagrams reasonably symmetrical, and allows a (± symmetrical) parabolic curve to be fitted. Symmetry might be sharpened by different scaling on the abscissa, or alternatively, by nonsymmetrical curves (e.g., gamma functions tailed to the left might be used to parameterize the patterns). To the right in Fig. 8.2 are parabolas of the form $\ln B = -c(\ln H - a)(\ln H - b)$, where a, b, and c are fitted constants: a and b represent the range of heights for which $B > 1$, and c is a

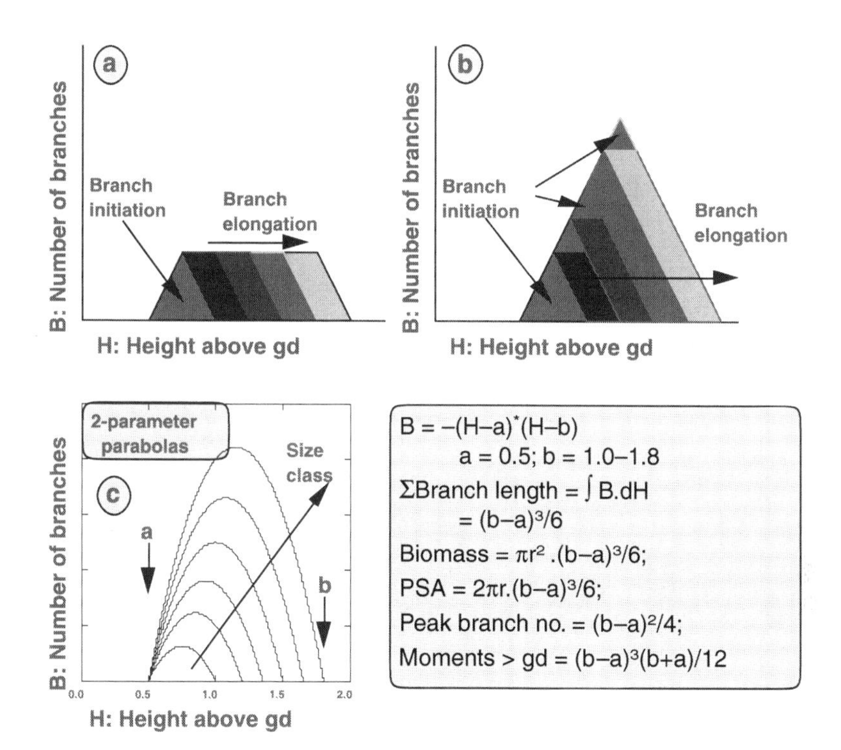

FIGURE 8.1. Diagrammatic representation of cactus growth in terms of branch number as a function of height above ground (both natural log scales). (a) Branch initiation occurs over a limited height range, hence growth is solely by branch elongation. (b) Branch initiation contributes continually to growth and supplements branch elongation. (c) Fitted parabolic curves summarize the branching pattern in terms of three parameters, *a* (fixed) at first branch height (0.5) and *b* (variable) increases with different size classes. Various summary statistics can be derived from the fitted representation.

multiplier that increases the elevation of the curve over the abscissa. In these figures, adjusted R^2 averages 0.82 for the largest four size classes of the four species.

Another feature of the branching patterns is the degree to which curves of smaller-size classes are encompassed wholly within those of larger-size classes, as in the *Pachycereus* spp., or extend above the outer curves to the left. The latter (noncongruent) curves indicate that smaller individuals are more branched at lower heights than are larger individuals (see *Harrisia*, and especially *Pilosocereus gaumeri*). A reasonable explanation for noncongruence is that small individuals lose lower branches as they grow, but an alternative explanation—that individuals reaching larger sizes are those that begin branching higher off the ground—is also plausible.

Derivative Measures

An advantage of fitting curves to branching patterns is that it allows derivation of simple variables that characterize the plant's branching morphology. Thus in Fig. 8.2, first and last branch heights (*a, b*) are given from the fitted curve of the largest-size class of

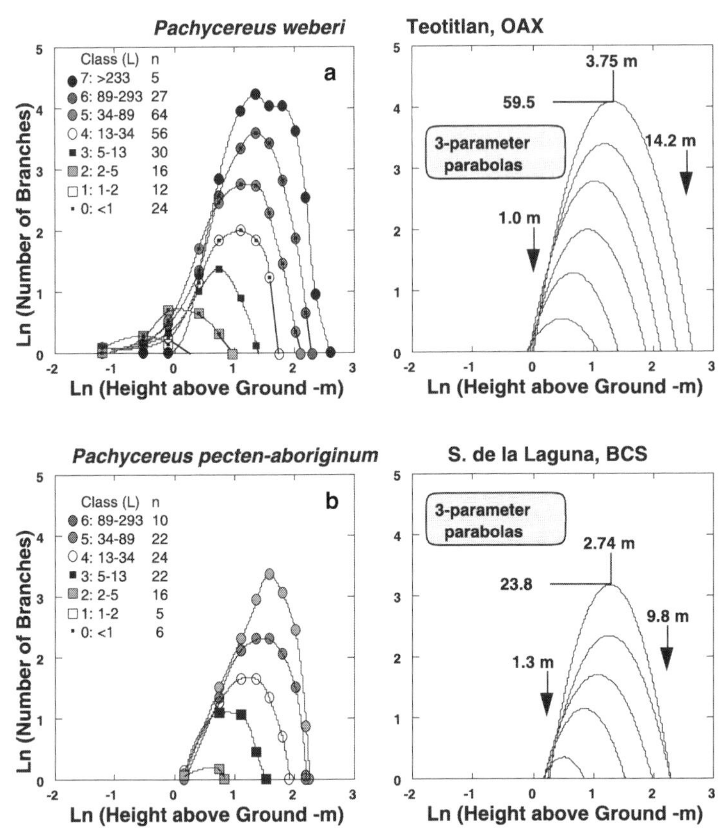

FIGURE 8.2. Raw data (left) and fitted curves (right) for four species of columnar cacti. (a) *Pachycereus weberi* population is divided into size classes 0–7 based on total stem length (L) as indicated, with sample sizes (percentage of individuals per size class). Fitted curves are three-parameter parabolas, but note that height where branch number B declines to one (14.2 m), and the number of branches (59.5) and height above ground (3.75 m) at peak branching in the largest size class (class 7). (b) Similar data and representation for congeneric *P. pecten-aboriginum* in Baja California Sur, (c) *Harrisia gracilis* in Jamaica, and (d) *Pilosocereus gaumeri* in Yucatan.

the species, as are height at peak branching $H = \exp[(b - a)/2]$ and peak branch number $N = \exp[c(b - a)^2/4]$. These are useful summary statistics that can be derived from fitted curves or directly from the branching diagrams; other derived measures are shown in Fig. 8.1. Derived measures involving integration, such as total branch length, are more conveniently obtained from linear than from log-scaled axes.

Form Functionals

Comparative summaries of branching patterns that reflect the ontogeny of branching are conveniently depicted by growth curves that join the mean peak branch number and mean height at peak branching for successive size classes. Such growth curves

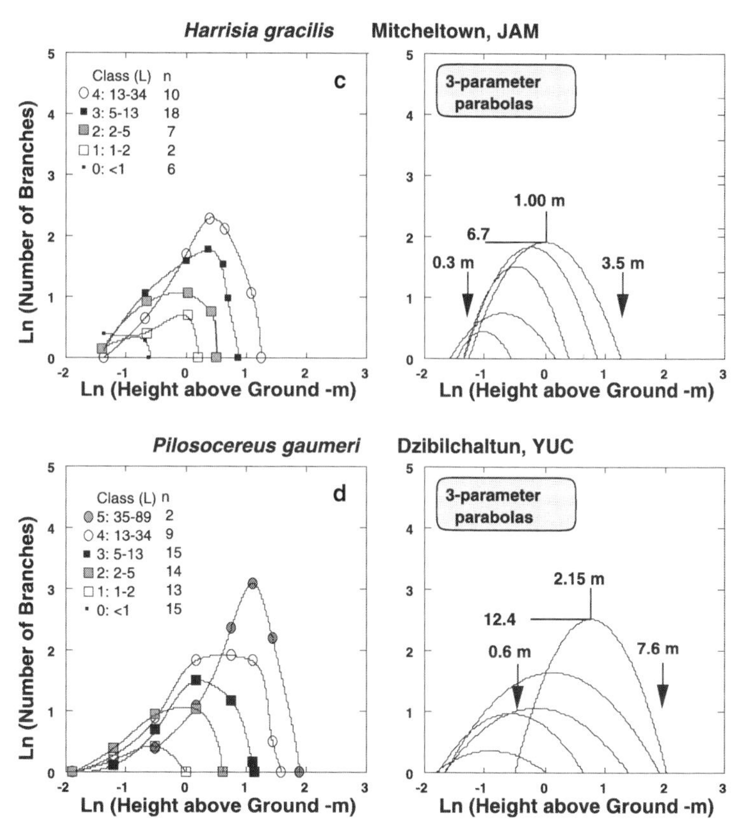

FIGURE 8.2. *Continued*

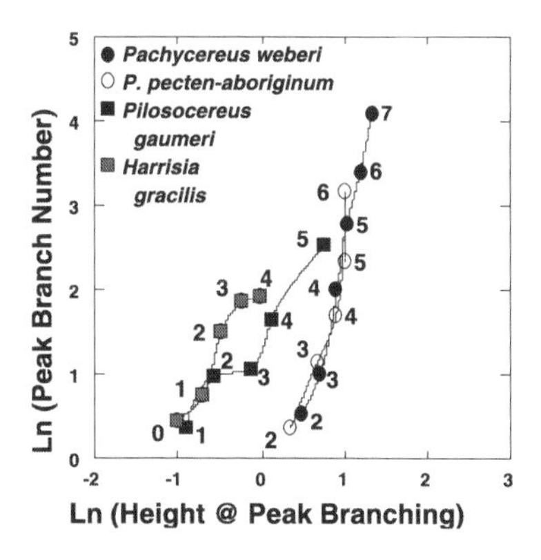

FIGURE 8.3. Summary representation of branching diagrams in terms of form functionals, where peak numbers over height for successive size classes are connected in each of the four species of Fig. 8.2. Note that the two *Pachycereus* species are very similar (although *P. weberi* reaches a greater size [class 7] than does *P. pecten-aboriginum* [6]), and the remaining two species differ from each other and from the *Pachycereus* species, although converging somewhat at the upper size classes.

are shown in Fig. 8.3 for the four species of Fig. 8.2. The figure illustrates the strong similarity between the two species of *Pachycereus* and the dissimilar growth, with lower or earlier branching, in *Harrisia* and *Pilosocereus gaumeri*. Note, however, that in general, larger individuals become relatively more similar among species, as indicated by the proximity of size class 5 in three species that attain that size.

Columnar Cacti in the Sonoran Desert

Species Segregation by Branching Pattern

About eight species of columnar cacti occur in the Sonoran Desert, but local species richness seldom exceeds four species, with the additional taxa accounted for by ecological equivalents across the Gulf of California (e.g., *Stenocereus alamosensis* for *S. gummosus*, *Carnegiea gigantea* in part for *Pachycereus pringlei*), species marginally present (*Bergerocactus emoryi* in northwestern Baja California), or whose occurrence in the desert is sporadic (e.g., *Myrtillocactus cochal*). Branching patterns in communities across the Sonoran Desert were reported in Cody (1984). A representative four-species community for Baja California is that near La Paz (Fig. 8.4a), and a comparable community on mainland Mexico occurs near Puerto Libertad, Sonora (Fig. 8.4b). Several different four-species combinations may be found at different sites within the desert, but a conspicuous feature of all sites is the widely divergent branching patterns of the coexisting species.

Specific branching patterns covary with a range of other morphological and ecological attributes of the cacti involved and are attributable to the evolution of species into different structural niches (Cody, 1985). In brief, species are segregated along a morphological/ecological gradient from those of greater stem diameter (which are taller, less branched, and shallow rooted) to narrow-stemmed species (which are shorter, more branched, and deep rooted). Because photosynthetic area per unit volume (PSA/V) scales inversely as stem diameter (d^{-1}), only narrow-stemmed species with higher PSA/V succeed in higher rainfall areas of low, coastal scrub and chaparral (in the Mediterranean climate region of northwestern Baja California). In these areas, water storage capacity is of reduced advantage and wide-stemmed species with shallow roots are at an advantage on steep, rocky slopes, where precipitation runoff is rapid and water storage between precipitation events is at a premium (Cody, 1984, 1985).

Niche Shifts with Species Composition and Habitat Change

With changes in local species composition, the branching patterns of individual species also change such that interspecific segregation is maintained. Thus in the presence of the taller, less- and higher-branched *Carnegiea*, *Pachycereus* at Puerto Libertad is shorter, more- and lower-branched than where *Carnegiea* is absent (e.g., at La Paz). *Lophocereus schottii* is the shortest species at Puerto Libertad where it is lowest- and most-branched, but *Stenocereus gummosus* fills that role at La Paz, where *L.*

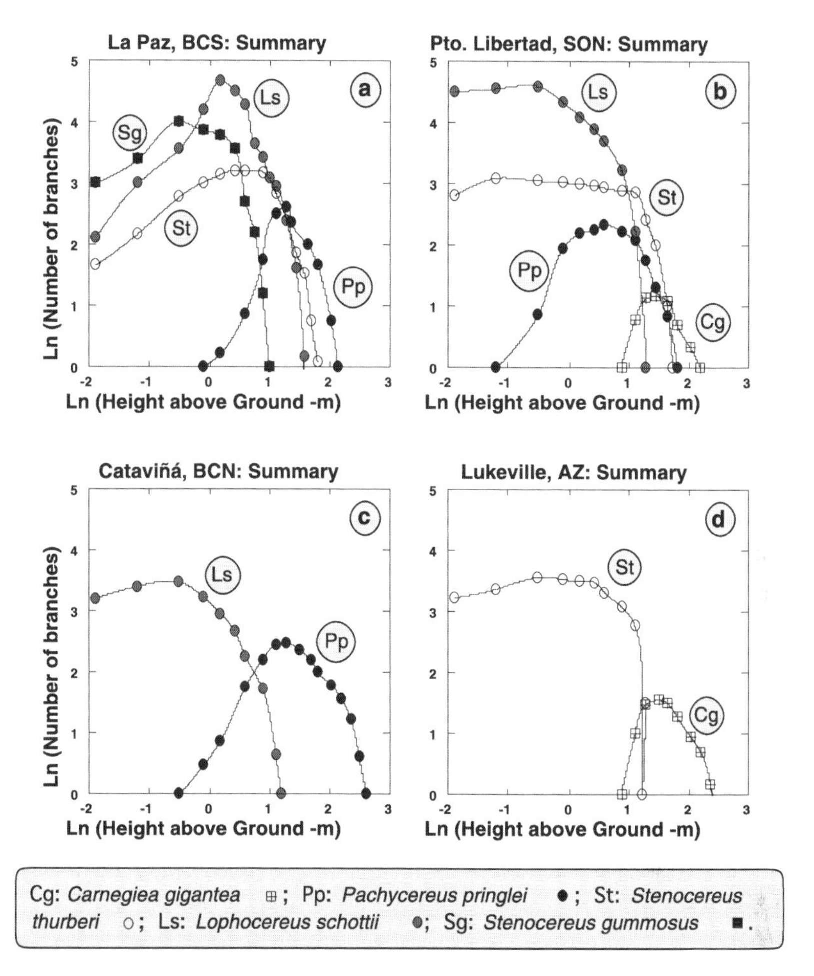

FIGURE 8.4. Summary curves of branching patterns at four Sonoran Desert sites, with four coexisting species (upper) and two species (lower). Note widely divergent branching patterns in coexisting species at a site, and variation in pattern within species among sites.

schottii is taller and much less branched at low heights (Fig. 8.4a,b). In two-species communities such as those at Cataviñá, Baja California Norte and Lukeville, Arizona (Fig. 8.4c,d), the larger species at each site are taller, or more branched, or branched lower (i.e., they have expanded branching niches) than where they co-occur with several other species. Although the two lower- and most-branched species at these two sites are different species, their branching patterns are nearly identical.

Changes in branching pattern in the widespread *Pachycereus pringlei* across the Sonoran Desert (Baja California and Sonora) are shown in Fig. 8.5. For ten desert sites on the peninsular and landbridge islands (those with peninsular connections during the Pleistocene), peak branch number and height at peak branching are enclosed by

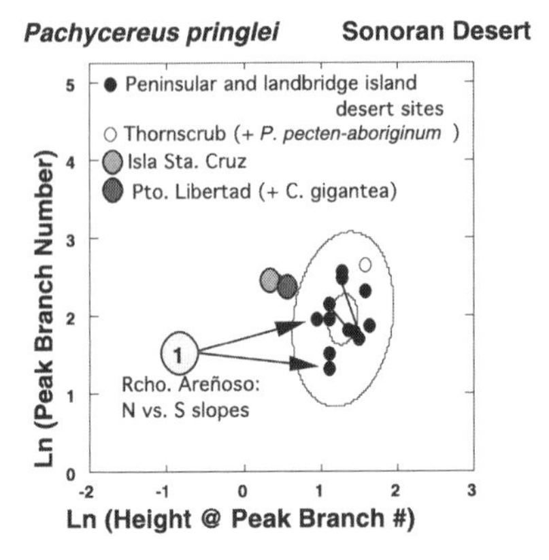

FIGURE 8.5. *Pachycereus pringlei* is widely distributed throughout the Sonoran Desert; its branching pattern is here represented by peak branch number in the highest size class and the height at which the peak occurs. In 11 peninsular desert sites (solid dots), variation is described by the ellipses (outer: 95% confidence ellipse; inner: 95% centroid). Two pairs of adjacent sites are shown within this distribution where site conditions (substrate, slope, and aspect) differ between the paired sites and result in differences in branching pattern (but still within the 95% ellipse). A thornscrub site is also within the ellipse, but two other sites are not: where *P. pringlei* is the only species present and where *P. pringlei* is not the tallest and least-branched species in the assemblage.

95% confidence ellipses around the mean (inner ellipse in Fig. 8.5) and the distribution (outer ellipse). All data, except for an additional site near La Paz are derived from Cody (1984). Some variation is attributable to local changes in habitat; the two pairs of points in the figure joined by lines represent shifts in branching pattern on Isla San José between sandy flats and rocky slopes and at La Paz between sandy bajada and rocky bench habitats, respectively. Another pair of data points is indicated by arrows in the figure: at Rancho Areñoso, Baja California Norte, near the northern limit of the peninsula range of *P. pringlei,* branching patterns change substantially between north- and south-facing slopes, with smaller branch numbers on north-facing slopes, where the species is likely limited by radiation and temperature (Nobel, 1980; Cody, 1984).

South of La Paz, at Rancho Los Divisadores, precipitation increases and thornscrub plus deciduous forest components, including *Pachycereus pecten-aboriginum,* are added to the vegetation. There *Pachycereus pringlei* has the highest recorded peak branch number (Fig. 8.5). But the most dramatic shifts in branching patterns may be seen in desert sites, where species composition differs from the usual two-to-four species set. At Puerto Libertad, *Carnegiea gigantea* is the tallest and least-branched of the four-species set; on smaller, rocky islands in the Gulf of California (e.g., Isla Santa Cruz), *P. pringlei* occurs essentially alone. In both situations, peak branching in *P. pringlei* occurs much lower on the plant, significantly beyond the usual desert pattern (i.e., the outer 95% confidence ellipse of Fig. 8.5). These shifts, corresponding to a change in species sequence and a much-reduced species number, respectively, considerably exceed changes in branching pattern that correspond with habitat shifts (e.g., from sandy flats to rocky slopes, Fig. 8.5; see also Cody et al., 1983).

Beyond these observations, which may be regarded as indicative of broader, first-order generalities, definitive studies on the determinants of branching pattern have yet to be conducted. The role of genetic vs. environmental controls (and among the latter, the influence of edaphic and other abiotic factors) and biotic influences (such as coexisting species) have not been resolved with precision, but Cornejo and Simpson (1997) give a detailed analysis of the effects of varying climatic factors on branching patterns in 26 taxa of North American columnar cacti. They emphasize the contribution of phylogenetic constraints to the conservation of these patterns.

Columnar Cacti in Thornscrub and Tropical Deciduous Forest

Moving up the Moisture Gradient

With increasing precipitation and a higher incidence of summer rainfall, Sonoran Desert habitats in western Mexico are replaced by thornscrub and eventually by dry woodlands and tropical deciduous forests that become taller with less open canopies and more evergreen trees (Gentry, 1942; Rzedowski and McVaugh, 1966; Rzedowski, 1978; Murphy and Lugo, 1986; Van Devender and Friedman, 1998). Similar habitats occur on the northern Yucatán Peninsula along a moisture gradient increasing from north to south, and in low, south-coastal sites on Caribbean islands. The lower and scrubbier of these habitats generally support two species of columnar cactus, although nonarborescent forms (e.g. scandent *Acanthocereus* and *Nyctocereus* spp.), shrubby *Opuntia* spp., and other cacti (e.g., *Ferocactus, Mammillaria*) may also be present. In taller woodlands to the south (e.g., south-central Mexico), arborescent platyopuntias (e.g., *Opuntia excelsa*) reach canopy height, resembling the treelike platyopuntias of the Galapagos Islands.

Thornscrub from Western Mexico to Jamaica

In tall thornscrub to scrubby woodland sites in southern Baja California and eastern Sonora, the two representative columnar cacti are *Pachycereus pecten-aboriginum* and *Stenocereus thurberi.* Summary branching patterns are shown in Fig. 8.6. Unlike conspecifics (*S. thurberi*) and congenerics (*P. pringlei*) in adjacent desert, the two species show convergently similar patterns in branching morphology, although *S. thurberi* is more branched at lower heights above the ground. The largest cacti are emergent from the vegetation canopy, which is composed mainly of leguminous trees (e.g., *Lysiloma, Acacia, Erythrina* spp.) along with *Bursera* spp. The bulk of the largest cacti's PSA lies within the canopy, at heights at or slightly above those at which canopy density is highest. Dashed lines in Fig. 8.6 show vegetation density as a function of height above ground, where vegetation density (ordinate at right) is measured as the reciprocal of the horizontal distance at which the vegetation reduces vision to half-obscurity (see MacArthur and MacArthur [1961] for methodology). Similar patterns are produced from sites in the northern (coastal) Yucatán, at Dzibilchaltun, and on the

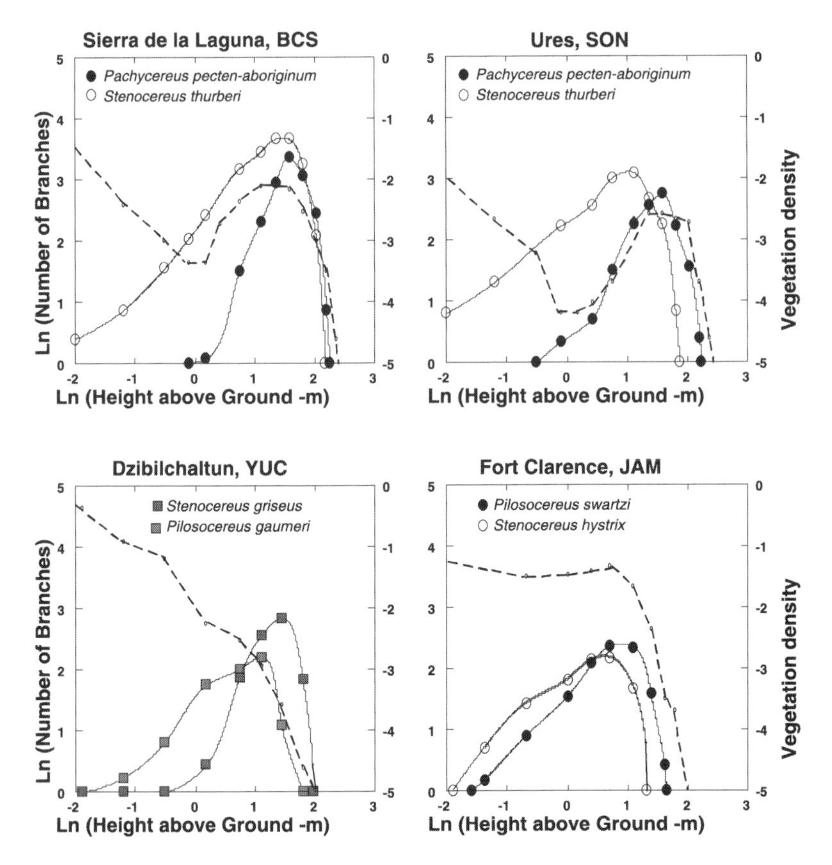

FIGURE 8.6. Branching patterns at four thornscrub sites in different regions, each with two coexisting species of columnar cacti. The vegetation profile at teach site is represented by the dashed line, with ordinate at right. Units are ln (distance to semiobscurity)–1, with distance measured in feet.

south coast of Jamaica at Fort Clarence. At the Yucatán site, the two species of columnar cactus are *S. griseus* and *Pilosocereus gaumeri;* in Jamaica, *S. hystrix* and *Pilosocereus swartzi.* The vegetation is lower and denser at the flat Yucatán site, and taller in Jamaica, where the site is located on a steep south-facing slope that admits more radiation into the vegetation and where branching in the cacti begins somewhat lower to the ground.

Variation in branching patterns with the height of the vegetation in which the cacti grow is explored further with *Stenocereus hystrix* in Jamaica. At each of two locations, the aforementioned Fort Clarence on a rocky hillside and at the Salt River (a flat, estuarine site), three sites of varying vegetation density were selected, and branching patterns relative to the vegetation profile were measured. Results are shown in Fig. 8.7. Within each site, vegetation height and density increase from left to right, with vegetation taller at Fort Clarence (upper graphs) than Salt River (lower); dashed lines in the figure show the decline in vegetation density with increasing

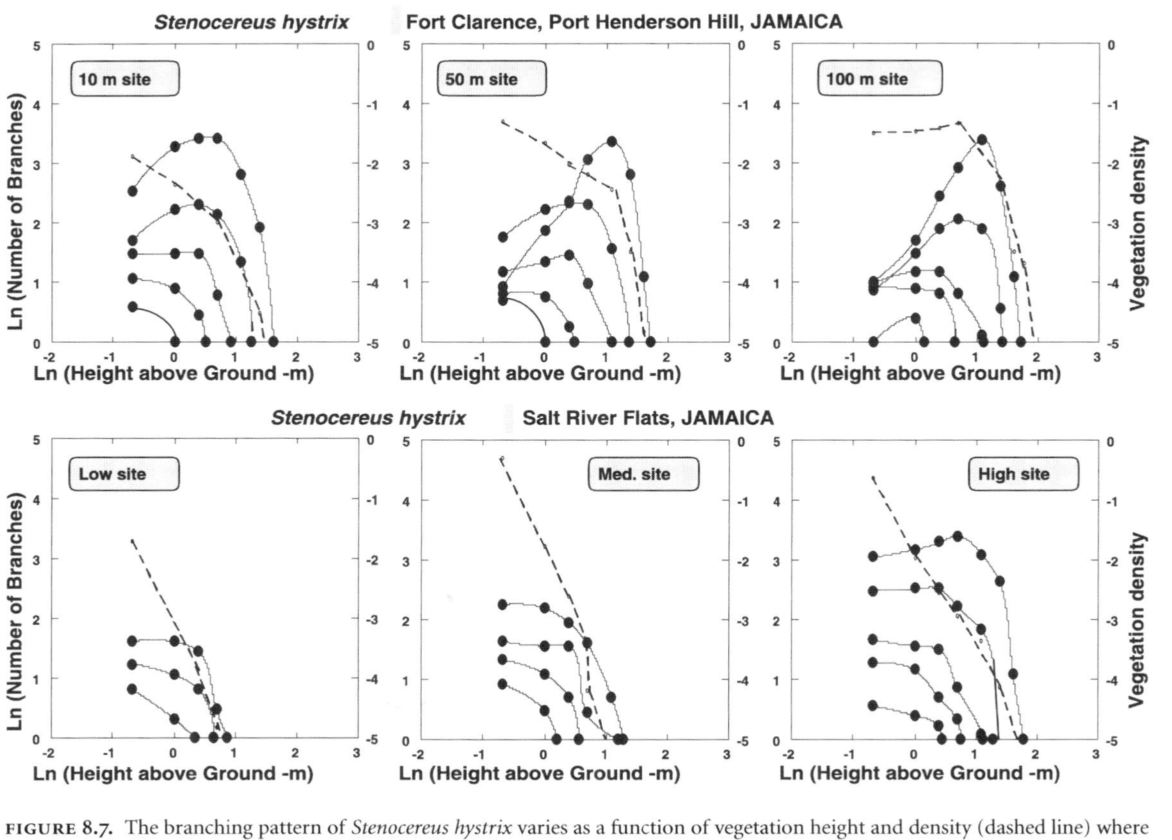

FIGURE 8.7. The branching pattern of *Stenocereus hystrix* varies as a function of vegetation height and density (dashed line) where it grows. Branching occurs earlier and lower on the plant where the vegetation is low and open and is delayed (occurs higher) in taller, denser vegetation.

height above ground, reaching canopy top at about 2 m to about 8 m. At Fort Clarence, the 10-m site (referring to elevation above coastline) represents low and open vegetation, where *S. hystrix* is extensively branched at much lower heights than it is in the taller, denser vegetation at the 100-m site. Vegetation is overall lower and more open at Salt River, where *S. hystrix* is again well ramified near the ground. Thus in taller, denser vegetation, the cactus grows taller and branches are initiated higher off the ground.

Tropical Deciduous Forest, Chamela

The dry, deciduous forest around the Universidad Nacional Autónoma de México's Estacion Biológica at Chamela, Jalisco, is taller and denser than at sites previously discussed. Columnar cacti there are represented by two tall species (*Pachycereus pecten-aboriginum* and *Stenocereus queretaroensis*), which are canopy emergents, and two additional species (*Pilosocereus purpusii* and *S. standleyi*) that rarely reach canopy height (Cody, 1986). Their branching patterns are shown in Fig. 8.8a–d in the standard format, showing ontogeny over increasing size classes; summary curves (Fig. 8.8e) and statistics are also given. The two larger canopy species both reach size class 7, at which their mean stem lengths L exceed 300 m; they are very similar to each other in branching pattern, overall size, stem diameter d, and branch number B at peak branching and the height H at which the peak occurs (Fig. 8.8). The subcanopy species are smaller in biomass and stem diameter, but relatively similar to each other in peak branching height, although *S. standleyi* is the more branched of the two. The subcanopy species differ additionally in habitat and growth form, with *S. standleyi* typically situated on ridgetops, with wide-spreading branches, and *Pilosocereus purpusii* on the sides and at the bottoms of arroyos, with a narrow and erect growth form.

Topographical influences, especially slope and aspect, on branching pattern in these cacti were discussed in Cody (1986). Topography influences the height and density (as well as composition) of the forest canopy, and this in turn is related to the incidence and extent of branching in the cacti. Based on a subset of the *Pachycereus pecten-aboriginum* data—individuals with total stem length of 45–120 m ($n = 21$)—the height above ground at which a second branch is produced from the initially unbranched young stem increases with canopy height (Fig.8.9a). Likewise the third, fifth, and seventh branches are produced higher under taller canopies (Fig. 8.9b–d), but the influence of the canopy on subsequent branching declines (r^2 and slope b both reduced); by the production of the tenth branch (Fig. 8.9e), there is no significant effect of canopy height. At the top of the plant, the height by which branch number declines to five is also related to canopy height (higher in taller canopies; Fig. 8.9f), but the effect is not as pronounced as that on branch initiation lower on the plant. In these deciduous forests, branching patterns are clearly a reflection of the vegetation at individual growth sites, with variation environmentally rather than genet-

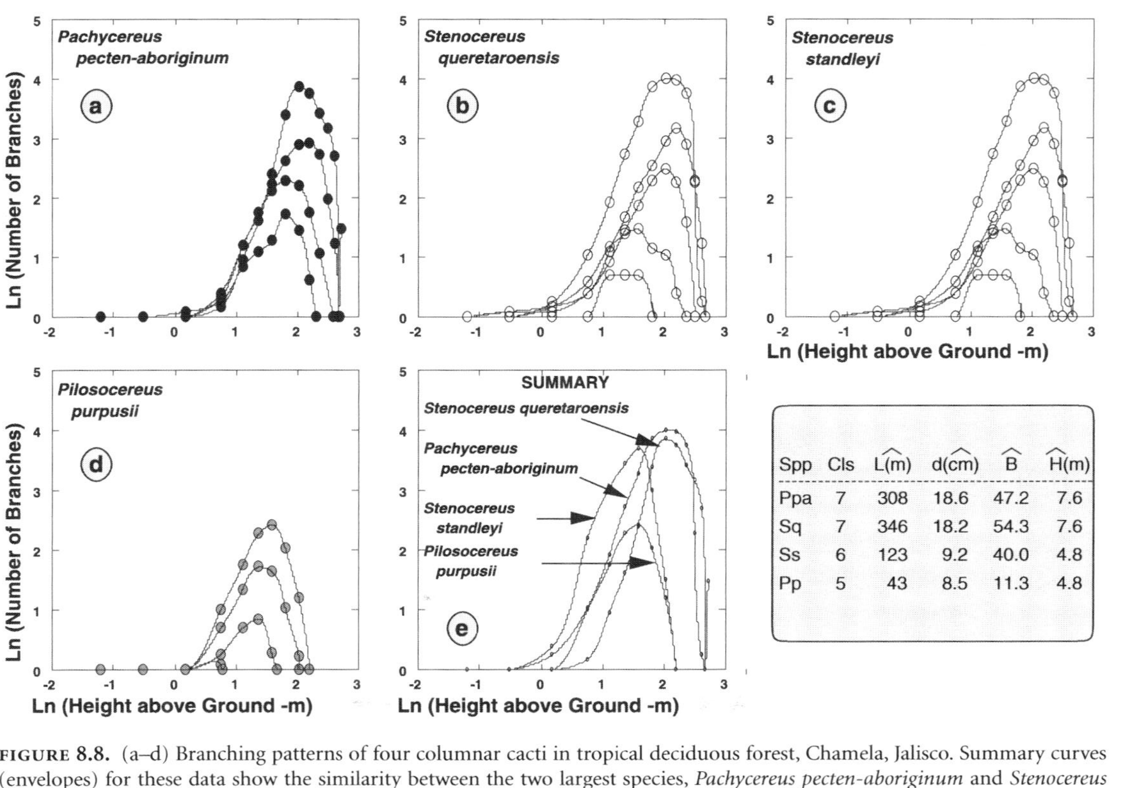

Spp	Cls	$\widehat{L}$(m)	$\widehat{d}$(cm)	$\widehat{B}$	$\widehat{H}$(m)
Ppa	7	308	18.6	47.2	7.6
Sq	7	346	18.2	54.3	7.6
Ss	6	123	9.2	40.0	4.8
Pp	5	43	8.5	11.3	4.8

FIGURE 8.8. (a–d) Branching patterns of four columnar cacti in tropical deciduous forest, Chamela, Jalisco. Summary curves (envelopes) for these data show the similarity between the two largest species, *Pachycereus pecten-aboriginum* and *Stenocereus queretaroensis.* Two shorter, subcanopy species differ from each other and from the canopy emergents. Summary statistics are shown in the lower right, where *L* represents the mean total stem length of the largest size class, *d* the mean stem diameter, *B* the peak number of branches, and *H* the height at which peak branch number occurs.

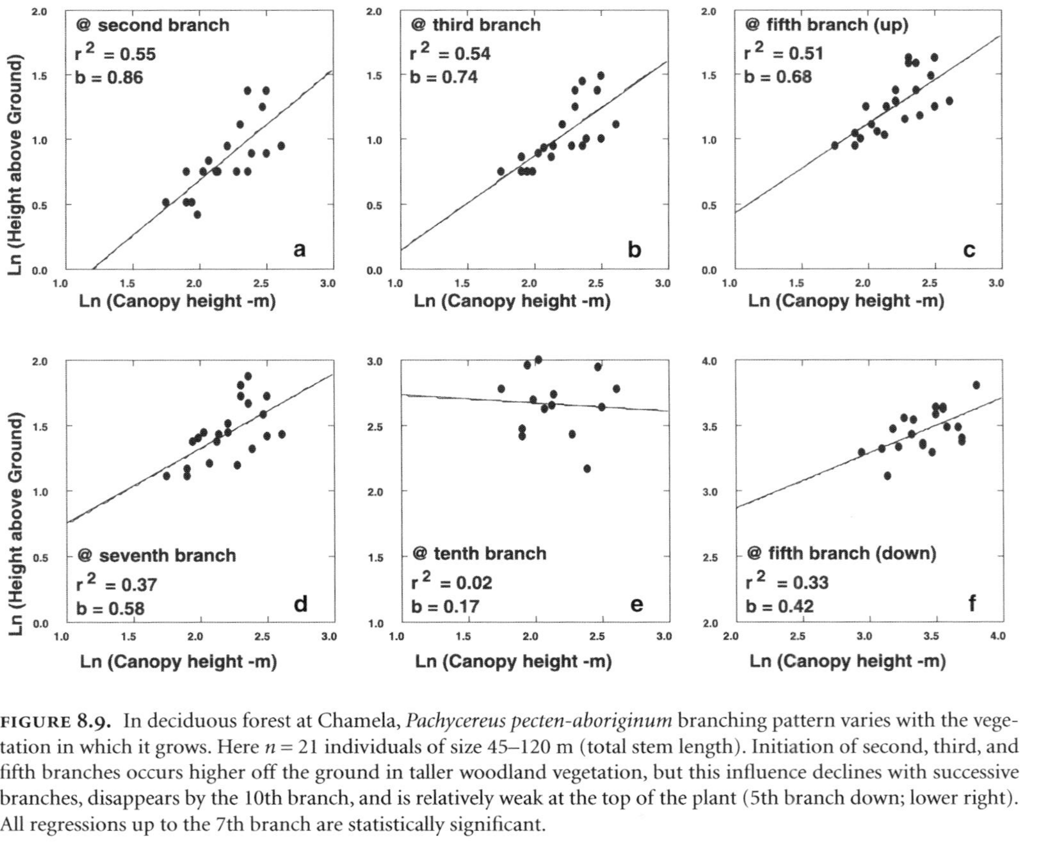

FIGURE 8.9. In deciduous forest at Chamela, *Pachycereus pecten-aboriginum* branching pattern varies with the vegetation in which it grows. Here $n = 21$ individuals of size 45–120 m (total stem length). Initiation of second, third, and fifth branches occurs higher off the ground in taller woodland vegetation, but this influence declines with successive branches, disappears by the 10th branch, and is relatively weak at the top of the plant (5th branch down; lower right). All regressions up to the 7th branch are statistically significant.

ically determined. Such response plasticity is presumably adaptive to the specific conditions of the growth site, perhaps to the PAR (photosynthetically active radiation) gradient within the vegetation.

Columnar Cacti in the Teotitlan Region, Oaxaca

From open semiarid desert scrub habitats at around 500 m elevation some 80 km south of Teotitlan, a striking habitat gradient extends south as Highway 131 gains elevation to the watershed at 1,440 m. A dense oak (*Quercus* spp.) woodland covers the hillsides above this point, but between 500 and 1,400 m, the scrubby desert passes into increasingly taller and denser tropical deciduous woodland or forest (in which *Acacia, Bursera,* and *Erythrina* are prominent). The vegetation again decreases in stature as a chaparral-like scrub replaces the woodland toward the 1,400-m watershed. As many as a dozen columnar cactus species are seen along this 70-km gradient (as well as a number of other noncolumnar species). Some are relatively restricted to certain habitats and elevations (e.g., *Neobuxbaumia tetetzo* to steep rocky slopes at lower elevations; *Mitrocereus fulviceps* to the higher reaches, where it is emergent from the chaparral), but several others (e.g., *Escontria chiotilla, Myrtillocactus geometrizans, Pachycereus weberi*) are more generally distributed and span a range of habitats from scrubby desert to open woodland and deciduous forest. A series of eight sites was established along this gradient, at each of which species composition and branching morphologies were recorded (see Table 8.1 for site locations). The aforementioned widely-distributed taxa plus *Stenocereus stellatus* are common at sites 1–5 and *Pilosocereus collinsii* is common on the upper half of the gradient; additional species of *Pachycereus* and *Stenocereus* are more sporadic in occurrence.

Five species are common in the taller forest (e.g., sites 5 and 6) and their branching patterns and a summary diagram are shown in Fig. 8.10. The three tallest species are canopy emergents: *Pachycereus weberi, Escontrilla chiotilla, and Myrtillocactus geometrizans.* All reach size class 7, in which mean total stem lengths L are 320, 307, and 286 m, respectively. The remaining two species (*Pilosocereus collinsii* and *Stenocereus*

TABLE 8.1
Data Collection Sites along a Habitat Gradient in the Teotitlan Region, Oaxaca

Site Number	*Elevation (m)*	*Distance along Gradient*[a] *(km)*
1	550	86
2	530	91
3	590	102
4	540	113
5	650	135
6	760	140
7	1,225	146
8	1,400	155

[a]Distances as measured by milestones along Highway 131.

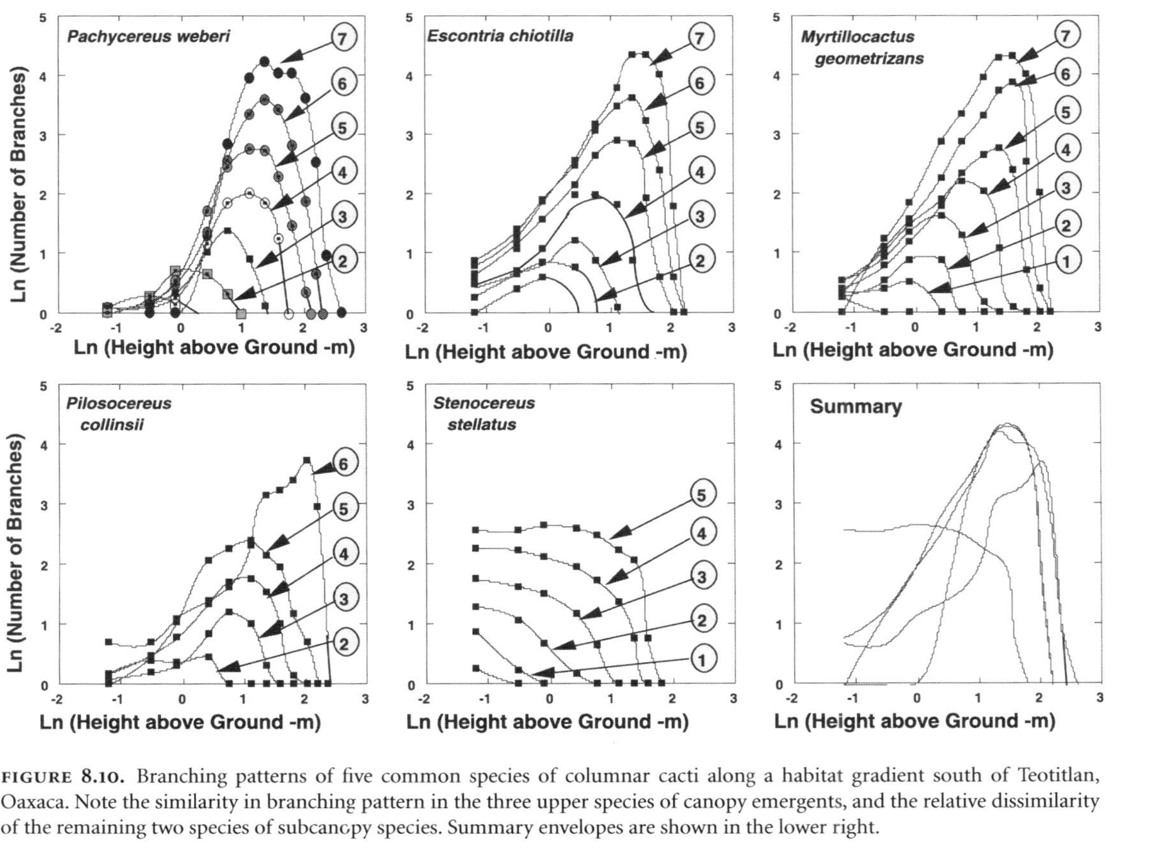

FIGURE 8.10. Branching patterns of five common species of columnar cacti along a habitat gradient south of Teotitlan, Oaxaca. Note the similarity in branching pattern in the three upper species of canopy emergents, and the relative dissimilarity of the remaining two species of subcanopy species. Summary envelopes are shown in the lower right.

stellatus) are both subcanopy species, although the former frequently extends into the woodland canopy; *P. collinsii* reaches size class 6, $L = 188$ m total stem length, and *Stenocereus stellatus* attains size class 5, $L = 45$ m total stem length. These two species are strongly parallel in growth form to the two subcanopy congenerics at Chamela, *Pilosocereus purpusii* and *S. standleyi,* with both *Pilosocereus* species narrowly erect and both *Stenocereus* species more broadly spreading beneath the canopy. Outside of the forest habitats, *Neobuxbaumia tetetzo* and *Mitrocereus fulviceps* are tall, but modestly branched, reaching size class 4, $L = 20$ m and class 6, $L = 143$ m, respectively.

The three canopy emergents are very similar in overall branching patterns (Fig. 8.10), and it seems appropriate to consider them to be convergently similar, as their phylogenetic positions within the tribe are distinct (Cota and Wallace, [1997]; Cornejo and Simpson [1997] place *Pachycereus* in subtribe Pachycereinae and *Escontria* and *Myrtillocactus* in subtribe Stenocereinae of tribe Pachycereeae). I interpret their similarities in branching patterns as common responses to growth in the closed canopy woodland they share; the largest *Pilosocereus* are also similar in branching pattern, with similar peak branch height and number. However, the branching patterns of the smaller individuals of all species appear more divergent among species, and thus similarities among the larger plants become more apparent with ontogeny. Fig. 8.11 attempts to show this convergence over branching ontogeny, with 75% confidence ellipses around successive size classes 1–7. Note that sample size (i.e., number of species) is reduced at the larger-size classes, an effect that would otherwise expand those confidence ellipses; however, it is the larger-size classes that are closely similar, and divergence occurs among species in the smaller-size classes. Because the ellipses are drawn and compared on log-scaled axes, in line with the universal appli-

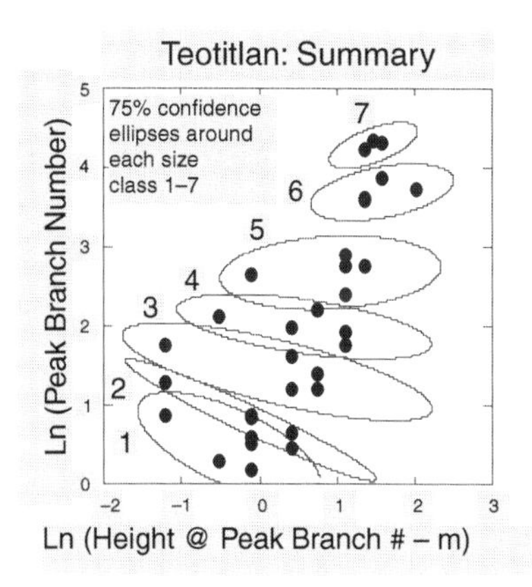

FIGURE 8.11. On the Teotitlán habitat gradient, the largest size classes of the five most common species (see Fig. 8.10) tend to converge in peak branch number and height at peak branch number, whereas successively smaller size classes (with logarithmic scaling) are conspicuously less similar among species. Fewer species contribute to the 75% confidence ellipses of the largest size classes, yet their areas are the smallest.

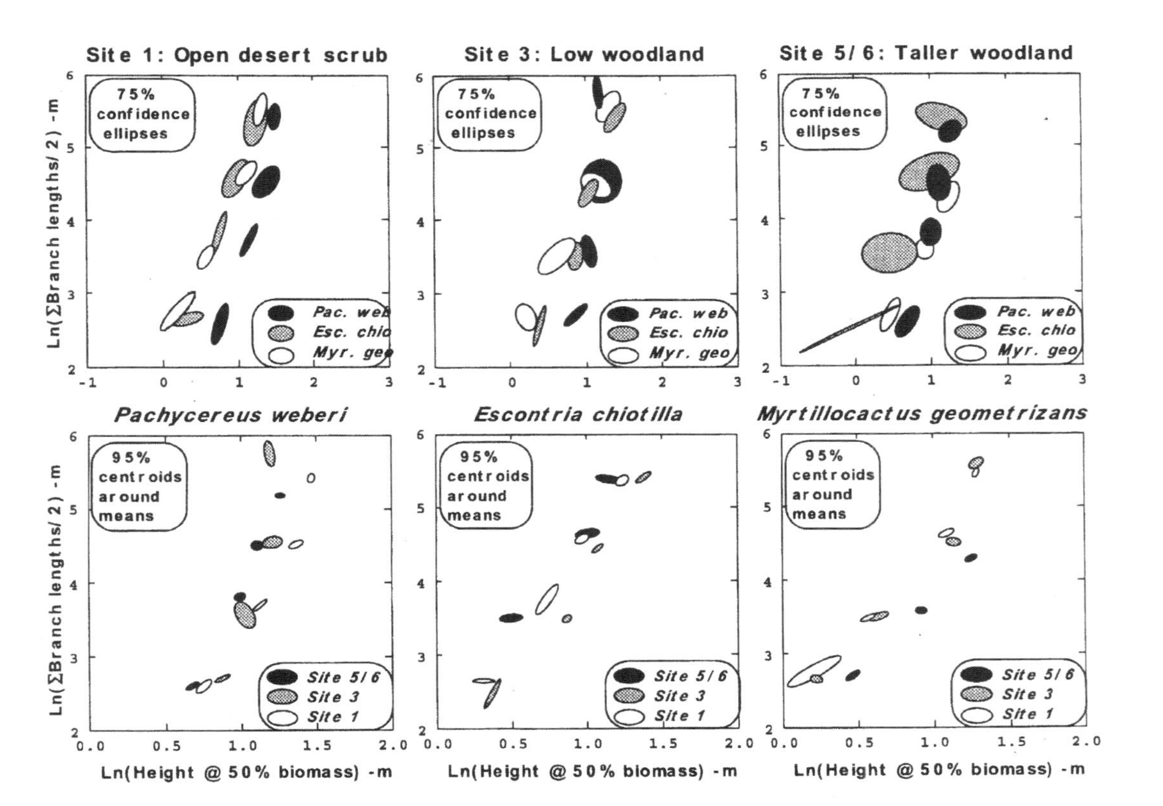

FIGURE 8.12. Upper row: for size classes 3, 4, 5, and 6/7 of the three most common species present at each of three sites (1, 3, 5/6) on the Teotitlán habitat gradient, 75% confidence ellipses enclose the distributions of half-height (abscissa: height where 50% biomass is above the 50% below) versus half-biomass (ordinate: half the total stem length of the individual cactus). The branching patterns represented by these distributions tend to be broader (more variable) in taller woodland, significantly so between the open desert scrub of site 1 and the tall woodland of sites 5/6. Lower row: 95% centroids, confidence limits around the means are shown for each of the three most common species as a function of sites along the gradient. In general, all species show significant differences in branching patterns among sites (as shown by nonoverlapping centroids), but the shifts vary in direction with individual species.

cation of such scaling to size metrics (from organs to populations), evidence for GF divergence among smaller individuals on the species set remains tentative.

Differences in branching pattern, within and among species, are expected with changes in vegetation among sites along the habitat gradient. To investigate such differences, I computed for each individual the "half-height": the height at which half the total stem length or biomass lies above and half below. Then for each size class of each species, a joint distribution of half-height and "half-biomass" (i.e., total stem length/2) was obtained, and variation displayed by the 75% confidence ellipse of this distribution. Results are shown in Fig. 8.12, upper row, for sites 1, 3, and 5/6, representing progression along the habitat gradient. I use the three most common species in this analysis (*Pachycereus weberi, Escontria chiotilla,* and *Myrtillocactus geometrizans*), as their sample sizes (i.e., per species per size group) are most uniform across species and sites, with uniformity enhanced by combining the two largest-size classes 6 and 7. By species as listed above, mean sample sizes (per size group) over sites are 8.33 ± 4.54 standard deviation (SD), 9.00 ± 5.32 SD, and 6.64 ± 2.54 SD, respectively; by sites ordered 1, 3, and 5/6, mean sample sizes over species are 7.08 ± 4.32 SD, 6.41 ± 2.94 SD, and 10.81 ± 4.57 SD, respectively. Inspection of Fig. 8.12 shows a tendency in taller forest sites for a greater variation within species and increased overlap in distributions among species, whereas in lower, open vegetation (site 1) there appears to be reduced variation among species, and more distinction among them. I quantified this by measuring mean areas of the 75% confidence ellipses at each site: these are (in the ln units common to both axes) at site 1, 0.074 ± 0.039 SD; site 3, 0.093 ± 0.063 SD; and site 5/6, 0.1351 ± 0.099. By *t*-test, more closely adjacent sites do not differ significantly, but ellipses at site 1 are significantly smaller that those at site 5/6 ($t = 1.872$, $df = 21$, $p < 0.05$).

To examine variation in interspecific overlap among sites, I computed the ratio of planar area occupied by ellipses to the total area of all species' ellipses, where the numerator is reduced over the denominator by interspecific overlaps. These figures give 8.7% overlap at site 1, and 15.2% and 12.0% at woodland sites 3 and 5/6, respectively. These analyses may indicate that there is greater variation in growing site conditions in the woodlands as compared with the open desert scrub, and that species respond to this variation with corresponding flexibility in GF ontogeny.

Last, I consider the species-specific responses to variations in growing conditions among sites, again using the three most common species of columnar cacti. Centroids (95% confidence intervals around the joint mean of half-height and half-biomass) are plotted in Fig. 8.12 (lower row) for size classes 3, 4, 5, and 6/7. Note that individuals may fall within a particular size class by having more branches lower, or fewer branches high off the ground. Significant differences in branching patterns within species among sites are reflected in the nonoverlap of the 95% centroids. Note that, in general, each species differs significantly from site to site, but the differences vary in direction among species and no general pattern emerges. Rather, each species appears to respond to changes in vegetation in its own unique way, as indicated by the fact that each species in the lower row of Fig. 8.12 reaches taller half-heights (right-shifted ellipses) at different sites. *Pachycereus weberi* shows delayed branching (per height) at site 1, and branches earlier (lower) in the taller and denser vegetation of sites 3 and 5/6. *Escontrilla chiotilla* branches later (higher off the ground) in the low, dense woodland at site 3, whereas *Myrtillocactus geometrizans* shows higher (later) branching at site 5/6, where it is less common and reaches only size class 5 (vs. class 6/7 at the lower elevation sites). This sort of analysis would clearly benefit from more extended and detailed study, with measurement of abiotic variables at specific growing sites; interpretation of the species-specific GF variation among sites of different vegetational characteristics should prove a fertile area for further research.

Discussion

Growth form variations in columnar cacti are relatively easily represented and are useful for showing differences within and among species. In the Sonoran Desert, coexisting species segregate into distinct structural niches as represented by divergent branching patterns and correlated differences in other morphological and ecological attributes. In the denser vegetation of thornscrub and in tropical deciduous forest approaching canopy closure, divergence in branching patterns is far less apparent, and indeed species of similar overall biomass appear convergent rather than divergent in morphology. In two-species thornscrub communities (Fig. 8.6), in the two larger cacti in taller, deciduous forest in Jalisco (Fig. 8.8), and among the three larger cacti represented on the Teotitlan gradient (Fig. 8.10), there are strong similarities in branching patterns. These species are apparently constrained to occupy similar structural niches by the nature of the vegetation in which they grow, and evidence is presented above that branching patterns are indeed influenced by the surrounding vegetation. Earlier (Cody, 1985), I suggested that, given an apparent lack of structural niche diversification in some columnar cacti in deciduous forests, species may segregate into different regeneration niches, sensu Grubb (1977). Some evidence for this comes from the fact that the largest-size classes of these cactus species are most similar in branching pattern, whereas the smaller-size classes tend to be rather more distinct.

Even in tall woodland, however, some segregation by species into different structural niches occurs, and the taller woodland and forest sites (Chamela, Fig. 8.8; Teotitlan, Fig. 8.10) support subcanopy columnar cacti that differ in branching pattern both from the convergently similar canopy emergents and from each other.

The data presented and discussed in this chapter are very much of the low-tech, paper-and-pencil variety, but serve, I hope, to illustrate that morphological attributes in columnar cacti are easily quantified, and that patterns of variation in these attributes, both within and among species, are interesting. Refinements of the patterns and unequivocal determination of their causes requires further research, and perhaps the data in hand will point to some directions in which further progress is likely. At one level is the extent to which morphology is environmentally determined (i.e., growth-site specific), and whether it is little or greatly constrained by phylogenetic or genetic factors. At another level, an elucidation of the environmental influences is needed; a particularly interesting classification is that between biotic and abiotic factors. Evidence points to a considerable plasticity in branching pattern over sites, with or without genetic change, and there is evidence that a major role in the evolution of local morphotypes is attributable to the number and specific identity of coexisting species. Coexisting species may influence branching patterns in a number of ways, including competitive effects from similarly branched species for water uptake or storage or light interception and commensal or quasimutualistic effects, among which the most pertinent is perhaps the role of nurse plants. Nurse plants are clearly important agencies for seedling establishment in columnar cacti (chapters 11 and 15, this volume), as well as in noncolumnar cacti (Cody, 1993), and nurse plants of different identity, size, or GF would be expected to influence the developing morphology of their juvenile cactus protégés in close proximity. Further, cactus GF is notoriously responsive to variation in abiotic conditions, but the exact role of such potentially important factors as precipitation, temperature, and PAR, and their variation among sites as well as seasonal variations within sites remains to be determined. This is a productive area for future studies.

Summary

Columnar cacti are among the most spectacular of plant growth forms, and within the tribe Pachycereeae and its close relatives is found a wide morphological diversity that, in leafless stem succulents, is easily described. The more conspicuous aspects of GF in columnar cacti vary and can be described primarily by branching patterns and secondarily by such stem characteristics as width and rib number. In this chapter, I view branching patterns as a facet of adaptive morphology and a signal of the structural niche occupied by the species in the vegetation. As with niche parameters in the usual classical sense, we might expect that adaptive shifts in branching pattern will occur within species among habitats as both abiotic and biotic components of selective regimes change. Among coexisting species, differences in branching pattern signify a diversification of the species' roles within the vegetation, the evolution of different resource utilization strategies, or a segregation into different structural niches.

In this chapter, I review a methodology for the quantification of branching patterns in columnar cacti, and I use this methodology as a comparative tool to describe variation within and among species. Segregation by branching patterns in coexisting species of desert and other semiarid, open habitats is described and compared with that in species of taller and more close-canopied scrub and woodland. For the latter habitats, examples are drawn from thornscrub and subtropical deciduous woodlands and forest of western Mexico, the Yucatán peninsula, and Jamaica. In the Teotitlan region of Oaxaca, a habitat gradient from more open desert to relatively closed woodland occurs along an elevational gradient, along which I quantify and examine species' distributions, branching patterns, and their variations.

Resumen

Las cactáceas columnares tienen formas de crecimiento (FC) muy espectaculares. En la tribu Pachycereeae y sus parientes se puede encontrar una extensa diversidad de morfologías. En estas plantas de tallos suculentos y sin hojas es muy fácil describirlas. Los aspectos de FC más evidentes en cactáceas columnares son variables, y se les puede describir ante todo, por patrones de ramificación, en segundo lugar por características del tallo tal como ancho y el número de costillas. En este capitulo los patrones de ramificación (PR) son contemplados como una faceta de la morfología adaptiva, también como señales del nicho estructural ocupado por la especie en la vegetación. De una manera similar con los parámetros del nicho en su sentido clásico, anticipamos que los cambios adaptivos en el PR de una especie ocurritán dentrode las especies entre hábitats en la medida que los componentes tauto abióticos como bióticos de los regímenes selectives cambien. Dentro de especies coexistentes, las diferencias en PR significan una diversificación de las funciones de las varias especies en la vegetación la evolución de diversas estrategias de utilización de recursos, o una segregación de las especies en sus nichos estructurales.

En este capítulo presento una metodología para la cuantificación del PR, la cual uso como un instrumento para describir la variación dentro y entre las differentes especies. Describo la segregación por PR entre especies coexistentes en los desiertos y en otros habitat semiáridos y abiertos en comparación con el PR de las especies de habitat más altos y densos, como los bosques. Esos hábitats sou ejemplificados, por los matorrales espinosos y los bosques subtropicales decíduos del oeste de México, la peninsula de Yucatán, y Jamaica. En el región de Teotitlán, Oaxaca, existe una gradiente de habitat que va del desierto más abierto a los bosques relativamente cerrados, que corresponde a un gradiente de elevación sobre el nivel del mar. Aquí las distribuciones de las especies, sus PR y variaciones entre ellas son cuantificadas y examinadas.

Acknowledgments

The field work on which this chapter is based has been generously supported by the Faculty Grants Program of the University of California, Los Angeles, and by Resources for the Future. Over the years, numerous undergraduate and graduate students have assisted with data collection, and their assistance is gratefully acknowledged here. My thanks also to editors and reviewers Dennis Cornejo, Ted Fleming, Alfonso Valiente-Banuet, and Thomas Van Devender for much useful input to the manuscript.

REFERENCES

Cody, M. L. 1984. Branching patterns in columnar cacti. In *Being Alive on Land,* eds. N. Margaris, M. Arianoutsou-Farragitaki, and W. Oechel, T:VS 14:201–36. The Hague: Junk.

———. 1985. Structural niches in plant communities. In *Community Ecology,* eds. J. M. Diamond and T. J. Case, 381–405. San Francisco: Harper and Row.

———. 1986. Distribution and morphology of columnar cacti in tropical deciduous woodland, Jalisco, México. *Vegetatio* 66:137–45.

———. 1989. Growth form diversity in desert plants. *Journal of Arid Environments* 17:199–209.

———. 1991. Niche theory and plant growth form. *Vegetatio* 97:39–55.

———. 1993. Do cylindropuntia cacti need or use nurse plants in the Mojave Desert? *Journal of Arid Environments* 24:139–54.

Cody, M. L., H. J. Thompson, and R. Moran. 1983. The plants. In *Island Biogeography in the Sea of Cortez,* eds. T. J. Case and M. L. Cody. Berkeley: University of California Press.

Cornejo, D. O., and B. B. Simpson. 1997. Analysis of form and function in North American columnar cacti (tribe Pachycereeae). *American Journal of Botany* 84:1482–1501.

Cota, J. H., and R. S. Wallace. 1997. Chloroplast DNA evidence for divergence in *Ferocactus* and its relationships to North American columnar cacti (Cactaceae: Cactoideae). *Systematic Botany* 22:529–42.

Crawley, M. J., ed. 1986. *Plant Ecology.* Oxford: Blackwell Scientific.

Gentry, H. S. 1942. *Rio Mayo Plants: a Study of the Flora and Vegetation of the Valley of the Rio Mayo, Sonora.* Publication 527. Washington, D.C.: Carnegie Institute of Washington.

Givnish, T. 1975. Cited in May, R. 1975. Stability in ecosystems: some comments. In *Unifying Concepts in Ecology,* eds. W. H. van Dobben and R. H. Lowe-McConnell, 161–68. The Hague: Junk.

Grubb, P. 1977. The maintenance of species-richness in plant communities: the importance of the regeneration niche. *Biological Reviews* 52:107–55.

Hunt, R. 1982. *Plant Growth Curves.* London: Edward Arnold.

MacArthur, R. H., and J. W. MacArthur. 1961. On bird species diversity. *Ecology* 42:594–98.

Murphy, P. G., and A. E. Lugo. 1986. Ecology of tropical dry forests. *Annual Review of Ecology and Systematics* 17:67–88.

Nicklas, K. J. 1994. *Plant Allometry: the Scaling of Form and Process.* Chicago: University of Chicago Press.

Nobel, P. N. 1980. Morphology, surface temperature, and northern limits of columnar cacti in the Sonoran Desert. *Ecology* 61:1–7.

Raunkier, C. 1934. *The Life Form of Plants.* Oxford: Clarendon Press.

Ricklefs, R. E., and D. Schluter, eds. 1993. *Species Diversity in Ecological Communities.* Chicago: University of Chicago Press.

Rosenzweig, M. L. 1995. *Species Diversity in Space and Time.* Cambridge: Cambridge University Press.

Rzedowski, J. 1978. *Vegetacion de México.* Mexico: Editorial Limusa.

Rzedowski, J., and R. McVaugh. 1966. La vegetation de Nueva Galicia. *University of Michigan Herbarium Contributions* 9:1–123.

Tilman, D. 1988. *Plant Strategies and the Dynamics and Structure of Plant Communities.* Monographs in Population Biology 26. Princeton: Princeton University Press.

Tomlinson, P. B., and M. H. Zimmerman, eds. 1978. *Tropical Trees as Living Systems.* Cambridge: Cambridge University Press.

Van Devender, T. R., and S. L. Friedman. 1998. From tropical deciduous forest to the Sonoran Desert: gradients in space and time. Unpublished paper in possession of the authors.

Whittaker, R. H. 1977. Evolution of species diversity in land communities. *Evolutionary Biology* 10:1–67.

CHAPTER 9

Physiological Ecology of Columnar Cacti

PARK S. NOBEL

Introduction

Much of the early ecophysiological research on columnar cacti was done at the Desert Botanical Laboratory of the Carnegie Institution of Washington, located near Tucson, Arizona. Approximately 100 publications were produced on environmental responses of cacti by 1940, when the Laboratory closed (McGinnies, 1981). For instance, root growth and the chlorophyll-containing tissue (chlorenchyma) were examined for *Carnegiea gigantea* (saguaro), the tall columnar cactus that is characteristic of the Sonoran Desert (Cannon, 1908, 1916). The osmotic pressure of its stem tissue increases from 0.6 MPa under wet conditions to 1.2 MPa after two months of drought, consistent with the accompanying water loss from its stem (MacDougal and Cannon, 1910). The ribs of cacti provide the accordion-like flexibility that is necessary for appreciable changes in stem water storage (Spalding, 1905; MacDougal and Spalding, 1910). The diameter of *C. gigantea* increases during the daytime and decreases at night (MacDougal, 1924), which is opposite the daily pattern for trees, an observation that anticipated modern studies on water relations and Crassulacean acid metabolism (CAM) for cacti. Stomates of *C. gigantea* tend to be closed during the daytime and open at night when temperatures are lower (MacDougal and Working, 1933), and stem-tissue acidity is highest in the morning (Long, 1915); other characteristics of CAM plants are discussed in Nobel (1988). Research at the Desert Botanical Laboratory also indicated that various cacti can tolerate stem temperatures above 55°C but succumb to freezing temperatures (Coville and MacDougal, 1903; MacDougal and Working, 1921). Of 65 species of columnar cacti in the Sonoran Desert, only *C. gigantea*, *Lophocereus schottii* (senita), and *Stenocereus thurberi* (organ pipe cactus) extend north of the frost limit and only for relatively short distances (Shreve, 1911). Based on damage following a cold wave in January 1937, low temperature was indeed proposed to limit the distribution of these three species (Turnage and Hinckley, 1938).

Following World War II, numerous technological advances occurred in instrumentation for studying the physiological responses of plants in their native habitats, the domain of physiological ecology. For instance, sensors were developed that could measure net CO_2 exchange per unit stem surface area; computers increased in processing speed, facilitating modeling efforts; and environmental chambers became available for studying plant responses under environmentally controlled conditions. This chapter will examine the influences of temperature, water, and light on net CO_2 uptake and hence growth of columnar cacti and their survival.

Thermal Tolerances

Columnar cacti tolerate subzero stem temperatures poorly but tolerate high temperatures extremely well (Table 9.1). The three prominent columnar cacti of the northern Sonoran Desert and two species of columnar *Trichocereus* from arid regions of Chile have an average low-temperature tolerance of –7.6°C when maintained at day/night air temperatures of 15°C/5°C and a low-temperature hardening as day/night air temperatures are reduced by 10°C/10°C of only 0.6°C. Such responses are based on the uptake of the vital stain neutral red into the vacuoles of chlorenchyma cells. Indeed, "death" is difficult to determine for columnar cacti, as a lethal freezing episode in the field may not kill the plants for a few years (Turnage and Hinckley, 1938; Steenbergh and Lowe, 1976; Nobel, 1988). Therefore, a laboratory method had to be devised that predicted eventual plant death, for which the uptake of neutral red proved an extremely reliable indicator (Onwueme, 1979; Didden-Zopfy and Nobel, 1982). In particular, when less than half of the cells accumulate the vital stain neutral red, tissue necrosis occurs for the affected region, which can lead to the death of the entire plant.

In contrast to death occurring at a relatively modest low temperature near –8°C, the three columnar cacti tested and presumably others occurring in similar environments are shown to be extremely tolerant of high temperatures. Indeed, *Carnegiea gigantea, Lophocereus schottii,* and *Stenocereus thurberi* have an average stem high-temperature tolerance of 61.6°C for plants maintained at day/night air temperatures of 45°C/35°C, caused in large part by a remarkable high-temperature hardening of 4.4°C as the day/night air temperatures are raised by 10°C/10°C (Table 9.1).

What are the consequences of the observed tolerances to extreme temperatures for columnar cacti? The low-temperature sensitivity clearly controls the natural geographical distribution of columnar cacti and influences the regions where certain species, such as those in the genus *Stenocereus,* may be cultivated for their fruits. Moreover, various morphological attributes, such as apical pubescence, shading of the stem by spines, and stem diameter can influence stem temperature and hence the environments in which particular species may survive. The high-temperature tolerance of columnar cacti is relatively unprecedented among vascular plants, as 55°C denatures most enzymes; yet even the exceptional high-temperature tolerance of columnar cacti can still be ecologically limiting. In particular, soil surface tempera-

TABLE 9.1
Tolerance of Extreme Temperatures by Columnar Cacti[a]

Species	*Low-Temperature Tolerance for Plants at Day/Night Air Temperatures of 15°C/5°C (°C)*	*Low-Temperature Hardening[b] (°C)*	*High-Temperature Tolerance for Plants at Day/Night Air Temperatures of 45°C/35°C (°C)*
Carnegiea gigantea	−8.4	0.5	61.3
Lophocereus schottii	−6.6	0.5	62.0
Stenocereus thurberi	−8.8	0.3	61.5
Trichocereus candicans	−6.9	1.0	—
T. chilensis	−7.4	0.9	—

[a] Stem segments were maintained for one hour at a series of temperatures. Uptake of the vital stain neutral red (3-amino-7-dimethyl-amino-2-methylphenazine [HCl]) was then measured for chlorenchyma cells in thin slices of tissue, and the temperature leading to a 50% decrease in the fraction of cells taking up the vital stain (a "tolerance") was determined graphically. Data are from Smith et al. (1984) and Nobel (1982, 1984b, 1988).

[b] Hardening refers to the change in such a temperature tolerance when the day/night air temperatures were reduced by 10°C/10°C (low-temperature hardening) or raised by 10°C/10°C (high-temperature hardening).

tures, which strongly influence stem temperatures and hence survival of small seedlings, can be 70°C or even higher in deserts (Coville and MacDougal, 1903; Nobel, 1988), so "nurse plants" are necessary during the early years for many columnar cacti (Steenbergh and Lowe, 1977). Indeed, seedlings of desert succulents can be more sensitive to high temperatures than are adult plants (Nobel, 1984a).

Plant Influences on Stem Temperature

Besides abiotic influences on stem temperature caused by air temperature, radiation (both short and long wavelength), and wind speed, various morphological features also affect stem temperatures. Spines, which are generally ascribed an antiherbivory function for cacti, modify the thermal environment adjacent to the stems. For instance, computer models indicate that the minimum temperature at the stem apex increases by about 5°C and the maximum temperature decreases by about 9°C when simulated plant morphology is varied from the spineless to the fully spine-covered condition for *Carnegiea gigantea* (Table 9.2). Also, the minimum apical stem temperature increases by about 6°C and the maximum temperature decreases by 11°C when the simulated plant morphology is varied from the absence of pubescence to a 10-mm thickness of apical pubescence (Table 9.2). Because the meristem producing the cell divisions necessary for stem extension growth is located at the stem apex, moderation of apical temperatures is crucial for the growth and survival of columnar cacti. In addition, minimal apical temperatures increase and maximal temperatures decrease by about 3°C as the stem diameter of *C. gigantea* increases sixfold (Table 9.2), consistent with statistical analyses for columnar cacti of a specific height

TABLE 9.2

Influence of Morphology on Apical Stem Temperatures of *Carnegiea gigantea*[a]

Condition	*Minimum Daily Temperature (°C)*	*Maximum Daily Temperature (°C)*
No spines	5.3	39.3
50% spine shading of apex	7.6	33.8
100% spine shading of apex	10.2	30.0
No pubescence	5.3	39.3
1-mm thickness of pubescence	6.9	36.4
10-mm thickness of pubescence	10.8	28.0
10-cm diameter	4.5	39.7
30-cm diameter	5.3	39.3
60-cm diameter	7.9	36.9

[a] The control condition was a plant with a stem that is 30 cm in diameter without apical spines or pubescence; the minimum and maximum air temperatures were 7.0°C and 38.0°C, respectively. Data are from Nobel (1978, 1980a, 1988).

(Cornejo and Simpson, 1997). Thus the apex tends to be warmer in the winter and cooler in the summer for the larger stems with appreciable spine coverage and a deep insulating layer of pubescence.

At 30° north latitude in Sonora, Mexico, the apex of *Carnegiea gigantea* is 9% shaded by spines, that of *Lophocereus schottii* is 20% shaded, and the apex of *Stenocereus thurberi* is 60% shaded, whereas at their northernmost limits in Arizona, shading increases to 41% for *C. gigantea* at 35° north latitude, remains unchanged for *L. schottii* at 32° north latitude, and increases to 71% for *S. thurberi* at 32° north latitude (Nobel, 1980a). Selection pressure with respect to avoiding low apical temperatures coupled with phenotypic plasticity in spine expression is thus evident for *C. gigantea.* Although the mean depth of apical pubescence does not change for the three columnar cacti over these latitudinal ranges, it is greatest for *C. gigantea* (10 mm) and virtually absent for the other two species, which are not distributed as far northward, where freezing temperatures are more common.

Seeds of many species of columnar cacti germinate under the canopy of another plant, referred to as a nurse plant (see chapter 15). For instance, shading by *Cercidium floridum* (blue paloverde) and *C. microphyllum* (foothill or little-leaf paloverde) reduces air and soil temperatures, which is crucial for the germination and survival of seedlings of *Carnegiea gigantea* during the hot parts of a year (Shreve, 1931; Turner et al., 1966; Despain, 1974; Steenbergh and Lowe, 1976). Shading by nurse plants also favors the establishment of *Stenocereus thurberi* in Arizona (Parker, 1987) and of *Neobuxbaumia tetetzo,* which predominates on the north side of nurse plants in various regions of Mexico (Valiente-Banuet and Ezcurra, 1991). Nurse plants can also protect the apical meristem of *Carnegiea gigantea* from a cold nighttime sky in the winter, which helps the plants avoid freezing injury (Steenbergh and Lowe, 1976; Nobel, 1980b, 1988). Yet the effects of nurse plants are complex. Shading of the soil by nurse plants lowers the rate of water evaporation from the soil, but the nurse plants also take up soil water. Litter and hence nutrient availability can be higher under nurse plants, but nurse plants also take up minerals. For instance, soil nitrogen levels tend to be lower under shrubs acting as nurse plants for the columnar cacti *Cephalocereus hoppenstedtii* and *N. tetetzo* than for bare soil in the semiarid Zapotitlán Valley in south-central Mexico (Valiente-Banuet et al., 1991). Nurse plants also reduce the photosynthetic photon flux (PPF, wavelengths of 400–700 nm) incident on the shaded cactus seedlings, thereby lowering their net rates of photosynthesis and hence growth. For instance, *Cercidium floridum* and *C. microphyllum,* even when leafless in the winter, reduce the PPF available for seedlings of *Carnegiea gigantea* under their canopies by 31% (Lowe and Hinds, 1971).

Water Relations

One key to the success of columnar cacti in deserts is their ability to survive drought, an aspect of which is the storage of water in their massive stems. The volume of water storage per unit of surface area across which transpiration can occur equals V/A, whose units are length^3/length^2; V/A indicates the average stem depth for water storage. For a cylindrical stem, ignoring the area of the top, $V/A = \pi r^2 l / 2\pi r l$ (where r is the radius and l is the length along the cylinder axis) or $r/2$, which equals 75 mm for a stem 30 cm in diameter, as is typical for many columnar cacti. Compare this with water storage for flat leaves, which have an average thickness of about 400 μm, so that V/A is 200 μm or 0.2 mm. Clearly, cacti and most other CAM plants can store tremendous amounts of water per unit surface area, which greatly increases the time during which stored water can be used for transpiration. However, seedlings of columnar cacti have a much lower V/A than do adult plants and so are much more vulnerable to drought than are adult plants (Nobel, 1988). For columnar cacti, water storage occurs primarily in the parenchyma, which varies widely in thickness and underlies the chlorenchyma, which is usually 1 to 5 mm in thickness. During drought, water moves from the large parenchyma cells, which are generally devoid of chloroplasts, to the green chlorenchyma cells, thereby sustaining biochemical reactions that support CO_2 fixation (Nobel, 1988).

Besides water storage in the stem, several other attributes of columnar cacti are crucial for their water relations and hence success in arid and semiarid regions. Deferring for the moment a consideration of daily gas exchange pattern for CAM plants, facilitation of water uptake by the roots and limitations on water loss by the roots and stems are crucial. The shallowness of roots of columnar cacti has long been noted; most roots occur at a depth of less than 0.4 m (Cannon, 1911, 1916), which allows them to respond rapidly to the light rainfalls that often occur in deserts (Rundel and Nobel, 1991; Nobel, 1996a). Although this has not been extensively studied for columnar cacti in the field, extrapolation from results for other cacti indicate that new root growth follows rapidly after rainfall and that existing roots can take up water, which is the predominant mode for water uptake for up to one week after rainfall interrupts drought (Szarek et al., 1973; Nobel and Sanderson, 1984; Nobel, 1988).

Although the roots of cacti represent only a small part of the plant biomass (generally 9–14% by dry weight; Nobel, 1988), a major problem is the prevention of appreciable water movement from a cactus filled with water to a drying soil. Over the short term (up to 14 days of drought), water loss from the plant to the soil is reduced by shrinkage of roots (especially young ones) away from the soil and hence the development of root-soil air gaps that conduct water poorly; over the longer term, the greatly reduced hydraulic conductivity of the soil decreases water loss from the roots to a drying soil (Nobel and Cui, 1992a,b). Water loss from the stems of columnar cacti to the air is decreased by a potentially high cuticular resistance (low cuticular conductance), which is a consequence of the waxy stem cuticle and the tight closing of stomata. In particular, the cuticle for *Carnegiea gigantea* and other columnar cacti is relatively thick (Gibson and Nobel, 1986), which is correlated with decreased water loss by cuticular transpiration when the stomata are closed. Also, a low stomatal count per unit area coupled with a high stomatal resistance during drought can greatly reduce the accompanying stem water loss through stomata (Nobel, 1988). Indeed, detached, unrooted stem segments of cacti can live for a few years without any source of external water (Nobel, 1988; P. S. Nobel, unpubl. data; S. R. Szarek and I. P. Ting, unpubl. data).

CAM

In addition to adaptations for arid conditions based on stem water storage, a small and shallow root system for water uptake, and a high cuticular resistance against water loss, cacti use CAM photosynthesis, whose predominantly nocturnal stomatal opening greatly reduces daily water loss per unit surface area while allowing considerable net CO_2 uptake (Nobel, 1988). In particular, the water-vapor content of air at saturation (which is virtually the case within the stems of columnar cacti) increases essentially exponentially with temperature. For example, saturated air contains 6.8 $g{\cdot}m^{-3}$ of water vapor at 5°C, 17.3 $g{\cdot}m^{-3}$ at 20°C, and 39.7 $g{\cdot}m^{-3}$ at 35°C. The water-vapor content of the air surrounding plants generally is far below the saturation value, especially in deserts, and does not change much during the course of a day unless a major change in the weather occurs. For air containing 4.0 $g{\cdot}m^{-3}$ of water vapor (relative humidity of 59% at 5°C, 23% at 20°C, and 10% at 35°C), the stem-to-air difference in water-vapor content is 2.8 $g{\cdot}m^{-3}$ at 5°C, 13.3 $g{\cdot}m^{-3}$ at 20°C, and 35.7 $g{\cdot}m^{-3}$ at 35°C. The rate of water loss (transpiration) for the same degree of stomatal opening is then 4.8-fold higher at 20°C than at 5°C and 2.7-fold higher at 35°C than at 20°C ($13.3/2.8 = 4.8$, $35.7/13.3 = 2.7$). If the temperatures were 15°C cooler at night than during the daytime, which is realistic for the deserts where columnar cacti occur, transpiration would be three- to fivefold lower at night than during the daytime for the same degree of stomatal opening, which underscores the importance of nocturnal stomatal opening and CAM for water conservation by columnar cacti.

Has CAM been established as the photosynthetic mode for columnar cacti? The most convincing evidence for CAM is substantial nocturnal CO_2 uptake by the photosynthetic organs, which has been clearly demonstrated for *Carnegiea gigantea, Pachycereus pringlei,* and *Stenocereus queretaroensis* (Fig. 9.1). Moreover, both the maximum net CO_2 uptake rates and the daily pattern are similar for these three species of columnar cacti. Because photosynthesis cannot occur without light, the CO_2 taken up at night cannot be immediately fixed into photosynthetic products such as the sugars glucose and sucrose. Rather, the CO_2 is incorporated into phosphoenolpyruvate (PEP) using the enzyme PEP carboxylase, which leads to the formation of an organic acid such as malate. The continued uptake of CO_2 throughout the night (Fig. 9.1) results in the continued accumulation of such organic acids in the chlorenchyma. This accumulation has long been recognized for *C. gigantea* (Long, 1915; Richards, 1915). Indeed, nocturnal acidification of the chlorenchyma has been used to demonstrate CAM in various columnar cacti (Table 9.3). A sophisticated but indirect method for determining whether a particular species uses CAM is to determine the ratio of various carbon isotopes in the plant tissues. In particular, differences in the enzymes involved in the initial carbon fixation and the subsequent biochemical processing of the fixed carbon leads to unique isotopic signatures for the three photosynthetic pathways—CAM (which is the only one with nocturnal stomatal opening), C_3, and C_4 (Nobel, 1988). Thus isotopic analysis of carbon by mass spectroscopy for tissue samples that can be readily obtained in the field or from herbarium specimens indicates the photosynthetic pathway. Isotopic analysis has been used to demonstrate CAM for various columnar cacti (Table 9.3). So far 23 species of columnar cacti have been shown to use CAM and no contrary cases of C_3 or C_4 metabolism have apparently been reported for mature plants (young organs on plants whose adult form exhibits CAM often exhibit substantial net CO_2 uptake during the daytime in the C_3 mode for a matter of weeks, but the organs then switch to predominantly CAM behavior; Nobel, 1988).

Environmental Influences on CO_2 Uptake

Net CO_2 uptake by a CAM plant does not respond to the instantaneous value of the PPF, as most uptake occurs at night in the absence of light (Fig. 9.1). Therefore, total daily PPF must be considered relative to the total daily net CO_2 uptake to determine the PPF responses of CAM plants (Nobel, 1988). And because most net CO_2 uptake by CAM plants occurs at night, the temperatures that have the major effect on total

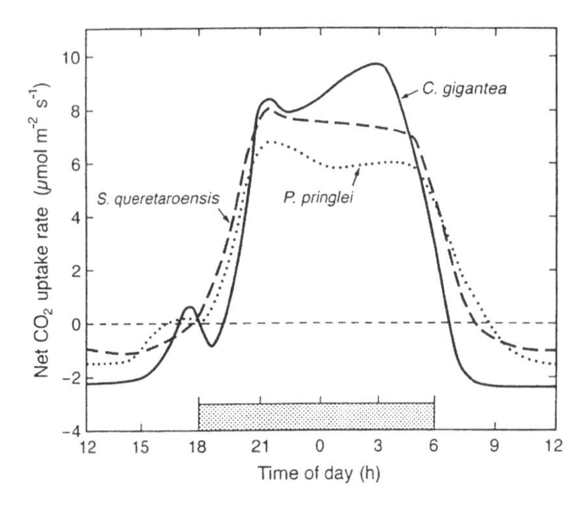

FIGURE 9.1. Daily patterns of CO_2 uptake for columnar cacti. Data are from Despain et al. (1970) for *Carnegiea gigantea,* Franco-Vizcaíno et al. (1990) for *Pachycereus pringlei,* and Nobel and Pimienta-Barrios (1995) for *Stenocereus queretaroensis.* Stippled bar indicates nighttime.

daily net CO_2 uptake are those at night. Rainfall, or more precisely, soil water status is the third main environmental variable affecting net CO_2 uptake; total daily net CO_2 uptake is maximal under wet conditions and decreases steadily during drought. Among columnar cacti, the effects of PPF, temperature, and drought on total daily net CO_2 uptake has apparently been determined only for *Stenocereus queretaroensis* (Fig. 9.2). Total daily net CO_2 uptake for this species increases approximately linearly with total daily PPF up to about 20 mol·m^{-2}·day^{-1} (Fig. 9.2a). Total daily net CO_2 uptake is maximal for mean nighttime temperatures of about 15°C, decreasing by approximately 40% at 5°C and 25°C (Fig. 9.2b). Total daily net CO_2 uptake for *S. queretaroensis* decreases by about 35% after one month of drought and by 92% after two months of drought (Fig. 9.2c).

For *Carnegiea gigantea,* the instantaneous rate of nocturnal net CO_2 uptake is maximal at about 14°C for plants grown with mean nighttime temperatures of 10°C and at about 21°C for plants grown with mean nighttime temperatures of 30°C (Nobel and Hartsock, 1981). The maximal rate is 74% lower for plants at the higher temperature, consistent with the substantial reduction in total daily net CO_2 uptake for *Stenocereus queretaroensis* at such high nighttime temperatures (Fig. 9.2b). Note that *S. queretaroensis* occurs natively and is cultivated in central Jalisco, Mexico, where the minimum nighttime temperatures annually average about 12°C and the mean nighttime temperatures average about 16°C; the latter temperature is essentially optimal for net CO_2 uptake by this columnar cactus. Increasing the CO_2 concentration in the atmosphere increases net CO_2 uptake by *S. queretaroensis;* specifically, doubling the current atmospheric value of 360 μmol CO_2 mol^{-1} increases daily net CO_2 uptake by 36%, similar to effects on certain other CAM plants (Nobel, 1996b).

TABLE 9.3
Summary of Columnar Cacti that Use CAM[a]

	Method for Determining CAM		
Species	*Nocturnal Net CO_2 Uptake*	*Nocturnal Acid Accumulation*	*CAM-like Carbon Isotope Ratio*
Acanthocereus tetragonus		Díaz and Medina (1984)	
Carnegiea gigantea	Despain et al. (1970)	Richards (1915), Long (1915)	
Cephalocereus royenii		Ting (1976)	
Cereus emoryi			Mooney et al. (1974)
C. peruvianus			Bender et al. (1973)
C. repandus		Díaz (1983)	
C. sylvestrii		Seeni and Gnanam (1980)	Troughton et al. (1974)
C. validus	Nobel et al. (1984)	Nobel et al. (1984)	
Eulychnia acida			Mooney et al. (1974)
E. castanea			Mooney et al. (1974)
E. iquiquensis			Mooney et al. (1974)
E. spinibarbis			Mooney et al. (1974)
Lophocereus schottii	Mooney et al. (1974)	Smith et al. (1984)	Mooney et al. (1974)
Myrtillocactus cochal	Mooney et al. (1974)		Mooney et al. (1974)
Neobuxbaumia tetetzo		Altesor et al. (1992)	
Pachycereus pringlei	Franco-Vizcaíno et al. (1990)	Franco-Vizcaíno et al. (1990)	Troughton et al. (1974)
Stenocereus griseus		Díaz and Medina (1984)	
S. gummosus	Mooney et al. (1974)	Nobel (1980c)	Mooney et al. (1974)
S. queretaroensis	Nobel and Pimienta-Barrios (1995)		
S. thurberi		Ting (1976)	
Subpilosocereus ottonis		Lüttge et al. (1989)	
Trichocereus chilensis		Nobel (1981)	Mooney et al. (1974)
T. coquimbanus			Mooney et al. (1974)

[a] References are those for which CAM was first cited for the species.

Vegetative and Reproductive Growth and Productivity

A surprising finding for *Stenocereus queretaroensis* in Jalisco is the seasonal offset of the times for stem net CO_2 uptake, stem extension growth, and reproduction. In particular, about 90% of the annual rainfall occurs from June to September (Fig. 9.3a). Although the cloudiness accompanying the summertime rainfall reduces the PPF incident on the stems (Pimienta-Barrios et al., 1998), the predicted monthly net CO_2 uptake by *S. queretaroensis* (Fig. 9.2) is greatest from June to October, primarily because of the availability of soil water at that time of the year. Yet the maximum monthly stem extension occurs at the end of this period and into the ensuing drought (Fig. 9.3b); specifically, 80% of the stem extension for *S. queretaroensis* occurs from

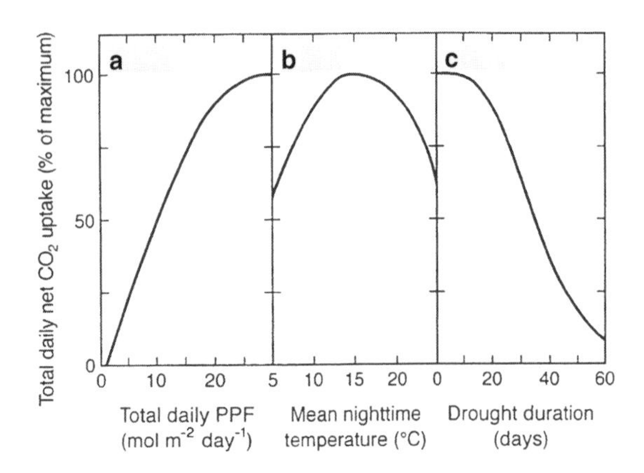

FIGURE 9.2. Responses of total daily net CO_2 uptake for *Stenocereus queretaroensis* to (a) PPF, (b) nighttime temperature, and (c) drought duration. Adapted from Nobel and Pimienta-Barrios (1995) and P. S. Nobel (unpubl. obs).

September through December. The delayed stem growth is apparently genetically fixed, reflecting adjustments to the shade under closed canopies during the rainy season for tropical deciduous forests that are the native habitat of *S. queretaroensis.* Flower buds appear and fruits develop during the driest time of the year, February to May (Fig. 9.3c), which facilitates germination during the summer rainy season. New roots appear as extensions of main roots and as lateral branches on existing roots during the summer rainy period, a time when the stems and reproductive organs are not visually increasing in size.

The offsets of vegetative and reproductive growth for *S. queretaroensis* have major impacts on its carbohydrate contents. For instance, the amount of sugars in its stems increases during the summer, when net CO_2 uptake is favored (Nobel and Pimienta-Barrios, 1995; Pimienta-Barrios et al., 1998). The sugar content decreases nearly threefold when the stem extension occurs in the autumn. Moreover, reducing sugars such as glucose and fructose, which are directly involved in the construction of new tissues, have an opposite seasonal pattern, representing only 10% of the total sugars in June but 90% in December. Thus, major net CO_2 uptake for *S. queretaroensis* occurs when environmental conditions are optimal, which is coincident with the development of new roots; stem extension growth occurs subsequently, followed by reproductive development during the driest time of the year (Fig. 9.3), leading to major decreases in glucans and sugars stored in the stems.

Stenocereus queretaroensis is cultivated for its fruits (the plants are often called "pitayos" and the fruits "pitayas"). Other columnar cacti cultivated for fruits in Mexico include *Myrtillocactus geometrizans, S. fricii, S. griseus,* and *S. stellatus* (Pimienta-Barrios and Nobel, 1994; Mizrahi et al., 1997; see also chapter 7). Fruits of *Carnegiea gigantea, Pachycereus pecten-aboriginum, P. pringlei, S. gummosus,* and *S. thurberi* are collected from the wild. Limited cultivation for fruits also occurs for the columnar

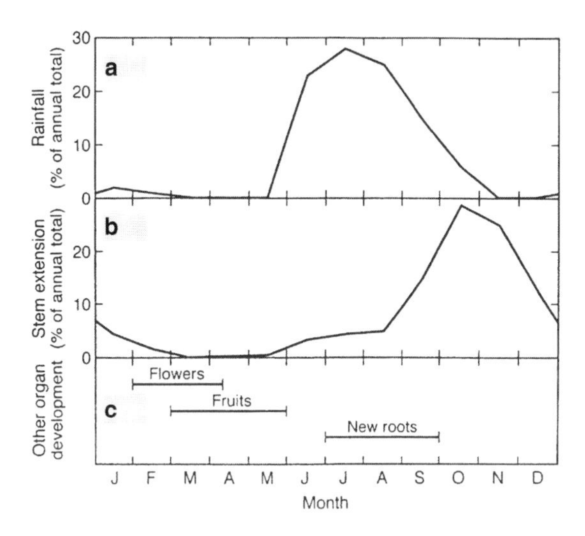

FIGURE 9.3. Monthly rainfall in the Sayula Basin of Jalisco, Mexico: (a) during the mid-1990s; (b) monthly stem extension for *Stenocereus queretaroensis;* and (c) development of other organs. Data are from Pimienta-Barrios (1995) and Pimienta-Barrios et al. (1998).

cacti *Cereus jamacaru, Cereus peruvianus,* and *Escontria chiotilla* (Mizrahi et al., 1997). Five of these columnar cacti have been cultivated in the Negev Desert of Israel (Nerd et al., 1993); *Cereus peruvianus* grows most easily, but it and *S. griseus* can be damaged by low temperatures (the other species grown are *P. pringlei, S. gummosus,* and *S. thurberi*). In any case, the possibility for expanded fruit production by columnar cacti should stimulate research on their ecophysiological responses, ranging from daily net CO_2 uptake in the field to seasonal carbohydrate sequestering and seasonal growth of various organs.

Various factors interact to determine the productivity of columnar cacti. For instance, branching patterns affect the interception of PPF. The majority of branches for *Carnegiea gigantea* occurs on the south-facing half of the main stems, which can increase the stem surface area for PPF interception by a plant (Geller and Nobel, 1986). Ribs also increase the stem surface area available for PPF interception. The fractional rib depth (the radial distance from a rib crest to the trough between adjacent ribs divided by the stem-crest radius) varies from 0.14 for *C. gigantea* to 0.35 for *Lophocereus schottii* for the relatively dry conditions prevailing for most of the year (Geller and Nobel, 1984). Ribs of *C. gigantea* tend to be closer on the southern side of stems, leading to more surface area where the ambient PPF is higher (Spalding, 1905; Walter and Stadelmann, 1974; Geller and Nobel, 1984). As mentioned above, spines moderate stem temperatures and thereby extend the ranges of columnar cacti (they can also deter herbivores), but the reduction by spines of PPF reaching the chlorenchyma reduces net CO_2 uptake per unit stem surface area (Fig. 9.2a).

Conclusions

Columnar cacti are not especially tolerant of freezing temperatures, death often occurring by −8°C, but they are extremely tolerant of high temperatures, including those above 60°C. Their stem morphology influences their thermal responses, which in turn influences their geographical distribution. For instance, increased apical pubescence, increased apical shading by spines, and increased stem diameter can extend the range of *Carnegiea gigantea* to colder regions compared with other columnar cacti of the Sonoran Desert. Nurse plants reduce the extreme temperatures experienced by the stems but also reduce their light absorption. The stem volume to surface-area ratio, which indicates the average depth of stem available for water storage and which affects the length of drought than can be tolerated, is high for columnar cacti. Columnar cacti also have shallow roots, which facilitates responses to light desert rainfalls, and low cuticular transpiration, which limits water loss.

All 23 species of columnar cacti investigated with respect to photosynthetic pathway exhibit CAM, in which the majority of net CO_2 uptake occurs at night, thereby decreasing plant water loss compared with daytime stomatal opening. Thus, nighttime temperatures are generally more influential on daily net CO_2 uptake than are daytime temperatures, with optimal temperatures near 15°C for *Stenocereus queretaroensis*. The maximal rate of net CO_2 uptake is about 8 $\mu mol \cdot m^{-2} \cdot s^{-1}$ for *S. queretaroensis*, with similar values for *Carnegiea gigantea* and *Pachycereus pringlei*. The total daily net CO_2 uptake for *S. queretaroensis* tends to increase linearly with the total daily PPF incident on the stems, decreases during drought, and can be increased by increasing the atmospheric CO_2 concentration. Various columnar cacti are cultivated for fruits, such as *S. queretaroensis* in Jalisco. Nearly all of its stem extension occurs in the autumn after the summer rainy season, and the production of flowers and fruit development occurs in the spring at the end of the dry season, so that vegetative and reproductive growth are out of phase with seasonal water acquisition and daily net CO_2 uptake. Many interesting ecophysiological aspects await discovery for columnar cacti, which presumably all exhibit CAM but have unique anatomical, morphological, and physiological responses to environmental conditions, perhaps especially with respect to seasonal carbohydrate sequestration and accumulation in various organs.

Summary

Columnar cacti tolerate high temperatures well but freezing temperatures poorly, so low temperatures can limit their distribution. Their massive stems can store considerable amounts of water, and their shallow roots are well adapted to light desert rainfalls. Net CO_2 uptake occurs primarily at night, as is characteristic of species utilizing CAM, resulting in water conservation because temperatures are lower at night than during the day. For *Stenocereus queretaroensis*, which like certain other columnar cacti is cultivated for fruits, flower and fruit production occur before the summer rainy season and major stem extension after it, with interesting physiological and ecological consequences.

Resumen

Las cactáceas columnares toleran sin problema alguno las temperaturas altas, sin embargo, la tolerancia de estas plantas a las heladas es pobre. De manera que su distribución puede verse limitada por las temperaturas bajas. Sus tallos masivos son capaces de almacenar cantidades considerables de agua y sus raíces someras están bien adaptadas a las lluvias ligeras de los desiertos. La asimilación neta de CO_2 es nocturna, lo cual es característico de las especies que utilizan el metabolismo ácido de las Crasuláceas. Este atributo les permite conservar agua debido a que las temperaturas nocturnas son menores que las diurnas. En el caso de *Stenocereus queretaroensis,* especie que al igual que otras cactáceas columnares es cultivada por sus frutos, la producción de flores y frutos ocurre antes de la temporada de lluvias en el verano y el crecimiento de los tallos ocurre después del mismo, lo cual tiene interesantes consecuencias fisiológicas y ecológicas.

REFERENCES

Altesor, A., E. Ezcurra, and C. Silva. 1992. Changes in the photosynthetic metabolism during the early ontogeny of four cactus species. *Acta Oecologia* 13:777–85.

Bender, M. M., I. Rouhani, H. M. Vines, and C. C. Black, Jr. 1973. $^{13}C/^{12}C$ ratio changes in Crassulacean acid metabolism plants. *Plant Physiology* 52:427–30.

Cannon, W. A. 1908. *The Topography of the Chlorophyll Apparatus in Desert Plants.* Publication 98. Washington, D.C.: Carnegie Institution of Washington.

———. 1911. *The Root Habits of Desert Plants.* Publication 131. Washington, D.C.: Carnegie Institution of Washington.

———. 1916. Distribution of the cacti with especial reference to the role played by the root response to soil temperature and soil moisture. *American Naturalist* 50:435–42.

Cornejo, D. O., and B. B. Simpson. 1997. Analysis of form and function in North American columnar cacti (Tribe Pachycereeae). *American Journal of Botany* 84:1482–1501.

Coville, F. V., and D. T. MacDougal. 1903. *Desert Botanical Laboratory of the Carnegie Institution.* Publication 6. Washington, D.C.: Carnegie Institution of Washington.

Despain, D. G. 1974. The survival of saguaro (*Carnegiea gigantea*) seedlings on soils of differing albedo and cover. *Journal of the Arizona Academy of Science* 9:102–7.

Despain, D. G., L. C. Bliss, and J. C. Boyer. 1970. Carbon dioxide exchange in Saguaro seedlings. *Ecology* 51:912–14.

Díaz, M. 1983. Estudios fisioecológicos de cactáceas bajo condiciones naturales. Master's thesis, Instituto Venezolano de Investigaciones Científicas, Caracas, Venezuela.

Díaz, M., and E. Medina. 1984. Actividad CAM de cactáceas en condiciones naturales. In *Eco-Fisiologia de Plantas CAM,* ed. E. Medina, 98–113. Caracas, Venezuela: Centro Internacional de Ecología Tropical.

Didden-Zopfy, B., and P. S. Nobel. 1982. High temperature tolerance and heat acclimation of *Opuntia bigelovii. Oecologia* 52:176–80.

Franco-Vizcaíno, E., G. Goldstein, and I. P. Ting. 1990. Comparative gas exchange of leaves and bark in three stem succulents of Baja California. *American Journal of Botany* 77:1272–78.

Geller, G. N., and P. S. Nobel. 1984. Cactus ribs: influence on PAR interception and CO_2 uptake. *Photosynthetica* 18:482–94.

———. 1986. Branching patterns of columnar cacti: influences on PAR interception and CO_2 uptake. *American Journal of Botany* 73:1193–1200.

Gibson, A. C., and P. S. Nobel. 1986. *The Cactus Primer.* Cambridge, Mass.: Harvard University Press.

Long, E. R. 1915. Acid accumulation and destruction in large succulents. *The Plant World* 18:261–72.

Lowe, C. H., and D. S. Hinds. 1971. Effect of paloverde (*Cercidium*) trees on the radiation flux at ground level in the Sonoran Desert in the winter. *Ecology* 52:916–22.

Lüttge, U., E. Medina, W. J. Cram, H. S. Lee, M. Popp, and J.A.C. Smith. 1989. Ecophysiology of xerophytic and halophytic vegetation of a coastal alluvial plain in northern Venezuela. II. Cactaceae. *New Phytologist* 111:245–51.

MacDougal, D. T. 1924. *Growth in Trees and Massive Organs of Plants—Dendrographic Measurements.* Publication 350. Washington, D.C.: Carnegie Institution of Washington.

MacDougal, D. T., and W. A. Cannon. 1910. *The Conditions of Parasitism in Plants.* Publication 129. Washington, D.C.: Carnegie Institution of Washington.

MacDougal, D. T., and E. S. Spalding. 1910. *The Water-Balance of Succulent Plants.* Publication 141. Washington, D.C.: Carnegie Institution of Washington.

MacDougal, D. T., and E. B. Working. 1921. Another high-temperature record for growth and endurance. *Science, New Series* 54:152–53.

———. 1933. *The Pneumatic System of Plants, Especially Trees.* Publication 441. Washington, D.C.: Carnegie Institution of Washington.

McGinnies, W. G. 1981. *Discovering the Desert: Legacy of the Carnegie Desert Botanical Laboratory.* Tucson: University of Arizona Press.

Mizrahi, Y., A. Nerd, and P. S. Nobel. 1997. Cacti as crops. *Horticultural Reviews* 18:291–319.

Mooney, H., J. H. Troughton, and J. A. Berry. 1974. Arid climates and photosynthetic systems. *Carnegie Institution Year Book* 73:793–805.

Nerd, A., E. Raveh, and Y. Mizrahi. 1993. Adaptation of five columnar cactus species to various conditions in the Negev Desert of Israel. *Economic Botany* 47:304–11.

Nobel, P. S. 1978. Surface temperatures of cacti—influences of environmental and morphological factors. *Ecology* 59:986–96.

———. 1980a. Morphology, surface temperatures, and northern limits of columnar cacti in the Sonoran Desert. *Ecology* 61:1–7.

———. 1980b. Morphology, nurse plants, and minimum apical temperatures for young *Carnegiea gigantea. Botanical Gazette* 141:188–91.

———. 1980c. Interception of photosynthetically active radiation by cacti of different morphology. *Oecologia* 45:160–66.

———. 1981. Influences of photosynthetically active radiation on cladode orientation, stem tilting, and height of cacti. *Ecology* 62:982–90.

———. 1982. Low-temperature tolerance and cold hardening of cacti. *Ecology* 63:1650–56.

———. 1984a. Extreme temperatures and thermal tolerances for seedlings of desert succulents. *Oecologia* 62:310–17.

———. 1984b. PAR and temperature influences on CO_2 uptake by desert CAM plants. *Advances in Photosynthesis Research IV* 3:193–200.

———. 1988. *Environmental Biology of Agaves and Cacti.* New York: Cambridge University Press.

———. 1996a. Ecophysiology of roots of desert plants, with special emphasis on agaves and cacti. In *Plant Roots: The Hidden Half.* 2nd ed., eds. Y. Waisel, A. Eshel, and U. Kafkafi, 823–44. New York: Marcel Dekker.

———. 1996b. Responses of some North American CAM plants to freezing temperatures and doubled CO_2 concentrations: implications of global climate change for extending cultivation. *Journal of Arid Environments* 34:187–96.

Nobel, P. S., and M. Cui. 1992a. Hydraulic conductances of the soil, the root-soil air gap, and the root: changes for desert succulents in drying soil. *Journal of Experimental Botany* 43:319–26.

———. 1992b. Shrinkage of attached roots of *Opuntia ficus-indica* in response to lowered water potentials—predicted consequences for water uptake or loss to soil. *Annals of Botany* 70:485–91.

Nobel, P. S., and T. L. Hartsock. 1981. Shifts in the optimal temperature for nocturnal CO_2 uptake caused by changes in growth temperature for cacti and agaves. *Physiologia Plantarum* 53:523–27.

Nobel, P. S., and E. Pimienta-Barrios. 1995. Monthly stem elongation for *Stenocereus queretaroensis:* Relationships to environmental conditions, net CO_2 uptake and seasonal variations in sugar content. *Environmental and Experimental Botany* 35:17–24.

Nobel, P. S., and J. Sanderson. 1984. Rectifier-like activities of roots of two desert succulents. *Journal of Experimental Botany* 35:727–37.

Nobel, P. S., U. Lüttge, S. Heuer, and E. Ball. 1984. Influence of applied NaCl on Crassulacean acid metabolism and ionic levels in a cactus, *Cereus validus. Plant Physiology* 75:799–803.

Onwueme, I. C. 1979. Rapid, plant-conserving estimate of heat tolerance in plants. *Journal of Agricultural Science, Cambridge* 92:527–36.

Parker, K. C. 1987. Site-related demographic patterns of organ pipe cactus populations in southern Arizona. *Bulletin of the Torrey Botanical Club* 114:149–55.

Pimienta-Barrios, E., and P. S. Nobel. 1994. Pitaya (*Stenocereus* spp., Cactaceae): an ancient and modern fruit crop of Mexico. *Economic Botany* 48:76–83.

———. 1995. Reproductive characteristics of pitayo (*Stenocereus queretaroensis*) and their relationships with soluble sugars and irrigation. *Journal of the American Society for Horticultural Science* 120:1082–86.

———. 1998. Vegetative, reproductive, and physiological adaptations to aridity of pitayo (*Stenocereus queretaroensis,* Cactaceae). *Economic Botany* 52:401–11.

Pimienta-Barrios, E., G. Hernandez, A. Domingues, and P. S. Nobel. 1998. Growth and development of the arborescent cactus *Stenocereus queretaroensis* in a subtropical semiarid environment, including effects of gibberellic acid. *Tree Physiology* 18:59–64.

Richards, H. M. 1915. *Acidity and Gas Interchange in Cacti.* Publication 209. Washington, D.C.: Carnegie Institution of Washington.

Rundel, P. W., and P. S. Nobel. 1991. Structure and function in desert root systems. In *Plant Root Growth—An Ecological Perspective,* ed. D. Atkinson, 349–78. Oxford: Blackwell Scientific.

Seeni, S., and A. Gnanam. 1980. Photosynthesis in cell suspension cultures of the CAM plant *Chamaecereus sylvestrii* (Cactaceae). *Physiologia Plantarum* 49:465–72.

Shreve, F. 1911. The influence of low temperatures on the distribution of the giant cactus. *The Plant World* 14:136–46.

———. 1931. Physical conditions in sun and shade. *Ecology* 12:96–104.

Smith, S. D., B. Didden-Zopfy, and P. S. Nobel. 1984. High-temperature responses of North American cacti. *Ecology* 65:643–51.

Spalding, E. S. 1905. Mechanical adjustment of the sahuaro (*Cereus giganteus*) to varying quantities of stored water. *Bulletin of the Torrey Botanical Club* 32:57–68.

Steenbergh, W. F., and C. H. Lowe. 1976. *Ecology of the Saguaro. I. The Role of Freezing Weather in a Warm-Desert Population.* National Park Service Scientific Monograph Series. Number 1. Washington, D.C.: U.S. Government Printing Office.

———. 1977. *Ecology of the Saguaro. II. Reproduction, Germination, Establishment, Growth, and Survival of the Young Plant.* National Park Service Scientific Monograph Series. Number 8. Washington, D.C.: U.S. Government Printing Office.

Szarek, S. R., H. B. Johnson, and I. P. Ting. 1973. Drought adaptation in *Opuntia basilaris.* Significance of recycling carbon through Crassulacean acid metabolism. *Plant Physiology* 52:539–41.

Ting, I. P. 1976. Crassulacean acid metabolism in natural ecosystems in relation to annual CO_2 uptake patterns and water utilization. In *CO_2 Metabolism and Plant Productivity,* eds. R. H. Burris and C. C. Black, 251–68. Baltimore: University Park Press.

Troughton, J. H., P. V. Wells, and H. A. Mooney. 1974. Photosynthetic mechanisms in ancient C_4 and CAM species. *Carnegie Institution Year Book* 73:812–16.

Turnage, W. V., and A. L. Hinckley. 1938. Freezing weather in relation to plant distribution in the Sonoran Desert. *Ecological Monographs* 8:529–50.

Turner, R. M., S. M. Alcorn, G. Olin, and J. A. Booth. 1966. The influence of shade, soil, and water on saguaro seedling establishment. *Botanical Gazette* 127:95–102.

Valiente-Banuet, A., and E. Ezcurra. 1991. Shade as a cause of the association between the cactus *Neobuxbaumia tetetzo* and the nurse plant *Mimosa luisana* in the Tehuacán Valley, Mexico. *Journal of Ecology* 79:961–71.

Valiente-Banuet, A., A. Bolongaro-Crevanna, O. Briones, E. Ezcurra, M. Rosas, H. Nuñez, G. Barnard, and E. Vazquez. 1991. Spatial relationships between cacti and nurse shrubs in a semi-arid environment in central Mexico. *Journal of Vegetation Science* 2:15–20.

Walter, H., and E. Stadelmann. 1974. A new approach to the water relations of desert plants. In *Desert Biology,* Vol. 2, ed. G. W. Brown, 214–302. Sydney: Academic Press.

PART III

Population and Community Ecology and Conservation

CHAPTER 10

Pollination Biology of Four Species of Sonoran Desert Columnar Cacti

THEODORE H. FLEMING

Introduction

Characterized by relatively warm winter temperatures and winter and summer rains, the Sonoran Desert of northwestern Mexico and southwestern United States is floristically the richest of North America's four deserts. Located between latitudes 24–34° north in the Mexican states of Sonora, Baja California Norte and Sur, and south-central Arizona in the United States, this desert includes the northernmost distributions of columnar cacti of the tribe Pachycereeae. Four of the northern members of this tribe are widely distributed in the Sonoran Desert, and their reproductive biology has been studied intensively in the past decade (Fleming et al., 1994; Murawski et al., 1994; Fleming et al., 1996, 1998, 2001). These species include *Carnegiea gigantea* (saguaro), *Lophocereus schottii* (senita), and *Pachycereus pringlei* (cardón) of subtribe Pachycereinae and *Stenocereus thurberi* (organ pipe) of subtribe Stenocereinae. Individually and collectively, these plants define the Sonoran Desert in many people's minds, and they are arguably among its most important species ecologically. In addition to providing floral and fruit resources for nectar- and fruit-eating animals, these plants provide shelter and, as is the case for cactophilic *Drosophila,* mating and incubation sites for many other species of animals.

To judge from the relatively recent geological appearance of the Sonoran Desert (Axelrod, 1979; chapter 1, this volume), it is likely that these four species of columnar cacti are evolutionarily young. Their phylogenetic affinities are clearly with species presently located much farther south in Mexico. The closest relative of *Carnegiea gigantea,* for example, is *Neobuxbaumia mezcalaensis* of Puebla and Oaxaca (Gibson and Horak, 1978; chapter 3, this volume). The closest relatives of *Lophocereus schottii* and *Pachycereus pringlei* are *P. marginatus* and *P. grandis,* respectively, of central Mexico (Gibson and Horak, 1978; Cota and Wallace, 1997). Only the closest relative of *Stenocereus thurberi*—*S. martinezii* of Sinaloa—is located close to the Sonoran Desert (Gibson, 1990). In addition to sharing many nonreproductive anatomical characteristics with their relatives, three of the four species (saguaro, cardón, and organ pipe) share a legacy of pollination by nocturnal vertebrates (i.e., bats). But because of their northern distributions, these species occur in areas where their potential vertebrate pollinators are seasonal migrants. It has been hypothesized that these migrant species might be temporally less reliable as pollinators than the permanently resident pollinators of more southern latitudes (Valiente-Banuet et al., 1996). If this is true, then we might expect to find more generalized pollination systems in vertebrate-pollinated columnar cacti in the Sonoran Desert compared with their southern relatives.

In this chapter, I will provide an overview of the pollination biology of the four species of Sonoran Desert columnar cacti. Questions that I will address in this review include: (1) Are the pollination systems of the three vertebrate-pollinated species more generalized than those of southern species? (2) To what extent has competition for vertebrate pollinators influenced the flowering biology of these three species? And (3) What conditions have favored the evolution of the highly specialized senita/senita-moth pollination mutualism?

Pollination Biology

FLORAL ANTHESIS AND MORPHOLOGY

As is typical of columnar cacti, each of the four species produce one-day flowers. Flowers of cardón, organ pipe, and senita open at sunset and, except during cool weather, usually close well before 1200 the next morning. During warm weather, flowers of senita generally close before sunrise, thereby preventing diurnal pollinators from gaining access to them. Flowers of organ pipe generally close before 0900, and those of cardón generally close before 1100. Order of closing in these species is positively correlated with their flower mass (Table 10.1). Flowers of the fourth species—saguaro—open well after sunset (between 2100 and 2300) and close the next afternoon. Thus, rank order of these species in terms of length of exposure to diurnal pollinators is saguaro > cardón > organ pipe > senita.

TABLE 10.1

Reproductive Characteristics of Four Species of Sonora Desert Columnar Cacti[a]

Parameter[b]	*Cardón*	*Saguaro*	*Organ Pipe*	*Senita*
Flower mass (g)	40.9 ± 1.7	33.4 ± 1.2	16.8 ± 1.1	1.6 ± 0.1
Fruit mass (g)	78.9 ± 9.4	52.6 ± 1.9	73.0 ± 5.3	4.5 ± 0.3
Seed mass (mg)	5.2 ± 0.2	0.7 ± 0.2	1.7 ± 0.1	3.6 ± 0.1
Seeds per fruit	1,329.0 ± 141	1,350.0 ± 99	351.0 ± 77	150.2 ± 8.5
Flowers per season	872	394	106	3,399
Fruits per season	170	246	24	850
Seeds per season	225,930	332,100	8,424	127,500

[a] Data from Fleming et al. (1996, 2001) and Sosa (1997).

[b] Mean ± standard error.

Flowers of saguaro, cardón, and organ pipe conform to a chiropterophilous (i.e., bat) pollination syndrome. In addition to nocturnal anthesis, their floral characteristics include funnelform shape, large size (their mass [wet weight] averages 16–42 g), light-colored tepals, hundreds of anthers producing about 0.5 g of pollen (i.e., about 1.5 million pollen grains) in saguaro and cardón, copious nectar production, and up to 1,700 ovules in their ovaries. In strong contrast, flowers of senita are small and delicate (Table 10.1). They bear relatively few anthers that produce modest amounts of pollen; they produce little or no nectar; and their ovaries contain fewer than 200 ovules.

Nectar Characteristics

Although their nectar contains similar percentages of sugar to cactus flowers pollinated by hawkmoths and hummingbirds, cacti pollinated by bats produce a greater total volume of nectar and more calories per flower (Fig. 10.1). Scogin (1985) reported similar frequencies of sucrose-dominated vs. hexose-dominated species in cacti pollinated by hawkmoths and bats. By producing 1.1–2.0 mL of nectar con-

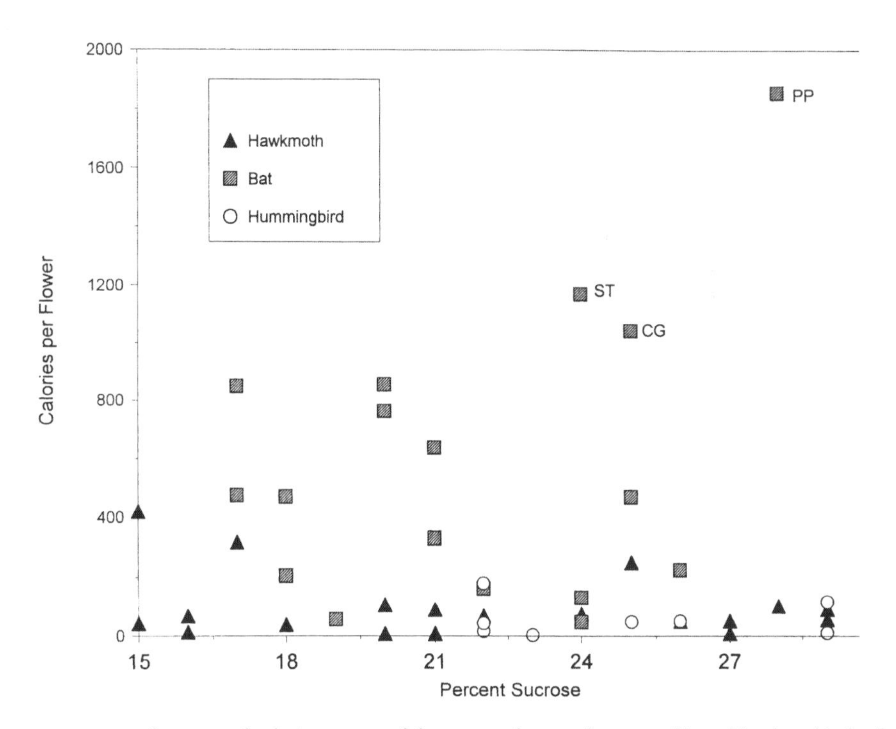

FIGURE 10.1. Sucrose and caloric content of the nectar of cactus flowers pollinated by three kinds of animals. Three species of Sonoran Desert columnar cacti include *Carnegiea gigantea* (CG; saguaro), *Pachycereus pringlei* (PP; cardón), and *Stenocereus thurberi* (ST; organ pipe). Data are from Scogin (1985); Fleming et al. (1996); and Nassar et al. (1997).

taining 25–30% sucrose equivalents of sugar, however, the three species of Sonoran Desert vertebrate-pollinated columnar cacti offer substantially larger energetic rewards than other cacti that have been studied (Fig. 10.1).

Nectar production in bat-pollinated columnar cacti tends to be unimodal, with peak secretion rates occurring before 2400. In four Venezuelan species studied by Nassar et al. (1997), nectar production ceased by sunrise. A similar situation occurred in three of those species studied by Petit and Freeman (1997) on Curaçao. Cardón and organ pipe of the Sonoran Desert conform to this pattern, except that nectar production continues past sunrise (Fig. 10.2). In contrast, saguaro has a bimodal nectar production schedule with one peak occurring around 0200 and another occurring well after sunrise, at about 0800 (Fig. 10.2). In all columnar cacti that have been studied, nectar sugar concentration changes during the night, with highest concentrations occurring around the time of peak secretion. Temporal changes in sugar concentration are relatively modest in cardón and organ pipe (i.e., from 26% to 28% and 23.5% to 25%, respectively), but they are substantial in Venezuelan cacti (e.g., from

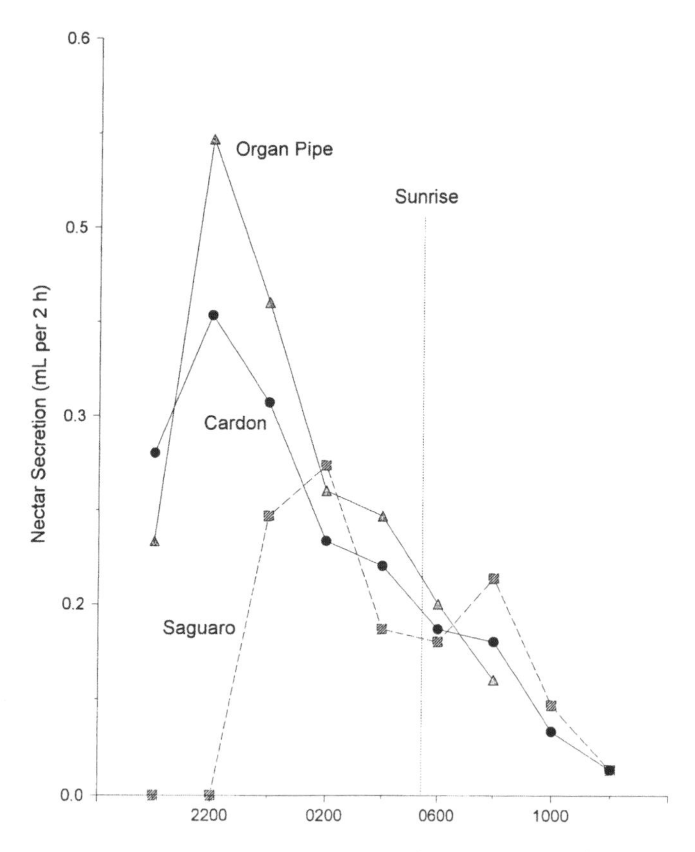

FIGURE 10.2. Nectar secretion curves for three species of Sonoran Desert columnar cacti. Redrawn from Fleming et al. (1996).

16.6–17.5% to 22% in *Pilosocereus moritzianus* and *Stenocereus griseus;* Fleming et al., 1996; Nassar et al., 1997).

Flowering Seasons

Columnar cacti in the Sonoran Desert are spring/summer bloomers, as is the case in most species of Mexican columnar cacti (Valiente-Banuet et al., 1996). Flowering begins in late March or early April in cardón, organ pipe, and senita and one or two weeks later in saguaro (Fig. 10.3). Cardón and saguaro have unimodal flowering curves that overlap extensively; their flowering seasons last for about two months. In contrast, organ pipe has an extended flowering season with peak numbers of flowers occurring in June, well after the flowering peaks of cardón and saguaro (Fig. 10.3a). The extended flowering season of organ pipe is caused by a polymorphism in the initiation of flowering. At our study site near Bahia Kino, Sonora, about 25% of the individuals of organ pipe begin flowering in early April, nearly 1.5 months before the rest of the population. Because of this polymorphism, the flowering season of organ pipe is about 50% longer than that of cardón and saguaro. Senita has a longer flowering season than the other three species, and its flowering season is multimodal, with individuals producing pulses of flowers at about monthly intervals (Fig. 10.3b).

Individuals of the four species of Sonoran Desert columnar cacti differ substantially in the number of flowers they produce daily and seasonally (Table 10.1). Senita produces nearly an order of magnitude more flowers per season (over 3,000) than the other three species. Of the bat-pollinated species, individuals of cardón produce over 800 flowers (with substantial annual variation) per season, whereas those of the much smaller organ pipe produce about 100 flowers per season.

Fruit and Seed Production

Like their flowers, fruits of cardón, saguaro, and organ pipe are relatively large, weighing nearly 80 g (wet weight) in the case of cardón , whereas those of senita are an order of magnitude smaller. Fruits of cardón and saguaro contain an average of 1,400 seeds; those of organ pipe and senita contain far fewer seeds. Seasonal fruit production per plant ranges from about 24 in organ pipe to about 850 in senita. Annual seed production per plant varies from about 8,400 in organ pipe to over 300,000 in saguaro (Table 10.1).

Factors Limiting Fruit Set

Factors limiting fruit set include resources a plant allocates to reproduction, pollen availability, and destruction of immature fruits by animals (usually larval insects). Each of these factors is known to affect fruit production in at least one of the four species of Sonoran Desert cacti. We have conducted hand-pollination experiments to determine whether fruit set in these cacti is resource- or pollen-limited (Fleming et

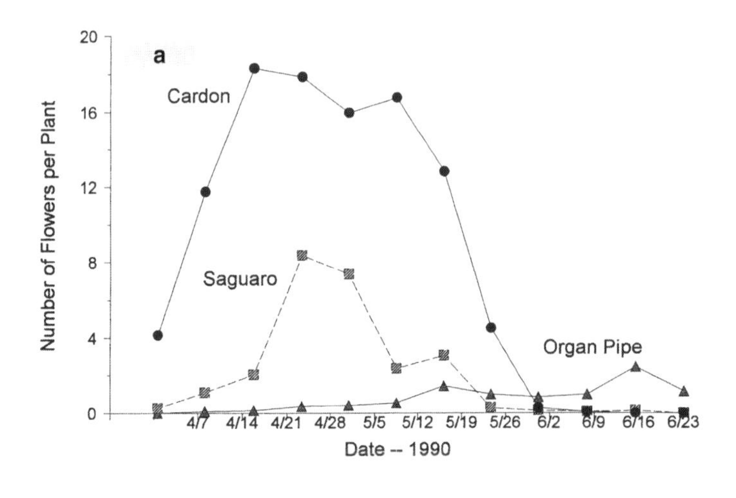

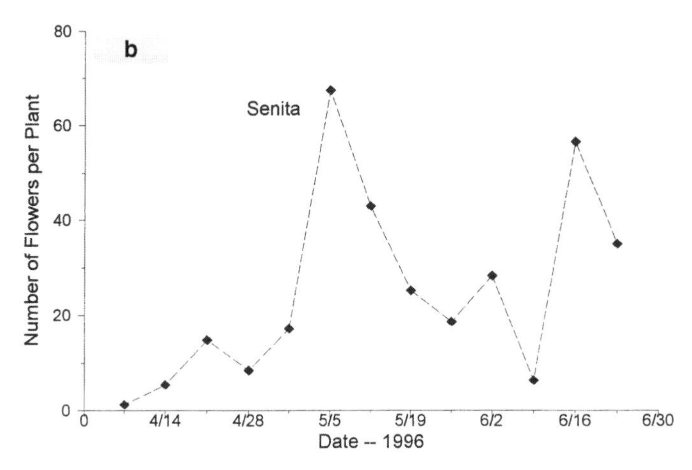

FIGURE 10.3. Flowering phenology of four species of Sonoran Desert columnar cacti. Data are mean number of flowers on 20 plants counted weekly. Redrawn from Fleming et al. (1996) and Holland and Fleming (1999).

al., 1996, 2001; Fleming and Holland, 1998, unpubl. data). Results indicate that fruit set is resource-limited in saguaro and senita. Fruit set resulting from open pollination is high in saguaro (about 65%; Fig. 10.4) and is similar to fruit set resulting from hand outcrossing. Similarly, fruit set averages about 46% in both open-pollinated and hand-outcrossed flowers of senita. In contrast, fruit set is pollen-limited in cardón (but only in females of this trioecious species; see section on gynodioecy and trioecy) and organ pipe. Fruit set in females of cardón is about twice as high (75% vs. 35%) in hand-outcrossed flowers compared with open-pollinated flowers. It is three times higher (90% vs. 30%) in organ pipe. Fruit set in cardón hermaphrodites is resource-limited.

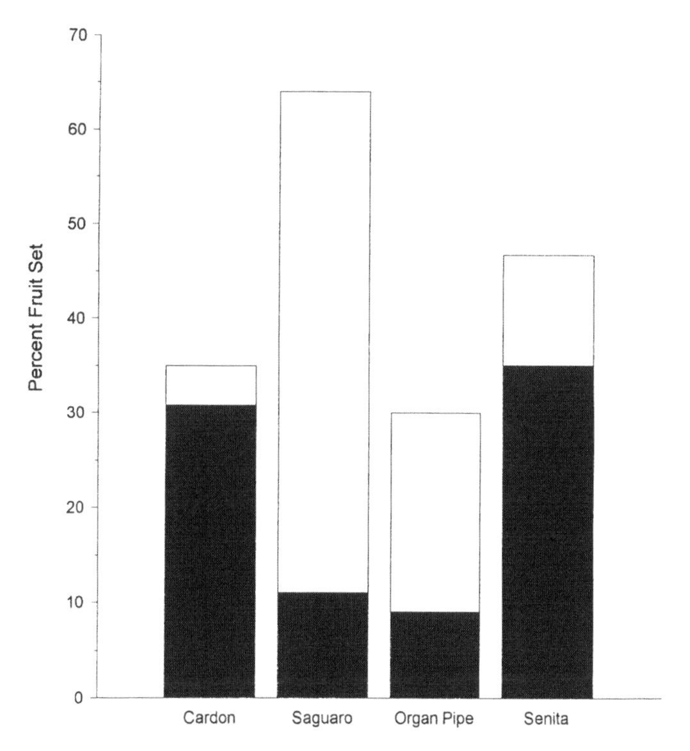

FIGURE 10.4. Contribution of nocturnal pollinators (shaded portion of each bar)—bats (in cardón, saguaro, and organ pipe) and the senita moth (in senita)—to open-pollinated fruit set in four species of Sonoran Desert columnar cacti. Data represent mean values based on two to six years of experimental data per species. Data are from Fleming et al. (2001).

In addition to aborting about 54% of its pollinated flowers because of resource (probably water) limitation, senita loses developing fruit to two species of larval seed predators. Senita's major seed predators are larvae of the senita moth, *Upiga virescens* (Pyralidae), which destroy about 30% of the fruit resulting from pollination by adult females of *U. virescens.* Larvae of another pyralid moth, *Cactobrosis fernandialis,* also destroy a few developing fruit (Fleming and Holland, 1998). Larvae of *C. fernandialis* also attack fruit of other columnar cacti (T. Fleming, unpubl. obs.)

Reduced fruit set owing to pollen limitation appears to be uncommon in other columnar cacti. Valiente-Banuet et al. (1996) reported that 550 of 560 flowers (98%) of *Neobuxbaumia tetetzo* set fruit. Nassar et al. (1997) conducted hand-pollination experiments with four species of Venezuelan cacti and found fruit set to be the same or higher in open-pollinated flowers than in hand-pollinated flowers in two species (*Stenocereus griseus* and *Subpilocereus horrispinus*). Fruit set in the other two species (*Pilosocereus moritzianus* and *Subpilocereus repandus*) was slightly, but not significantly, higher in hand-pollinated flowers than in open-pollinated flowers. In contrast to these studies and to our results for saguaro, McGregor et al. (1962) reported that fruit set in saguaro was substantially higher in hand-pollinated flowers than in open-pollinated flowers. Fruit set in the two treatments in 1959 was 99% and 48%, respectively; in 1960, it was 71% and 54%. These results suggest that factors limiting fruit set in saguaro vary geographically (or perhaps temporally).

Breeding Systems

HERMAPHRODITISM

As is the case in most angiosperms, hermaphroditism is likely the most common breeding system in the Cactaceae. Most of the columnar cacti that have been examined are self-incompatible hermaphrodites. These species include three of the four Sonoran Desert cacti we have studied (saguaro, organ pipe, and senita) as well as *Neobuxbaumia tetetzo* and *N. macrocephala* from the Tehuacán Valley, Mexico (Valiente-Banuet et al., 1997). Of the four Venezuelan species studied by Nassar et al. (1997), only *Pilosocereus moritzianus* was self-compatible.

GYNODIOECY AND TRIOECY

Cardón has a nonhermaphroditic breeding system featuring gynodioecy (separate females and hermaphrodites) and trioecy (separate males, females, and hermaphrodites) in different parts of its range (Fleming et al., 1998). Trioecy, which is an extremely rare breeding system in plants, occurs in the northern part of cardón's range in Sonora and in the southern part of its range in Baja California. Elsewhere, populations of cardón are gynodioecious. Males and females occur in relatively high frequencies (≥20%) in most trioecious populations. Hermaphrodites are about twice as common as females in gynodioecious populations.

Geographic variation in the form of cardón's breeding system appears to be associated with geographic variation in the abundance of the nectar-feeding bat, *Leptonycteris curasoae,* which is cardón's main pollinator (see section on pollinator relationships). Trioecious populations of cardón occur near maternity roosts of these wide-ranging bats. Outside the flight range of these bats, most populations of cardón are gynodioecious. A theoretical analysis indicates that when the abundance of pollinators varies geographically, trioecy can be an evolutionarily stable breeding system rather than a transitional state from hermaphroditism to dioecy (Maurice and Fleming, 1995). Recent work in Sonora, however, indicates that geographic variation in pollinator abundance cannot be the only factor affecting variation in the form of cardón's breeding system (F. Molina, M. Cervantes, T. Fleming, and S. Buchmann, unpubl. data). Geographic variation in cytoplasmic factors involved in sex determination are also likely to be involved in this variation.

Unlike most cacti, hermaphrodites in cardón are self-compatible, and their selfing rate is about 64% (Murawski et al., 1994). For males and females to persist in populations of self-fertilizing hermaphrodites, they must produce substantially more pollen and viable ovules, respectively, per breeding season than hermaphrodites if sex determination involves only nuclear genes. For example, in the absence of inbreeding depression in the progeny of hermaphrodites, relative fertilities of males and females must be twice that of hermaphrodites for trioecy or gynodioecy to persist. If inbreeding depression is substantial, then the relative fertilities of males and females need not be twice that of hermaphrodites (Charlesworth and Charlesworth, 1978). Detailed study of reproduction in cardón at Bahia Kino indicates that males and females annually produce about 1.6 times more pollen and seeds, respectively, than hermaphrodites but that the growth and survivorship of outcrossed progeny of females is no greater than that of the selfed progeny of hermaphrodites (Fleming et al., 1994; Sosa and Fleming, 1999). Therefore, fertility differences and inbreeding depression apparently cannot explain the persistence of gynodioecy and trioecy in cardón. If sex determination involves cytoplasmic as well as nuclear genes, however, then segregation ratios, rather than relative fertilities, will determine the frequencies of males and females in nonhermaphroditic species (Couvet et al., 1986). We clearly need data on the genetic basis of sex determination in cardón to resolve this issue.

How did cardón's nonhermaphroditic breeding system evolve? Murawski et al. (1994) have suggested that a ploidy event (i.e., chromosomal duplication) was the key step in the evolution of gynodioecy and trioecy from hermaphroditism. Unlike its closest relatives and other columnar cacti, which are diploid (Pinkava and McLeod, 1971; Pinkava et al., 1977), cardón is an autotetraploid that exhibits tetrasomic inheritance. This kind of genetic system produces low levels of individual homozygosity and reduces the effects of inbreeding depression (Husband and Schemske, 1997). The reduced cost of selfing in autotetraploids, in turn, probably resulted in relaxed selection for the maintenance of self-incompatibility, especially in desert plants occurring at low population densities, where selfing might be advantageous. Once the ability to self-fertilize was established in hermaphroditic populations of cardón, the stage was set for successful invasion into these populations by mutant individuals bearing a male sterility allele (i.e., females). Females may be able to produce more seeds per season and per lifetime than hermaphrodites because they produce smaller flowers and do not incur the cost of pollen production. If females outproduced hermaphrodites and their outcrossed progeny were initially more fit than the selfed progeny of hermaphrodites, then they could have persisted in populations, and cardón would have a gynodioecious breeding system. Finally, in areas of abundant pollinators, female sterile mutants (i.e., males) that produced more pollen per season and lifetime than hermaphrodites could successfully invade gynodioecious populations. In areas of low pollinator abundance, hermaphrodites would be more successful at passing on their genes through male and female function and would outcompete males for representation in future generations.

One other columnar cactus—*Neobuxbaumia mezcalaensis*—has a nonhermaphroditic breeding system. It is androdioecious (separate males and hermaphrodites), and hermaphrodites are self-incompatible (Valiente-Banuet et al., 1997).

Conditions that permit males to persist in populations of hermaphrodites are currently unknown.

Pollinator Relationships

Vertebrate-Pollinated Species

Based on timing of floral anthesis and flower morphology, Valiente-Banuet et al. (1996) concluded that at least 42 of 70 species in tribe Pachycereeae are primarily bat pollinated. Sixteen species have diurnal anthesis and are pollinated by birds, and four appear to be hawkmoth pollinated. Based on their nocturnal anthesis and flower morphologies, saguaro, cardón, and organ pipe clearly fall in the bat-pollinated pollination syndrome. But floral characteristics alone are not sufficient for identifying the most important pollinator(s) of a plant. Pollinator exclusion experiments are needed to make this determination rigorously.

The potential pollinators of saguaro, cardón, and organ pipe include nocturnal and diurnal vertebrates and insects. Nocturnal visitors include moths and three species of cactus-visiting bats: the phyllostomids *Choeronycteris mexicana* and *Leptonycteris curasoae* and the vespertilionid *Antrozous pallidus.* Of these bats, *L. curasoae* is by far the most common visitor to flowers of columnar cacti in the Sonoran Desert. *C. mexicana* is uncommon in the coastal lowlands of Sonora, and *A. pallidus* is an opportunistic visitor to flowers of columnar cacti and paniculate agaves (Herrera et al., 1993). Moths rarely visit these flowers. Diurnal flower visitors include native bees and honeybees as well as a variety of nectar-seeking birds, including white-winged doves, woodpeckers, hummingbirds, orioles, verdins, and house finches (Haughey, 1986; Schmidt and Buchmann, 1986; Fleming et al., 1996).

We have conducted pollinator exclusion experiments for three to five years per species at two different sites: Bahia Kino, Sonora, and Organ Pipe Cactus National Monument, Arizona. Results for Bahia Kino (Fig. 10.4) indicate that bats account for most fruit set only in cardón. In most years, bats are relatively minor pollinators of saguaro, a result also reported for two years at a different location in Arizona by McGregor et al. (1962). Only in 1995, when the flowering season of cardón was delayed one month by unusually cool weather, did bats account for most fruit set in saguaro at Bahia Kino. White-winged doves and honeybees are major pollinators of saguaro in the northern Sonoran Desert. At Bahia Kino, hummingbirds and honeybees account for over 67% of fruit set in organ pipe, with bats accounting for the remainder. In contrast, at Organ Pipe Cactus National Monument, where hummingbirds are scarce during the spring, bats accounted for 88% of control fruit set in organ pipe in 1997. In summary, in the Sonoran Desert bats are consistently important pollinators of cardón, but their importance varies temporally in saguaro and geographically in organ pipe. Diurnal pollinators are actually (saguaro, organ pipe) or potentially (cardón) important pollinators of the three species.

Senita/Senita-Moth Mutualism

The closest relative of senita is *Pachycereus marginatus,* a hummingbird-pollinated plant of central Mexico (Gibson and Horak, 1978). The pollination biology of senita includes an evolutionary shift to an obligate mutualism involving a pyralid moth, whose larvae destroy senita seeds and fruits. Senita moths account for most, but not all, of fruit set in senita (Fig. 10.4); two or three species of halictid bees (in the genera *Agapostemon, Augochlorella, Dialictis*) account for the remaining fruit set. The senita/senita-moth pollination system shares many features (e.g., reduced nectar production, active pollination by female moths, limited seed destruction by moth larvae) with the yucca/yucca-moth mutualism (Fleming and Holland, 1998). Comparison of the proportion of fruits produced by pollination by senita moths with the proportion of fruits destroyed by moth larvae indicates that the benefit:cost ratio of this interaction to the plant is about 3:1, a ratio that is similar to those in two other highly coevolved pollination mutualisms: the fig/fig-wasp and yucca/yucca-moth interactions.

Evolutionary Considerations

Shifts in Pollinator Importance

Bat pollination is widespread in the tribe Pachycereeae and probably represents the ancestral condition in this and perhaps other tribes (e.g., Cereeae, Browningieae) of Cactaceae. Studies in central Mexico, Venezuela, and Curaçao indicate that bats account for nearly 100% of fruit set in night-blooming columnar cacti (Sosa and Soriano, 1993; Petit, 1995; Valiente-Banuet et al., 1996, 1997; Nassar et al., 1997). This is not the case at the northern and southern limits of these cacti. For example, Sahley (1996) reported that bats, hummingbirds, and diurnal insects were important pollinators of *Weberbauerocereus weberbaueri* in the Andes of southwestern Peru. Like cardón, *W. weberbaueri* is a self-compatible, autotetraploid cactus that exhibits considerable intrapopulational variation in flower morphology. This variation probably represents relaxed selection for pollinator specialization owing to extensive year-to-year variation in the abundance of the migratory nectar-feeding bat, *Platalina genovensium* (Phyllostomidae). Furthermore, unlike other columnar cacti, about 44% of the flowers of *W. weberbaueri* initiate fruit development without pollinator visits, which further supports the hypothesis that pollinators of this species are either scarce or unreliable in the Peruvian Andes. El Niño events strongly affect flowering behavior, pollinator abundance, and reproductive success in this species (Sahley, 1996).

The four columnar cacti that we have studied represent the northernmost members of the tribe Pachycereeae. Major vertebrate pollinators of saguaro, cardón, and organ pipe in the Sonoran Desert include the bat *Leptonycteris curasoae,* the white-winged dove, and two or three species of hummingbirds. Each of these species is a seasonal migrant to the Sonoran Desert (Haughey, 1986; Cockrum, 1991; Baltosser and Scott, 1996; Wilkinson and Fleming, 1996). Fleming et al. (1996, 2001) have documented extensive year-to-year and site-to-site variation in the relative abundances and visitation rates of these pollinators to flowers of saguaro, cardón, and organ pipe. Spatial and temporal variation in the abundance of pollinating bats should favor a "bet-hedging" pollination strategy, in which flowers of nocturnally blooming cacti remain open and secrete nectar after sunrise. This nectar secretion schedule attracts diurnal pollinators and ensures that some fruit set will occur in the absence of bats.

Of the four species of Sonoran Desert cacti, senita has undergone the greatest change from an ancestral condition of diurnal hummingbird pollination to nocturnal pollination by a small moth. The selective pressures that caused this shift are currently unknown. Nor is it yet known how active pollination behavior evolved in the senita moth. It seems reasonable to postulate, however, that a shift in time of flower opening was the first step in the evolution of this specialized mutualism, although the conditions favoring this shift are not yet clear. Based on similarities between the yucca and senita mutualisms, Fleming and Holland (1998) proposed that a self-incompatible breeding system and resource-limited fruit set were also important during the evolution of these mutualisms.

Pollinator Redundancy and Complementarity

Our pollinator exclusion experiments allow us to quantify the extent to which nocturnal and diurnal pollinators are "redundant" or "complementary" in terms of overall fruit set in the three species of vertebrate-pollinated cacti. Redundancy (sensu Lawton, 1994) occurs when the sum of fruit set effected by nocturnal and diurnal pollinators separately exceeds open-pollinated fruit set. Complementarity (sensu Tilman, 1980) occurs when the sum of fruit set by nocturnal and diurnal pollinators equals open-pollinated fruit set. The concepts of redundancy and complementarity have important conservation implications. In a redundant pollination system, loss of one group of pollinators (e.g., bats) will not necessarily result in reduced fruit set because other pollinators (e.g., diurnal species) can compensate for this loss. In a complementary pollination system, however, loss of one group of pollinators will result in significantly reduced fruit set, because no compensation can occur.

As I have used them here, the concepts of complementarity and redundancy refer only to quantitative aspects of fruit set. They do not refer to qualitative (i.e., genetic) aspects of fruit set. If pollinators differ in the distance they typically transport pollen (e.g., between nearest neighbors vs. between distant neighbors) and/or in the number of different pollen genotypes they deposit on a stigma, then the loss of pollinators could alter mating patterns within populations and gene flow between populations in both complementary and redundant pollination systems. As an extreme example, consider the potential genetic consequences of visits by bats and honeybees. Compared with sedentary honeybees, which are major diurnal pollinators of the three vertebrate-pollinated cacti, wide-ranging *Leptonycteris* bats probably transport pollen much longer distances between plants and deposit pollen from a greater number of individuals on each stigma they contact. If bats were to disappear from the Sonoran Desert, it is likely that pollen flow would be greatly reduced within and between populations. Elsewhere, my colleagues and I will present data quantifying the qualitative effects of bats, birds, bees, and senita moths on fruit set in the four species of cacti. The macrogeographic genetic consequences of bat-, bird-, and insect-mediated gene flow in these species are discussed in chapter 6.

In the species we have studied, cardón, saguaro, and senita have redundant pollination systems, whereas that of organ pipe is complementary. In cardón, bats and diurnal pollinators account for about 88% and 53% of open-pollinated fruit set, respectively, at Bahia Kino. If bats were to disappear, fruit set in cardón would still be substantial. Similarly, fruit set effected by bats and diurnal pollinators in saguaro is 45% and 90% of open-pollinated fruit set, respectively. Loss of bats would have little effect on fruit set in this species. In contrast, maximum fruit set in organ pipe depends on the joint effect of bats and diurnal pollinators, particularly hummingbirds. The sum of fruit set resulting from visits by bats and diurnal pollinators equaled open-pollinated fruit set in five of six years at Bahia Kino. Loss of either group would result in substantially reduced fruit set in this species.

Pollinator exclusion experiments in senita indicate that visits by diurnal halictid bees result in about 34% of open-pollinated fruit set and are thus redundant with the senita moth, whose visits result in 83% of open-pollinated fruit set. Flower closing in senita is temperature dependent (J. N. Holland and T. Fleming, unpubl. data); these delicate flowers remain open after sunrise only on cool mornings. Thus, bees are redundant with senita moths only under certain climatic conditions. During warm weather, senita moths are the exclusive pollinators of senita flowers. Copollinators are not present in the yucca/yucca-moth mutualism, and this raises the question: Is senita still in transition from a pollination system that included both diurnal (the presumed ancestral condition) and nocturnal insects to a nocturnal-only pollination system? That some individuals of senita still produce nectar (in contrast with total loss of nectar production in yucca) provides some support for the "in-transition" hypothesis. Nonetheless, it is also possible that senita will remain polymorphic in nectar production as a hedge against becoming entirely dependent on a single pollinator (presuming that nectar is more important for attracting bees than moths).

Competition among Columnar Cacti for Pollinators

Substantial overlap in flowering times appears to characterize some, but not all, populations of night-blooming columnar cacti. Flowering overlap appears to be extensive, for example, among species living in the Tehuacán Valley, Mexico (Valiente-Banuet et al., 1996, 1997). In contrast, flowering overlap is relatively low in three species of co-occurring species in the dry intermontane valleys of southwestern Venezuela and western Colombia (Soriano et al., 1991; Ruiz et al., 1997). Whenever two or more species with the same pollination syndrome flower concurrently, the question arises: Are these species likely to compete for pollinator visits? The answer to this question depends on at least two factors: (1) the abundance of effective pollinators and (2) the extent to which fruit set is pollen-limited. Competition for pollinators is likely to occur only when pollinators are scarce relative to the number of open flowers and/or when fruit set is pollen-limited.

In the northern Sonoran Desert, flowering overlap is relatively high between the three vertebrate-pollinated species (Fig. 10.3), the abundance of pollinators varies seasonally, and fruit set is pollen-limited in two species (in females of cardón and in organ pipe). This situation would seem to promote significant interspecific competition between these species. Do these species compete for pollinators, and if not, how is competition avoided? Because cardón and organ pipe have the most similar flower-opening and nectar-secretion schedules, they would seem to be the most likely competitors for flower visits by *Leptonycteris* bats. Seasonal differences in times of peak flowering (Fig. 10.3) probably reduce some, but not all, of this competition. But some individuals of organ pipe flower in April, during the blooming peaks of both cardón and saguaro. Because individuals of organ pipe produce considerably fewer flowers per night than the other two species, they would appear to be inferior competitors for bat visits. Do early-blooming individuals of organ pipe experience lower fruit set owing to interspecific competition for pollinators than late-blooming individuals?

In fact, detailed analysis of flowering and fruit set in organ pipe indicates that early-flowering plants produce more fruit per season than do late-flowering plants (T. Fleming and K. Conway, unpubl. data). These results indicate that, contrary to expectations, early-flowering individuals of organ pipe are not necessarily at a competitive disadvantage to cardón and saguaro. Not only do they appear to attract pollinators, but they apparently do not suffer from reduced fruit set when their stigmas receive heterospecific pollen. If bats are relatively nonselective in their visits to cactus flowers, individuals of organ pipe should frequently receive heterospecific pollen because of their much lower flower density. Hand-pollination experiments have revealed, however, that fruit set in organ pipe is not zero when its flowers receive heterospecific pollen. Whereas fruit set in cardón and saguaro was zero when their stigmas received organ pipe pollen, fruit set in organ pipe was 27% with cardón pollen (compared with 47% with organ pipe pollen) and 13% with saguaro pollen (compared with 54% with organ pipe pollen) (Alcorn et al., 1959; T. Fleming and K. Conway, unpubl. data). The mechanism that allows organ pipe flowers to produce fruits containing apparently viable seeds when they receive heterospecific pollen is currently unknown but probably involves agamospermy (seed set without fertilization; Negron-Ortiz, 1998). Such a mechanism obviously reduces organ pipe's competitive disadvantage when it blooms with cardón and saguaro.

Because it produces fewer calories per flower and fewer flowers per night (Fig. 10.1, Table 10.1), saguaro should also be at a competitive disadvantage to cardón when both species bloom together in April and May. A shift in its flower opening and closing times as well as its nectar secretion schedule apparently allows saguaro to avoid competing for bat visits with cardón (and perhaps with organ pipe). By opening later at night and closing in mid-afternoon, saguaro flowers are exposed to diurnal pollinators longer than they are exposed to nocturnal pollinators. This shift to greater reliance on diurnal pollinators results in higher and less variable fruit set from year to year in saguaro than in cardón and organ pipe (Fleming et al., 2001). Along with physiological adaptations (chapter 9, this volume), this shift also allows saguaro a more northerly distribution than that achieved by other species of Sonoran Desert columnar cacti.

Conclusions

Columnar cacti of the Pachycereinae and Stenocereinae moved north from their ancestral ranges as the climate of northwestern Mexico, Baja California, and southwestern Arizona became drier. During this process, three species—cardón, saguaro, and organ pipe—have undergone subtle changes in their pollination biology (primarily in terms of times of flower opening and closing) and currently depend less on bats for maximum fruit set than their southern relatives. In this sense, these species have more generalized pollination systems than their relatives. A reasonable hypothesis to explain this shift is that bats are less reliable pollinators at the northern edge of their range than they are farther south. A further consequence of bat scarcity in the north is potentially strong competition for bat pollination, which has had an important effect on timing of flower anthesis in saguaro and on the effect of heterospecific pollen on stigmas of organ pipe. Cardón has evolved a nonhermaphroditic breeding system that involves gynodioecy and trioecy. Again, the distribution and abundance of *Leptonycteris* bats appear to have had an influence on this process. A fourth species of cactus—senita—has undergone a major shift from hummingbird pollination to pollination by a specialized moth. Reliance on a nonmigratory insect rather than on migratory vertebrates seems to have favored the evolution of this specialized mutualism.

Summary

In this chapter I describe the pollination biology of four common species of Sonoran Desert columnar cacti of tribe Pachycereeae: *Carnegiea gigantea* (saguaro), *Pachycereus pringlei* (cardón), and *Lophocereus schottii* (senita) of the subtribe Pachycereinae and *Stenocereus thurberi* (organ pipe) of subtribe Stenocereinae. In terms of floral anthesis, flower morphology, and nectar characteristics, three of the species (saguaro, cardón, and organ pipe) clearly display affinities with vertebrate-pollinated cacti of southern Mexico. Senita has a highly derived pollination system and is coevolved with a specialized lepidopteran (moth) pollinator. Three of the four species have hermaphroditic, self-incompatible breeding systems whereas cardón has a self-compatible, geographically variable breeding system involving gynodioecy and trioecy. Flowering begins in the spring in each of these species and lasts for about two months in saguaro and cardón and four to five months in senita and organ pipe. Competition for pollinators probably has influenced timing of peak flowering in organ pipe and time of flower opening and closing in saguaro. Major pollinators include birds and bees (in saguaro and organ pipe), bats (in cardón), and the pyralid moth *Upiga virescens* (in senita). Observations over several years and at two sites in the Sonoran Desert indicate that the importance of particular pollinators varies geographically in organ pipe and temporally in saguaro. Except for the highly specialized pollination system of senita, the pollination biology of these northern cacti appears to be somewhat more generalized regarding pollinator relationships than that of columnar cacti in southern Mexico and Venezuela.

Resumen

En este capítulo se describe la biología de la polinización de cuatro especies de cactáceas columnares. Estas son comunes del desierto Sonorense, pertenecientes a la tribu Pachycereeae: *Carnegiea gigantea* (saguaro), *Pachycereus pringlei* (cardón), y *Lophocereus schottii* (senita) de la subtribu Pachycereinae y *Stenocereus thurberi* (organ pipe) de la subtribu Stenocereinae. En términos de una antésis floral, morfología floral, y características del néctar, tres de las especies (saguaro, cardón, y organ pipe) muestran claras afinidades con cactus polinizados por vertebrados del sur de México. Senita tiene un sistema de polinización altamente derivado y ha coevolucionado con un lepidóptero especializado (polilla). Tres de las cuatro especies tienen un sistema reproductivo caracterizado, por flores hermafroditas y auto-incompatibilidad, mientras que el cardón tiene un sistema reproductivo auto-compatible, que varía geográficamente e incluye las condiciones ginodióica y trióica. La floración se inicia en la primavera en todas las especies y dura alrededor de 2 meses en saguaro y cardón y 4–5 meses en senita y organ pipe. La competencia por polinizadores ha influído probablemente en el tiempo al que se alcanza el máximo de floración en organ pipe y el tiempo de apertura y cierre floral en saguaro. Los principales polinizadores incluyen a las aves y abejas (en saguaro y organ pipe), murciélagos (en cardón), y la polilla pirálida *Upiga virescens* (en senita). Observaciones durante varios años y en dos localidades differentes en el desierto de Sonora indican que la importancia de polinizadores particulares varía geograficamente en organ pipe y temporalmente en saguaro. Excepto por el sistema de polinización altamente especializado de senita, la biología de la polinización de estas cactáceas, parece ser algo más generalizada en referencia a relaciones con polinizadores mantienen los que cactus columnares en el sur de México y Venezuela.

Acknowledgments

I thank several landowners for permission to study bats and cacti on their property at Bahia Kino and the Mexican government for research permits. The U.S. National Park Service provided permission to work at Organ Pipe Cactus National Monument. I was assisted by and collaborated with many people, including M. Horner, M. Tuttle, C. Sahley, V. Sosa, S. Maurice, N. Holland, J. Hamrick, J. Nason, F. Molina, and S. Buchmann in my field and laboratory studies. Financial support for our studies was provided by the National Geographic Society, National Fish and Wildlife Foundation, and the U.S. National Science Foundation.

REFERENCES

Alcorn, S. M., S. E. McGregor, G. D. Butler, and E. B. Kurtz. 1959. Pollination requirements of the saguaro (*Carnegiea gigantea*). *Cactus and Succulent Journal* 31:39–41.

Axelrod, D. I. 1979. Age and origin of the Sonoran desert vegetation. *Occasional Papers, California Academy of Sciences* 132:1–74.

Baltosser, W. H., and P. E. Scott. 1996. Costa's hummingbird (*Calypte costae*). In *The Birds of North America,* No. 251, eds. A. Poole and F. Gill, 1–32. Philadelphia and Washington, D.C.: The Academy of Natural Sciences and the American Ornithologist's Union.

Charlesworth, B., and D. Charlesworth. 1978. A model for the evolution of dioecy and gynodioecy. *American Naturalist* 112:975–97.

Cockrum, E. L. 1991. Seasonal distribution of northwestern populations of the long-nosed bats, *Leptonycteris sanborni* Family Phyllostomidae. *Anales del Instito Biologia Universidad Nacional Autonoma de México, Serie Zoologica* 62:181–202.

Cota, J. H., and R. S. Wallace. 1997. Chloroplast DNA evidence for divergence in *Ferocactus* and its relationships to North American columnar cacti (Cactaceae: Cactoideae). *Systematic Botany* 22:529–42.

Couvet, D., F. Bonnemaison, and P-H. Gouyon. 1986. The maintenance of females among hermaphrodites: the importance of nuclear-cytoplasmic interactions. *Heredity* 57:325–30.

Fleming, T. H., and J. N. Holland. 1998. The evolution of obligate mutualisms: the senita cactus and senita moth. *Oecologia* 114:368–75.

Fleming, T. H., S. Maurice, S. Buchmann, and M. D. Tuttle. 1994. Reproductive biology and the relative fitness of males and females in a trioecious cactus, *Pachycereus pringlei. American Journal of Botany* 81:858–67.

Fleming, T. H., M. D. Tuttle, and M. A. Horner. 1996. Pollination biology and the relative importance of nocturnal and diurnal pollinators in three species of Sonoran Desert columnar cacti. *Southwestern Naturalist* 41:257–69.

Fleming, T. H., S. Maurice, and J. L. Hamrick. 1998. Geographic variation in the breeding system and the evolutionary stability of trioecy in *Pachycereus pringlei* (Cactaceae). *Evolutionary Ecology* 12:279–89.

Fleming, T. H., C. T. Sahley, J. N. Holland, J. Nason, and J. L. Hamrick. 2001. Sonoran Desert columnar cacti and the evolution of generalized pollination systems. *Ecological Monographs* 71:511–30.

Gibson, A. C. 1990. The systematics and evolution of subtribe Stenocereinae. 8. Organ pipe cactus and its closest relatives. *Cactus and Succulent Journal* 62:13–24.

Gibson, A. C., and K. E. Horak. 1978. Systematic anatomy and phylogeny of Mexican columnar cacti. *Annals of the Missouri Botanical Garden* 65:999–1057.

Haughey, R. A. 1986. Diet of desert-nesting western white-winged doves, *Zenaida asiatica mearnsi.* Master's thesis, Arizona State University, Tempe.

Herrera, L. G., T. H. Fleming, and J. S. Findley. 1993. Geographic variation in the carbon composition of the pallid bat, *Antrozous pallidus,* and its dietary implications. *Journal of Mammalogy* 74:601–6.

Holland, J. N., and T. H. Fleming. 1999. Mutualistic interactions between *Upiga virescens* (Pyralidae), a pollinator seed-consumer, and *Lophocereus schottii* (Cactaceae). *Ecology* 80:2074–84.

Husband, B. C., and D. W. Schemske. 1997. The effect of inbreeding in diploid and tetraploid populations of *Epilobium angustifolium* (Onagraceae): implications for the genetic basis of inbreeding depression. *Evolution* 51:737–46.

Lawton, J. H. 1994. What do species do in ecosystems? *Oikos* 71:367–74.

McGregor, S. E., S. M. Alcorn, and G. Olin. 1962. Pollination and pollinating agents of the saguaro. *Ecology* 43:259–67.

Maurice, S., and T. H. Fleming. 1995. The effect of pollen limitation on plant reproductive systems and the maintenance of sexual polymorphisms. *Oikos* 74:55–60.

Murawski, D., T. H. Fleming, K. Ritland, and J. Hamrick. 1994. The mating system of an autotetraploid cactus, *Pachycereus pringlei. Heredity* 72:86–94.

Nassar, J., N. Ramirez, and O. Linares. 1997. Comparative pollination biology of Venezuelan columnar cacti and the role of nectar-feeding bats in their sexual reproduction. *American Journal of Botany* 84:918–27.

Negron-Ortiz, V. 1998. Reproductive biology of a rare cactus, *Opuntia spinossima* (Cactaceae), in the Florida Keys: why is seed set very low? *Sexual Plant Reproduction* 11:208–12.

Petit, S. 1995. The pollinators of two species of columnar cacti on Curaçao, Netherlands Antilles. *Biotropica* 27:538–41.

Petit, S., and C. E. Freeman. 1997. Nectar production of two sympatric species of columnar cacti. *Biotropica* 29:175–83.

Pinkava, D. J., and M. G. McLeod. 1971. Chromosome numbers in some cacti of western North America. *Brittonia* 23:171–76.

Pinkava, D. J., L. A. McGill, and T. Reeves. 1977. Chromosome numbers in some cacti of western North America III. *Bulletin of the Torrey Botanical Club* 104:105–10.

Ruiz, A., M. Santos, P. Soriano, J. Cavelier, and A. Cadena. 1997. Relaciones mutualisticas entre el murcielago *Glossophaga longirostris* y las cactaceas columnares en la zona arida de La Tatacoa, Colombia. *Biotropica* 29:469–79.

Sahley, C. T. 1996. Bat and hummingbird pollination of an autotetraploid columnar cactus, *Weberbauerocereus weberbaueri* (Cactaceae). *American Journal of Botany* 83:1329–36.

Schmidt, J. O., and S. L. Buchmann. 1986. Floral biology of the saguaro (*Cereus giganteus*). I. Pollen harvest by *Apis mellifera. Oecologia* 69:491–98.

Scogin, R. 1985. Nectar constituents of the Cactaceae. *Southwestern Naturalist* 30:77–82.

Soriano, P. J., M. Sosa, and O. Rossel. 1991. Feeding habits of *Glossophaga longirostris* Miller (Chiroptera: Phyllostomidae) in an arid zone of the Venezuelan Andes. *Revista Biologica Tropical* 39:263–68.

Sosa, M., and P. J. Soriano. 1993. Solapamiento de dieta entre *Leptonycteris curasoae* y *Glossophaga longirostris* (Mammalia: Chiroptera). *Revista Biologica Tropical* 41:529–32.

Sosa, V. 1997. Dispersal and recruitment ecology of columnar cacti in the Sonoran Desert. Ph.D. diss., University of Miami, Coral Gables, Florida.

Sosa, V., and T. H. Fleming. 1999. Seedling performance in a trioecious cactus, *Pachycereus pringlei:* effects of maternity and paternity. *Plant Systematics and Evolution* 218:145–51.

Tilman, D. 1980. Resources: a graphical-mechanistic approach to competition and predation. *American Naturalist* 116:362–93.

Valiente-Banuet, A., M. C. Arizmendi, A. Martinez-Rojas, and L. Dominquez-Canesco. 1996. Geographical and ecological correlates between columnar cacti and nectar-feeding bats in Mexico. *Journal of Tropical Ecology* 12:103–19.

Valiente-Banuet, A., M. C. Arizmendi, A. Martinez-Rojas, and P. Davila. 1997. Pollination of two columnar cacti (*Neobuxbaumia mezcalaensis* and *Neobuxbaumia macrocephala*) in the Tehuacán Valley, central Mexico. *American Journal of Botany* 84:452–55.

Wilkinson, G. S., and T. H. Fleming. 1996. Migration and evolution of lesser long-nosed bats, *Leptonycteris curasoae,* inferred from mitochondrial DNA. *Molecular Ecology* 5:329–39.

CHAPTER 11

Biotic Interactions and Population Dynamics of Columnar Cacti

Alfonso Valiente-Banuet
María del Coro Arizmendi
Alberto Rojas-Martínez
Alejandro Casas
Héctor Godínez-Alvarez
Carlos Silva
Patricia Dávila-Aranda

Introduction

Recent studies indicate that the most diverse and densest forests of columnar cacti occur within Mexico (Valiente-Banuet et al., 1996, 1997a). A total of 70 species of columnar cacti belonging to the tribes Pachycereeae and Cereeae have been described for this country, many of which constitute the physiognomically and structurally dominant elements of tropical deciduous forests as well as arid shrublands, which we refer here to as "columnar cactus forests" (Valiente-Banuet et al., 1995, 1996). The distribution of these forests is concentrated in the south-Pacific drainage in south-central Mexico, which includes the Tehuacán Valley and the Balsas River Basin (Fig. 11.1). This region contains the highest diversity of columnar cacti, including 45 of the 70 species reported for Mexico. Two continuous distribution belts derive from this region: one along the Pacific coast and the other along the coast of the Gulf of Mexico. Both belts generally have fewer species of columnar cacti than the south-Pacific drainage has (Valiente-Banuet et al., 1996).

Floral and faunal inventories conducted in areas of high columnar cactus diversity indicate that a considerable number of species inhabit these communities. In the Tehuacán Valley, for instance, Arizmendi and Espinosa de los Monteros (1996) reported a total of 91 species of birds, ten of which are endemic to columnar cactus forests, and Rojas-Martínez and Valiente-Banuet (1996) found 34 species of bats. For the same region, Valiente-Banuet et al. (2000) reported nine vegetation types that are physiognomically and structurally dominated by columnar cacti and the occurrence of nearly 350 species of trees and shrubs in these plant communities.

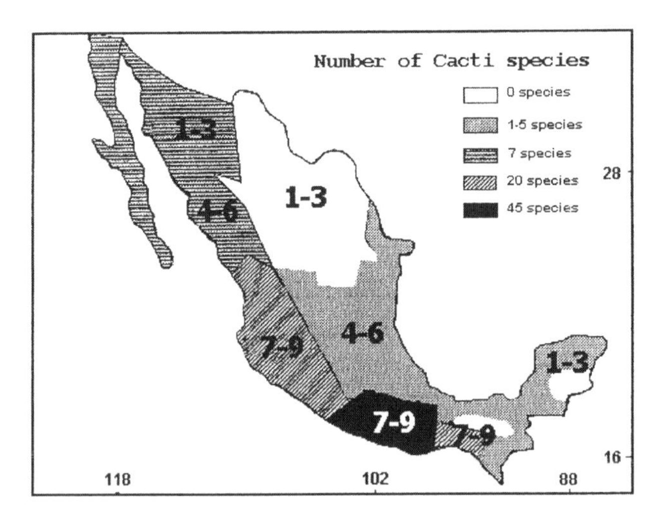

FIGURE 11.1. Distribution map of species of columnar cacti and nectar-feeding bats (annotated values in the map) in Mexico. Modified from Valiente-Banuet et al. (1996).

Many of these species inhabiting columnar cactus forests exert a considerable influence on pollination, seed dispersal, and establishment of columnar cacti, as well as predators and competitors with these plants, indicating that biotic interactants may play important roles in the establishment and growth of columnar cacti (Fleming et al., 1996; Nassar et al., 1997; Valiente-Banuet et al., 1997a,b; Godínez-Alvarez and Valiente-Banuet, 1998). These studies provide an ecological context for determining the relative importance of biotic interactions in the population dynamics of cacti. This chapter summarizes our present knowledge about how abiotic and biotic factors interact to influence the distribution and abundance of particular species, leading to particular patterns of survival, fecundity, and growth. To do this, we analyze ecological processes occurring at the different life stages of the life cycle of columnar cacti. To determine their relative importance in the population dynamics, we analyze cactus demography using matrix models.

Ecological Processes Occurring during the Life Cycle

Columnar cacti are long-lived plants whose lifespans reach hundreds of years (Shreve, 1910; Steenbergh and Lowe, 1977, 1983). At different stages (e.g., seed, seedling, juvenile, adult), the cacti are exposed to a considerable number of mortality factors, including seed and seedling predation, drought, and winter cold (Steenbergh and Lowe, 1977). Traditionally, early stages of the life cycle have been considered to be the most critical for maintaining viable populations (Steenbergh and Lowe, 1969, 1977; Valiente-Banuet and Ezcurra, 1991), and therefore the ecological processes occurring during these earliest life-cycle stages are considered here.

Seed Germination

In all of the few studies that have analyzed seed germination under field conditions, water and temperature are considered the most important limiting factors for this process. In some studies, it has been shown that imbibition of seeds enhances seed germination of *Carnegiea gigantea, Stenocereus thurberi,* and *S. griseus* (Alcorn and Kurtz, 1959; McDonough, 1964; Martínez-Holguín, 1983). Dubrowsky (1996, 1998) reported that the seeds of the columnar cacti *C. gigantea, Pachycereus pecten-aboriginum, S. thurberi,* and *S. gummosus,* when subjected to hydration-dehydration cycles of different lengths germinated faster and accumulated higher biomass compared with untreated seeds. This "seed hydration memory," in which seeds have the ability to retain during dehydration the physiological changes promoted during hydration, has important ecological implications because it may facilitate seed germination under conditions of unpredictable rainfall (Dubrowsky, 1996).

Rojas-Aréchiga et al. (1997) have reported that germination of columnar cactus seeds occurs over a wide range of temperatures, ranging from 15°C to 40°C (Nobel, 1988; Rojas-Aréchiga and Vazquez-Yanes, 2000). In genera such as *Cephalocereus, Neobuxbaumia,* and *Pachycereus,* optimal temperatures for germination lie between 20°C and 30°C (Rojas-Aréchiga et al., 1997).

Seedling Establishment and Growth

In columnar cactus forests, the establishment phase of perennial plants occurs under unpredictable conditions of precipitation and in soils with high temperatures and low water content. In these areas, columnar cacti recruit beneath the canopies of perennial "nurse plants" (Turner et al., 1966; Steenbergh and Lowe, 1969, 1977; Valiente-Banuet and Ezcurra, 1991; Valiente-Banuet et al., 1991a; Arriaga et al., 1993), which modify the environment beneath their canopies and enhance seedling establishment (Turner et al., 1966; Valiente-Banuet and Ezcurra, 1991). Nurse plants provide protection against direct solar radiation, lower soil temperatures, and higher soil moisture for seed germination and early seedling survival (Turner et al., 1966; Franco and Nobel, 1989; Valiente-Banuet and Ezcurra, 1991). In marginal populations in the northern distribution of columnar cacti, nurse plants provide protection against winter temperatures, decreasing the risk of frost injury (Brum, 1973).

In all cases studied, a clumped distribution of cacti beneath the canopy of a nurse plant has important ecological consequences related to intra- and interspecific competition. The saguaro, *Carnegiea gigantea,* effectively competes intraspecifically, which affects the apical growth and reproductive potential of individuals (McAuliffe and Janzen, 1986). Interspecific competition between columnar cacti and their nurse plants has also been reported for *Carnegiea gigantea* and *Cercidium microphyllum* (McAuliffe, 1984), and for *Neobuxbaumia tetetzo* and *Mimosa luisana* (Valiente-Banuet et al., 1991b; Flores-Martínez et al., 1994). In both cases, cacti outcompete their nurse plants by affecting their growth and decreasing their production of vegetative and reproductive parts.

Sexual Reproduction in Columnar Cacti

Columnar cacti as a group rely heavily on animals for pollination (Gibson and Nobel, 1986; Valiente-Banuet et al., 1996). Of the 70 species of columnar cacti found in Mexico, 72% have a bat-flower syndrome characterized by nocturnal anthesis, large whitish, bowl-shaped flowers with a strong odor, large quantities of pollen, large or many anthers, and abundant nectar. This syndrome is common in most of the species of *Pachycereus* and *Stenocereus* and in all the species of *Carnegiea, Mitrocereus,* and *Neobuxbaumia.* About 6% of the species have solitary diurnal bowl-shaped flowers pollinated by diurnal insects such as bees; these occur in the genera *Escontria, Myrtillocactus,* and *Polaskia.* Another 6% (in the genus *Machaerocereus*) have strongly scented night-blooming flowers with white or whitish perianths and long, slender nectar-containing floral tubes that are visited by hawkmoths. The remaining 16% (mostly in the genus *Stenocereus*) have red, tubular hummingbird flowers (Valiente-Banuet et al., 1996).

Not surprisingly, the geographic distribution of columnar cacti overlaps with the distribution of nectar-feeding bats in North America (Fig. 11.1; Valiente-Banuet et al., 1996), and different studies have indicated that bats are the most effective pollinators of giant columnar cacti in south-central Mexico and Venezuela (Valiente-Banuet et al., 1996, 1997a,b; Sosa and Soriano, 1992; Nassar et al., 1997; Casas et al., 1999). Three species of nectar-feeding bats, *Leptonycteris curasoae, L. nivalis,* and *Choeronycteris mexicana,* have been reported as specialized nectar and pollen consumers of plants (Alvarez and González, 1970; Cockrum, 1991; Valiente-Banuet et al., 1996, 1997a,b).

In the Sonoran Desert at latitude 29° north, pollination studies conducted with chiropterophilous cacti such as *Carnegiea gigantea, Pachycereus pringlei,* and *Stenocereus thurberi* (McGregor et al., 1959; Alcorn et al., 1959, 1961, 1962; Fleming et al., 1996; chapter 10, this volume) indicate that they are pollinated by a wide spectrum of animals, including birds, bats, and bees. Fruit set by diurnal (birds and bees) and nocturnal (bats) pollinators of *C. gigantea* were, respectively, 68% and 40%; of *S. thurberi* were 21% and 8%; and 28% and 31% for *P. pringlei.* These results contrast with the patterns reported for the tropics, in which all species studied are self-incompatible and are pollinated exclusively by bats with fruit sets ranging between 75% and 98% in Mexico and between 46% and 76% in Venezuela (Sosa and Soriano, 1992; Valiente-Banuet et al., 1996, 1997a,b; Nassar et al., 1997; Casas et al., 1999). This apparent dichotomy found within and outside the tropics among columnar cacti with bat-pollinated flowers has been explained as a consequence of the predictability of pollinators throughout the year (Valiente-Banuet et al., 1997a,b). Such predictability is lower in extratropical areas, where nectar-feeding bats are seasonal migrants to Arizona and northern Sonora from the tropical deciduous forests of Sonora and probably Sinaloa (Rojas-Martínez et al. [1999]; but see Wilkinson and Fleming [1996] for a different view on this topic). In contrast, *Leptonycteris* species have resident populations within the tropics, and their year-round presence has been reported for south-central Mexico (Rojas-Martínez, 1996; Rojas-Martínez et al., 1999).

Most of the species of columnar cacti studied within the tropics show a mutually dependent relationship with bats, in which the plants can only produce seeds by visitation of bats, and bats depend (for most of the year) on resources from plants such as Cactaceae, Bombacaceae, and Agavaceae for food (Valiente-Banuet et al., 1997b; Rojas-Martínez et al., 1999; chapter 13, this volume). Nectar production and flower function are exclusively nocturnal in most of the species of the intertropical deserts (Venezuela and Mexico), whereas *Carnegiea gigantea, Pachycereus pringlei,* and *Stenocereus thurberi* have longer anthesis periods than any other columnar cacti in the tropics (Table 11.1). Columnar cacti of Venezuela have the shortest anthesis times (ca. 12 hours), whereas these times in south-central Mexico are 13–15 hours and 19–23 hours in extratropical deserts. Diurnal pollination of the flowers therefore occurs only in northern Mexico and the southwestern United States (Alcorn et al., 1959, 1961; McGregor et al., 1959,1962; Fleming et al., 1996; Valiente-Banuet et al., 1996). Interestingly, specialization for bat pollination seems to be related to flower function rather than to morphology because flowers from the northern and tropical columnar cacti share such morphological characteristics as total length and perianth width (Table 11.1).

Although flowers in columnar cacti are hermaphroditic, hints of the evolution of breeding systems towards dioecy have been found in different taxa. Trioecy occurs in *Pachycereus pringlei,* which has males, seed-producing females, and hermaphrodites in some of its populations (Fleming et al., 1994; chapter 10, this volume), and androdioecy occurs in *Neobuxbaumia mezcalaensis* with male (female sterile) and hermaphrodite individual plants (Table 11.2; Valiente-Banuet et al., 1997b). Both trioecy and androdioecy are uncommon sexual systems, and theoretically dioecy could be established via androdioecy only if males (female steriles) have more than twice the pollen fertility of hermaphrodites. If these conditions are not met, then it is not possible for male plants to invade hermaphroditic populations (Charlesworth and Charlesworth, 1978).

TABLE 11.1

Ranges of Floral Characteristics of Columnar Cacti in Northwestern Mexico, South-Central Mexico, and Northwestern Venezuela[a]

Area	*Flower Length (cm)*	*Perianth Width (cm)*	*Nectar-Sugar Concentration (% sucrose wt/wt)*	*Maximum Duration of Anthesis (h)*	*Nectar Production per Flower (ml)*	*Percentage of Species with Diurnal Nectar Production*
Northwestern Mexico	7.9–10.2	1.9–2.5	23.6–26.4	19–23	1.1–2.0	100 ($n = 3$)
South-central Mexico	4.3–10.2	1.6–4.0	17.6–24.9	12–15	0.38–2.4	50 ($n = 8$)
Northwestern Venezuela	4.3–5.9	2.1–2.5	18–20.7	11–12	0.67–1.09	0 ($n = 4$)

[a] Data are from Alcorn et al. (1959, 1961, 1962); Casas et al. (1999); Fleming et al. (1994, 1996); Nassar et al. (1997); and Valiente-Banuet et al. (1996, 1997a,b).

TABLE 11.2

Summary of the Pollinators and Reproductive Systems of Columnar Cacti in Northwestern Mexico, South-Central Mexico, and Northwestern Venezuela[a]

Cactus Species	*Locality and Latitude*	*Pollinators (in Order of Importance)*	*Reproductive System*
Carnegiea gigantea	NW Mexico, 28°50′ N	bees, bats, and birds	perfect flowers
Pachycereus pringeli	NW Mexico, 28°50′ N	bats, birds, and bees	trioecious
Stenocereus thurberi	NW Mexico, 28°50′ N	birds, bats, and bees	perfect flowers
Pilosocereus chrysacanthus	south-central Mexico, 18°20′ N	bats	perfect flowers
Neobuxbaumia mezcalaensis	south-central Mexico, 18°20′ N	bats, birds	androdioecious
N. tetetzo	south-central Mexico, 18°20′ N	bats	perfect flowers
N. macrocephala	south-central Mexico, 18°20′ N	bats	perfect flowers
Pachycereus weberi	south-central Mexico, 18°20′ N	bats	perfect flowers
Stenocereus stellatus	south-central Mexico, 18°20′ N	bats	perfect flowers
Subpilocereus horrispinus	NW Venezuela, 10°56′ N	bats	perfect flowers
Subpilocereus repandus	NW Venezuela, 10°56′ N	bats	perfect flowers
Stenocereus griseus	NW Venezuela, 10°56′ N	bats	perfect flowers
Pilosocereus moritzianus	NW Venezuela, 10°56′ N	bats	perfect flowers
Pilosocereus lanuginosus	NW Venezuela, 10°56′ N	bats	perfect flowers

[a] Data are from Alcorn et al. (1959, 1961, 1962); Casas et al. (1999); Fleming et al. (1994, 1996); Nassar et al. (1997); and Valiente-Banuet et al. (1996, 1997a,b).

SEED DISPERSAL

Because seedlings of columnar cacti do not survive in open spaces, directed seed dispersal beneath the canopies of trees and shrubs seems to be an important step in the population dynamics of cacti (Valiente-Banuet and Ezcurra, 1991). Information about the mechanisms of cactus-seed dispersal under natural conditions is scarce and most of the studies on the topic deal only with fruit removal or qualitative aspects of dispersal. Considering the information presented below, seed dispersal by animals in columnar cactus forests deserves investigation to determine if this process has important consequences for the population dynamics of cacti.

Fleshy fruits are found in practically all the columnar cactus species and they serve as attractants for different groups of animals (e.g., birds and mammals) that consume the pulp and seeds (Steenbergh and Lowe, 1977; Silva, 1988; León de la Luz and Cadena, 1991; Soriano et al., 1991; Silvius, 1995; Valiente-Banuet et al., 1996; Godínez-Alvarez and Valiente-Banuet, 2000). In south-central Mexico, fruits of Mexican columnar cacti belonging to the genera *Cephalocereus, Mitrocereus,* and *Neobuxbaumia* have the same white color and odor as the flowers (Valiente-Banuet et al., 1995, 1996) and are consumed while still on the plant by a wide variety of animals, including the bats *Artibeus intermedius, A. jamaicensis, Choeronycteris mexicana, Glossophaga soricina, Leptonycteris curasoae, L. nivalis,* and *Sturnira lilium;* and such birds as *Campylorhynchus bruneicapillus, Carpodacus mexicanus, Guiraca caerulea, Melanerpes hypopolius, Passerina versicolor, Phainopepla nitens, Toxostoma curvirostre,* and *Zenaida asiatica* (Valiente-Banuet et al., 1996; unpubl. data).

In some cases, passage of seeds through vertebrate guts increases the rate of germination in *Stenocereus gummosus* (León de la Luz and Cadena, 1991), whereas in other species, including *Carnegiea gigantea* and *S. griseus,* rate of seed germination is not affected or even decreases (Steenbergh and Lowe, 1977; Silvius, 1995). Laboratory experiments in which seed ingestion by birds and bats was simulated indicate that germination percentages were higher than 70% and that the results from these treatments do not differ from those for controls (Godínez-Alvarez and Valiente-Banuet, 1998).

In the Sonoran Desert, the bat *Leptonycteris curasoae* can remove 10–80% of the available seeds of the columnar cactus *Pachycereus pringlei* per night (Fleming and Sosa, 1994). This bat is an important disperser of columnar cacti seeds whenever it defecates or spits out the seeds when resting in night roosts beneath trees and shrubs (Hirshfeld et al., 1977; Howell, 1979; Cockrum, 1991; Valiente-Banuet et al., 1996; Godínez-Alvarez and Valiente-Banuet, 2000). In the Tehuacán Valley, such plant species as *Caesalpinia melanadenia, Cercidium praecox,* and *Mimosa luisana* are used as night roosts by bats, and some of these plants have a significantly higher number of young individuals of columnar cacti growing beneath their canopies than is expected by chance seed dispersal (Valiente-Banuet et al., 1991a; Godínez-Alvarez and Valiente-Banuet, 2000).

Demographic Models

Comparative demographic analyses of related taxa such as columnar cacti may be a useful method for analyzing the effects of different environments on life-history parameters (Silvertown et al., 1993). The life cycle of a plant can be described by a life-cycle graph (Hubbell and Werner, 1979), from which a population projection matrix may be derived (Caswell, 1989; Silvertown et al., 1993). Populations where individuals are size- or stage-classified are best described by a Lefkovitch matrix, because fecundity, survival, and growth in most plants are more closely related to stage of growth than age (Caswell, 1989).

Matrix models represented as $n_{t+1} = An_t$, include a stage projection square matrix A whose entries a_{ij} are based on the quantitative description of the life cycle (i.e., representing the transitions or contributions from individuals in the *i*th category to the *j*th category after one time step), and a population vector n_t representing the number of individuals in the population for each stage class *i* at time *t* (Lefkovitch, 1965). The stage projection matrix A contains three principal parts: (1) the first row refers to fecundity values for all reproductive stages, (2) the main diagonal defines probabilities of stasis or proportions of individuals in stage class *i* that remain in the same class after one time interval, and (3) the first lower subdiagonal defines transition probabilities or proportion of individuals of stage class *i* that pass to the next class.

The solution of the matrix projection model by the power method (Caswell, 1989) or the repeated multiplication of the matrix A by the vector n_t allows one to estimate the finite rate of increase (the largest positive eigenvalue), the stable stage distribution (the right-hand eigenvector), and stage-specific reproductive values (the left-hand eigenvector; De Kroon et al., 1986; Caswell, 1989; Silvertown et al., 1993).

The stable stage distribution and the vector of reproductive values are used to calculate the elasticity e_{ij} of each element a_{ij} of the matrix. Elasticity is a measure of the relative change in the finite rate of increase resulting from small changes in the matrix elements ($e_{ij} = \partial \log \lambda / \partial \log a_{ij}$), and represents the relative contribution of a given matrix element to the species fitness as measured by λ (De Kroon et al., 1986; Caswell, 1989). Therefore, elasticity values, which sum to unity (or 100%), may be summed across regions of the matrix related to fecundity, growth, and survival to compare the relative importance of each of these demographic parameters defining the life history of the species (Godínez-Alvarez et al., 1999).

The results of demographic studies conducted using the matrix approach (Silvertown et al., 1993; Silva, 1996; Godínez-Alvarez et al., 1999) indicate that the population dynamics of *Carnegiea gigantea* and *Pachycereus pringlei* in the Sonoran Desert and *Neobuxbaumia tetetzo* in the Tehuacán Valley are similar. In all of these columnar cacti, population structures are characterized by a monotonic decrease in the number of individuals with increasing height categories (Fig. 11.2) and cohort survival patterns that can be described as type III curves, with high mortality occurring during the early years of life (i.e., seeds and seedlings) and decreasing at older ages (Fig. 11.3). These studies also show that fecundity increases with size, except in *P. pringlei* in which senescence may occur (Fig. 11.4). An increase in fecundity with respect to size has been reported for other species of columnar cacti, such as *C. gigantea* (Steenbergh and Lowe, 1977), *Stenocereus thurberi,* and *Lophocereus schottii* (Parker, 1987, 1989). Of the many thousands of seeds that columnar cacti such as *C. gigantea* and *N. tetetzo* produce, only four to eight in 1,000 seeds reach a site where natural germination can occur (Steenbergh and Lowe, 1977; Valiente-Banuet and Ezcurra, 1991). The small number of seedlings emerging has been associated mainly with seed predation by such granivores as birds, rodents, and ants. In addition, most of the succulent, weakly rooted seedlings are eaten or uprooted by foraging animals or die from drought or winter cold (Steenbergh and Lowe, 1977; Valiente-Banuet and Ezcurra, 1991). Indeed, at the northern distribution limit of columnar cacti, populations of saguaro are periodically subjected to catastrophic freeze-kill throughout their populations (Steenbergh and Lowe, 1969).

Elasticity Analysis

Matrix models can be a useful tool for determining the role of biotic interactions in population dynamics of species of columnar cacti by determining whether an inter-

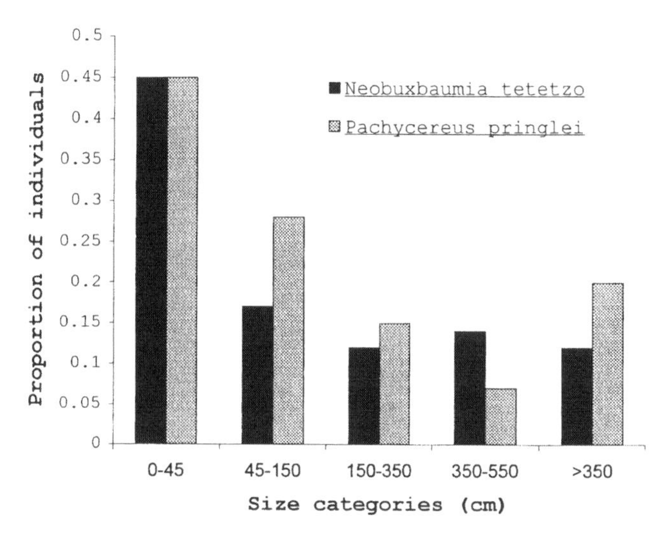

FIGURE 11.2. Size category distributions for *Pachycereus pringlei* in Baja California Sur (Silva, 1996) and *Neobuxbaumia tetetzo* (Godínez-Alvarez et al., 1999).

action affects those life stages that significantly affect the population's finite rate of increase (Godínez-Alvarez et al., 1999). Results of an elasticity analysis conducted on *Neobuxbaumia tetetzo* (Fig. 11.5) indicate that the highest elasticity values occur between the stages of seedling and mature plant (individuals with heights of the principal trunk between 2 and 250 cm), whereas those for growth and reproduction are

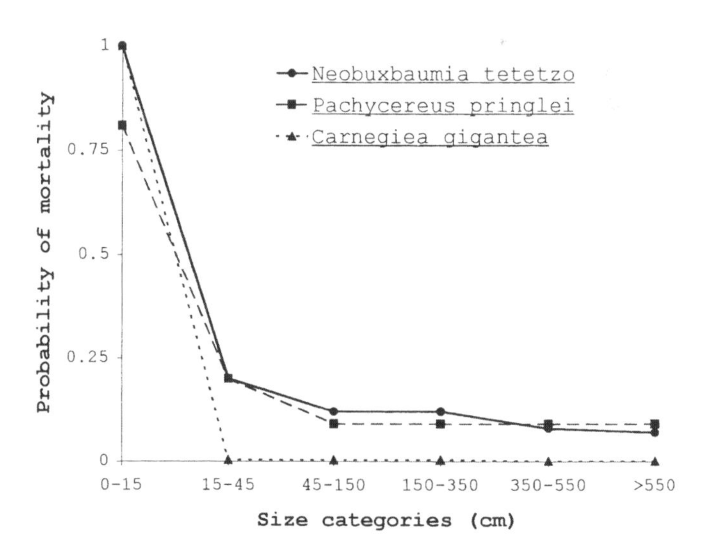

FIGURE 11.3. Probability of mortality for different size categories of *Pachycereus pringlei* in Baja California (Silva, 1996), *Carnegiea gigantea* (Steenbergh and Lowe, 1983), and *Neobuxbaumia tetetzo* (Godínez-Alvarez et al., 1999).

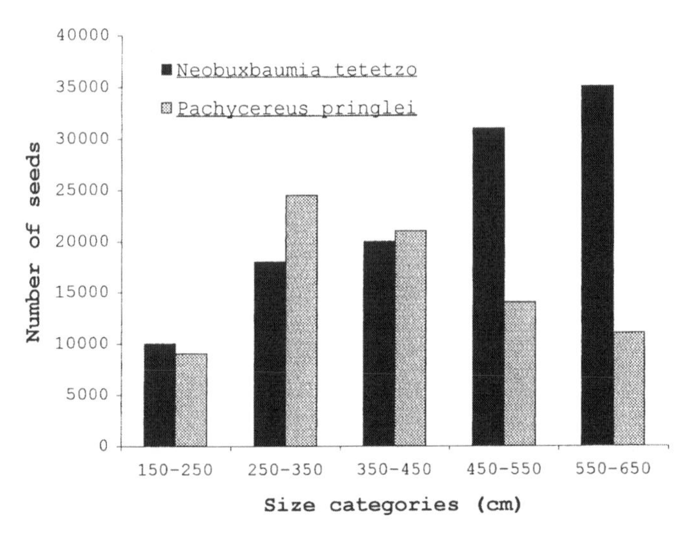

FIGURE 11.4. Total number of seeds ha^{-1} produced per size category for individuals of *Pachycereus pringlei* in Baja California (Silva, 1996) and *Neobuxbaumia tetetzo* (Godínez-Alvarez et al., 1999).

relatively lower (Fig. 11.5). In a similar way, by summing the elasticity values of the main life-history parameters of the life cycle (i.e., survival, fecundity, and growth), of the columnar cacti *Carnegiea gigantea, Neobuxbaumia tetetzo,* and *Pachycereus pringlei,* survival again is the most important parameter, with elasticity values ranging from 86% (in *P. pringlei*) to 99.7% (in *C. gigantea*) (Fig. 11.6). These results mean that survival, mostly during the earliest life stages, exerts the highest effect on the finite rate of increase of the populations studied. Therefore, germination, seedling es-

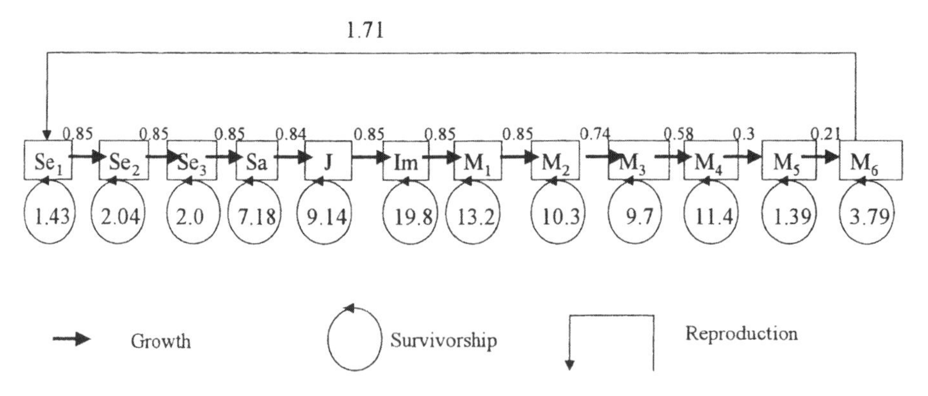

FIGURE 11.5. Life-cycle diagram with elasticity values (in percentages) for *Neobuxbaumia tetetzo* growing in the Tehuacán Valley, Mexico. Size categories are as follows: 0–2 cm, Se_1; 2–8 cm, Se_2; 8–15 cm, Se_3; 15–45 cm, Sa; 45–100 cm, J; 100–150 cm, Im; 150–250 cm, M_1; 250–350 cm, M_2; 350–450 cm, M_3; 450–550 cm, M_4; 550–650; M_5; ≥650 cm, M_6. Values corresponding to seed germination and seedling establishment are incorporated in the reproduction of mature individuals. Life-cycle stages from Se_1 to Im grow in association with perennial nurse plants. Modified from Godínez-Alvarez et al. (1999).

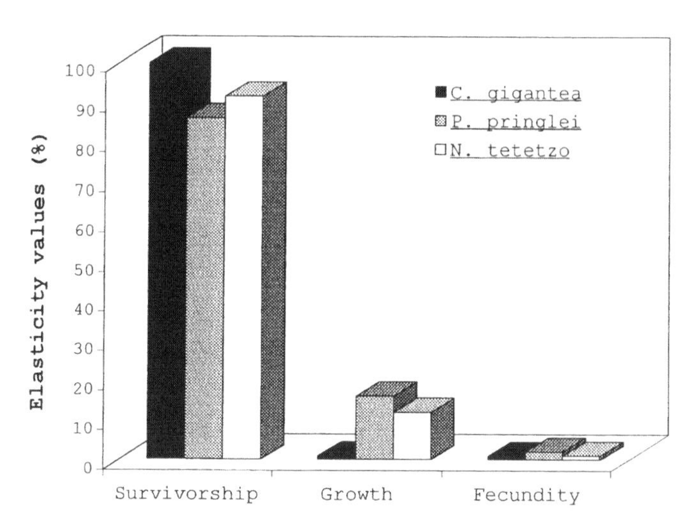

FIGURE 11.6. Summed elasticity values for the main life-history parameters (Silvertown et al., 1993); *Pachycereus pringlei* (Silva, 1996), and *Neobuxbaumia tetetzo* (Godínez-Alvarez et al., 1999).

tablishment, and survival of young plants appear to be the most critical phases of the life cycle and, as pointed out above, seed and seedling predation, as well as the ameliorating effect of the physical environment produced by perennial nurse plants, are interactions that strongly affect the population dynamics of columnar cacti (Steenbergh and Lowe 1969, 1977; Valiente-Banuet and Ezcurra, 1991).

As would be expected for long-lived species in which there is a trade-off between the importance of fecundity and survival, the population dynamics of columnar cacti depend mostly on the survival parameter. However, it is evident that seed production (pollination) and seed transportation directed to nurse plants (dispersal) are also important interactions because many species of columnar cacti reproduce only by seeds (Steenbergh and Lowe, 1977, 1983; Silvertown et al., 1993; Silva, 1996; Godínez-Alvarez et al., 1999).

Cyclical Relationships between Columnar Cacti and Their Nurse Plants

The establishment and growth of columnar cacti beneath other plants may eventually affect the original nurse plant. McAuliffe (1984) found that the establishment of *Carnegiea gigantea* is facilitated by *Cercidium microphyllum*, which in the long term is competitively replaced by the former. Valiente-Banuet et al. (1991b) presented evidence that the columnar cactus *Neobuxbaumia tetetzo* replaces its legume nurse plant *Mimosa luisana* by competition. In both cases, competition for water is the most likely mechanism to explain the phenomenon, as has been found by other authors (Flores-Martínez et al., 1994). Thus, the nature of the interaction between columnar cacti and nurse plants changes through time, starting with commensalism (0, +) and ending with competition (–, –) and/or competitive exclusion (Valiente-Banuet et al., 1991b). These properties of the nurse phenomenon have allowed nurse systems to be described as Markovian (McAuliffe, 1988), in which communities are patch-structured and vegetation dynamics are different in each patch. The regional space defined as the entire community is therefore composed of vegetation patches with different occupation stages, starting with open spaces that may be colonized by nurse plants, which in turn are colonized by cacti species (McAuliffe, 1988).

The dynamics in each patch are structured by positive (pollination, seed dispersal, and nurse phenomenon) and negative (competition and seed predation) interactions that occur differentially over the landscape in each independent patch (McAuliffe, 1988; Valiente-Banuet and Ezcurra, 1991). Therefore, the differential outcomes of the interactions occurring in each patch define a cycle of community dynamics, as has been reported for other arid ecosystems (Yeaton, 1978; Yeaton and Romero-Manzanares, 1986; Yeaton and Esler, 1990).

Conclusions

Forests of columnar cacti occupy a considerable area of Mexico, overlapping in distribution with important groups of animals with which they interact. In these environments, positive and negative plant-to-plant and animal-to-plant interactions play an important role in the maintenance of cactus diversity. These biotic interactions—positive and negative—affect the most sensitive life stages and the finite rate of increase of cacti. Thus these interactions constitute key processes in the maintenance of this group of plants. Consequently, to protect these highly diverse columnar cacti forests, conservation practices need to be directed towards protecting biotic interactions and not merely protecting plant species.

Summary

Columnar cacti have critical stages during their life cycle, whose alteration may produce dramatic changes in the dynamics of entire populations. Critical life-cycle stages include seedling establishment and early growth stages that only occur successfully beneath canopies of nurse plants. Considering this situation and given that many species of columnar cacti reproduce only by seeds, pollination and seed dispersal are critical interactions for ensuring the maintenance of columnar cactus forests. These interactions are analyzed in this chapter in the context of plant demography. We have found important differences in patterns of pollination interactions associated with geographical distribution and we examined hypotheses for explaining such differences.

Resumen

En su ciclo de vida, las plantas presentan estadíos "sensibles" en los cuales cualquier modificación que se presente puede afectar la dinámica de las poblaciones. Para el caso de las cactáceas columnares las fases más sensibles de su ciclo de vida están relacionadas con el establecimiento de las plántulas y la sobrevivencia de los primeros estadíos, que sólo ocurre exitosamente debajo de la sombra de arbustos llamados plantas nodriza. Bajo estás condiciones, y considerando que muchas de las especies de cactáceas columnares solo se reproducen, por semillas, la polinización y dispersión de semillas son interacciones clave para el mantenimiento de estos bosques de cactáceas. Tales interacciones se analizan aquí en el contexto de la demografía. Se muestran importantes diferencias en los patrones de polinización biótica, asociadas a la ubicación geográfica y se analizan hipótesis para explicarlas.

ACKNOWLEDGMENTS

Our studies on biotic interactions were supported by Fondo Mexicano para la Conservación de la Naturaleza (FMCN, Project A-1-97-36) and Dirección General de Asuntos del Personal Académico de la Universidad Nacional Autónoma de México, Project IN 207798.

REFERENCES

Alcorn, S. M., and E. B. Kurtz. 1959. Some factors affecting the germination of seeds of the saguaro cactus (*Carnegiea gigantea*). *American Journal of Botany* 46:526–29.

Alcorn, S. M., S. E. McGregor, G. D. Butler, and E. B. Kurtz. 1959. Pollination requirements of the saguaro (*Carnegiea gigantea*). *Cactus and Succulent Journal* 31:39–41.

Alcorn, S. M., S. E. McGregor, and G. Olin. 1961. Pollination of saguaro cactus by doves, nectar-feeding bats and honey bees. *Science* 132:1594–95.

———. 1962. Pollination requirements of the organpipe cactus. *Cactus and Succulent Journal* 34:134–38.

Alvarez, T., and Q. L. González. 1970. Análisis polínico del contenido gástrico de murciélagos Glossophaginae de México. *Anales del la Escuela Nacional de Ciencias Biológicas, México* 18:137–65.

Arizmendi, M. C., and A. Espinosa de los Monteros. 1996. Avifauna de los bosques de cactáceas columnares en el Valle de Tehuacán. *Acta Zoológica Mexicana* (n.s.) 67:25–46.

Arriaga, L., Y. Maya, S. Díaz, and J. Cancino. 1993. Association between cacti and nurse perennials in a heterogeneous tropical forest in northwestern Mexico. *Journal of Vegetation Science* 4:349–56.

Brum, G. D. 1973. Ecology of the saguaro (*Carnegiea gigantea*): phenology and establishment in marginal populations. *Madroño* 22:195–204.

Casas, A., A. Valiente-Banuet, A. Rojas-Martínez, and P. Dávila. 1999. Reproductive biology and the process of domestication of the columnar cactus *Stenocereus stellatus* in central Mexico. *American Journal of Botany* 86:534–42.

Caswell, H. 1989. *Matrix Population Models. Construction, Analysis and Interpretation.* Sunderland, Mass.: Sinauer Associates.

Cockrum, E. L. 1991. Seasonal distribution of north-western populations of the long nosed bats family Phyllostomidae. *Anales del Instituto de Biología. Universidad Nacional Autónoma de México. Serie Zoología* 62:181–202.

Charlesworth, B., and D. Charlesworth. 1978. A model for the evolution of dioecy and gynodioecy. *American Naturalist* 112:975–97.

De Kroon, H., A. Plaiser, J. van Groenendael, and H. Caswell. 1986. Elasticity: the relative contribution of demographic parameters to population growth rate. *Ecology* 67:1427–31.

Dubrowsky, J. G. 1996. Seed hydration memory in Sonoran Desert cacti and its ecological implication. *American Journal of Botany* 83:624–32.

———. 1998. Discontinuous hydration as a facultative requirement for seed germination in two cactus species of the Sonoran Desert. *Journal of the Torrey Botanical Society* 125:33–39.

Fleming, T. H., and V. J. Sosa. 1994. Effects of nectarivorous and frugivorous mammals on reproductive success of plants. *Journal of Mammalogy* 75:845–51.

Fleming, T. H., S. Maurice, S. L. Buchmann, and M. D. Tuttle. 1994. Reproductive biology and relative male and female fitness in a trioecious cactus, *Pachycereus pringlei* (Cactaceae). *American Journal of Botany* 81:858–67.

Fleming, T. H., M. D. Tuttle, and M. A. Horner. 1996. Pollination biology and the relative importance of nocturnal and diurnal pollinators in three species of Sonoran Desert columnar cacti. *The Southwestern Naturalist* 41:257–69.

Flores-Martínez, A., E. Ezcurra, and S. Sánchez-Colón. 1994. Effect of *Neobuxbaumia tetetzo* on growth and fecundity of its nurse plant *Mimosa luisana. Journal of Ecology* 82:325–30.

Franco, A. C., and P. S. Nobel. 1989. Effect of nurse plants on the microhabitat and growth of cacti. *Journal of Ecology* 77:870–86.

Gibson, A. C., and P. S. Nobel. 1986. *The Cactus Primer.* Cambridge, Mass.: Harvard University Press.

Godínez-Alvarez, H., and A. Valiente-Banuet. 1998. Germination and early seedling growth of Tehuacán Valley cacti species: the role of soils and seed ingestion by dispersers on seedling growth. *Journal of Arid Environments* 39:21–31.

Godínez-Alvarez, H., A. Valiente-Banuet, and B. L. Valiente. 1999. The role of biotic interactions in the population dynamics of the long-lived columnar cactus *Neobuxbaumia tetetzo* in the Tehuacán Valley. *Canadian Journal of Botany* 77:203–8.

Godínez-Alvarez, H., and A. Valiente-Banuet. 2000. Fruit-feeding behavior of the bats *Leptonycteris curasoae* and *Choeronycteris mexicana* in flight cage experiments: consequences for dispersal of columnar cactus seeds. *Biotropica* 32:552–56.

Hirshfeld, J. R., Z. C. Nelson, and G. Bradley. 1977. Night roosting behavior in four species of desert bats. *The Southwestern Naturalist* 22:427–33.

Howell, D. J. 1979. Flock foraging in nectar-feeding bats: Advantages to the bats and to the host plants. *The American Naturalist* 114:23–49.

Hubbell, S. P., and P. A. Werner. 1979. On measuring the intrinsic rate of increase of populations with heterogeneous life histories. *American Naturalist* 113:277–93.

Lefkovitch, L. P. 1965. The study of population growth in organisms grouped by stages. *Biometrics* 21:1–18.

León de la Luz, J. L., and R. D. Cadena. 1991. Evaluación de la reproducción por semilla de la pitaya agria (*Stenocereus gummosus*) en Baja California Sur, México. *Acta Botánica Mexicana* 14:75–87.

Martínez-Holguín, E. 1983. Germinación de semillas de *Stenocereus griseus* (Haw.) Buxbaum (Pitayo de Mayo). *Cactáceas y Suculentas Mexicanas* 28:51–57.

McAuliffe, J. R. 1984. Sahuaro-nurse tree associations in the Sonoran Desert: competitive effects of sahuaros. *Oecologia* 64:319–21.

———. 1988. Markovian dynamics of simple and complex desert plant communities. *The American Naturalist* 131:459–90.

McAuliffe, J. R., and F. J. Janzen. 1986. Effects of intraspecific crowding on water uptake, water storage, apical growth, and reproductive potential in the sahuaro cactus, *Carnegiea gigantea. Botanical Gazette* 147:334–41.

McDonough, W. T. 1964. Germination responses of *Carnegiea gigantea* and *Lemairocereus thurberi. Ecology* 45:155–59.

McGregor, S. E., S. M. Alcorn, E. B. Kurtz, and G. D. Butler. 1959. Bee visitors to Saguaro flowers. *Journal of Economic Entomology* 52:1002–4.

McGregor, S. C., S. M. Alcorn, and G. Olin. 1962. Pollination and pollinating agents of the Saguaro. *Ecology* 43:259–67.

Nassar, J. M., N. Ramírez, and O. Linares. 1997. Comparative pollination biology of Venezuelan columnar cacti and the role of nectar-feeding bats in their sexual reproduction. *American Journal of Botany* 84:918–27.

Nobel, P. S. 1988. *Environmental Biology of Agaves and Cacti.* New York: Cambridge University Press.

Parker, K. C. 1987. Seedcrop characteristics and minimum reproductive size of organ pipe cactus (*Stenocereus thurberi*) in southern Arizona. *Madroño* 34:294–303.

———. 1989. Height structure and reproductive characteristics of senita, *Lophocereus schottii* (Cactaceae), in southern Arizona. *The Southwestern Naturalist* 34:392–401.

Rojas-Aréchiga, M., and C. Vazquez-Yanes. 2000. Cactus seed germination. A review. *Journal of Arid Environments* 44:85–104.

Rojas-Aréchiga, M., A. Orozco-Segovia, and C. Vazquez-Yanes. 1997. Effect of light on the germination of seven species of cacti from the Zapotitlán Valley in Puebla, Mexico. *Journal of Arid Environments* 36:571–78.

Rojas-Martínez, A. 1996. Estudio poblacional de tres especies de murciélagos nectarívoros considerados como migratorios y su relación con la presencia estacional de los recursos florales en el Valle de Tehuacán y la Cuenca del Balsas. Master's thesis, Universidad Nacional Autónoma de México, Mexico.

Rojas-Martínez, A., and A. Valiente-Banuet. 1996. Análisis comparativo de la quiropterofauna del Valle de Tehuacán-Cuicatlán, Puebla, Oaxaca. *Acta Zoologica Mexicana* (n.s.) 67:1–23.

Rojas-Martínez, A., A. Valiente-Banuet, M. C. Arizmendi, A. Alcántara-Egúren, and H. T. Arita. 1999. Seasonal distribution of the long-nosed bat (*Leptonycteris curasoae*) in North America: Does a generalized migration pattern really exist? *Journal of Biogeography* 26:1065–78.

Shreve, F. 1910. The rate of establishment of the giant cactus. *The Plant World* 13:235–39.

Silva, P. C. 1996. Demografía comparativa de *Pachycereus pringlei* en dos unidades geomórficas contrastantes del paisaje en Baja California Sur, México. Master's thesis, Universidad Nacional Autónoma de México, Mexico.

Silva, W. R. 1988. Ornitocoria em *Cereus peruvianus* (Cactaceae) na Serra do Japi, Estado de Sao Paulo. *Revista Brasileira de Biologia* 48:381–89.

Silvertown, J. W., M. Franco, I. Pisanty, and A. Mendoza. 1993. Comparative plant demography: relative importance of the life-cycle components to the finite rate of increase in woody and herbaceous perennials. *Journal of Ecology* 81:465–76.

Silvius, K. M. 1995. Avian consumers of cardón fruits (*Stenocereus griseus:* Cactaceae) on Margarita Island, Venezuela. *Biotropica* 27:96–105.

Soriano, P., M. Sosa, and O. Rossell. 1991. Hábitos alimentarios de *Glossophaga longirostris* Miller (Chiroptera: Phyllostomidae) en una zona árida de los andes venezolanos. *Revista de Biología Tropical* 39:263–68.

Sosa, M., and P. Soriano. 1992. Los murciélagos y los cactus: una relación muy estrecha. *Carta Ecológica* 61:7–10.

Steenbergh, W. H., and C. H. Lowe. 1969. Critical factors during the first years of life of the saguaro (*Cereus giganteus*) at the Saguaro National Monument, Arizona. *Ecology* 50:825–34.

———. 1977. Ecology of Saguaro II. Reproduction, germination, establishment, growth, and survival of the young plant. *National Park Service Scientific Monograph Series.* No. 8. Washington, D.C.: U.S. Government Printing Office.

———. 1983. Ecology of Saguaro III. Growth and demography. *National Park Service Scientific Monograph Series.* No. 17. Washington, D.C.: U.S. Government Printing Office.

Turner, R. M., S. M. Alcorn, G. Olin, and J. A. Booth. 1966. The influence of shade, soil and water on saguaro seedling establishment. *Botanical Gazette* 127:95–102.

Valiente-Banuet, A., and E. Ezcurra. 1991. Shade as a cause of the association between the cactus *Neobuxbaumia tetetzo* and the nurse plant *Mimosa luisana* in the Tehuacán Valley, Mexico. *Journal of Ecology* 79:961–71.

Valiente-Banuet, A., O. Briones, A. Bolongaro-Crevenna, E. Ezcurra, M. Rosas, H. Núñez, G. Barnard, and E. Vázquez. 1991a. Spatial relationships between cacti and nurse shrubs in a semi-arid environment in central Mexico. *Journal of Vegetation Science* 2:15–20.

Valiente-Banuet, A., F. Vite, and A. Zavala-Hurtado. 1991b. Interaction between the cactus *Neobuxbaumia tetetzo* and the nurse shrub *Mimosa luisana. Journal of Vegetation Science* 2:12–14.

Valiente-Banuet, A., P. Dávila, M. C. Arizmendi, A. Rojas-Martínez, and A. Casas. 1995. Bases ecológicas del desarrollo sustentable en zonas áridas: el caso de los bosques de cactáceas columnares en el Valle de Tehuacán y Baja California Sur, México. In *IV Curso sobre Desertificación y Desarrollo Sustentable en América Latina y el Caribe,* ed. G. M. Anaya and C.S.F. Díaz, 20–36. Mexico: United Nations Environmental Program/Food and Agriculture Organization Mexico/Colegio de Postgraduados en Ciencias Agrícolas.

Valiente-Banuet, A., M. C. Arizmendi, A. Rojas-Martinez, and L. Domínguez-Canseco. 1996. Ecological relationships between columnar cacti and nectar feeding bats in Mexico. *Journal of Tropical Ecology* 12:103–19.

Valiente-Banuet, A., A. Rojas-Martínez, M. C. Arizmendi, and P. Dávila. 1997a. Pollination biology of two columnar cacti (*Neobuxbaumia mezcalaensis* and *Neobuxbaumia macrocephala*) in the Tehuacán Valley, central Mexico. *American Journal of Botany* 84:452–55.

Valiente-Banuet, A., A. Rojas-Martínez, A. Casas, M. C. Arizmendi, and P. Dávila. 1997b. Pollination biology of two winter-blooming giant columnar cacti in the Tehuacán Valley, central Mexico. *Journal of Arid Environments* 37:331–41.

Valiente-Banuet, A., A. Casas, A. Alcántara, P. Dávila, N. Flores-Hernández, A.M.C. Arizmendi, J. L. Villaseñor, J. Ortega, J. A. Soriano. 2000. La vegetación del Valle de Tehuacán-Cuicatlán. *Boletín de la Sociedad Botánica de México* 67:24–74.

Wilkinson, G. S., and T. H. Fleming. 1996. Migration and evolution of lesser long-nosed bats *Leptonycteris curasoae,* inferred from mitocondrial DNA. *Molecular Ecology* 5:329–39.

Yeaton, I. R. 1978. A cyclical relationship between *Larrea tridentata* and *Opuntia leptocaulis* in the Northern Chihuahuan Desert. *Journal of Ecology* 66:651–56.

Yeaton, I. R., and A. Romero-Manzanares. 1986. Organization of vegetation mosaics in the *Acacia schaffneri-Opuntia streptacantha* association, Southern Chihuahuan Desert, Mexico. *Journal of Ecology* 74:211–17.

Yeaton, I. R., and K. J. Esler. 1990. The dynamics of a succulent karoo vegetation. A study of species association and recruitment. *Vegetatio* 88:103–13.

CHAPTER 12

The Role of Bats and Birds in the Reproduction of Columnar Cacti in the Northern Andes

Pascual J. Soriano
Adriana Ruiz

Introduction

The Andes of northern South America contain a set of dry valleys with climatic characteristics that result from orographic rainfall shadows (Sarmiento, 1975). These arid valleys constitute an archipelago of small arid enclaves separated from each other by wet vegetation formations. They occur from the Cordillera de Mérida, in western Venezuela, to Ecuador through the three Andean mountain chains of Colombia. During the Pleistocene, these enclaves were connected with the current arid region of the Caribbean coast (Sarmiento, 1975, 1976) and at present constitute refuges for the flora and fauna of the Andean dry ecosystems (Hernández et al., 1992). They therefore represent an important genetic reservoir where allopatric speciation may be occurring (Hernández et al., 1992).

Ecologically, these dry valleys can be classified into two types according to their climatic characteristics and altitude (Sarmiento, 1976). The first type includes enclaves found above 1,800 m, characterized by meso- and microthermic climates (annual mean temperatures below 20°C) and vegetation types with few features in common with those of the Caribbean coast. This lack of relationship is evidenced by the absence of columnar cacti, among other aspects. The second type comprises enclaves located below 1,800 m, which are characterized by warm climates (annual mean temperatures above 24°C) and which are similar in ecology, flora, and vegetation physiognomy to the Caribbean coast (Sarmiento, 1972, 1975). At least eleven units of the second type can be recognized in the northern Andes (Fig. 12.1). We will focus on these latter enclaves owing to the presence of columnar cacti, whose ecological relationships are the subject of the present chapter.

The plant communities of these medium- and low-altitude valleys show close floristic relationships to the vegetation of the Caribbean coast, although they contain

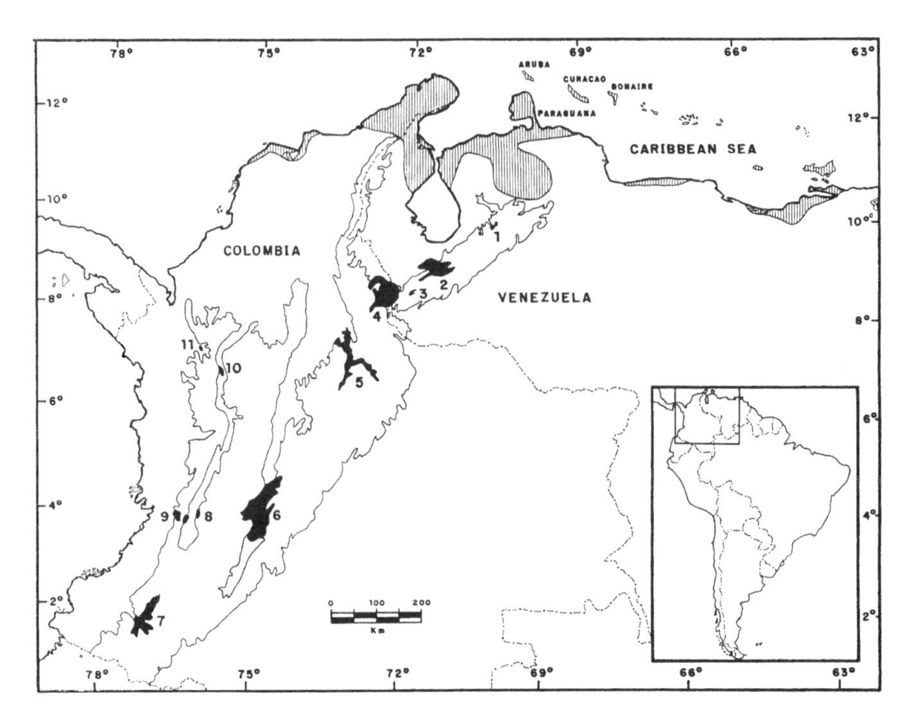

FIGURE 12.1. Arid zones in northwestern South America. Caribbean coastal dry zone (barred areas) and the eleven Andean enclaves (black areas). (1) La Puerta; (2) Lagunillas; (3) La Quinta; (4) Cúcuta-Ureña; (5) Chicamocha; (6) La Tatacoa; (7) Patía; (8) Cauca River; (9) Dagua; (10) Santa Fé de Antioquia; (11) Sucio River. Modified from Sarmiento (1976) and Cavelier et al. (1996).

fewer plant species (Sarmiento, 1975). Columnar cacti are characteristic elements of these landscapes, where some individuals reach heights of 7–9 m and emerge above the tree canopy. Their flowers and fruits are located mainly at the tips of branches and provide access to bats and birds (Fig. 12.2). These vertebrates directly influence the reproductive biology of these Cactaceae, with which they have established different degrees of dependency in pollination (zoogamy) and seed dispersal (zoochory).

During the past decade, there has been an increase in the number of studies dealing with the interactions between birds, bats, and tropical columnar cacti. These studies have documented interactions between Cactaceae and phyllostomid bats of the subfamily Glossophaginae in certain arid enclaves of the Venezuelan and Colombian Andes and on some Caribbean islands, as well as the importance of birds in the process of seed dispersal (Bosque, 1984; Silva, 1988; Wendelken and Martin, 1988; Soriano et al., 1991; Sosa and Soriano, 1993, 1996; Petit, 1995; Silvius, 1995; Nassar, et al., 1997; Rengifo, 1997; Ruiz et al., 1997; Naranjo 1998).

One of the most important species in these interactions is *Glossophaga longirostris* (Fig. 12.3a), a nectarivorous, pollinivorous, frugivorous bat (Goodwin and Greenhall, 1961; Pirlot, 1964; Alvarez and González, 1970; Gardner, 1977; Soriano et al.,

FIGURE 12.2. Panoramic view of the Lagunillas enclave at the middle Chama River basin. The branches of columnar cacti emerge from the open canopy of the woody plants. Photo by P. J. Soriano.

1991; Sosa and Soriano, 1996; Ruiz et al., 1997) that inhabits arid environments in northern South America (Webster and Handley, 1986). It is a nonmigratory species (Soriano et al., 1991; Sosa and Soriano, 1996; Ruiz et al., 1997). In contrast, another glossophagine species involved in these interactions is *Leptonycteris curasoae* (Fig. 12.3b), which is a migrant species in North America (Hayward and Cockrum, 1971; Cockrum, 1991; Fleming et al., 1993a) and whose presence in the Andean enclaves is restricted to certain periods of the year (Sosa and Soriano, 1993).

Our knowledge of the relationships between birds and columnar cacti is restricted to species lists of fruit consumers and indirect measures of fruit consumption, such as visitation rates (Ellner and Shmida, 1981; Bosque, 1984; Silva, 1988; Wendelken and Martin, 1988; Silvius, 1995). Comparisons between birds and bats with respect to their efficiency as potential seed dispersers and their role in germination have not been made, although Rengifo (1997) and Naranjo (1998) provide experimental evidence for the effect on germination of seeds passed through the gut of certain species of birds and the bat *Glossophaga longirostris*, respectively.

The purpose of this chapter is to summarize available information on the ecological relationships between columnar cacti and their main pollinators and seed dispersers in the arid Andean enclaves of Colombia and Venezuela. We will describe the adaptive strategies that can be deduced from the floral and fruit features of the cacti that inhabit these enclaves. We will also compare the ecological role of birds and bats as seed dispersers of these columnar cacti. Additionally, we will discuss the different

a

b

FIGURE 12.3. The main pollinator and seed disperser bats in the Andean enclaves. (a) *Glossophaga longiorostris*, a resident species. (b) *Leptonycteris curasoae*, a migrant species. Photos by P. J. Soriano.

configurations of the bat-cactus system in the Andean enclaves (especially the occurrence of *Glossophaga longirostris* and *Leptonycteris curasoae*) and the ecological implications of the presence or absence of these bats. For this purpose, we will emphasize enclaves located in the Cordillera Oriental (Colombia) and the Cordillera de Mérida (Venezuela).

The Andean Arid Enclaves

We give special emphasis to seven of the 11 enclaves where columnar cacti occur in the Andes of Colombia and Venezuela (Fig. 12.1); four are located in the Cordillera de Mérida (Venezuela) and other three in the Cordillera Oriental and Cordillera Occidental (Colombia):

1. *La Puerta.* This enclave comprises a small swath restricted to the mountain hillsides of the middle Motatán river basin, between the towns of La Puerta (Trujillo

State, Venezuela) and Timotes (Mérida State, Venezuela). Its approximate area is 27 km^2.

2. *Lagunillas.* The enclave extends from the town of Estanques, located in the middle Chama river basin, to the town of El Morro in the higher Nuestra Señora river basin in Mérida State, Venezuela. It occupies an area of 262 km^2 and represents the largest enclave in the Cordillera de Mérida.
3. *La Quinta.* The enclave is located between the confluence of the La Grita and Del Valle rivers, to the west of La Grita (Táchira State, Venezuela). The approximate area is 29 km^2.
4. *Cúcuta-Ureña.* The enclave is located at the Venezuelan and Colombian border in Táchira State (Venezuela) and Norte de Santander Department (Colombia). This enclave occupies an area of 1,472 km^2 with the driest zone occupying 13% of the total area.
5. *Chicamocha.* This enclave is located between the middle Chicamocha river valley and the Suárez river valley, Santander Department (Colombia). It occupies an area of 1,400 km^2; the driest portion represents 8% of the total area.
6. *La Tatacoa.* This enclave is located in the high Magdalena river valley, between the towns of Honda (Tolima Department, Colombia) and Garzón (Huila Department, Colombia). It occupies an area of 11,185 km^2, and its driest zone represents 12% of the total area. This is the largest enclave in the northern Andes.
7. *Patía.* It is located in the middle Patía river valley in the Cordillera Occidental of the Colombian Andes, Cauca and Nariño Departments. This enclave occupies about 1,127 km^2.

The large enclaves (La Tatacoa, Patía, Lagunillas, Chicamocha, and Cúcuta-Ureña) contain a moisture gradient; different community types occur along this gradient, from very arid to wetter zones (premontane thornshrub and tropical dry forest, respectively, sensu Holdridge; Espinal and Montenegro, 1963; Ewel et al., 1976). The driest zones containing the highest density of columnar cacti occupy very small areas relative to the total area of these enclaves. For most of these enclaves, dry forests have been replaced by cultivation and extensive grazing by cows and goats. Grazing has increased the processes of erosion and desertification.

These dry forests have a discontinuous canopy 4–8 m in height. Most of them share genera of woody plants, such as *Acacia, Bursera, Caesalpinia, Cercidium, Croton, Guazuma, Jatropha,* and *Prosopis.* They also share certain genera of columnar cacti, such as *Stenocereus, Subpilocereus, Pilosocereus,* and *Cereus* (Fig. 12.4a,d,g,j). Other Cactaceae present in these forests belong to the genera *Acanthocereus, Hylocereus, Mammillaria, Melocactus, Opuntia,* and *Rhipsalis.* Although the cactus composition of these forests is known (Ponce, 1989; Cavelier et al., 1996; chapter 16, this volume), we lack taxonomic studies to determine the extent of species shared between them.

FIGURE 12.4. Representatives of the four genera of columnar cacti (individual, flower, and fruit) present in the Andean enclaves. (a, b, c) *Stenocereus griseus;* (d, e, f) *Subpilocereus repandus;* (g, h, i) *Pilosocereus tillianus;* and (j, k, l) *Cereus hexagonus.* Photos by P. J. Soriano.

FIGURE 12.4. *Continued*

Reproductive Strategies of Columnar Cacti: Pollination and Seed Dispersal

Morphological Features of Flowers and Fruits of Columnar Cacti

Flowers and fruits of the columnar cacti of the enclaves described in the previous section possess well-defined functional and anatomical features that favor chiropterogamy as a pollination mechanism (Valiente-Banuet et al., 1996, 1997; Nassar et al., 1997). Features of chiropterochory and ornithochory as strategies for seed dispersal can also be recognized in their fruits (Greenhall, 1957; Snow, 1981; van der Pijl, 1982; Soriano et al., 1991). Flowers of the genera *Pilosocereus, Stenocereus,* and *Subpilocereus* possess morphological and functional characteristics that clearly favor bat pollination (Fleming, 1989; Nassar et al., 1997). Major morphological features shared by these flowers are funnel-shaped corolla tubes with thick fleshy or leathery walls, greenish floral tubes, white to rose-colored petals, numerous stamens, and production of large quantities of pollen and nectar (Fig. 12.4b,e,h). Upon maturation, these flowers emit an odor reminiscent of decomposed vegetables.

Nocturnal anthesis is another important feature of these Cactaceae, which helps restrict potential pollinators to nocturnal species. The anthesis is crepuscular and synchronous (between 1830 and 1930) in all columnar cacti in the northern Andean enclaves (Nassar et al., 1997; Ruiz et al., 2000). Their flowers remain open during the entire night and part of the next morning, with the exception of *Stenocereus griseus,* which closes before sunrise. These closing times are critical for excluding birds but do not exclude them completely. Petit (1995) and Nassar et al. (1997) have demonstrated that the bat species *Glossophaga longirostris* and *Leptonycteris curasoae* are the most effective pollinators of *Stenocereus griseus* and *Subpilocereus repandus* in Curaçao and northern Venezuela, respectively. Additionally, columnar cacti show a strong trend towards self-incompatibility, which increases their dependency on mobile pollinators, as has been established by Petit (1995) and Nassar et al. (1997) in *Stenocereus griseus, Subpilocereus horrispinus,* and *Subpilocereus repandus,* which are also present in the Andes. This set of chiropterogamic features and other evidence suggest that these Cactaceae depend mainly on glossophagine bats for fruit and seed set.

In contrast, the flowers of *Cereus hexagonus* (Fig. 12.4k) show features that favor pollination by moths of the family Sphingidae (P. Soriano, pers. obs.). Thus the floral tube is greenish in color and very long and narrow (about 20 cm in length); the perianth is 8–10 cm in diameter with white petals; and anthesis is nocturnal. Long floral tubes allow moths to reach the nectaries with their long proboscides but prevent access by bats.

Morphological features of fruits of the Cactaceae that favor chiropterochory and ornithochory as mechanisms of seed dispersal include size, shape, and pulp color. Fruits of *Stenocereus griseus* are spherical and 5 cm in diameter with a thin pericarp (1–2 mm) and numerous spiny areoles (Fig. 12.4c). When unripe, they are reddish green, but when ripe, the areoles tend to be easily detached from the pericarp, and the endocarp becomes red (Soriano et al., 1991). This species also has a morph whose pericarp is greenish yellow with white pulp. The seeds are small (1.5 × 1 mm). The genus *Subpilocereus* has ellipsoid to oval fruits 4–8 cm in length and 1.5–3.5 cm wide without areoles (Fig. 12.4f). They are dehiscent with a thick pericarp (5–8 mm) of green color with purple stains. A white pulp with abundant small seeds (3 × 2 mm) is revealed upon opening the fruit. The fruits of *Pilosocereus* spp. are nearly spherical (5 cm long by 3–4 cm wide), without areoles, and bluish-green in color (Fig. 12.4i). They are dehiscent, with a thick pericarp (4–5 mm). The white pulp contains numerous small seeds (1 × 0.7 mm). Different species of this genus have populations that exhibit white pulp (*P. tillianus,* in Lagunillas), red pulp (*Pilosocereus* sp., La Tatacoa), or white and red morphs in the same population (*Pilosocereus* sp., La Quinta). *Cereus hexagonus* produces spineless oval fruits 10–13 cm long and 5–8 cm in diameter; upon maturation the pericarp changes from green to red (Fig. 12.4l). There are varieties with red or white pulp. The seeds are small (3 × 2 mm) and black. These characteristics attract birds and bats as potential dispersal agents (Greenhall, 1957; Silva, 1988).

Seed-Dispersal Efficiency of Birds and Bats

Bird communities in Andean enclaves are as yet poorly studied; nevertheless, studies at Lagunillas (Soriano et al., 1999) show that the community of fruit-eating birds associated with two columnar cacti (*Stenocereus griseus* and *Subpilocereus repandus*) contains 19 species (Table 12.1). This species richness is equivalent to that of five other neotropical arid localities (Bosque, 1984; Silva, 1988; Santos, 1995; Silvius, 1995; Wendelken and Martin, 1988). Each of these areas shares taxonomic and functional similarities, including one species each in the families Columbidae, Mimidae, Parulidae, Picidae, Psittacidae, and Tyrannidae and at least two Emberizidae. Experiments and observations indicate that these species belong to three functional groups: seed dispersers, seed predators, and pulp consumers (Silvius 1995; Soriano et al., 1999). This appears to be the basic functional structure in cactus-rich arid communities.

In the Lagunillas community, birds remove twofold more seeds of *Stenocereus griseus* than *Subpilocereus repandus,* perhaps because of differences in their dispersal syndromes (Soriano et al., 1999). The main cactus-fruit eater is *Melanerpes rubricapillus* (Picidae), which transports 78% and 39% of seeds consumed by birds of *Subpilocereus repandus* and *Stenocereus griseus,* respectively (Table 12.1). Next in significance is the seed predator *Saltator albicollis* (responsible for 11% and 18% of the seed transport for the two species of cacti, respectively). Each cactus species or morph has a slightly different bird assemblage (Table 12.1). The bird assemblage associated with the red morph of *Stenocereus griseus* is richer (15 species) than that associated with the white morph (11 species) or with that of *Subpilocereus repandus* (ten species). The assemblages of *Subpilocereus repandus* and the white morph of *Stenocereus griseus* are the most similar (eight species in common). As a result, birds transport a greater num-

TABLE 12.1

Avian Consumption of the Fruit of the Cacti *Subpilocereus repandus* and *Stenocereus griseus* in the Lagunillas enclave, Mérida State, Venezuela[a]

				S. griseus		
Family	*Species*	*Function[b]*	*S. repandus*	*Red*	*White*	*Both*
Phasianidae	*Colinus cristatus*	Pd			—	—
Columbidae	*Leptotila verreauxi*	Pd		4.02		4.02
Psittacidae	*Forpus passerinus*	Pd		1.79		1.79
Trochilidae	*Phaethornis hispidus*	PE	0			
Picidae	*Melanerpes rubricapillus*	Di	78.75	26.55	13	39.55
Tyrannidae	*Pitangus sulphuratus*	Di			0.01	0.01
	Tyrannus melancholicus	Di(?)		0.01		0.01
Turdidae	*Turdus nudigenis*	Di		6.73		6.73
Mimidae	*Mimus gilvus*	Di		8.36	3.33	11.69
Emberizidae	*Saltator albicollis*	Pd	10.78	8.67	9.12	17.79
	Tiaris bicolor	Pd	0.03	0.69	0.39	1.08
	Sicalis flaveola	Pd		—		
	Zonotrichia capensis	Pd	0.03			
	Euphonia laniirostris	Di	0.93	0.67	2.16	2.83
	Piranga rubra	Di(?)		—		—
	Tachyphonus rufus	Di/Pd	6.86	0.52	0.65	1.17
	Thraupis episcopus	Di	2.59	2.96	2	4.96
Parulidae	*Coereba flaveola*	Cp	0	0	0	0
Fringillidae	*Carduelis psaltria*	PE	0.03	3.46	5	8.46
	TOTALS		100	64.45	35.68	100

[a] Numerical data indicate the percentage of seeds transported or predated by bird species (determined by defecation analysis), based on the total number of seeds removed by the avian community observed during the fruiting periods of each cactus species ($n = 697$ obs.). Data are from Soriano et al. (1999).

[b] — = consumption data not available; PE = pulp eater; Pd = predator; Di = disperser.

ber of seeds of the red morph than the white morph (Table 12.1). However, not all birds that eat fruits of these two species of cacti have a positive effect on their reproductive success because only seven of 19 birds appear to be legitimate seed dispersers.

Recent studies of some columnar cacti indicate that the phenotypic characteristics (e.g., color) exhibited by fruits influence the seed dispersal efficiency of birds and bats. The red color of pulp represents an adaptation to attract birds as dispersal agents, as these vertebrates select their food by color. In contrast, odor is more important for food choice in bats (van der Pijl, 1982; Debussche and Isenmann, 1989; Willson and Whelan, 1990; Willson et al., 1990). The red morph of *Stenocereus griseus* fruits should favor detection by birds. Rengifo (1997) provides experimental evidence demonstrating in the laboratory the preference for the red morph of *S. griseus* by the birds *Melanerpes rubricapillus, Mimus gilvus,* and *Thraupis episcopus,* whereas *Tachiphonus rufus* preferred the white morph. The general trend of field data shows that seed dispersers prefer the red morph and seed predators prefer the white morph (Table 12.1). Additional experimental evidence is needed to clarify this point.

Field experiments provide quantitative data about the role of bats and birds in the cactus-disperser system by using fruits of *Stenocereus griseus* and *Subpilocereus repandus* and measuring the consumption index by each of these groups (Rengifo, 1997; Naranjo, 1998). (The consumption index is defined as the ratio of pulp removed to pulp offered; it ranges in value between 0 and 1, indicating no pulp consumption or complete depletion, respectively.) The consumption of *Subpilocereus repandus* fruits by bats was greater than consumption by birds (0.79 ± 0.05 and 0.48 ± 0.06, $t = 2.0639$; $df = 24$; $p < 0.05$). However, for *Stenocereus griseus,* consumption by bats and birds was very similar (0.9 ± 0.03 and 0.8 ± 0.05, $t = 2.2009$; $df = 11$; $p > 0.05$), because birds are attracted by the red fruit coloration, but bats do not discriminate by color and consume as many red as white fruits.

The high density of *Stenocereus griseus* at Lagunillas and La Tatacoa (Santos, 1995; Sosa and Soriano, 1996) in comparison with other species of cacti agrees with the prediction of Fleming et al. (1993b) that the fitness of plants whose fruit syndromes allow dispersal by two or more taxa is higher than those dispersed by a single agent. Silvius (1995) observed that the dehiscence of fruits of *S. griseus* shows similar day and night frequencies, which increases resource availability and reduces competition between both groups.

In addition to number of seeds transported by seed dispersers, the quality of the treatment produced by the passage of seeds through animal intestines is an important component of dispersal. Seeds can be affected in different ways: At one extreme, their germination probability is enhanced, and at the other, the seeds are destroyed in the process (Schupp, 1993). Seeds of Cactaceae contain substances that inhibit the germination process (Williams and Arias, 1978), which makes them dependent on the effective elimination of this substance by dispersers. Passage of seeds of *Stenocereus griseus* through the digestive tracts of birds and bats does not produce seed scarification. It does eliminate the seed coat and increases percentage of germination (Table 12.2; Rengifo, 1997; Naranjo, 1998). This passage reduces mean germination times (between three and five days) and increases percentage of germination compared with unwashed seeds in *Stenocereus griseus* and *Subpilocereus repandus* (Table 12.2).

In the case of *Pilosocereus tillianus,* we lack data for birds, but seed passage through the digestive tract of *Glossophaga longirostris* does not significantly affect mean germination time (Naranjo, 1998). This suggests that this bat is not needed to release the seed inhibitor. In addition, such factors as erratic flowering and fruiting (Sosa and Soriano, 1996) and reproduction mainly involving vegetative dispersal (Rico et al., 1996) probably hinder the establishment of a close dispersal association with a bat species.

Now that data are available for evaluating some dispersal components, such as seed removal and the effect of seed passage through the digestive tract of birds and bats, there is a need to study aspects of the quality of dispersal: movement patterns of dispersers (seed shadows), patterns of seed deposition, nutrients provided to the seeds by the feces of dispersers, and viability of seeds in the seed bank. Postdispersal effects, including seed predation, establishment, recruitment, and nurse-plant effects have been studied in Mexico but not in the Andean enclaves (Valiente-Banuet and Ezcurra, 1991; Valiente-Banuet et al., 1991a; Valiente-Banuet et al., 1991b).

TABLE 12.2

Percent Germination of Two Species of Columnar Cactus Seeds Passed by the Digestive Tract of Bats and Birds and Two Controls (Unwashed and Washed Seeds)[a]

Treatment	*Stenocereus grisseus*		*Subpilocereus repandus*	
	Germination (%)	D_{max} [b]	*Germination (%)*	D_{max}
Bats				
Glossophaga longirostris ($n = 10$)	88	0.28*	86	0.12*
Birds				
Melanerpes rubricapillus ($n = 8$)	72	0.23*	83	0.07[NS]
Mimus gilvus ($n = 4$)	86	0.23*	—	—
Thraupis episcopus ($n = 8$)	76	0.25*	82	0.10*
Tachyphonus rufus ($n = 2$)	SD	SD	87	0.09*
Washed seeds	98		91	
Unwashed seeds	71		75	

[a] $n = 500$ seeds per treatment. SD = Seeds damaged by the passage; significance: * $p < 0.001$; NS = not significant. Data are from Rengifo (1997) and Naranjo (1998).

[b] D_{max} of the Kolmogorov-Smirnov two-sample test gives the efficiency of elimination of the inhibitor, measured as the maximum difference between animals and unwashed seeds treatment cumulative frequency distributions.

Adaptive Responses of *Glossophaga longirostris* and *Leptonycteris curasoae* to Environmental Conditions

INTERDEPENDENCE BETWEEN GLOSSOPHAGINES AND COLUMNAR CACTI

Studies conducted in some enclaves of the northern Andes show that the diet of *Glossophaga longirostris* varies geographically (Soriano et al., 1991; Sosa and Soriano, 1993, 1996; Ruiz et al., 1997, Cadena et al., 1998). In Lagunillas, *G. longirostris* is associated with three species of columnar cacti (*Pilosocereus tillianus, Stenocereus griseus,* and *Subpilocereus repandus*) and acts as their pollinator and seed disperser. The latter two species provide a food supply for the bats nearly all year-round. They have asynchronous flowering and fruiting peaks, which reduces interspecific competition for seed dispersers and pollinators. During the annual scarcity period, this bat eats fruits of *Chlorophora tinctoria* (Moraceae) (Soriano et al., 1991; Sosa and Soriano, 1993, 1996). At La Tatacoa, Ruiz et al. (1997) showed that *G. longirostris* eats mainly fruits and pollen of the cactus *Stenocereus griseus,* fruits of *Muntingia calabura* (Elaeocarpaceae), nectar and pollen of *Helicteris baruensis* (Sterculiaceae), and, in smaller amounts, pollen and fruits of the cactus *Pilosocereus* sp. Fruits of *M. calabura* constitute the principal food item in its diet during the annual scarcity period, in the same way as *Piper* spp. and *Solanum* spp. have been reported as eaten by other bat species in wet forests (Gilbert, 1980; Fleming, 1985; Marinho-Filho, 1991). At Chicamocha, in addition to columnar cacti, *G. longirostris* eats fruits of eight other plant species, mainly from the genera *Ambrosia, Cecropia,* and *Muntingia* (Cadena et al., 1998). In these enclaves, seasonal variations in diet are correlated mainly with the availability of resources and they do not reflect food preference by *G. longirostris* (Soriano et al., 1991; Sosa and Soriano, 1996; Ruiz et al., 1997)

At Lagunillas, where *Leptonycteris curasoae* is found during part of the year, it consumes mainly nectar and pollen of the families Cactaceae and Agavaceae and in smaller amounts, fruits of *Stenocereus griseus* and *Subpilocereus repandus* (Sosa and Soriano, 1993). In contrast, at Chicamocha, this species was recorded eating pollen and fruit of *Stenocereus griseus* (Cadena et al., 1998). This information agrees with data obtained by Aranguren (1995) in the Peninsula of Paraguaná and that of other authors for arid zones of North America (Alcorn et al., 1961; Hayward and Cockrum, 1971; Gardner, 1977; Fleming, 1989), confirming that this species is a pollinator and seed disperser of Cactaceae. Although at Lagunillas, the diet of the two bats (*Glossophaga longirostris* and *L. curasoae*), as measured by similarity in fecal-sample content, overlaps 82% (quantitative index of Sørensen; Magurran, 1988), the diet of *G. longirostris* includes fruit of other cacti (e.g., *Pilosocereus tillianus*) and other noncactus species not eaten by *L. curasoae* (Sosa and Soriano, 1993). High dietary overlap does not necessarily indicate interspecific competition, because *L. curasoae* is found only in the enclave during the period of high abundance of flowers and fruits of *Stenocereus griseus* and flowers of *Subpilocereus repandus* (December–April; Sosa, 1991; Sosa and Soriano, 1993). Availability of resources probably does not represent an environmental constraint during this period.

The reproductive pattern of *Glossophaga longirostris* is bimodal polyestry (Sosa and Soriano, 1996; Ruiz et al., 1997), which is synchronized with the production of flowers and fruits of columnar cacti. At Lagunillas, one of the annual birth peaks coincides with high availability of flowers and fruits of *Stenocereus griseus* and *Subpilocereus repandus,* but the second peak does not seem to correspond to a resource peak (Fig. 12.5a). At La Tatacoa, lactation peaks coincide with high levels of fruits of *Stenocereus griseus* (Fig. 12.5b). The differences found in diet and timing of reproductive periods between these two populations of *G. longirostris* appear to be adjustments to differences in the spatial and temporal distribution of food resources. In Curaçao, *G. longirostris* has a monoestrous pattern, which coincides with the strong overlap in the flowering periods of *Stenocereus griseus* and *Subpilocereus repandus* (Petit, 1997).

In contrast, the reproductive pattern of *Leptonycteris curasoae* is seasonal monoestry (Smith and Genoways, 1974). On Peninsula de Paraguaná, as in most of its geographic distribution, *L. curasoae* exhibits a single peak of parturition and lactation in the period May–July, which can extend until August (Hayward and Cockrum, 1971; Martino et al., 1998; but see chapter 14, this volume). During this period, *L. curasoae* is absent from Lagunillas (Sosa and Soriano, 1993), and preliminary surveys seem to indicate that this species abandons the Andes then (Ruiz and Soriano, 1998; Sánchez and Cadena, 1999; Soriano et al., 2000a). The single reproductive peak is possibly the consequence of the high energetic cost involved in migration and will be discussed below.

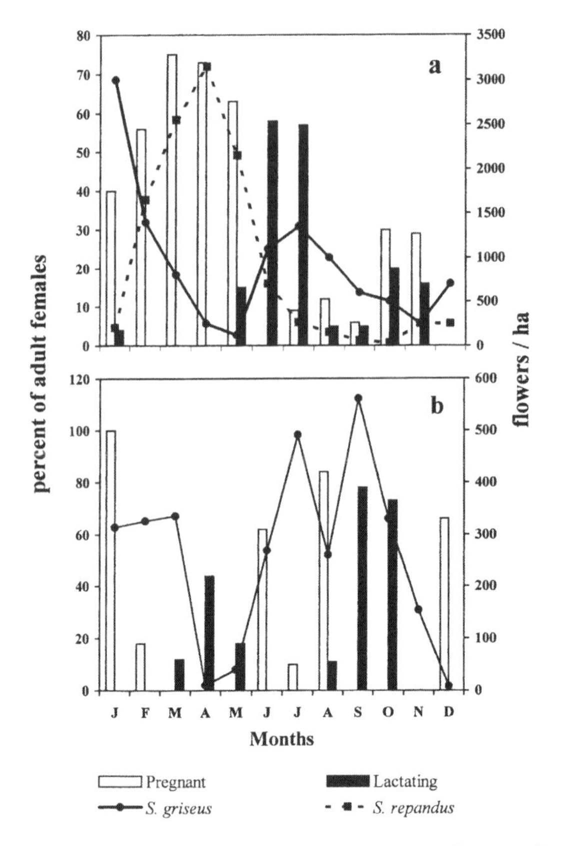

FIGURE 12.5. Reproductive pattern of *Glossophaga longirostris* versus the main floral resources in (a) Lagunillas and (b) La Tatacoa. Data are from Sosa and Soriano (1996) and Ruiz et al. (1997).

CONFIGURATION OF THE BAT-CACTUS SYSTEM IN THE ANDEAN ENCLAVES

The seven enclaves considered here show differences in size, species richness, and abundance of columnar cacti, as well as in the presence or absence of resident and/or migrant bat species (Table 12.3). Recent preliminary data have confirmed the presence of *Glossophaga longirostris* in Chicamocha as well as the presence of *Leptonycteris curasoae* in Chicamocha, La Quinta and La Puerta (Ruiz and Soriano, 1998; Sánchez and Cadena, 1999; Soriano et al., 2000a). These data show that the configuration of the bat-cactus system in a given enclave depends on particular combinations of variables, which include different species of Cactaceae, other plants, and other bat species. At least four different configurations are known.

1. *Configuration with a migrant species.* The exclusive presence of a migrant species in an enclave indicates that the availability of food is not sufficient to support bats for at least part of the year. In such a situation, only one species is able to use the

TABLE 12.3
Columnar Cacti and Glossophagine Bats in the Andean Enclaves[a]

		Columnar Cacti[b]				*Glossophagine Bats*				
Enclave	*Area (km²)*	*S. g.*	*S. r.*	*S. h.*	*P. spp.*	*G. l.*	*L. c.*	*G. s.*	*A. g.*	*C.g.*
La Puerta	27				++		X		X	
La Quinta	29	+			++		X	X	X	
Cúcuta-Ureña	191	+			++		X	X		
La Tatacoa	1342	+++			+	X		X	X	
Lagunillas	262	+++	++		++	X	X			
Chicamocha	112	+++		+	++	X	X		X	
Patía	1127				++				X	X

[a] Data are from Cadena et al. (1998); Ruiz and Soriano (1998); and Soriano et al. (2000a). Among the bats, *Leptonycteris curasoae* is a migratory species.

[b] *S. g., Stenocereus griseus; S. r., Subpilocereus repandus; S. h., Subpilocereus* cf. *horrispinus; P.* spp., *Pilosocereus* spp.; *G. l., Glossophaga longirostris; L. c., Leptonycteris curasoae; G. s., Glossophaga soricina; A. g., Anoura geoffroyi;* C. g., *Choeroniscus godmani;* X, present; + , rare; ++, abundant; +++, very abundant.

seasonal resources of the enclave and abandons it in the unfavorable period. To do this, the bat must be highly mobile, as has been reported for *L. curasoae* (Sahley et al., 1993). This seems to be the case at La Puerta, La Quinta, and Cúcuta-Ureña, where, due to scarcity of columnar cacti (Table 12.3) and/or noncomplementary phenological patterns, high resource levels do not occur year-round and therefore the presence of a resident species is not possible. In these enclaves, *L. curasoae* is present only seasonally. The small size of the two first enclaves is an additional factor that contributes to the absence of a resident species. Although the Cúcuta-Ureña enclave is relatively large (Table 12.3), its low species richness and paucity of columnar cacti discourages the presence of resident species.

2. *Configuration with a resident species.* This situation is typified by La Tatacoa (Table 12.3), which has a large area, in which phenological patterns of cacti species guarantee an adequate supply of resources all year to support a resident population of bats, in contrast with Cúcuta-Ureña and La Quinta (Ruiz et al., 1997, 2000). Large area and a significant population of one of the cacti seem to be the main advantages of La Tatacoa. The absence of a migrant species seems to be a consequence of the isolation of the enclave. In fact, whereas the distance separating most enclaves is about 60–100 km (Fig. 12.1), La Tatacoa is at least 220 km from the nearest enclave (Chicamocha). Fleming (1997) suggested that *L. curasoae* can migrate up to 480 km on its fat reserves before it must replenish them. This means that its maximum flight distance in one night is around 160 km. Therefore, an isolation distance 220 km must constitute, for La Tatacoa, a barrier to *L. curasoae.*
3. *Configuration with a resident and a migrant species.* This configuration includes the same conditions as the last configuration to maintain the resident species as well as proximity to other enclaves that allows the arrive of migrant species. The situation is exemplified by the enclaves of Lagunillas and Chicamocha (Table

12.3). This table shows that those enclaves in which *Stenocereus griseus* is absent or scarce are those in which a resident bat is absent. *S. griseus* provides resources during a significant portion of the year because of its bimodal phenological pattern and abundant flower and fruit production. Thus the presence of *S. griseus* in high densities seems to be especially important for the presence of resident species.

4. *Configuration with exclusively opportunistic species.* This type is characterized by the absence of bat species strongly linked to columnar cacti, and the only glosophagine species are those representatives from wetter, bordering ecologic units. Marginal presence of these "alien" species occurs in all types described before. For instance, *Glossophaga soricina* acts marginally in Cúcuta-Ureña as an opportunistic species, and in a similar way, *Anoura geoffroyi* can contribute to cactus pollination in the enclaves of La Puerta and La Quinta (Table 12.3). However, this configuration type occurs as the only possibility in those enclaves where, simultaneously, the resource supply is not guaranteed all year-round and isolation from neighbor enclaves prevents visitation by migrant species. Such is the case at the Patía enclave in the Cordillera Occidental of Colombia, where *A. geoffroyi, Choeroniscus godmani,* and *Phyllostomus discolor* were recorded eating pollen of Cactaceae (Cadena et al., 2000). Although these bat species do not belong to this arid ecosystem, they occasionally use the resources offered by its columnar cacti.

Movements of *Leptonycteris curasoae*

In North America, seasonal movements of *Leptonycteris curasoae* follow the flowering patterns of the plants that it consumes (Cockrum, 1991; Fleming et al., 1993a; Fleming, 1995). The same pattern emerges along that region of the Caribbean coast where its populations might travel back and forth to the Andean enclaves. Thus, on the Península de Paraguaná, populations of *L. curasoae* decrease in the period September–April (Martino et al., 1998), when the floral supply declines in the coastal region (Petit, 1997). At Lagunillas, this species' highest abundance coincides with the greater availability of columnar cacti flowers (Fig. 12.6); when this resource declines in May–July, *L. curasoae* disappears (Sosa and Soriano, 1993, 1996). Likewise, at Chicamocha this species is in highest abundance when columnar cacti exhibit their peaks of flower and fruit production; it disappears from the enclave in April–July (Cadena et al., 1998; Sánchez and Cadena, 1999). Although data are not available for resource supplies in the other enclaves under consideration, we speculate that the trough in flower production observed at Lagunillas might be a synchronous and general pattern over all Andean enclaves, as these enclaves display lower cacti richness and reduced population (Table 12.3).

Although the cactus nectar consumed by *Leptonycteris curasoae* has high caloric content (Helversen and Reyer, 1984; Petit and Freeman, 1996), when a reduction in flower supply occurs, the energy obtained could be lower than the investment in flying long distances from roosting to foraging sites. The amount of nectar that a single

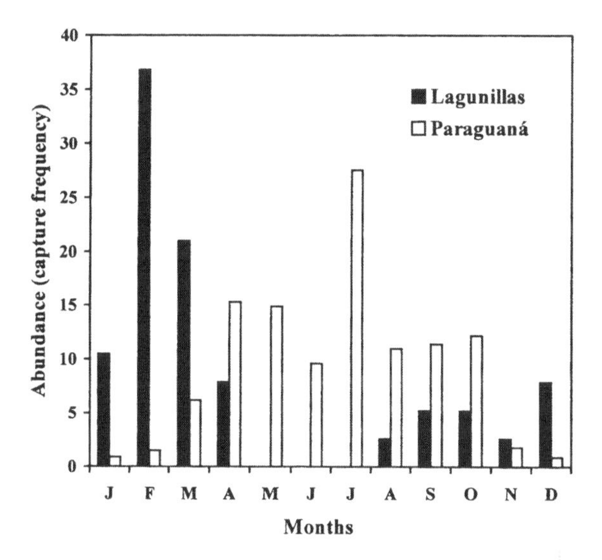

FIGURE 12.6. Monthly relative frequency of captures of *Leptonycteris curasoae* in Lagunillas and Paraguaná. Data are from Sosa and Soriano (1996) and Martino et al. (1998). Although the Paraguaná data come from one cave and are not directly comparable with Lagunillas data, banded individuals in Paraguaná have been found on the mainland (A. Martino, pers. com.).

flower can produce is about 0.6–1.1 mL/night, of which less than 0.1 mL is harvested in a single visit (Arends et al., 1995; Nassar et al., 1997). This low volume of nectar in relation to the energetic demand of the bat promotes visits to many plants with open flowers in the same night, which implies a great investment in flight time (Arends et al., 1995; Nassar et al., 1997). The energy available in the nectar of Cactaceae averages 2.78 kJ/flower, and this glossophagine needs 49–61 kJ/day, which is equivalent to 200 cactus flower visits per night; the caloric requirement may double for lactating females (Petit and Pors, 1995). Increasing the investment in foraging for resources also increases the cost:benefit ratio, which can be energetically unfavorable for the animal. Thus *L. curasoae* may migrate to other localities where resources are more abundant to reduce the energetic expense involved in food search.

The high energetic expenditure that migratory movements demand of *L. curasoae* in northern South America restricts its reproduction to a single annual event, which takes place outside of the Andean enclaves (Hayward and Cockrum, 1971, Bradbury, 1977; Cockrum, 1991; Sosa and Soriano, 1993; Martino et al., 1998). However, gregarious behavior in large maternity colonies (which can surpass 20,000 individuals in hot caves) is very important for economizing energy and increasing the growth rate of the juveniles and thus reducing the lactation period for females (Tuttle and Stevenson, 1982; Arends et al., 1995). The Andean enclaves do not seem to possess adequate refuges that permit the establishment of successful maternity colonies.

Conservation of the Andean Arid Zones

The dry zones of Colombia and Venezuela include relatively large areas ecologically connected with the plains of the Caribbean (coastal regions of northern Colombia and Venezuela) and with the Andean arid enclaves. However, these zones have been little studied (Hernández et al., 1992) when compared with the tropical wet forests in the lowlands. They have also received little attention in conservation plans. With the exception of a few hectares on the margins of Macuira National Park (on the Península de la Guajira in Colombia) and the Médanos de Coro National Monument (in Venezuela), no dry areas are under protection in the national park systems of either country. It is important to conduct studies to provide the necessary information to prepare management and conservation plans for these areas. These plans should take the following into account:

1. The inter-Andean dry valleys represent an important reservoir of plant and animal species adapted to xeric conditions. Given the current isolation of these dry enclaves, they may represent important areas of speciation and endemicity. For instance, the area of Villavieja in La Tatacoa is the type locality of the subspecies of bat *Glossophaga longirostris reclusa* (Webster and Handley, 1986). Likewise, the population of this bat in Lagunillas constitutes another endemic subspecies (*Glossophaga longirostris maricelae;* Soriano et al., 2000b). Similarly, the columnar cactus *Pilosocereus tillianus* is endemic to Lagunillas, and it is very probable that each enclave contains populations undergoing the speciation process. Some of the columnar cacti could represent currently unrecognized endemics, as is the case in a new, undescribed species of Cactaceae, *Pilosocereus* sp., located in La Tatacoa.
2. The Andean enclaves constitute a refuge for some populations of *Leptonycteris curasoae,* allowing it to maintain its population levels during the annual period of resource shortage along the Caribbean coast. This bat is a key element in the pollination of columnar cacti in the enclaves, where resident species are absent. The geographical arrangement of the enclaves, aligned throughout the Cordillera de Mérida and the Cordillera Oriental similar to the steps of a staircase, allows the passage of *L. curasoae* from the Caribbean coast to Chicamocha. This arrangement renders the enclave cacti vulnerable, because the disappearance or severe disturbance of intermediate enclaves would sever the passage and prevent *L. curasoae* from pollinating the Cactaceae in farther enclaves. This effect would be particularly pernicious in those enclaves lacking resident species. The persistence of the Andean enclaves can only be guaranteed through implementation of conservation policies in both countries.

Summary

Most of the Andean enclaves in Colombia and Venezuela contain species of columnar cacti from the genera *Pilosocereus, Stenocereus,* and *Subpilocereus.* All of these enclave species possess flowers that exhibit syndromes of chiropterogamy, and bats play the main role in their pollination. The fruits of these cacti show certain morphological features that favor both chiropterochory and the ornitochory. In these environments, the presence of two species of Glossophaginae bats has been detected: *Glossophaga longirostris* and *Leptonycteris curasoae*. These bats fulfill the double roles of pollinator and seed dispersers. The first species forms resident populations in some of these environments, whereas the second seems to be a seasonal visitor, if present at all. In addition, at least two functional groups of fruit-consuming birds exist in the community: seed dispersers and seed predators. Using the available information on these environments, we make a comparison of the roles of both groups in the reproduction process of these cacti. The dispersion of seeds is a function that seems to be shared between birds and bats. We use the arid enclave of Lagunillas, Venezuela, as a study case for the comparison of the communities of birds associated with two species of columnar cacti that bear fruits of different color. We also describe and compare four possible configurations of the bat-cactus system from some Andean arid enclaves.

Resumen

En la mayoría de los enclaves andinos de Colombia y Venezuela se observa la presencia de especies de cactáceas columnares pertenecientes a los géneros *Pilosocereus, Stenocereus,* y *Subpilocereus.* Todas estas especies poseen flores que presentan síndromes de quiropterogamia y los murciélagos juegan un papel muy importante en su polinización. Los frutos de estas cactáceas muestran ciertos atributos morfológicos que favorecen la quiropterocoria y la ornitocoria. En estos ambientes, se ha detectado la presencia de dos especies de murciélagos glosofaginos (*Glossophaga longirostris* y *Leptonycteris curasoae*). Estos juegan el doble papel de polinizadores y dispersores de semillas de estas cactáceas. La primera especie tiene poblaciones residentes en algunos de estos ambientes. La segunda parece ser un visitante estacional de los mismos aunque en ocasiones está ausente. En relación a las aves asociadas al consumo de frutos de cactáceas se sabe que existen al menos dos grupos funcionales en la comunidad: los dispersores y los depredadore de semillas. Con base en la información disponible para uno de estos ambientes, se realizó una comparación de los papeles de ambos grupos en los procesos reproductivos de las cactáceas. Con respecto a la dispersión de las semillas, este papel parece estar compartido por las aves y los murciélagos. Usamos el enclave de Lagunillas (Venezuela)como estudio de caso par comparar las comunidades de aves asociadas a dos especies de cactáceas columnares que presentan frutos con diferente color. Así mismo, se describen y comparan cuatro posibles configuraciones del sistema murciélago-cactáceas en distintas enclaves áridos andinos.

ACKNOWLEDGMENTS

We are grateful to Alfonso Valiente-Banuet and Ted Fleming for inviting us to participate in the workshop Evolution, Ecology, and Conservation of Columnar Cacti and Their

Mutualists, which took place in the summer of 1998 at Tehuacán, Mexico, and thank them for their valuable comments that improved the first draft of this chapter. M. Acevedo helped us with the early English version. Part of the information we provide in this work was obtained with funds of the Consejo de Desarrollo Científico, Humanístico y Tecnológico de la Universidad de los Andes (CDCHT-ULA), Grant C-966-99-01-B (awarded to P. J. Soriano) and the Bat Conservation International Student Scholarship Grant (awarded to A. Ruiz).

REFERENCES

Alcorn, S. M., S. E. McGregor, and G. Olin. 1961. Pollination of saguaro cactus by doves, nectar-feeding bats, and honeybees. *Science* 133:1594–95.

Álvarez, T., and L. González. 1970. Análisis polínico del contenido gástrico de murciélagos Glossophaginae de México. *Anales de la Escuela Nacional de Ciencias Biológicas* 18:137–65.

Aranguren, J. 1995. Composición de la dieta de *Leptonycteris curasoae* en Paraguaná, Estado Falcón, Venezuela. Thesis, Universidad Nacional Experimental Francisco de Miranda, Coro, Venezuela.

Arends, A., F. J. Bonaccorso, and M. Genoud. 1995. Basal rates of metabolism of nectarivorous bats (Phyllostomidae) from semiarid thorn forests in Venezuela. *Journal of Mammalogy* 78:947–56.

Bosque, C. A. 1984. Structure and diversity of arid zone bird communities in Venezuela. Ph.D. diss., University of Washington, Seattle.

Bradbury, J. W. 1977. Social organization and comunication. In *Biology of Bats*. Vol. 3, ed. W. A. Wimsatt, 1–72. New York: Academic Press.

Cadena, A., J. Álvarez, F. Sánchez, C. Ariza, and A. Albesiano. 1998. Dieta de los murciélagos frugívoros en la zona árida del Río Chicamocha (Santander, Colombia). *Boletín de la Sociedad Biológica de Concepción, Chile* 69:69–75.

Cadena, A., F. Sánchez, J. Alvarez, C. Ariza, and A. Albesiano. 2000. Diversidad de murciélagos en zonas áridas colombianas. In *Libro de Resúmenes, Primer Congreso Colombiano de Zoología—Año 2000,* eds. P. Muñoz de Hoyos and J. Aguirre, 67. Bogotá, Colombia: Instituto de Ciencias Naturales, Universidad Nacional de Colombia.

Cavelier, J., A. Ruiz, and M. Santos. 1996. *El Uso de las Cactáceas como Bioindicadores del Proceso de Desertificación en Enclaves Secos Interandinos de Colombia.* Bogotá, Colombia: Instituto de Hidrología, Meteorología y Estudios Ambientales (IDEAM).

Cockrum, E. L. 1991. Seasonal distribution of northwestern populations of the long-nosed bats, *Leptonycteris sanborni* Family Phyllostomidae. *Anales del Instituto de Biología de la Universidad Nacional Autónoma de México, Serie Zoológica* 62:181–202.

Debussche, M., and P. Isenmann. 1989. Fleshy fruit characters and the choices of bird and mammal seed dispersers in a Mediterranean region. *Oikos* 56:327–38.

Ellner, S., and A. Shmida. 1981. Why are adaptations for long-range seed dispersal rare in desert plants? *Oecologia* 51:133–44.

Espinal, L. S., and E. Montenegro. 1963. *Formaciones Vegetales de Colombia. Memoria Explicativa Sobre el Mapa Ecológico.* Bogotá, Colombia: Instituto Geográfico Agustín Codazzi.

Ewel, J. J., A. Madriz, and J. A. Tossi. 1976. *Zonas de Vida en Venezuela.* Caracas, Venezuela: Ministerio de Agricultura y Cría.

Fleming, T. H. 1985. Coexistence of five sympatric *Piper* (Piperaceae) species in a tropical dry forest. *Ecology* 66:688–700.

———. 1989. Climb every cactus. *Bats* 7:3–7.

———. 1995. The use of stable isotopes to study the diets of plant-visiting bats. *Symposium of the Zoological Society of London* 67:99–110.

———. 1997. Energetics and nectar corridors in *Leptonycteris curasoae.* In *Abstracts of the Seventh International Theriological Congress,* ed. R. Medellín, 105. Acapulco, Mexico.

Fleming, T. H., R. A. Nuñez, and L. Sternberg. 1993a. Seasonal changes in the diets of migrant and non-migrant nectarivorous bats as revealed by carbon stable isotope analysis. *Oecologia* 94:72–75.

Fleming, T. H., D. L. Venable, and L. G. Herrera. 1993b. Opportunism vs. specialization: the evolution of dispersal strategies in fleshy-fruit plants. *Vegetatio* 107/108:107–20.

Gardner, A. L. 1977. Feeding habits. In *Biology of Bats of the New World Family Phyllostomatidae.* Part 2, ed. R. J. Baker, J. K. Jones, and D. C. Carter, 293–350. Texas Tech University Museum Special Publication No. 12. Lubbock, Texas: Texas Tech Press.

Gilbert, L. E. 1980. Food web organization and the conservation of neotropical diversity. In *Conservation Biology: an Evolutionary—Ecological Perspective,* ed. M. E. Soulé and B. A. Wilcox, 11–33. Sunderland, Mass.: Sinauer Associates.

Goodwin, G. G., and A. M. Greenhall. 1961. A review of the bats of Trinidad and Tobago. Descriptions, rabies infection, and ecology. *Bulletin of the American Museum of Natural History* 122:187–302.

Greenhall, A. M. 1957. Food preferences of Trinidad fruit bats. *Journal of Mammalogy* 38:409–10.

Hayward, B. J., and E. L. Cockrum. 1971. The natural history of the western long-nosed bat *Leptonycteris sanborni. Publication of the Office of Research, Western New Mexico University* 1:74–123.

Helversen, O. V., and H. U. Reyer. 1984. Nectar intake and energy expenditure in a flower-visiting bat. *Oecologia* 63:178–84.

Hernández, J., T. Walschburger, R. Ortiz, and A. Hurtado. 1992. Origen y distribución de la biota suramericana y colombiana. In *La Diversidad Biológica de Iberoamérica,* ed. G. Halffter, 55–104. Volumen especial de Acta Zoológica Mexicana. Xalapa, Mexico: Instituto de Ecología, A. C. CYTED-D.

Magurran, A. E. 1988. *Ecological Diversity and Its Measurement.* Cambridge: Cambridge University Press.

Marinho-Filho, J. S. 1991. The coexistence of two frugivorous bat species and the phenology of their food plants in Brazil. *Journal of Tropical Ecology* 7:59–67.

Martino, A., A. Arends, and J. Aranguren. 1998. Reproductive pattern of *Leptonycteris curasoae* Miller (Chiroptera: Phyllostomidae) in northern Venezuela. *Mammalia* 62:69–76.

Naranjo, M. E. 1998. Efecto del murciélago *Glossophaga longirostris* en la germinación de tres cactáceas columnares de los Andes venezolanos. Bachelor's thesis, Universidad de los Andes, Mérida, Venezuela.

Nassar, J. M., N. Ramírez, and O. Linares. 1997. Comparative pollination biology of Venezuelan columnar cacti and the role of nectar-feeding bats in their sexual reproduction. *American Journal of Botany* 84:918–27.

Petit, S. 1995. The pollinators of two species of columnar cacti on Curaçao, Netherlands Antilles. *Biotropica* 27:538–41.

———. 1997. The diet and reproductive schedules of *Leptonycteris curasoae curasoae* and *Glossophaga longirostris elongata* (Chiroptera: Glossophaginae) on Curaçao. *Biotropica* 29: 214–23.

Petit, S., and E. Freeman. 1996. Nectar production of two sympatric species of columnar cacti on Curaçao. *Biotropica* 28:175–83.

Petit, S., and L. Pors. 1995. Survey of columnar cacti and carrying capacity for nectar-feeding bats on Curaçao. *Conservation Biology* 10:769–75.

Pirlot, P. 1964. Nota sobre la ecología de ciertos quirópteros de la región del Río Palmar. *Kasmera* 1:289–307.

Ponce, M. 1989. *Distribución de Cactáceas en Venezuela y su Ámbito Mundial.* Maracay, Venezuela: Universidad Central de Venezuela.

Rengifo, C. 1997. Efecto de las aves en la germinación de las cactáceas columnares *Stenocereus griseus* y *Subpilocereus repandus.* Bachelor's thesis, Universidad de los Andes, Mérida, Venezuela.

Rico, R., L. E. Rodríguez, R. Pérez, and A. Valero. 1996. Mapa y análisis de la vegetación xerófila de las lagunas de Caparú, cuenca media del Río Chama, Estado Mérida. *Plantula* 1:83–94.

Ruiz, A., and P. J. Soriano. 1998. Bat-cactus interrelations in the inter-Andean arid pockets of Colombia and Venezuela. In *International Workshop on the Evolution, Ecology, and Conservation of the Columnar Cacti and their Mutualists,* eds. A. Valiente-Banuet and T. H. Fleming. Tehuacán, México.

Ruiz, A., M. Santos, P. Soriano, J. Cavelier, and A. Cadena. 1997. Relaciones mutualísticas entre el murciélago *Glossophaga longirostris* y las cactáceas columnares en la zona árida de La Tatacoa, Colombia. *Biotropica* 29:469–79.

Ruiz, A., M. Santos, J. Cavelier, and P. J. Soriano. 2000. Estudio fenológico de cactáceas en el, enclave seco de La Tatacoa, Colombia. *Biotropica* 32:397– 407.

Sahley, C., M. A. Horner, and T. H. Fleming. 1993. Flight speeds and mechanical power outputs of the nectar-feeding bat, *Leptonycteris curasoae* (Phyllostomidae: Glossophaginae). *Journal of Mammalogy* 74:594–600.

Sánchez, F., and A. Cadena. 1999. Migración de *Leptonycteris curasoae* (Chiroptera: Phyllostomidae) en las zonas áridas de norte de Colombia. *Revista de la Academia Colombiana de Ciencias Exactas, Físicas y Naturales* 23:683–86.

Santos, M. 1995. Fenología de cuatro cactáceas y su relación con polinizadores y dispersores en el enclave seco de La Tatacoa, Huila, Colombia. Bachelor's thesis, Universidad de los Andes, Bogotá, Colombia.

Sarmiento, G. 1972. Ecological and floristic convergences between seasonal plant formations of tropical and subtropical South America. *Journal of Ecology* 60:367–410.

———. 1975. The dry formations of South America and their floristic conections. *Journal of Biogeography* 2:233–51.

———. 1976. Evolution of arid vegetation in tropical America. In *Evolution of Desert Biota,* ed. D. W. Goodall, 65–99. Austin: University of Texas Press.

Schupp, E. W. 1993. Quantity, quality and the effectiveness of seed dispersal by animals. *Vegetatio* 107/108:15–29.

Silva, W. R. 1988. Ornitocoria em *Cereus peruvianus* (Cactaceae) na Serra do Japi, Estado de Sâo Paulo. *Revista Brasileira de Biologia* 48:381–89.

Silvius, K. M. 1995. Avian consumers of cardón fruits (*Stenocereus griseus:* Cactaceae). *Biotropica* 27:96–105.

Smith, J. D., and H. H. Genoways. 1974. Bats of Margarita Island, Venezuela, with zoogeographic comments. *Bulletin Southern California Academy of Sciences* 73:64–79.

Snow, D. W. 1981. Tropical frugivorous birds and their food plants: a world survey. *Biotropica* 13:1–14.

Soriano, P. J., M. Sosa, and O. Rossell. 1991. Hábitos alimentarios de *Glossophaga longirostris* Miller (Chiroptera: Phyllostomidae) en una zona árida de los Andes venezolanos. *Revista de Biología Tropical* 39:267–72.

Soriano, P. J., M. E. Naranjo, C. Rengifo, M. Figuera, M. Rondón, and R. L. Ruiz. 1999. Aves consumidoras de frutos de cactáceas columnares del enclave semiárido de Lagunillas, Mérida, Venezuela. *Ecotropicos* 12:91–100.

Soriano, P. J., A. Ruiz, and J. Nassar. 2000a. Notas sobre la distribución e importancia ecológica de los murciélagos *Leptonycteris curasoae* y *Glossophaga longirostris* en zonas áridas andinas. *Ecotropicos* 13:91–95.

Soriano, P. J., M. R. Fariñas, and M. E. Naranjo. 2000b. A new subspecies of Miller's long-tongued bat (*Glossophaga longirostris*) from a semiarid pocket of the Venezuelan Andes. *Zeitschrift für Säugetierkunde* 65:1–6.

Sosa, M. 1991. Relaciones ecológicas entre el murciélago *Glossophaga longirostris* y las cactáceas columnares en el bolsón árido de Lagunillas, Mérida, Venezuela. Bachelor's thesis, Universidad de los Andes, Mérida, Venezuela.

Sosa, M., and P. J. Soriano. 1993. Solapamiento de dieta entre *Leptonycteris curasoae* y *Glossophaga longirostris* (Mammalia: Chiroptera). *Revista de Biología Tropical* 41:529–32.

———. 1996. Resource availability, diet and reproductive pattern of *Glossophaga longirostris* (Mammalia: Chiroptera) in an arid zone of the Venezuelan Andes. *Journal of Tropical Ecology* 12:805–18.

Tuttle, M., and D. Stevenson. 1982. Growth and survival of bats. In *Ecology of Bats*, ed. T. H. Kunz, 105–50. New York: Plenum Publishing.

Valiente-Banuet, A., and E. Ezcurra. 1991. Shade as a cause of the association between the cactus *Neobuxbaumia tetetzo* and the nurse plant *Mimosa luisana* in the Tehuacán Valley, Mexico. *Journal of Ecology* 79:961–71.

Valiente-Banuet, A., F. Vite, and J. A. Zavala-Hurtado. 1991a. Interaction between the cactus *Neobuxbaumia tetetzo* and the nurse shrub *Mimosa luisana*. *Journal of Vegetation Science* 2:11–14.

Valiente-Banuet, A., A. Bolongaro-Crevenna, O. Briones, E. Ezcurra, M. Rosas, H. Nuñez, G. Barnard, and E. Vazquez. 1991b. Spatial relationships between cacti and nurse shrubs in a semi-arid environment in central Mexico. *Journal of Vegetation Science* 2:15–20.

Valiente-Banuet, A., M. C. Arizmendi, A. Rojas, and L. Dominguez. 1996. Ecological relationships between columnar cacti and nectar-feeding bats in Mexico. *Journal of Tropical Ecology* 11:1–17.

Valiente-Banuet, A., A. Rojas-Martínez, M. C. Arizmendi, and P. Dávila. 1997. Pollination biology of two columnar cacti (*Neobuxbaumia mezcalaensis* and *Neobuxbaumia macrocephala*) in the Tehuacán Valley, central Mexico. *American Journal of Botany* 84:452–55.

van der Pijl, L. 1982. *Principles of Dispersal in Higher Plants*. Berlin: Springer-Verlag.

Webster, W. D., and C. O. Handley, Jr. 1986. Systematics of Miller's long-tongued bat, *Glossophaga longirostris*, with description of two new subspecies. *Occasional Papers, The Museum, Texas Tech University* 100:1–22.

Wendelken, P. W., and R. E. Martin. 1988. Avian consumption of the fruit of the cacti *Stenocereus eichlamii* and *Pilosocereus maxonii* in Guatemala. *The American Midland Naturalist* 119:235–43.

Williams, P. M., and I. Arias. 1978. Physio-ecological studies on plants species from the arid and semi-arid regions of Venezuela. I. The role of endogenous inhibitors in the germination of the seeds of *Cereus griseus* (Haw.) Br. and R. (Cactaceae). *Acta Científica Venezolana* 29:93–97.

Willson, M. F., D. A. Graff, and C. J. Whelan. 1990. Color preferences of frugivorous birds in relation to the colors of fleshy fruits. *Condor* 92:545–55.

Willson, M. F., and C. J. Whelan. 1990. The evolution of fruit color in fleshy-fruited plants. *The American Naturalist* 136:790–809.

CHAPTER 13

Columnar Cacti and the Diets of Nectar-Feeding Bats

Maria del Coro Arizmendi
Alfonso Valiente-Banuet
Alberto Rojas-Martínez
Patricia Dávila-Aranda

Introduction

Neotropical nectarivororus bats, mainly belonging to the subfamily Glossophagine (Phyllostomidae), are notable for their elongate rostra, long and protrusible tongues, and reduced dentition (Findley, 1993; chapter 5, this volume). They are specialized for obtaining food from flowers but they also consume other resources, such as fruits and insects (Alvarez and González, 1970; Howell, 1974a,b; Hevly, 1979; Findley, 1993). In arid lands of North America, three species of glossophagine bats—*Choeronycteris mexicana, Leptonycteris curasoae,* and *L. nivalis*—are specialized for consuming nectar, pollen, and fruits (Alvarez and González, 1970; Koopman, 1981; Heithaus, 1982; Arita and Santos del Prado, 1999). Other species, including those of *Anoura* and *Glossophaga,* consume nectar and pollen but also considerable amounts of fruit and insects (Arita and Santos del Prado, 1999). Other phyllostomids, such as species of *Artibeus,* are only occasional nectar/pollen eaters and can be considered to be predominantly frugivorous.

In the Americas, nectar-feeding bats are a diverse group comprising 36 species that mainly inhabit arid and semiarid lands (Koopman, 1981; Valiente-Banuet et al., 1996a,b; Arita and Santos del Prado, 1999). The importance of these bats as pollinators of many plant species has been extensively documented. In tropical America, several studies have shown that bats are the only effective pollinators of columnar cacti (Nassar, 1991; Soriano et al., 1991; Sosa and Soriano, 1992, 1996; Petit 1995; Rojas-Martínez, 1996; Valiente-Banuet et al., 1996a,b, 1997a,b; Nassar et al., 1997) and are thus responsible for the maintenance of cactus diversity at these sites. They are also considered to be important pollinators of certain plants of economic significance, including wild banana (*Musa paradisiaca*), gum (*Manilkara zapota*), and mango

(*Mangifera indica*) (Fleming, 1982). They are also the main pollinators of agaves, which are the basis for the production of tequila in Mexico (Arita, 1991).

The aim of this chapter is to summarize our present knowledge of the diets of nectarivorous neotropical bats and in doing so, explore the mutualistic relationship between bats and columnar cacti. The composition of their diets—in particular, the importance of cacti relative to other plants—throughout their geographic ranges is also documented.

Nectarivorous Bat Diets

The diets of glossophagine bats in arid parts of the neotropics is thought to be based primarily on columnar cacti and agaves and, to a lesser extent, on other plant species (Fleming, 1982; Arita, 1991; Valiente-Banuet et al., 1996b). Based on the current literature, 16 species of bats belonging to three subfamilies of Phyllostomidae (Nowak and Paradiso, 1983) consume nectar and pollen from more than 90 species of plants. These bats include Glossophaginae: *Anoura caudifer, Anoura geoffroyi, Choeronycteris mexicana, Glossophaga leachii, G. longirostris, G. soricina, Leptonycteris curasoae, L. nivalis,* and *Platalina genovensium;* Phyllostominae: *Micronycteris megalotis* and *Phyllostomus discolor;* and Sternodermatinae: *Artibeus jamaicensis, Artibeus intermedius, Artibeus lituratus, Chiroderma salvini,* and *Sturnina lilium.* Information on diet is obtained by analyses of stomach contents, observations of foraging activities, or pollen analyses of stomach contents, feces, and bodies of the bats (i.e., Alvarez and González, 1970; Heithaus et al., 1975; Gardner, 1977; Hevly, 1979; Fleming 1982; Quiroz et al., 1986; Soriano et al., 1991; Sosa and Soriano, 1992, 1993, 1996; Petit, 1995; Rojas-Martínez, 1996; Valiente-Banuet et al., 1996a,b, 1997a,b; Nassar et al., 1997).

Many of the plants known to occur in bat diets are cacti and agaves (22%; see Table 13.1). In addition, our data show that for most species of arid land bats, cacti and agaves represent between 75% and 100% of the total diet (Fig. 13.1). Other important plants that are consumed by flower-visiting bats include Bombacaceae (*Bombax ellipticum, Ceiba pentandra, Pseudobombax ellipticum,* and *Pseudobombax septinatum*), Bignoniaceae (*Crescentia alata, Crescentia amazonica,* and *Crescentia cujete*), Boraginaceae (*Cordia dodecandra* and *Cordia* sp.), Burseraceae (*Bursera* sp.), Caesalpinaceae (*Cassia* sp., *Mucuna andeana,* and *Mucuna* sp.), Caparidaceae (*Cleome speciosa* and *Crataeva* sp.), Leguminosae (*Bauhinia ungulata* and *Bauhinia* sp.), Mimosoideae (*Calliandra houstoniana, Enterolobium cyclocarpum,* and *Pithecellobium dulce*), Myrtaceae (*Psidium guajava*), and Sapotaceae (*Manilkara zapota* and *Sideroxylon capiri*).

To further explore the importance of columnar cacti and agaves in the diets of the 16 species of bats, we analyzed the composition of their diets using a hierarchical clustering analysis. Data used in this analysis included (1) the number of species of agaves, (2) the number of species of columnar cacti, (3) the number of other plant

TABLE 13.1

Reported Diets of Nectarivorous Bats in North, Central, and South America[a]

Plant Species	*Lc*	*Ln*	*Cm*	*Gs*	*Gle*	*Glo*	*Mm*	*Ag*	*Sl*	*Aj*	*Al*	*Ai*	*Cs*	*Ac*	*Pg*	*Pd*	*Locality*	*Source*
Abutilon		X															Baja California, Mexico	Hevly (1979)
Acacia sp.				X													Guerrero, Mexico	Alvarez and González (1970)
Agave schottii	X																Arizona and Sonora	Alcorn, et al. (1961)
A. palmeri	X	X															Arizona and New Mexico	Howell (1979), Howell and Roth (1981), Slauson (2000)
A. harvardiana		X															Texas	Kuban (1989)
Agave sp.				X													Guerrero, Mexico	Alvarez and González (1970)
Agave sp.				X													Michoacán, Mexico	Alvarez and González (1970)
Agave sp.				X													Morelos, Mexico	Alvarez and González (1970)
Agave sp.								X									Michoacán and Oaxala, Mexico	Alvarez and González (1970)
Agave sp.			X														Morelos and Guerrero, Mexico	Alvarez and González (1970)
Agave sp.	X																Guerrero, Mexico	Alvarez and González (1970)
Agave sp.	X																Arizona and Sonora	Hayward and Cockrum (1971); Howell (1974)
Agave sp.	X		X														Arizona and Sonora	Hevly (1979)
Agave sp.			X														Baja California, Mexico	Hevly (1979)
Agave sp.			X														Morelos, Mexico	Hevly (1979)
Agave sp.			X														Oaxaca, Mexico	Hevly (1979)
Agave sp.	X					X											Curaçao, Mexico	Petit (1997)

Agave sp.	X			X						Guerrero, Mexico	Quiroz et al. (1986)
Agave sp.	X				X					Venezuela	Sosa and Soriano (1993)
Alexa grandifolia				X						Brazil	Carvalho (1961)
Alnus sp.	X			X						Guerrero, Mexico	Quiroz et al. (1986)
Amaranthaceae				X						Guerrero, Mexico	Quiroz et al. (1986)
Annona cherimola				X						Guerrero, Mexico	Quiroz et al. (1986)
Asteraceae	X		X							Arizona and Sonora	Hevly (1979)
Asteraceae			X							Baja California, Mexico	Hevly (1979)
Asteraceae			X							Morelos, Mexico	Hevly (1979)
Asteraceae			X							Oaxaca, Mexico	Hevly (1979)
Bauhinia ungulata				X				X	X	Brazil	Fischer (1992)
B. ungulata	X			X						Guerrero, Mexico	Quiroz et al. (1986)
B. pauletia							X			Costa Rica	Heithaus et al. (1974)
B. pauletia				X						Costa Rica	Baker (1970)
B. pauletia						X				Costa Rica	Heithaus et al. (1974)
Bauhinia sp.				X						Costa Rica	Heithaus et al. (1975)
Bombacaceae				X						Costa Rica	Howell (1974)
Bombax ellipticum	X			X						Guerrero, Mexico	Quiroz et al. (1986)
Bombax sp.	X									Guerrero, Mexico	Alvarez and González (1970)
Bombax sp.							X			Costa Rica	Howell (1974)
Bougambilea spectabilis		X								Brazil	Carvalho (1961)
Bursera sp.	X			X						Guerrero, Mexico	Quiroz et al. (1986)
Calliandra houstoniana	X	X								Guerrero, Mexico	Quiroz et al. (1986)
Carnegiea gigantea	X									Arizona and Sonora	Alcorn et al. (1961)
C. gigantea	X									Arizona and Sonora	Alcorn et al. (1961)
C. gigantea		X								Arizona and Sonora	Beatty (1955)
C. gigantea	X									Sonora, Mexico	Fleming et al. (1996)

(continued)

TABLE 13.1 *(continued)*

Plant Species	*Lc*	*Ln*	*Cm*	*Gs*	*Gle*	*Glo*	*Mm*	*Ag*	*Sl*	*Aj*	*Al*	*Ai*	*Cs*	*Ac*	*Pg*	*Pd*	*Locality*	*Source*
C. gigantea	X																Arizona and Sonora	Hayward and Cockrum (1971), Howell (1974)
C. gigantea	X		X														Arizona and Sonora	McGregor et al. (1962)
Cassia sp.	X			X													Guerrero, Mexico	Quiroz et al. (1986)
Cecropia sp.							X										Brazil	Gardner (1977)
Cedrela sp.				X													Guerrero, Mexico	Quiroz et al. (1986)
Ceiba pentandra				X													Costa Rica	Heithaus et al. (1975)
C. pentandra									X								Costa Rica	Heithaus et al. (1975)
C. pentandra										X							Costa Rica	Heithaus et al. (1975)
C. pentandra											X						Costa Rica	Heithaus et al. (1975)
C. pentandra										X							México	Villa (1967)
Ceiba sp.				X													Guerrero, Mexico	Alvarez and González (1970)
Ceiba sp.				X													Michoacán, Mexico	Alvarez and González (1970)
Ceiba sp.				X													Morelos, Mexico	Alvarez and González (1970)
Ceiba sp.								X									Michoacán, Mexico and Oax., Mexico	Alvarez and González (1970)
Ceiba sp.			X														Morelos and Gue., Mexico	Alvarez and González (1970)
Ceiba sp.	X																Guerrero, Mexico	Alvarez and González (1970)
Ceiba sp.	X																Hidalgo, Mexico	Alvarez and González (1970)
Ceiba sp.			X														Morelos, Mexico	Hevly (1979)
Ceiba sp.			X														Oaxaca, Mexico	Hevly (1979)

Ceiba sp.					X	Costa Rica	Howell (1974b)
Ceiba sp.	X			X		Curaçao	Petit (1997)
Ceiba sp.	X		X			Guerrero, Mexico	Quiroz et al. (1986)
Cereus type	X	X				Arizona and Sonora	Hevly (1979)
Cereus type		X				Baja California, Mexico	Hevly (1979)
Cereus type		X				Morelos, Mexico	Hevly (1979)
Cereus type		X				Oaxaca, Mexico	Hevly (1979)
Cleome speciosa	X		X			Guerrero, Mexico	Quiroz et al. (1986)
Cnidoscolus aconitifolius	X					Guerrero, Mexico	Quiroz et al. (1986)
Columnar cacti	X		X			Guerrero, Mexico	Quiroz et al. (1986)
Combretum farinosum	X	X				Guerrero, Mexico	Quiroz et al. (1986)
Compositae	X		X			Guerrero, Mexico	Quiroz et al. (1986)
Compositae			X			Michoacán, Mexico and Oax., Mexico.	Alvarez and González (1970)
Conzanttia sp.			X			Guerrero, Mexico	Alvarez and González (1970)
Cordia dodecandra			X			Yucatán, Mexico	Quelch (1892), Gardner (1977)
Cordia sp.			X			Guerrero, Mexico	Alvarez and González (1970)
Cordia sp.			X			Michoacán, Mexico	Alvarez and González (1970)
Cordia sp.			X			Guerrero, Mexico	Quiroz et al. (1986)
Crataeva benthami			X			Brazil	Carvalho (1961)
Crescentia alata	X		X			Guerrero, Mexico	Quiroz et al. (1986)
Crescentia amazonica		X				Brazil	Carvalho (1961)
Crescentia cujete			X			Brazil	Carvalho (1961)
C. cujete			X			Costa Rica	Porsch (1939)
Crescentia sp.			X			Costa Rica	Heithaus et al. (1975)

(continued)

TABLE 13.1 *(continued)*

Plant Species	*Lc*	*Ln*	*Cm*	*Gs*	*Gle*	*Glo*	*Mm*	*Ag*	*Sl*	*Aj*	*Al*	*Ai*	*Cs*	*Ac*	*Pg*	*Pd*	*Locality*	*Source*
Crescentia sp.				X													Costa Rica	Heithaus et al. (1975)
Crescentia sp.									X								Costa Rica	Heithaus et al. (1975)
Crescentia sp.										X							Costa Rica	Heithaus et al. (1975)
Crescentia sp.											X						Costa Rica	Heithaus et al. (1975)
Crescentia sp.				X													Costa Rica	Howell (1974b)
Chenopodiaceae			X														Arizona and Sonora	Hevly (1979)
Chenopodiaceae			X														Baja California, Mexico	Hevly (1979)
Chenopodiaceae			X														Morelos, Mexico	Hevly (1979)
Chenopodiaceae			X														Oaxaca, Mexico	Hevly (1979)
Chenopodiaceae	X			X													Guerrero, Mexico	Quiroz et al. (1986)
Durio zibethinus				X													Honduras	Baker (1970)
Elizabetha paraense			X														Brazil	Carvalho (1961)
Enterolobium cyclocarpum	X																Guerrero, Mexico	Quiroz et al. (1986)
Eriobotrya japonica						X											Brazil	Gardner (1977)
Erythrina sp.	X			X													Guerrero, Mexico	Quiroz et al. (1986)
Eucalyptus sp.								X									Michoacán, Mexico and Oax., Mexico.	Alvarez and González (1970)
Eucalyptus sp.	X			X													Guerrero, Mexico	Quiroz et al. (1986)
Euphorbia sp.				X													Guerrero, Mexico	Quiroz et al. (1986)
Gomphrena sp.	X			X													Guerrero, Mexico	Quiroz et al. (1986)
Gramineae	X			X													Guerrero, Mexico	Quiroz et al. (1986)
Helicteris baruensis			X														Venezuela	Ruiz et al. (1997)
Heliocarpus terebinthinaceus	X			X													Guerrero, Mexico	Quiroz et al. (1986)
Hibiscus sp.	X			X													Guerrero, Mexico	Quiroz et al. (1986)
Hymenaea coubaril				X													Brazil	Carvalho (1961)
H. coubaril									X								Costa Rica	Heithaus et al. (1975)
H. coubaril										X							Costa Rica	Heithaus et al. (1975)

Hymenaea sp.			X					Costa Rica	Howell (1974b)
Inga sp.			X					Costa Rica	Howell (1974b)
Ipomoea	X							Arizona and Sonora	Hevly (1979)
Ipomoea		X						Morelos, Mexico	Hevly (1979)
Ipomoea sp.			X					Guerrero, Mexico	Alvarez and González (1970)
Ipomoea sp.			X					Morelos, Mexico	Alvarez and González (1970)
Ipomoea sp.		X						Morelos and Gue., Mexico	Alvarez and González (1970)
Ipomoea sp.	X							Guerrero, Mexico	Alvarez and González (1970)
Jambosa vulgaris				X				Brazil	Gardner (1977)
Kigelia aethropica			X					Costa Rica	Vogel (1958)
Lagenaria sp.	X		X					Guerrero, Mexico	Quiroz et al. (1986)
Leguminosae			X					Guerrero, Mexico	Quiroz et al. (1986)
Lemairocereus sp.			X					Michoacán, Mexico	Alvarez and González (1970)
Lemairocereus sp.		X						Morelos, Mexico	Alvarez and González (1970)
Lemairocereus sp.	X							Hidalgo, Mexico	Alvarez and González (1970)
Malvaceae			X					Guerrero, Mexico	Quiroz et al. (1986)
Mangifera indica			X					Guerrero, Mexico	Quiroz et al. (1986)
Manilkara zapota			X					Costa Rica	Heithaus et al. (1975)
M. zapota					X			Costa Rica	Heithaus et al. (1975)
M. zapota						X		Costa Rica	Heithaus et al. (1975)
M. zapota							X	Costa Rica	Heithaus et al. (1975)
Marcgravia sp.			X					Costa Rica	Vogel (1958)
Melia sp.			X					Guerrero, Mexico	Quiroz et al. (1986)

(continued)

TABLE 13.1 *(continued)*

Plant Species	*Lc*	*Ln*	*Cm*	*Gs*	*Gle*	*Glo*	*Mm*	*Ag*	*Sl*	*Aj*	*Al*	*Ai*	*Cs*	*Ac*	*Pg*	*Pd*	*Locality*	*Source*
Mimosoideae			X														Arizona and Sonora	Hevly (1979)
Mimosoideae			X														Morelos, Mexico	Hevly (1979)
Mucuna andeana				X													Costa Rica	Baker (1970)
Mucuna sp.				X													Costa Rica	Howell (1974b)
Musa paradisiaca							X										Brazil	Gardner (1977)
M. paradisiaca				X													Costa Rica	Howell (1974b)
Myrtillocactus sp.				X													Guerrero, Mexico	Alvarez and González (1970)
Myrtillocactus sp.				X													Michoacán, Mexico	Alvarez and González (1970)
Myrtillocactus sp.				X													Morelos, Mexico	Alvarez and González (1970)
Myrtillocactus sp.								X									Michoacán, Mexico and Oax., Mexico	Alvarez and González (1970)
Myrtillocactus sp.			X														Morelos, Mexico	Alvarez and González (1970)
Myrtillocactus sp.	X																Guerrero, Mexico	Alvarez and González (1970)
Myrtillocactus sp.	X																Hidalgo, Mexico	Alvarez and González (1970)
Neobuxbaumia macrocephala	X	X	X														Tehuacan	Valiente-Banuet et al. (1997a)
N. mezcalaensis	X	X	X														Tehuacan	Valiente-Banuet et al. (1997a)
N. tetetzo	X	X															Tehuacan	Valiente-Banuet et al. (1996b)
Ochroma lagopus				X													Costa Rica	Heithaus et al. (1975)
O. lagopus									X								Costa Rica	Heithaus et al. (1975)
O. lagopus										X							Costa Rica	Heithaus et al. (1975)

O. lagopus											X			Costa Rica	Heithaus et al. (1975)
Oenothera			X											Arizona and Sonora	Hevly (1979)
Operculina sp.				X										Guerrero, Mexico	Quiroz et al. (1986)
Pachycereus pringlei	X													Sonora, Mexico	Fleming et al. (1996)
Pachycereus tilianus					X									Venezuela	Sosa and Soriano (1993)
Pachycereus weberi	X	X	X	X					X	X		X	X	Tehuacán, Mexico	Valiente-Banuet et al. (1997b)
Parmentiera alata				X										Costa Rica	Porsch (1939)
Pilosocereus chrysacanthus	X	X	X			X								Tehuacán, Mexico	Valiente-Banuet et al. (1997b)
P. mortizianus	X	X												Venezuela	Nassar et al. (1997)
Pilosocereus tilianus				X										Venezuela	Sosa and Soriano (1996)
Pinus sp.	X			X										Guerrero, Mexico	Quiroz et al. (1986)
Pinus-Ulmus			X											Arizona and Sonora	Hevly (1979)
Pitacairnia sp.				X										Costa Rica	Howell (1974b)
Pithecellobium dulce	X	X												Guerrero, Mexico	Quiroz et al. (1986)
Poaceae	X		X											Arizona and Sonora	Hevly (1979)
Poaceae			X											Baja California, Mexico	Hevly (1979)
Poaceae			X											Morelos, Mexico	Hevly, 1979
Poaceae			X											Oaxaca, Mexico	Hevly (1979)
Proteaceae	X			X										Guerrero, Mexico	Quiroz et al. (1986)
Pseudobambax ellipticum								X						Costa Rica	Heithaus et al. (1975)
P. ellipticum										X				Morelos, Mexico	Eguiarte and Martínez del Río (1987)
P. septinatum		X												Costa Rica	Heithaus et al. (1975)
P. septinatum							X							Costa Rica	Heithaus et al. (1975)
Psidium guajava							X							San Luis Potosí	Dalquest (1953)
P. guajava							X							Brazil	Gardner (1977)
Roupala sp.				X										Michoacán, Mexico	Alvarez and González (1970)

(continued)

TABLE 13.1 *(continued)*

Plant Species	*Lc*	*Ln*	*Cm*	*Gs*	*Gle*	*Glo*	*Mm*	*Ag*	*Sl*	*Aj*	*Al*	*Ai*	*Cs*	*Ac*	*Pg*	*Pd*	*Locality*	*Source*
Ruelia geminiflora	X			X													Guerrero, Mexico	Quiroz et al. (1986)
Sclerocarpus divaricatus	X																Guerrero, Mexico	Quiroz et al. (1986)
Sideroxilon capiri	X			X													Guerrero, Mexico	Quiroz et al. (1986)
Solanum paniculatum						X											Brazil	Gardner (1977)
Stenocereus griseus	X					X											Venezuela	Nassar et al. (1997)
Stenocereus griseus	X				X												Curaçao	Petit (1995, 1997)
Stenocereus griseus	X				X												Venezuela	Sosa and Soriano (1993)
Stenocereus griseus					X												Venezuela	Sosa and Soriano (1996)
Stenocereus thurberi	X																Sonora, Mexico	Fleming et al. (1996)
Subpilocereus horrispinus	X	X															Venezuela	Nassar et al. (1997)
Subpilocereus repandus	X		X														Venezuela	Nassar et al. (1997)
S. repandus	X		X														Curaçao	Petit (1995, 1997)
S. repandus	X			X													Venezuela	Sosa and Soriano (1993)
S. repandus				X													Venezuela	Sosa and Soriano (1996)
Weberbauerocereus weberbaureri										X							Peru	Sahley (1996)
Willardia sp.				X													Guerrero, Mexico	Quiroz et al. (1986)
Yucca + Dasylirion			X														Arizona and Sonora	Hevly (1979)

[a] Abbreviations: Lc = *Leptonycteris curasoae*, Ln = *Leptonycteris nivalis*, Cm = *Choeronycteris mexicana*, Gle = *Glossophaga leachii*, Glo = *Glossophaga longirostris*, Mm = *Micronycteris megalotis*, Ag = *Anoura geoffroyi*, Sl = *Sturnira lilium*, Aj = *Artibeus jamaicensis*, Al = *Artibeus lituratus*, Ai = *Artibeus intermedius*, Cs = *Chiroderma salvini*, Ac = *Anoura caudifer*, Pg = *Platalina genovensium*, Pd = *Phyllostomus discolor*.

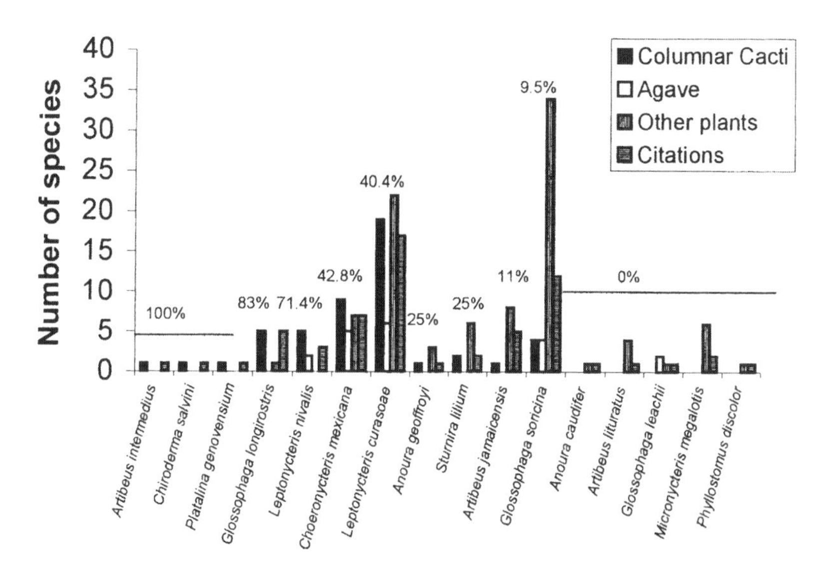

FIGURE 13.1. Bat diet composition in terms of the number of species of columnar cacti, agaves, or other plant species visited by each species. Also indictaed is the number of citations in which the diet of each bat was mentioned. Percentages above the groups of species represent the proportion of columnar cacti in relation to other plant species.

species, and (4) the number of citations mentioning the species' diet (Hierarchical Clustering using complete linkage; JMP, 1995; Fig. 13.2). Four major groups were defined by the analysis. The first group contains the best-studied bat species, which consume agaves, columnar cacti, and other plants (*Choeronycteris mexicana, Glossophaga longirostris, G. soricina,* and *Leptonycteris curasoae*). The second group includes those bats that are not as well studied and that consume only columnar cacti and plants other than cacti and agaves (*Anoura geoffroyi, Artibeus jamaicensis,* and *Sturnina lilium*). The third group contains bats that do not include columnar cacti in their diets (*Anoura caudifer, Artibeus lituratus, Glossophaga leachii, Micronycteris megalotis,* and *Phyllostomus discolor*). The final group includes bats that eat only columnar cacti (*Artibeus intermedius, Chiroderma salvini,* and *Platalina genovensium*) and the species that eats mainly columnar cacti and some agaves (*L. nivalis*). The diets of these 16 species of bats are described in more detail below.

Leptonycteris curasoae, one of the better-known bat species, is distributed widely, from southern Arizona and New Mexico in the United States to Colombia and Venezuela (Nowak and Paradiso, 1983). It is abundant throughout its geographic range (Arita and Santos del Prado, 1999) and is regarded as one of the species specializing in consuming cactus and agave resources. It occurs mainly in dry scrublands (Nowak and Paradiso, 1983). Based on our data, this species consumes 19 species of columnar cacti, six species of agaves, and 22 species of plants other than cacti or agaves (Table 13.1). However, based on their floral morphology, several of those 22

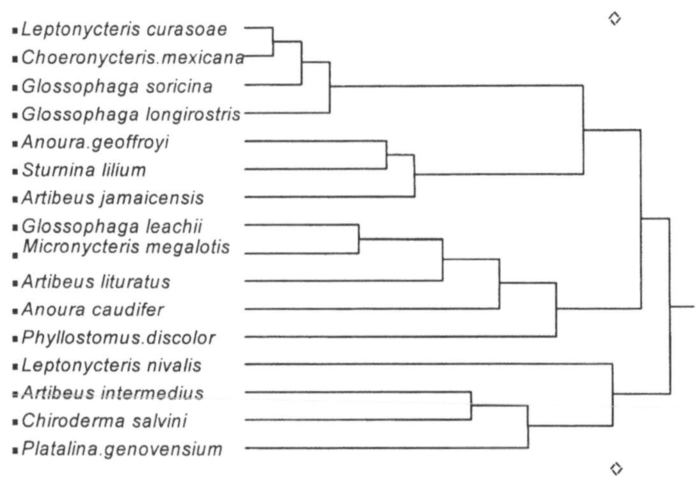

FIGURE 13.2. Hierarchical clustering analysis of bat diets (complete linkage; JMP 1985). Variables used are (1) number of species of agaves, (2) number of species of columnar cacti, (3) number of other plant species, and (4) number of citations mentioning each species' diet.

nonsucculent plants (e.g., Asteraceae, Poaceae) are probably accidental visits and not true food resources.

Leptonycteris nivalis is a species with a wide distribution (from Texas to Guatemala) but is scarce in the localities studied (Arita and Santos del Prado, 1999). It is often associated with pine-oak vegetation and is not common in desert scrublands. It has been reported to consume nectar and pollen from only five species of columnar cacti and two species of *Agave.*

Choeronycteris mexicana is also a widely distributed species (southwestern United States to Guatemala) that is considered to be locally scarce (Arita and Santos del Prado, 1999). It has been reported to feed on nine species of columnar cacti, five unidentified species of *Agave,* and seven other plant species, of which five are considered to be accidental visits.

Glossophaga soricina is a widespread (northern Mexico to northern Argentina), abundant species (Arita and Santos del Prado, 1999), which apparently is not associated with columnar cacti. It includes many plants in its diet (34 identified species, including four cacti and four unidentified species of agaves, plus a few nonsucculents that have not been identified). It occurs in moister tropical forests, where columnar cacti are not physiognomically dominant.

Glossophaga leachii has a restricted distribution (central Mexico to central Costa Rica) and it can be locally abundant (Arita and Santos del Prado, 1999). It has been reported to consume nectar and pollen only from *Pseudobombax ellipticum.* In Venezuela and Curaçao, another species of the same genus, *G. longirostris,* feeds on five species of columnar cacti (*Pilosocereus tilianus, P. mortizianus, Stenocereus griseus, Subpilosocereus horrispinus,* and *Subpilosocereus repandus*), two species of

Agave (both unidentified; one in Venezuela and one in Curaçao), and *Ceiba* sp. (Bombacaceae).

Anoura geoffroyi is a nectarivorous species that is widespread (Mexico to Brazil and northwestern Argentina) and abundant (Arita and Santos del Prado, 1999). It has been reported as consuming nectar and pollen of *Agave* spp., *Ceiba* sp., *Eucalyptus* sp., and *Myrtillocactus* sp. It is considered to be predominantly insectivorous, including in its diet beetles and moths that, due to their size, are not likely to be accidentally ingested while the bat is consuming nectar and pollen (Gardner, 1977; Sazima and Sazima, 1978). It is a common species in moist forests at high altitudes (Handley, 1976).

In addition, other bat species are occasional consumers of nectar and pollen. For example, a population of *Artibeus jamaicensis* studied in Costa Rica consumed pollen and nectar from ten species of plants from several families other than cacti and agaves (*Bahuinia pauletia, Bombax* sp., *Ceiba pentandra, Ceiba* sp., *Crescentia* sp., *Hymenaea coubaril, Hymenaea* sp., *Manilkara zapota, Ochroma lagopus,* and *Pseudobombax ellipticum*). However, the same species in Mexico has been reported to consume *Pachycereus weberi* and *Pilosocereus chrysacanthus,* both columnar cacti. It is a widespread species (Mexico to Brazil and Bolivia) that is primarily frugivorous, although it also consumes pollen, nectar, flower parts, and insects (Gardner, 1977).

Artibeus lituratus also has a wide distribution (from Mexico to northern Argentina). It has been reported to consume *Ceiba pentandra, Crescentia* sp., *Manilkara zapota,* and *Ochroma lagopus* in Central America (Heithaus et al., 1975), where it inhabits dry or moist forests (Handley, 1976) .

Specialized for consuming fruits, *Chiroderma salvini* is distributed from northwestern Mexico to Ecuador. It is a common species in montane evergreen forests (Gardner, 1977; Handley, 1976). However, it has recently been reported to visit *Pachycereus weberi* (Cactaceae) in a dry tropical deciduous forest in central Mexico (Valiente-Banuet et al., 1997a).

Micronycteris megalotis, an insectivorous species that is distributed from northeastern Mexico to Peru and Brazil, occasionally consumes fruits (Gardner, 1977). It inhabits dry deserts and moist tropical forests (Handley, 1976). It has been reported to consume *Psidium guajava* (Myrtaceae) in Mexico and *Cecropia* sp. (Moraceae), *Eriobotrya japonica, Jambosa vulgaris,* and *Psidium guajava* (Myrtaceae), *Musa paradisiaca* (Musaceae), and *Solanum paniculatum* (Solanaceae) in Brazil (Dalquest, 1953; Gardner, 1977).

Other nectarivorous species distributed in South America include (1) *Anoura caudifer* (from northern South America to Brazil and Peru), which consumes *Bauhinia ungulata* (Leguminoseae) in Brazil; (2) *Platalina genovensium,* endemic in arid regions of Peru, which consumes *Weberbauerocereus weberbaueri* (Cactaceae); and (3) *Phyllostomus discolor,* which is omnivorous and consumes fruits, pollen grains, nectar, and insects found in flowers (Gardner, 1977). It was reported to consume *Bauhinia ungulata* (Leguminosae) in Costa Rica and Brazil.

In general, our analysis of the dietary data indicates that phyllostomid bats consume a wide array of plant species, depending on seasonal availability of resources in each locality. Bats are capable of undertaking local or latitudinal migrations when resources are scarce. It is clear that most of these bat species also consume fruits and insects. In addition, those species specializing in nectar and pollen may accidentally eat insects that are found in flowers. In contrast, *Micronycteris megalotis* consumes large insects in a nonaccidental fashion. Our review clearly indicates that some species that were formerly considered as "typically frugivorous," such as *Artibeus jamaicensis,* consume pollen and nectar from many plant species throughout its range. It seems to be a pollen-nectar-fruit consumer, and its diet sometimes resembles those of more nectarivorous species. Perhaps nectarivorous species such as *Leptonycteris curasoae* that also eat large amounts of fruit (Valiente-Banuet et al., 1996a; Godínez-Alvarez et al., 1999) should be recognized by a more accurate term that reflects their diets of nectar, pollen, and fruit.

The proportion of nonsucculent plant species consumed by bats depends partly on the aridity of the habitat where they were studied (Table 13.1). For example, the diets of species that inhabit tropical dry forest, such as *Glossophaga soricina* in Costa Rica (Heithaus et al., 1975), include many species of plants other than cacti and agaves. But such bats as *Leptonycteris nivalis* or *Choeronycteris mexicana* that inhabit dry deserts feed extensively on columnar cacti or agaves. These variations in diet reflect differences between the floristic composition of tropical dry forests and arid scrublands. The former vegetation is not physionomically dominated by columnar cacti, whereas in the drylands of both North and South America, columnar cacti often represent the main floristic elements (Sosa and Soriano, 1993, 1996; Valiente-Banuet et al., 1996a,b, 1997 a,b; Nassar et al., 1997).

Conclusions

Reproductive mutualisms (both pollination and seed dispersal) have been considered by many authors to be key processes that are responsible for the maintenance of viable populations of plants (e.g., Faegri and van der Pijl, 1979; Boucher, 1985; Bronstein, 1994). Many plant species, especially those that are self-incompatible, need the services of pollen vectors that can effect outcrossing pollination. This is the case for most of the tropical columnar cacti that depend entirely on bats for fruit production (Valiente-Banuet et al., 1996a,b, 1997a,b).

The importance of this interaction is obvious for many plants that depend exclusively on bats for pollination, including many columnar cacti in central Mexico (Valiente-Banuet et al. 1996a,b, 1997a,b) and South America (Soriano et al., 1991; Sosa and Soriano, 1992, 1996; Nassar et al., 1997). However, bats are not necessarily obligate cactus feeders: On the contrary, they often feed sequentially in time and space on a wide array of plant species, moving altitudinally or latitudinally to meet their energetic requirements (Rojas-Martínez, 1996).

There are, however, few studies that include a year-round analysis of bat diets and plant phenologies, where the role of both mutualists is described (but see Soriano et al., 1991; Sosa and Soriano, 1992, 1996; Rojas-Martínez, 1996; Valiente-Banuet et al. 1996a,b, 1997a,b; Nassar et al., 1997, Rojas-Martinez et al., 1999).

Resource distributions measured on both temporal and spatial scales can influence foraging and reproductive behaviors of bats (Fleming, 1982; Heithaus et al., 1975). Such differential resource distribution can be the cause of local movements reported in tropical arid lands and may account in part for the latitudinal movements analyzed in chapter 11.

Summary

In North America, the geographic distribution of nectarivorous bats is coincident with the geographic and seasonal distribution of columnar cacti and agaves, suggesting a close relationship between both groups. In this chapter, we summarize studies of the diets of bats, to highlight the importance of columnar cacti in the diets of phyllostomid bats. Differences in the diets of bats are analyzed based on habitat preferences and overall diet composition. Nectar-feeding bats include in their diets many species of cacti and agaves, as well as a few other species. In contrast, *Glossophaga soricina* shows the reverse diet pattern: It consumes many different plants other than cacti and agaves and very few succulent species. This diet difference reflects the floristic composition of the habitats where this bat occurs.

Resumen

En Norte América se ha documentado que los murciélagos nectarívoros presentan patrones de distribución geográfica coincidiendo con la distribución de la floración de plantas como cactáceas columnares y agaves lo que sugiere una estrecha relación entre ambos grupos. En este capítulo analizamos la composición de la dieta de los murciélagos nectarívoros en América para tratar de documentar la importancia de las cactáceas columnares de la dieta de estos animales. Encontramos diferencias entre las diferentes especies de murciélagos de la familia Phyllostomidae. Podemos distinguir dos grandes grupos, el primero formado, por la mayoría de las especies de las cuáles existe una información, cuya dieta está compuesta en su mayoría por cactáceas columnares y agaves. El segundo formado por especies como *Glossophaga soricina* es presenta en patrón contrario consumiendo en su mayoría plantas que nos son cactus ó agaves. Se postula contrario que ésto debe ser un reflejo de la composición florística de los hábitats preferidos por las especies. Probablemente la diferencia se pueda correlacionar con la composición florística de los hábitats que las murciélagos usan preferentemente y con la distribución espacial y temporal de los recursos.

ACKNOWLEDGMENTS

Our studies have been supported by Fondo Mexicano para la Conservación de la Naturaleza (Project FMCN A-1-97/36) and by Dirección General de Asuntos del Personal Académico de la Universidad Nacional Autónoma de México, DGAPA-IN208398; DGAPA-IN207798. We thank T. H. Fleming for his editorial help with this chapter.

REFERENCES

Alcorn, S. M., S. E. McGregor, and G. Olin. 1961. Pollination of Saguaro cactus by doves, nectar-feeding bats and honey bees. *Science* 132:1594–95.

Alvarez, T., and L. Q. González. 1970. Análisis polínico del contenido gástrico de murciélagos Glossophaginae de México. *Anales de la Escuela Nacional de Ciencias Biológicas, México* 18:137–65.

Arita, H. T. 1991. Spatial segregation in long-nosed bats, *Leptonycteris nivalis* and *Leptonycteris curasoae,* in Mexico. *Journal of Mammalogy* 72:706–14

Arita, H. T., and D. E. Wilson. 1987. Long nosed bats and agaves: the tequila connection. *Bats* 5:3–5.

Arita, H. T., and K. Santos del Prado. 1999. Conservation biology of nectar feeding bats in Mexico. *Journal of Mammalogy* 80:31–41.

Baker, H. G. 1970. Two cases of bat pollination in Central America. *Revista de Biología Tropical* 17:187–97.

Beatty, L. D. 1955. Autecology of the long nosed bat *Leptonycteris nivalis* (Saussure). Master's thesis, University of Arizona, Tucson.

Boucher, D. H. 1985. *The Biology of Mutualism: Ecology and Evolution.* New York: Oxford University Press.

Bronstein, J. L. 1994. Our current understanding of mutualism. *Quarterly Review of Biology* 69:31–51.

Carvalho, C. T. De. 1961. Sobre os hábitos alimentares de Phillostomideos (Mammalia, Chiroptera). *Revista de Biología Tropical* 9:53–60.

Dalquest, W. W. 1953. Mexican bats of the genus *Artibeus. Proceedings of the Biological Society of Washington* 66:61–66.

Eguiarte, L., and C. Martínez del Río. 1987. El néctar y el polen como recursos: el papel ecológico de los visitantes a las flores de *Pseudobombax ellipticum* (H.B.K.) Dugand. *Biotropica* 19:74–82.

Faegri, K., and L. van der Pijl. 1979. *The Principles of Pollination Ecology.* 3rd ed. Oxford: Pergamon Press.

Findley, J. S. 1993. *Bats: a Community Perspective.* Cambridge: Cambridge University Press.

Fischer, E. A. 1992. Foraging of nectarivorous bats on *Bauhinia ungulata. Biotropica* 24:579–82.

Fleming, T. H. 1982. Foraging strategies of plant-visiting bats. In *Ecology of Bats,* ed. T. H. Kunz, pp. 287–325. New York: Plenum Press.

Fleming, T. H., M. D. Tuttle, and M. A. Horner. 1996. Pollination biology and the relative importance of nocturnal and diurnal pollinators in three species of Sonoran Desert columnar cacti. *The Southwestern Naturalist* 41:257–69.

Gardner, A. L. 1977. Feeding habits. In *Biology of Bats of the New World Family Phyllostomatidae.* Part II, eds. R. J. Baker, J. K. Jones, Jr., and D. C. Carter, 293–350. The Texas Tech University Museum Special Publication No. 13. Lubbock, Texas: Texas Tech University.

Godínez-Alvarez, H., A. Valiente-Banuet, and B. L. Valiente. 1999. Biotic interactions and the population dynamics of the long-lived columnar cactus *Neobuxbaumia tetetzo* in the Tehuacán Valley, Mexico. *Canadian Journal of Botany* 77:203–8.

Handley, C. O. 1976. Mammals of the Smithsonian Venezuela Project. *Brigham Young University Science Bulletin. Biology Series* 20:1–89.

Hayward, B., and E. L. Cockrum. 1971. The natural history of the western long-nosed bats *Leptonycteris sanborni. Western New Mexico University Research Science* 1:75–123.

Heithaus, R. E. 1982. Coevolution between bats and plants. In *Ecology of Bats,* ed. T. H. Kunz, 327–67. New York: Plenum Press.

Heithaus, R. E., P. A. Opler, and H. G. Baker. 1974. Bat activity and pollination of *Bauhinia pauletia:* plant pollinator coevolution. *Ecology* 55:412–19.

Heithaus, R. E., T. H. Fleming, and P. A. Opler. 1975. Foraging patterns and resource utilization in seven species of bats in a seasonal tropical forest. *Ecology* 56:841–54.

Hevly, R. H. 1979. Dietary habits of two nectar and pollen feeding bats in southern Arizona and northern Mexico. *Journal of Arizona-Nevada Academy of Sciences* 14:14–18.

Howell, D. J. 1974a. Bats and pollen: physiological aspects of the syndrome of chiropterophily. *Comparative Biochemistry and Physiology* 48:263–76.

———. 1974b. Feeding and acoustic behavior in glossophaginae bats. *Journal of Mammalogy* 55:293–308.

———. 1979. Flock foraging in nectar-feeding bats: advantages to the bats and the host plants. *The American Naturalist* 114:23–49.

Howell, D. J., and B. S. Roth. 1981. Sexual reproduction in agaves: the benefits of bats; the cost of semelparous advertising. *Ecology* 62:1–7.

JMP. 1995. *Statistics Made Visual.* Version 3.1.6.2. Cary, North Carolina: SAS Institute.

Johnsgard, P. A. 1983. *The Hummingbirds of North America.* Washington, D.C.: Smithsonian Institution Press.

Koopman, K. F. 1981. The distributional patterns of new world nectar-feeding bats. *Annals of the Missouri Botanical Garden* 68:352–69.

Kuban, J. F. 1989. The pollination biology of two populations of the Big Bend century plant *Agave harvardiana* Trel.: a multiple pollinator syndrome with floral specialization for vertebrate pollinators. Ph.D. diss. Syracuse University, Syracuse, New York.

McGregor, S. C., S. M. Alcorn, and G. Ollin. 1962. Pollination and pollinating agents of the saguaro. *Ecology* 43:259–67.

Nassar, J. M. 1991. Biología reproductiva de cuatro especies de cactáceas venezolanas (Ceereae: *Stenocereus griseus, Pilosocereus moritzianus, Subpilocereus repandus* y *S. horrispinus*) y estrategias de visita de los murciélagos asociados a éstas. Bachelor's thesis, Universidad Central de Venezuela, Caracas.

Nassar, J. M., N. Ramírez, and O. Linares, O. 1997. Comparative pollination biology of Venezuelan columnar cacti and the role of nectar feeding bats in their sexual reproduction. *American Journal of Botany* 84:918–27.

Nowak, R. M., and J. L. Paradiso. 1983. *Mammals of the World.* Vol. 1. Baltimore: The Johns Hopkins University Press.

Petit, S. 1995. The pollinators of two species of columnar cacti on Curaçao, Netherlands Antilles. *Biotropica* 27:538–41.

———. 1997. The diet and reproductive schedules of *Leptonycteris curasoae curasoae* and *Glossophaga longirostris elongata* (Chiroptera: Glossophaginae) on Curaçao. *Biotropica* 29: 214–23.

Porsch, O. 1939. Das Bestäubungsleben der Kakteenblüte II. *Cactaceae. Jahrbuch Deutsch, Kakteen-Gesellschaft* 1939:81–142.

Quiroz, D. L., M. S. Xelhuantzi, and M. C. Zamora. 1986. *Análisis Palinológico del Contenido Gastrointestinal de los Murciélagos* Glossophaga soricina *y* Leptonycteris yerbabuenae *de las Grutas de Juxtlahuaca, Guerrero.* Mexico: Instituto Nacional de Antropología e Historia.

Rojas-Martínez, A. 1996. Estudio poblacional de tres especies de murciélagos nectarívoros considerados como migratorios y su relación con la presencia estacional de los recursos florales en el Valle de Tehuacán. Master's thesis, Universidad Nacional Autónoma de México, Ciudad Universitaria, Mexico.

Rojas-Martínez, A., A. Valiente-Banuet, M. C. Arizmendi, A. Alcántara-Eguren, and H. Arita. 1999. Seasonal distribution of the long nosed bat *Leptonycteris curasoae* in North America: Does a generalized migration pattern really exist? *Journal of Biogeography* 26:1065–77.

Ruiz, A., M. Santos, P. J. Soriano, J. Cavelier, and A. Cadena. 1997. Relaciones mutualistas entre el murciélago *Glossophaga longirostris* y las cactáceas columnares en la zona árida de la Tatacoa, Colombia. *Biotropica* 29:469–79.

Sahley, C. T. 1996. Bat and hummingbird pollination of an autotetraploid columnar cactus, *Weberbauerocereus weberbaueri* (Cactaceae). *American Journal of Botany* 83:1329–36.

Sazima, M., and I. Sazima. 1978. Bat pollination of the passion flower, *Passiflora mucronata* in southwestern Brazil. *Biotropica* 10:100–109.

Slauson, L. A. 2000. Pollination biology of two chiropterophilous Agaves in Arizona. *American Journal of Botany* 87:825–36.

Soriano, P. J., M. Sosa, and O. Rossell. 1991. Hábitos alimentarios de *Glossophaga longirostris* Miller (Chiroptera: Phyllostomidae) en una zona árida de los Andes venezolanos. *Revista de Biología Tropical* 39:263–68.

Sosa, M., and P. J. Soriano. 1992. Los murciélagos y los cactus: una relación muy estrecha. *Carta Ecológica* 61:7–10.

———. 1993. Solapamiento de dieta entre *Leptonycteris curasoae* y *Glossophaga longirostris* (Mammalia :Chiroptera). *Revista de Biología Tropical* 41:529–32.

———. 1996. Resource availability, diet and reproduction in *Glossophaga longirostris* (Mammalia: Chiroptera) in an arid zone of the Venezuelan Andes. *Journal of Tropical Ecology* 12:805–18.

Valiente-Banuet, A., M. C. Arizmendi, and A. Rojas-Martínez. 1996a. Nectar-feeding bats in columnar cacti forests of Central Mexico. *Bats* 14:12–15.

Valiente-Banuet, A., M. C. Arizmendi, A. Rojas-Martínez, and L. Domínguez-Canseco. 1996b. Ecological relationships between columnar cacti and nectar feeding bats in Mexico. *Journal of Tropical Ecology* 11:103–19.

Valiente-Banuet, A., A. Rojas-Martínez, M. C. Arizmendi, and P. Dávila, P. 1997a. Pollination biology of two columnar cacti (*Neobuxbaumia mezcalaensis* and *Neobuxbaumia macrocephala*) in the Tehuacán Valley, Central Mexico. *American Journal of Botany* 84:452–55.

Valiente-Banuet, A., A. Rojas-Martínez, A. Casas, M. C. Arizmendi, and P. Dávila. 1997b. Pollination ecology of two winter-blooming giant columnar cacti in the Tehuacán Valley, Mexico. *Journal of Arid Environments* 37:331–41.

Villa, B. 1967. *Los Murciélagos de México.* Mexico: Universidad Nacional Autónoma de México.

Vogel, S. 1958. Fledermausblumen in Südamerika. *Österreichische Botanische Zeitschrift* 105: 491–530.

CHAPTER 14

Population Biology of the Lesser Long-Nosed Bat *Leptonycteris curasoae* in Mexico and Northern South America

THEODORE H. FLEMING
JAFET NASSAR

Introduction

According to Valiente-Banuet et al. (1996), a majority (42 of 70 species) of columnar cacti of tribe Pachycereeae produce flowers belonging to the chiropterophilous or bat-pollination syndrome. Features associated with this syndrome include nocturnal anthesis and robust, funnel-shaped flowers producing substantial amounts of nectar and pollen. These features ostensibly reflect the size, energetic requirements, and activity rhythms of flower-visiting bats. Fleming and Wallace (chapters 3 and 10, this volume) suggest that bat pollination likely represents the ancestral condition in this and certain other tribes of Cactaceae (e.g., Cereeae and Browningieae). If this is true, then nectar-feeding bats have played an especially important role in the evolution and ecology of columnar cacti. Dietary and other analyses indicate that several species of nectar-feeding bats of the Phyllostomidae (New World leaf-nosed bats), including *Anoura geoffroyi, Artibeus jamaicensis, Choeronycteris mexicana, Glossophaga longirostris, G. soricina, Leptonycteris curasoae, L. nivalis,* and *Platalina genovensium* visit cactus flowers regularly or occasionally (Alvarez and Gonzalez, 1970; Hayward and Cockrum, 1971; Gardner, 1977; Quiroz et al., 1986; Fleming et al., 1993; Petit, 1995; Sahley, 1996; Valiente-Banuet et al., 1996, 1997; Ceballos et al., 1997; Nassar et al., 1997). Of these species, the lesser long-nosed bat *L. curasoae* is probably the most common chiropteran visitor to flowers of columnar cacti in Mexico and northern South America. Thus knowledge about its population biology, including seasonal changes in its distribution and abundance as well as its diet and nightly foraging behavior, is important for understanding how nectar-feeding bats and columnar cacti have interacted evolutionarily and ecologically.

In this chapter, we review the population biology of *Leptonycteris curasoae* and its daily and seasonal interactions with columnar cacti. We emphasize *L. curasoae*'s annual migratory and reproductive cycles and how these cycles affect its interaction with columnar cacti as a pollinator and seed disperser.

Evolution of *Leptonycteris* Bats

A variety of evidence, including morphological, karyological, and biochemical data, indicates that bats of the genus *Leptonycteris* are fairly derived members of subfamily Glossophaginae (sensu Wilson and Reeder, 1993) (Koopman, 1981; Griffiths, 1982; Haiduk and Baker, 1982; Van den Bussche, 1992; Freeman, 1995; chapter 5, this volume). *Leptonycteris* bats differ morphologically from less-derived glossophagines in at least two sets of morphological traits:

1. They are about 40% larger in terms of forearm length (a standard measure of size in bats) and mass. Forearm lengths of adult males and females of *L. curasoae* average 54.3 mm and 54.0 mm, respectively; masses of adult males and nonpregnant females average 26.4 g and 24.9 g, respectively (Ceballos et al., 1997).
2. They have larger wing spans and wing areas, lower aspect ratios (i.e., relatively narrow wings), and higher wing loading (i.e., more mass per unit wing area) than most other glossophagines (Norberg and Rayner, 1987).

Sahley et al. (1993) have interpreted these two sets of traits, which are also found in two other species of arid zone nectarivorous bats (*Choeronycteris mexicana* and *Platalina genovensium*), as adaptations for long-distance foraging flights. *Leptonycteris* bats differ from other arid zone cactus-visiting bats in having only a moderately elongated snout and tongue, as is typical of most members of Glossophaginae (Fig. 14.1). Thus *Leptonycteris* bats have diverged most strongly from their ancestors in overall size and wing shape but not in the relative length of their jaws.

The genus *Leptonycteris* contains two species: *L. curasoae* (the lesser long-nosed bat; formerly *L. sanborni*) and *L. nivalis* (the greater long-nosed bat). As its common name implies, *L. nivalis* is slightly larger than its congener (forearm length in males and females average 56.3 mm and 56.0 mm, respectively; Arita and Humphrey, 1988). It is restricted in distribution primarily to upland regions of central and eastern Mexico and apparently does not occur in the coastal lowlands of western Mexico (Arita, 1991). Its northern distributional limits include Big Bend National Park, Texas (Easterla, 1972) and far southwestern New Mexico (Hoyt et al., 1994). The lesser long-nosed bat contains two subspecies, *L. curasoae yerbabuenae* and *L. curasoae curasoae* (Arita and Humphrey, [1988]; but see chapter 5, this volume). Subspecies *yerbabuenae* is widely distributed in arid and semiarid regions of Mexico (Fig. 14.2a); it is a seasonal resident in the southern quarter of Arizona and far southwestern New Mexico (Arita, 1991; Hoyt et al., 1994). Subspecies *curasoae* occurs in the arid and semiarid regions of northern Venezuela (including the dry Andean valleys of southwestern Venezuela) and northern Colombia. It also occurs on the

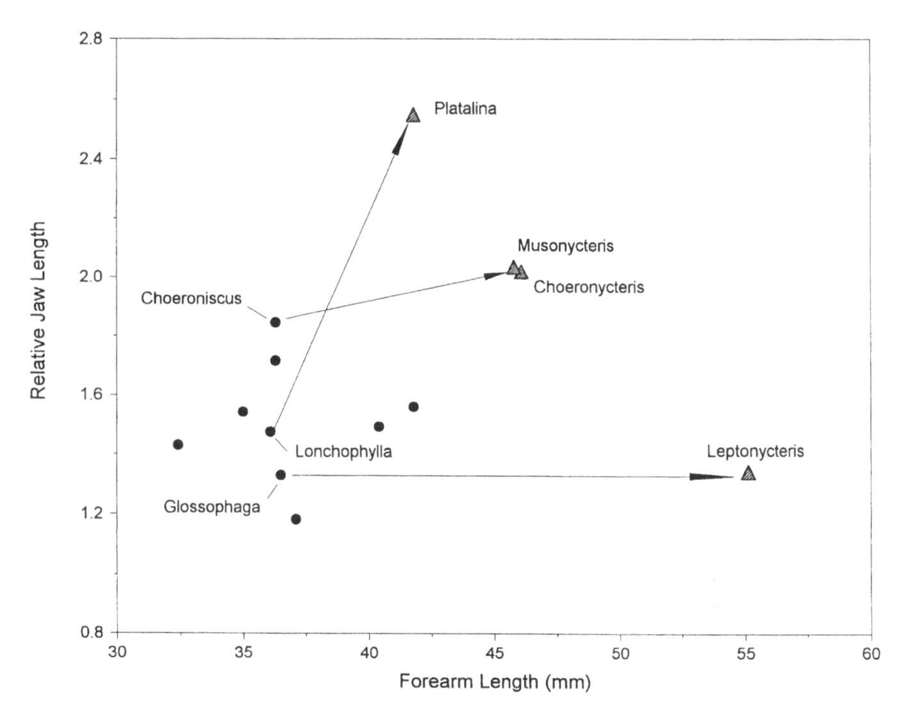

FIGURE 14.1. Ordination of 13 genera of glossophagine bats by forearm length (overall size) and relative jaw length (i.e., ratio of length to width of upper jaw). Arid zone, cactus-visiting genera are indicated by hatched triangles; mesic habitat genera by filled cricles. Arrows connect arid genera with close relatives to indicate likely direction of morphological evolution. Modified from Fleming (1995b).

Caribbean islands of Curaçao, Aruba, Bonaire, and Margarita (Fig. 14.2b; Cuervo-Diaz et al., 1986; Linares, 1998).

Wilkinson and Fleming (1996) used gene sequence data from the control region of mitochondrial DNA to determine phylogenetic relationships within *Leptonycteris. Glossophaga soricina,* a basal member of Glossophaginae, was used as an outgroup in this analysis. Results indicated that *Glossophaga* and *Leptonycteris* last shared a common ancestor about 2.4 million years ago, that the two species of *Leptonycteris* last shared a common ancestor about one million years ago, and that the two subspecies of *curasoae* separated about 0.54 million years ago. *L. nivalis* was determined to be the older of the two species of *Leptonycteris.* Wilkinson and Fleming (1996, pp. 335–36) concluded, "If these data accurately reflect evolutionary history, they suggest that these taxa [*Leptonycteris* spp.] are the products of climatic events occurring during the late Pliocene and Pleistocene." They suggested that a single species of *Leptonycteris* occurred in Mexico prior to the uplift of the Mexican plateau in the late Pliocene. Uplift and formation of the Sierra Madres apparently split this species into the two current species. They also noted (p. 336), "The presence of *L. curasoae* in Mexico and northern South America suggests that an arid or semi-arid corridor con-

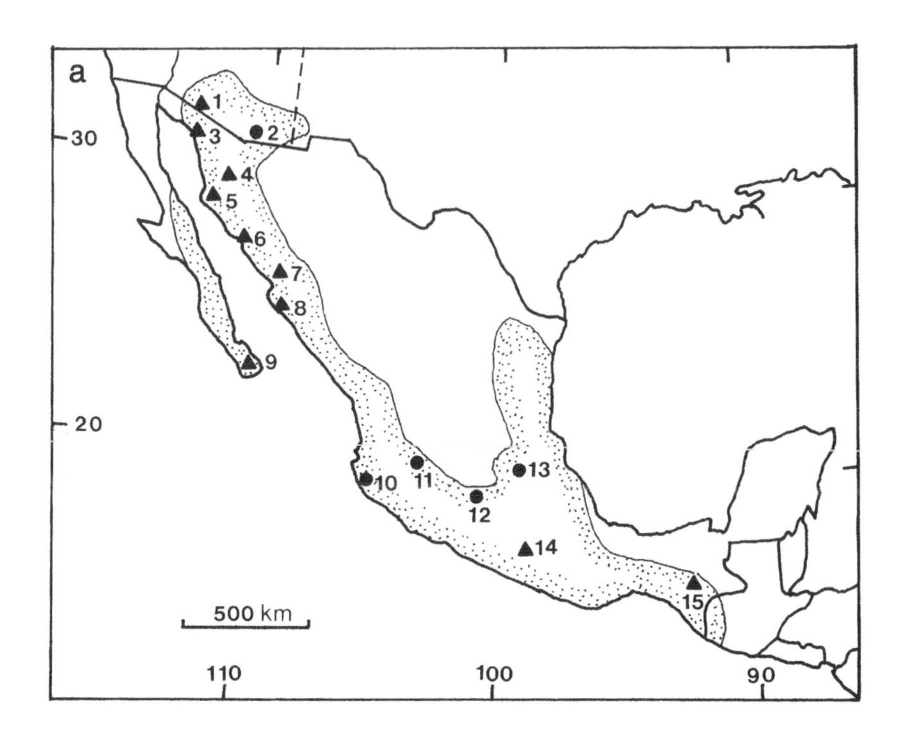

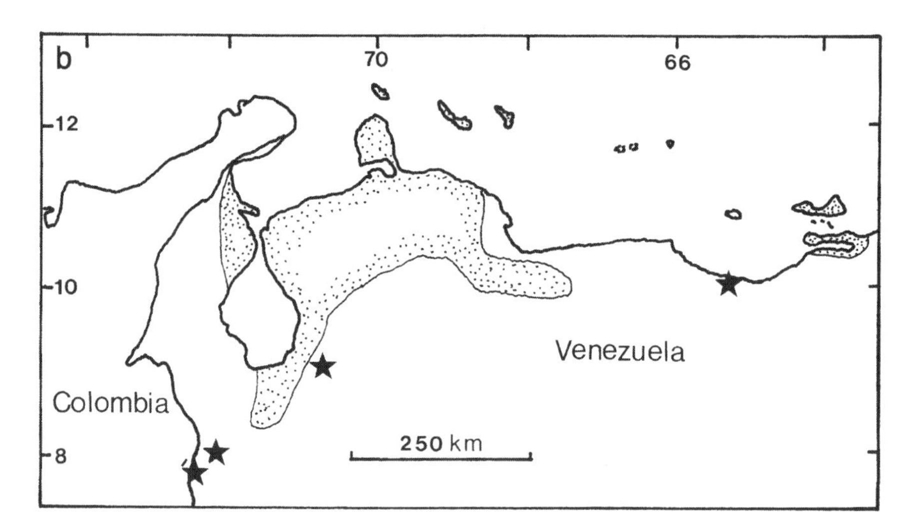

FIGURE 14.2. Maps of the geographic distributions of the two subspecies of *Leptonycteris curasoae*. (a) *L. c. yerbabuenae;* triangles represent maternity roosts. (b) *L. c. curasoae;* stars represent new locality records. Roosts in A include: (1) Mine in Organ Pipe Cactus National Monument, Arizona; (2) Patagonia Bat Cave, Arizona; (3) Lava tube in Pinacate Biosphere Reserve, Sonora; (4) Cueva del Tigre, Sonora; (6) Santo Domingo Mine, Sonora; (7) Cave near Buenavista, Baja California Sur; (8) San Andres Island, Jalisco; (9) Cueva la Mina, Jalisco; (10) Cueva Las Grutas, Michoacán; (11) Cueva de Xoxafi, Hidalgo; (12) Gruta Juxtlahuaca, Guerrero; (13) Cueva Tempisque, Chiapas. Data are from Arita (1991); Wilkinson and Fleming (1996); Linares (1998); and J. Nassar and P. Soriano (unpubl. data).

nected these regions during at least one Pleistocene glacial advance. Remnants of that corridor . . . are evident as far south as north-western Costa Rica today."

Estimates of the evolutionary ages of other cactus-visiting bats (i.e., *Choeronycteris* and *Platalina*) are not yet available, so we cannot construct a rigorous scenario relating the evolution of pollination syndromes in the Pachycereeae and other tribes with bat-pollinated species to the adaptive radiation of large, long-snouted phyllostomid bats. Current information, however, suggests that bat-cactus interactions are relatively recent—probably not more than a few million years old. This, in turn, raises two questions: How old are bat-pollinated columnar cacti, and if the answer is more than several million years old, then who were their early chiropteran pollinators—less-derived members of the Glossophaginae?

Seasonal Distribution and Abundance

In Mexico, it is likely that both species of *Leptonycteris* undertake seasonal migrations and that their geographic ranges contract and expand seasonally. The clearest evidence for this is the seasonal occurrence of these bats in certain caves and mines and large fluctuations in the sizes of *Leptonycteris* populations in other roosts in which these bats are year-round residents. Seasonal occupancy occurs in all known roosts of *L. curasoae* north of about mid-Sonora (Cockrum, 1991). A similar pattern holds for *L. nivalis* in Mt. Emory Cave, Big Bend National Park (Easterla, 1972). Data for bats roosting in a lava tube in Pinacate Biosphere Reserve, Sonora, are typical of occupancy patterns in northern maternity roosts of *L. curasoae* (Fig. 14.3). Pregnant females arrive at this roost in mid- to late April. Over 100,000 adults occupy this roost by mid-May, and females and young leave the roost by early September. A similar occupancy pattern occurs in a maternity roost of 12,000–15,000 adults in Organ Pipe Cactus National Monument, Arizona (V. Dalton and T. Tibbetts, pers. comm.). Two non-maternity roosts containing tens of thousands of *L. curasoae* and located in far south-central Arizona begin to fill up in mid-July and are empty by mid-September (Y. Petryszyn and D. Noel, pers. comm.). These roosts contain adult females and young of the year of both sexes. Caves and mines in the uplands of southeastern Arizona reach peak numbers (a few hundreds) in August and are empty by late September (Cockrum, 1991). Elsewhere in Mexico, seasonal occupancy by *L. curasoae* occurs in Cueva Las Grutas, Michoacán (in June through August; Zamacona [1991]), Cueva de Xoxafi, Hidalgo (in February or March through September; Alvarez and Gonzalez [1970]), and Gruta Juxtlahuaca, Guerrero (in August or September through February; Quiroz et al. [1986]) (Fig. 14.2a).

Even when they are occupied year-round, roosts located from southern Sonora south undergo strong seasonal fluctuations in size and sexual composition. Our most complete dataset comes from a sea cave located near Chamela, Jalisco (Figs. 14.2a, 14.3). The population of this roost ranges from about 5,000 bats in March to about

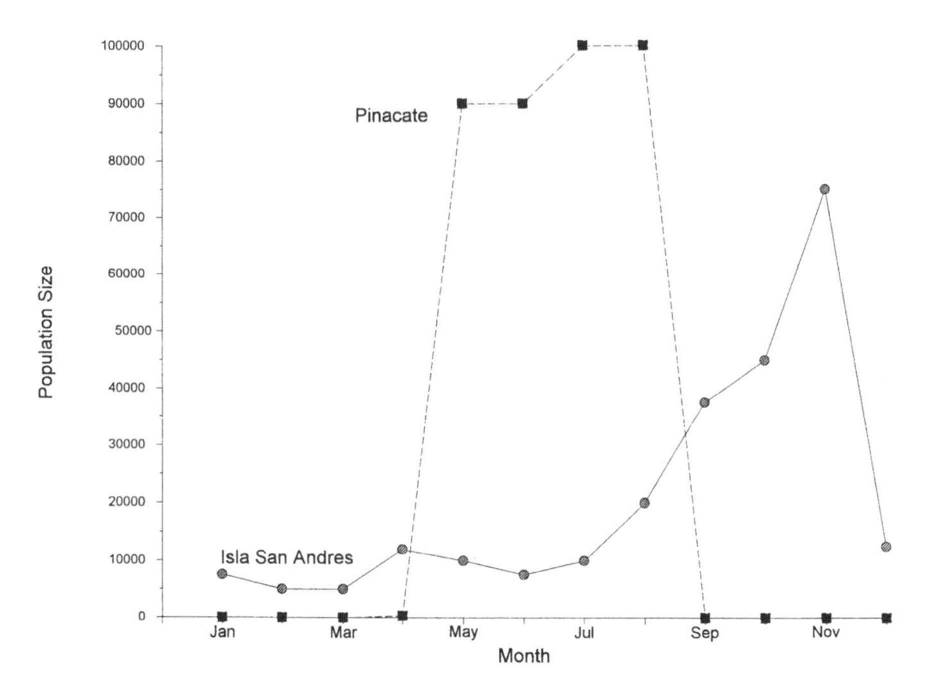

FIGURE 14.3. Seasonal changes in the sizes of two roosts of *L. curasoae* in Mexico. The Pinacate maternity roost is located in northern Sonora; the San Andres Island mating roost is located in coastal Jalisco. Data are from Ceballos et al. (1997) and W. Peachey (pers. comm.).

75,000 in November (Ceballos et al., 1997). Similarly, an abandoned mine near Alamos in southern Sonora (Fig. 14.2a) contained about 1,000 bats in October 1992 and about 20,000 pregnant females in February through at least late April 1993 (Ceballos et al., 1997).

In addition to marked changes in roost size, seasonal changes in sexual composition and sexual segregation are characteristic features of the population biology of *Leptonycteris* bats. For example, the roost near Chamela contains almost exclusively males from March through August, but by October, females are nearly as common as males (Ceballos et al., 1997). A similar shift from a male-biased sex ratio to a balanced sex ratio occurs between June and August in Cueva Las Grutas, Michoacán (Zamacona, 1991). In both of these cases, the appearance of females in male-dominated caves coincides with the onset of the mating season. In contrast, sexual segregation is strongly associated with the maternity period. Known maternity roosts of *L. curasoae* in the Sonoran Desert contain only adult females and their young. Adult males roost separately—usually at great geographic distances—from females during the maternity period.

As is the case in other species of highly gregarious, migratory bats (e.g., *Tadarida brasiliensis,* the Mexican free-tailed bat; McCracken et al. [1994]; McCracken and Gassel [1997]), migration patterns are likely to be complex in *Leptonycteris* bats (Rojas-

Martinez et al., 1999). This complexity raises such questions as: Who migrates (just females or both males and females)? How far do bats migrate? And why do bats migrate? Data from the sea cave near Chamela, Jalisco, can be used to illustrate this complexity. As mentioned above, the size and sexual composition of this roost undergo strong seasonal changes. Tens of thousands of males and females migrate to this cave in late summer. Where most of these bats come from is currently unknown. After mating in this cave in November and December, both males and females leave the cave. The genetic analysis conducted by Wilkinson and Fleming (1996) indicates that some of these females migrate up to 1,500 km to maternity roosts in northern Sonora and Arizona. But they do not appear in northern caves until late March or early April at the earliest, and their locations between mid-December and late March are unknown. Do they stay in central Mexico for some time before quickly migrating north or do they slowly move north, stopping at a series of roosts, over a period of several months? Where does the majority of males reside after they leave the Chamela cave? Based on the presence of *Ceiba* and *Bombax* pollen on their fur, Alvarez and Gonzalez (1970) speculated that individuals of *L. curasoae* arriving at the Cueva de Xoxafi, Hidalgo, in the spring are coming from subtropical forests in Morelos over 200 km to the south. It is likely that such seasonal habitat shifts are common in this bat (Herrera, 1997).

Results of their genetic analysis led Wilkinson and Fleming (1996) to hypothesize that lesser long-nosed bats migrate into the northern parts of their range from south-central Mexico along two pathways: a Pacific coastal route and an inland route along the western slopes of the Sierra Madres. Females from the Chamela cave, for example, are likely to migrate along the coastal route to arrive at their northern maternity roosts in the spring. In contrast, bats arriving in southern and southeastern Arizona later in the year are likely to migrate along the inland route. As discussed below, these two migration pathways undoubtedly differ in the flower resources they provide for migratory bats. Flowers of columnar cacti provide nectar and pollen along the coastal route, whereas flowers of agaves provide energy and protein along the inland route.

Although detailed studies have not yet been published, it is likely that the southern subspecies of *Leptonycteris curasoae* also undergoes seasonal movements in Venezuela. For example, Sosa and Soriano (1993) reported that *L. curasoae* was absent from their Lagunillas study site near Mérida between May and August and suggested that these bats undergo local movements to regions of higher food availability at that time. Marked changes in the size and sexual composition of roost populations—for example, in Cueva Piedra Honda on the Paraguaná Peninsula (Martino et al., 1997) and in Cueva El Convento on Margarita Island (Smith and Genoways, 1974)—also suggest that this species makes local seasonal movements in Venezuela. Soriano and Ruiz (chapter 12, this volume) further discuss movements of *L. curasoae* among arid Andean enclaves in Venezuela and Colombia.

At least two interrelated factors—the annual reproductive cycle and seasonal changes in resource availability—are likely to be responsible for seasonal changes in the distribution and local abundance of *Leptonycteris* bats. As discussed in more detail below, the mating period is a time when large numbers of bats of both sexes are concentrated in a few widely scattered roosts. After mating is completed, these roosts disband, and most bats move to other roosts. About two-thirds of the way through their gestation period, pregnant females (as well as a few yearling, nonreproductive females) form large maternity colonies located at great distances from their mating sites and away from males. They will remain in these roosts until they have weaned their young. Then adults and young abandon the maternity roosts and probably occupy a series of roosts en route back to their winter ranges, where mating will again take place.

Correlated with these large-scale movements are changes in the abundance of plant resources. Indeed, it is likely that seasonal changes in resource availability ultimately drive the annual reproductive and migratory cycles of *Leptonycteris curasoae* (and *L. nivalis*). For example, spring and summer are peak times in the availability of flowers and fruit, respectively, of Sonoran Desert columnar cacti (Steenbergh and Lowe, 1977; Turner et al., 1995; Fleming et al., 1996). Similarly, late summer and early fall are peak blooming times for bat-pollinated paniculate agaves in northern Mexico and Arizona (Gentry, 1982; Fleming et al., 1993). Resource-rich regions such as the Sonoran Desert undoubtedly provide a potent incentive for pregnant *Leptonycteris* bats to migrate north from tropical and subtropical regions in Mexico each spring. Large populations of blooming agaves in the uplands of northern Mexico and adjacent parts of the southwestern United States during the late summer and early fall provide a similar attraction to *Leptonycteris* bats of both species. In contrast, winter months are energetically barren in the north but are peak flowering times for bat-pollinated trees and shrubs in tropical dry forest (Frankie et al., 1974; Bullock and Solis-Magallanes, 1990). Concentration of *L. curasoae* in tropical and subtropical dry forests in the late fall, winter, and early spring coincides with increased resource (i.e., flower) levels in those habitats.

Reproduction and Physiology

Reproduction

As it is for most organisms, reproduction is a central feature in the population biology of *Leptonycteris* bats. Critical reproductive activities include sperm production and mating for males and mating, gestation, birth, and lactation for females. Despite the overwhelming importance of reproduction in the population biology of these bats, we currently have only a gross understanding of their annual reproductive cycle. It has long been known, for example, that females of *L. curasoae* migrate to the Sonoran Desert to give birth to a single pup between mid-May and early June and that they nurse their young in May and June (Hayward and Cockrum, 1971). A similar late-spring or early-summer birth pattern has been documented for *L. curasoae* on Curaçao and in Venezuela (Petit, 1997; Martino et al., 1998) and probably also occurs in *L. nivalis* in Mexico (Easterla, 1972; Moreno, 2000). But until recently, questions concerning when and where mating takes place and how many pregnancies a female undergoes per year have been unanswered.

As summarized in Ceballos et al. (1997), current reproductive data indicate that the annual reproductive cycle of *L. curasoae* in Mexico is geographically variable in terms of times of mating and birth. In mainland Mexico, there appear to exist two reproductive schedules: a northern spring-birth schedule and a southern winter-birth schedule. Different populations of this species probably adhere to one schedule or the other, so that males only mate at one time of the year and females give birth once a year.

Data on annual variation in testis size in males from two caves in south-central Mexico—San Andres Island near Chamela and Las Grutas near Ciudad Hidalgo (Fig. 14.2a)—provide a starting point for discussing these two reproductive schedules (Fig. 14.4). Year-round data from the Chamela cave indicate that testis size increases markedly in October through November (Ceballos et al., 1997). Although detailed histological studies are needed, these data strongly suggest that sperm production in males is highly seasonal. Males in this population appear to be reproductively competent only in late fall, and mating must take place between October and December. Babies conceived during this mating period are born in northern roosts in mid-May after a gestation period of about six months (Fig. 14.4). This gestation period is about two months longer than that of many tropical phyllostomids (e.g., *Carollia perspicillata, Glossophaga soricina;* Fleming [1988]); detailed histological studies are needed to determine whether gestation in *L. curasoae* involves delayed implantation or delayed development. A. Arends and A. Martino (pers. comm.) have observed mating in November and births in May in *L. curasoae* on the Paraguaná Peninsula of Venezuela.

Limited data from males occupying the Las Grutas cave in June–August indicate that testis size is large in June and small in July and August (Zamacona, 1991). Additional data (T. Fleming, unpublished) from a small colony in the Chiricahua Mountains of southeastern Arizona indicated that 13 of 14 adult males had large testes in mid-May 1993. Based on these data, we have constructed a hypothetical testis curve for late-spring mating males in Cueva Las Grutas. This curve is similar to the Chamela curve except for the timing of peak testis size (Fig. 14.4). Current information suggests that, like the Chamela cave, the Las Grutas cave is a mating site for *Leptonycteris curasoae.* Pups conceived there in June will be born elsewhere, probably in December in southern Mexico (Fig. 14.5). We currently know the locations of two winter-birth maternity roosts—Gruta Juxtlahuaca in Guerrero and Cueva Tempisque in Chiapas (Fig. 14.2a)—where pups are born in December and January (Ceballos et al., 1997). Banding and/or genetic studies are needed to determine whether females that mate in Cueva Las Grutas give birth in either of these two caves.

In summary, mating and maternity roosts of *L. curasoae* in mainland Mexico are geographically separated (Fig. 14.5). Current information suggests that mating takes place in central Mexico, in May–June or November–December, and that pups are born

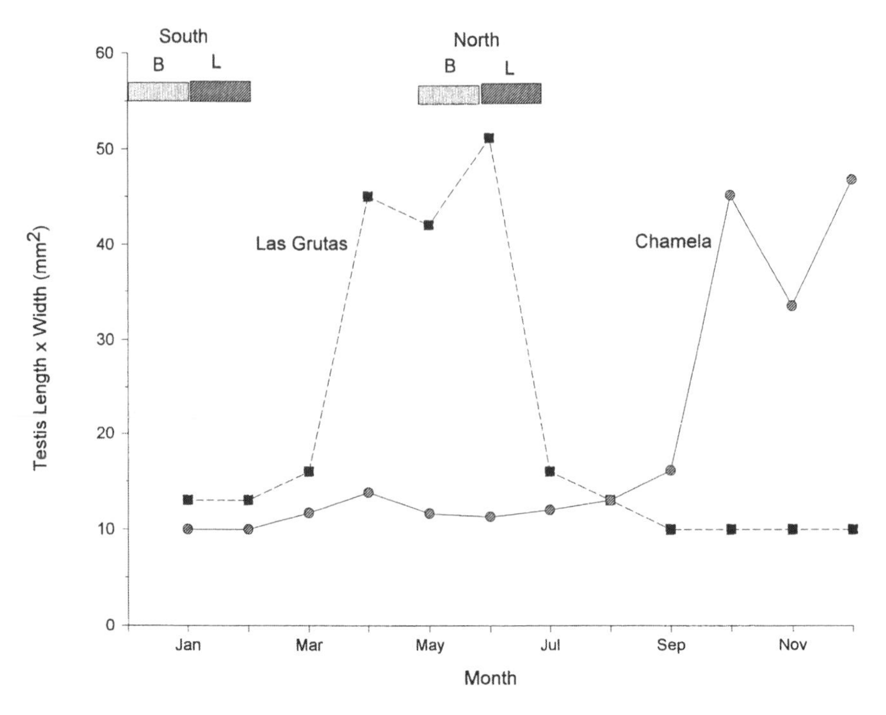

FIGURE 14.4. Seasonal changes in testis size of adult males of *L. curasoae* from two caves in south-central Mexico: Las Grutas, Michoacán, and San Andres Island, Jalisco. See text for construction of the Las Grutas curve. The rectangles indicate periods of birth (B) and location (L) in the northern spring-birth population and in the southern winter-birth population. Data are from Zamacona (1991); Ceballos et al. (1997); and T. Fleming (unpubl. data).

about six months later, in southern or northern maternity roosts, respectively. In both cases, periods of birth and lactation coincide with peaks in the availability of flowers and fruit in tropical dry forest (winter births) or the Sonoran Desert (spring births).

In addition to mainland Mexico, *Leptonycteris curasoae* is resident year-round in southern Baja California (Woloszyn and Woloszyn, 1982). Its reproductive schedule there appears to differ somewhat from the two mainland schedules. In mid-April 1993, one of us (Fleming) visited a cave near Buenavista, Baja California Sur, that contained about 30,000 adult females and their young. Most of the pups appeared to be about one month old and had probably been born in mid-March, about midway between the winter and spring birth peaks on the mainland. These observations indicate that *L. curasoae* probably has at least three different reproductive schedules in Mexico. The genetic consequences of these different schedules—Are there detectable reproductive demes in *L. curasoae?* Is this species undergoing a speciation event?—need to be investigated.

Details about the mating system of *Leptonycteris curasoae* (and other glossophagine bats) are currently unknown. Behavioral studies are needed to determine

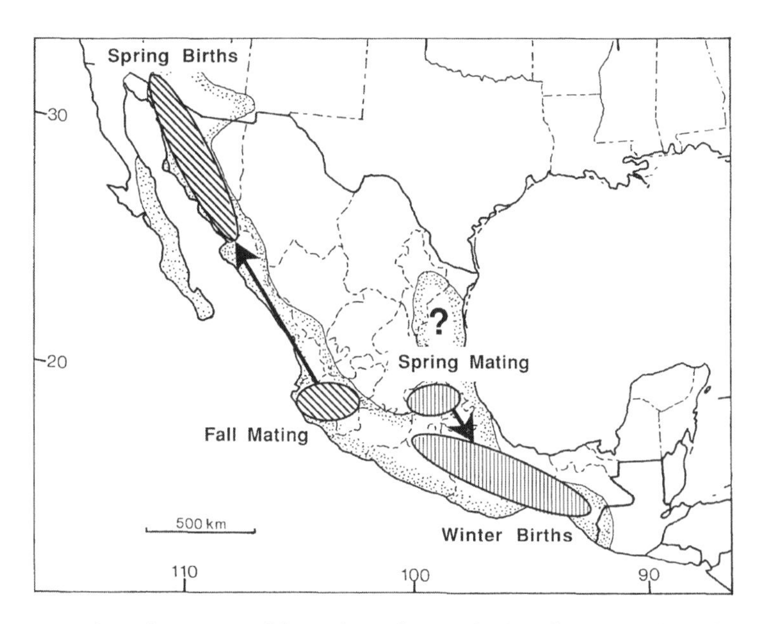

FIGURE 14.5. Schematic summary of the mating and maternity sites of *L. curasoae* in Mexico. Mating and maternity sites of eastern populations of this bat are currently unknown.

whether males mate with females in temporary harems or whether they attract females into small mating territories or form leks inside mating roosts (Altringham, 1996). Given the high density of bats within mating caves and the temporary nature of these aggregations, it is unlikely that lasting bonds form between males and their mates. The extent to which males compete among themselves for access to sexually receptive females and whether females exercise selectivity in their choice of mates are fascinating topics for future study.

Temperature Regulation and Metabolism

The roosting strategy of *Leptonycteris* bats differs from that of most other phyllostomid bats in two important respects. These bats occur in much larger and more densely packed colonies (tens of thousands to over 100,000 individuals), and their colonies are widely separated in space (e.g., Fig 14.2a). In contrast, most other phyllostomids roost in small colonies (maximum size is a few thousand) that are relatively uniformly distributed over the landscape (Fleming, 1993). The roosting ecology of *L. curasoae* also includes living in "hot" caves—caves that trap metabolic heat and moisture produced by the bodies of thousands of bats—often in association with large populations of several other species of hot-cave bats such as *Mormoops megalophylla, Pteronotus parnellii, P. davyi* (Mormoopidae), and *Natalus stramineus* (Natalidae) (Smith and Genoways, 1974; Arends et al., 1995; Ceballos et al., 1997; Martino et al., 1997).

Leptonycteris curasoae appears to select roosts based on their microclimatic conditions, and this roost selection profoundly influences the bat's physiological ecology.

Metabolic studies (Carpenter and Graham, 1967; Arends et al., 1995) indicate that nectarivorous glossophagine bats are normothermic endotherms (i.e., they maintain a high and constant body temperature over a wide range of ambient temperatures and do not undergo torpor) and have basal metabolic rates (BMR) that are slightly higher than expected for other mammals of equivalent mass. Relative to other glossophagines, however, *L. curasoae* has a lower BMR and a higher thermal conductance (i.e., rate of heat exchange with the environment as mediated by pelage), probably as a result of roosting in warm caves. In northern Venezuela and Arizona, for example, females roost during the maternity period within caves whose ambient temperature (T_a) is 33–34°C (Arends et al., 1995; Fleming et al., 1998). This T_a is within *L. curasoae*'s thermoneutral zone, so metabolic expenditures associated with thermoregulation in such caves are minimal. In contrast, two other glossophagines (*Choeroniscus godmani* and *Glossophaga longirostris*) roost in cool, shaded caves and mines at ambient temperatures several degrees below their thermoneutral zones in northern Venezuela. They need a higher BMR and better insulation from their pelage (hence, lower thermal conductance) to maintain high constant body temperatures (Arends et al., 1995). Females of *L. curasoae* thus gain several physiological benefits from roosting in warm, densely packed caves and mines, including reduced energy expenditure during the day, reduced evaporative water loss (caused by high rates of heat loss), and increased growth rates of their embryos and neonates.

Nutritional Physiology

As a nectarivorous and frugivorous bat, *Leptonycteris curasoae* consumes a diet that is rich in water but potentially poor in protein and salts. This diet produces unique physiological challenges not commonly faced by insectivorous bats or most other mammals. For example, the kidneys of most mammals are designed to conserve water and to excrete salts. But the lesser long-nosed bat is faced with the opposite problem: Its kidneys need to excrete excess water and retain as much of the salts and other solutes in the kidney filtrate as possible (Carpenter, 1969). Carpenter's detailed studies indicate that *L. curasoae* indeed differs from other desert-dwelling bats (e.g., *Eptesicus fuscus, Tadarida brasiliensis*) in kidney structure and function. Its kidney, for example, features a very small renal papilla (the site of maximum water reabsorption) and a thick medullary cortex, where resorption of salts and other solutes occurs. As a result, the bat excretes copious amounts of very dilute urine. It produces the most hypotonic urine (relative to the salt concentration of blood plasma) of any mammal. Despite a high rate of evaporative water loss during flight (it loses about 3.9% of its mass per hour of flight), *L. curasoae* is independent of free water, unlike insectivorous bats that must ingest free water to excrete excess nitrogen associated with their protein-rich diets. Because of this free-water independence, *Leptonycteris* bats do not need to roost or feed near scarce (non-nectar) water sources in the desert.

Howell (1974) addressed the question, Where does a nectar-feeding bat such as *Leptonycteris curasoae* get its protein? When *Leptonycteris* bats visit cactus and agave flowers, their faces and necks become covered with pollen. While they rest in night roosts between foraging bouts, these bats vigorously groom pollen off themselves (and each other) with their long tongues. Howell therefore postulated that the pollen ingested by this bat contains enough protein to maintain a positive nitrogen balance. Her measurements indicated that, compared with most pollen, the protein content of pollen of saguaro cacti and *Agave palmeri* is very high, averaging 43.7% and 22.9%, respectively. She calculated that by ingesting about one gram of pollen per day, this bat can maintain a positive nitrogen balance. Results of laboratory dietary studies indicate that bats fed synthetic diets of agave or saguaro nectar containing pollen maintained a positive nitrogen balance, whereas bats fed synthetic diets that lacked pollen quickly went into negative nitrogen balance and died. And so although nectar is a protein-poor food source, pollen is protein rich. Its ingestion and digestion eliminates the need for *Leptonycteris* bats to supplement their diets with insects, as hummingbirds usually do (Feinsinger, 1987).

Diet and Foraging Behavior

Diet

Four different methods have been used to study the diet of *Leptonycteris curasoae:* three direct methods (identification of stomach contents, fecal pellets, and pollen on fur) and one indirect method (stable-isotope analysis). Direct methods reveal the identity of recently ingested or visited food sources, whereas the indirect method allows one to determine the quantitative importance of different kinds of plants (e.g., CAM vs. C3 plants) in an animal's diet (Fleming, 1995a).

Plant species and genera known to occur in the diet of *Leptonycteris curasoae* are listed in Table 14.1 (see also chapters 5 and 13, this volume). Only two plant families (Agavaceae and Cactaceae) occur in the diet of this bat in temperate Arizona, whereas a broader range of families occurs in its diet in the subtropical and tropical parts of its range. In terms of number of species, Agavaceae and Cactaceae undoubtedly are the most important families in its diet in most parts of its range.

Carbon stable-isotope analysis is a powerful technique for quantifying the relative importance of CAM plants (i.e., Agavaceae and Cactaceae) vs. C3 plants (other plant families) in the diet of *Leptonycteris curasoae.* Such analyses have now been conducted throughout the ranges of both subspecies of the lesser long-nosed bat (Fleming et al., 1993; Ceballos et al., 1997; Petit, 1997; J. Nassar et al., unpubl. data). Results indicate that the relative importance of CAM plants varies geographically. Data from Mexico (Fig. 14.6) illustrates the range of this variation. Bats living on Baja California, for example, appear to feed primarily on CAM plants year-round. Winter-blooming *Agaves* help promote this CAM specialization (Fleming et al., 1993). In contrast, bats living in the sea cave near Chamela, Jalisco, where cactus and agave densities are low, feed primarily on C3 plants year-round. And bats that migrate season-

TABLE 14.1
Major Food Plants of *Leptonycteris curasoae*[a]

Location	*Plant Family*	*Species*	*Food Type*[b]
Arizona	Agavaceae	*Agave palmeri*	N, P
		A. Parryi	N, P
		A. deserti	N, P
		A. schottii	N, P
	Cactaceae	*Carnegiea gigantea*	N, P, F
		Stenocereus thurberi	N, P, F
Mexico	Agavaceae	many *Agave* species	N, P
	Cactaceae	*Carnegiea gigantea*	N, P, F
		Stenocereus spp.	N, P, F
		Pachycereus spp.	N, P, F
		Neobuxbaumia spp.	N, P, F
		other columnar cacti	N, P, F
	Bombacaceae	*Bombax ellipticum*	N, P
		Ceiba spp.	N, P
	Convolvulaceae	*Ipomoea arborescens*	N, P
	Leguminosae	*Bauhinia ungulata*	N, P
	Bignoniaceae	*Crescentia alata*	N, P
Venezuela and Curaçao	Agavaceae	*Agave cocuy*	N, P
	Cactaceae	*Stenocereus griseus*	N, P, F
		Pilosocereus lanuginosus	N, P, F
		Subpilocereus repandus	N, P, F
	Bombacaceae	*Ceiba pentandra*	N, P
	Myrtaceae	*Syzygium cumini*	F
	Sapotaceae	*Manilkara zapota*	F

[a] Data are from Alvarez and Gonzalez (1970); Hayward and Cockrum (1971); Howell and Roth (1981); Quiroz et al. (1986); Sosa and Soriano (1993); Fleming et al. (1996); Valiente-Banuet et al. (1996, 1997); Nassar et al. (1997); and Petit (1997).

[b] N = nectar, P = pollen, F = fruit.

ally from south-central Mexico into the Sonoran Desert and other northern localities probably undergo an annual "carbon cycle," in which they feed on a mixture of CAM and C3 plants in southern parts of their range and concentrate on CAM plants during migration and in the northern parts of their range. Data from Curaçao (Petit, 1997) and Venezuela (J. Nassar et al., unpubl. data) indicate that, like bats living in Baja California, the southern subspecies feeds primarily on CAM plants year-round. Thus the lesser long-nosed bat can be either a CAM or a C3 feeding specialist, depending on the local availability of flowers and fruit. It is a CAM specialist in the most arid parts of its range and is a C3 specialist or a CAM-C3 generalist (or sequential specialist) in the more mesic parts.

Nectar and pollen probably form the bulk of the diet of *Leptonycteris curasoae*, but it also consumes fruit. It is currently thought that columnar cacti produce most of the fruit eaten by lesser long-nosed bats in Arizona, Mexico, and Venezuela/Curaçao (Hayward and Cockrum, 1971; Sosa and Soriano, 1993; Petit, 1997; Sosa, 1997). Roosts housing *Leptonycteris* bats often contain deep layers of cactus seeds on the floor. Cactus fruits are particularly important food (and water) sources during lactation and postlactation periods in the Sonoran Desert. Insects have also been re-

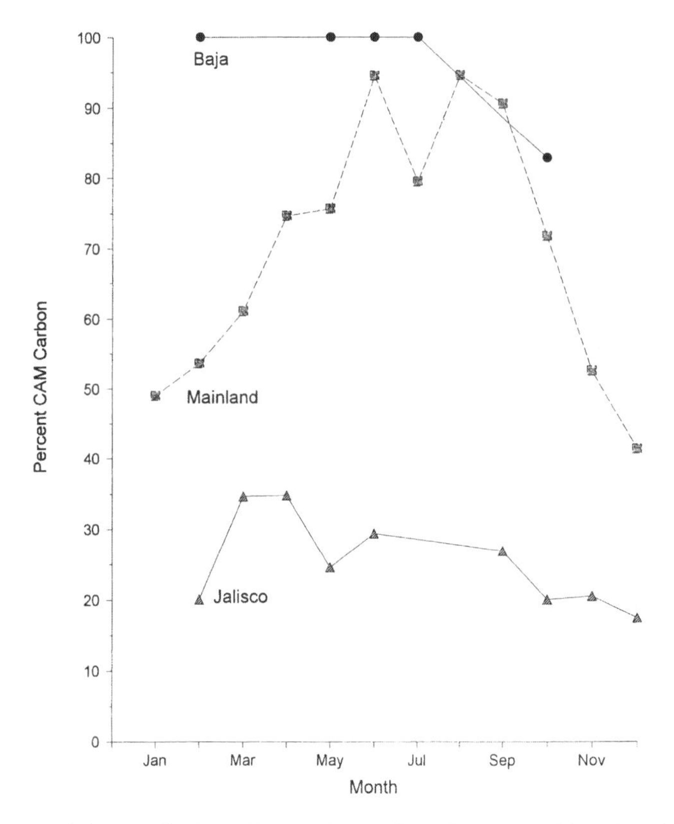

FIGURE 14.6. Relative contribution of CAM carbon to the carbon composition of Mexican populations of *L. curasoae.* Data from "mainland" include monthly samples from various locations in mainland Mexico. Data are from Fleming et al. (1993) and Ceballos et al. (1997).

ported in stomach contents or fecal samples (e.g., Howell, 1974; Sosa and Soriano, 1993), but they generally appear to be minor dietary items. Plant products provide most of the energy and nutrients assimilated by these bats.

Foraging Behavior and Energetics

Lesser long-nosed bats leave their day roosts around sunset and usually return to them at least one hour before sunrise (Fleming et al., 1998). They spend several hours in flight each night and sometimes (perhaps often) fly substantial distances between their day roosts and feeding areas. Horner et al. (1998) conducted a detailed foraging study of this bat using radio telemetry and light tagging techniques near Bahia Kino, Sonora. They discovered that females commuted 25–30 km from maternity roosts located on Tiburon Island in the Gulf of California to feed at cactus flowers on the Mexican mainland. Individuals returned to the same feeding areas on successive nights; these areas averaged about 0.43 km^2 in size but sometimes exceeded 2 km^2. Individual feeding areas overlapped, and bats sometimes foraged in groups of two to four individuals. Night-to-night compositional stability of these groups is unknown but is likely to be low, given the absence of socially cohesive behavior (e.g., allogrooming) among bats in their day roosts (Fleming et al., 1998). Bats visited flowers on many cactus plants and generally visited each flower five or fewer times. This flower visitation pattern probably results in a mixture of short- (<100 m) and long-distance (>500 m) pollen movements.

Although thousands of *Leptonycteris* bats roost together by day, these bats spread out over a large area (approximately 2,830 km^2 for a roost with a 30-km foraging radius) at night, and local density within cactus patches tends to be low. For example, T. Fleming et al. (2001) censused the relative abundance of *L. curasoae* in cactus patches at Bahia Kino for two seasons and at Organ Pipe Cactus National Monument for one season. They observed an average of 1.3 bats in an area of about 2.5 ha; this represents a density of about 0.5 bats/ha (ca. 50 bats/km^2). Such a low density should translate into very low flower visitation rates. In support of this expectation, Fleming et al. (1996) reported that visitation rates to flowers of cardón (*Pachycereus pringlei*) averaged 0.7 and 0.3 per hour in 1989 and 1990, respectively. Visitation rates to flowers of saguaro (*Carnegiea gigantea*) and organ pipe (*Stenocereus thurberi*) were even lower, both at Bahia Kino and at Organ Pipe Cactus National Monument (Fleming et al., 1996, T. Fleming, unpubl. data). In contrast, T. Fleming and F. Molina (unpubl. data) observed an extremely high bat density and high visitation rates to flowers of cardón (approximately one visit every ten minutes) in a remnant cactus patch near Guaymas, Sonora, in April 1998. At this site, *Leptonycteris* bats deposited about 20,000 pollen grains on each cardón stigma, whereas at sites where bat density was low, they deposited only a few thousand pollen grains on cardón stigmas (but still enough to fertilize all ovules in a fruit; T. Fleming and F. Molina, unpubl. data).

Different studies report differences in times of peak flower visitation by *Leptonycteris* bats. Nassar et al. (1997), for instance, found that bats (both *L. curasoae* and *Glossophaga longirostris*) visit cactus flowers most frequently at times of peak nectar secretion (generally before 2400), whereas Horner et al. (1998) noted that peak visitation occurred between 2400 and 0200, about two to four hours after peak nectar secretion (Fig. 14.7). A similar late visitation peak at flowers of saguaro and organ pipe cacti also occurs at Organ Pipe Cactus National Monument (T. Fleming, unpubl. data). In contrast, near Guaymas, T. Fleming and F. Molina (unpubl. data) found that *L. curasoae* begins to visit cardón and saguaro flowers frequently as soon as they open. What causes different populations of *L. curasoae* to differ in flower visitation behavior is currently unknown.

Horner et al. (1998) used their radiotracking data to construct a daily time and energy budget for *Leptonycteris curasoae.* They estimated that this bat spends about five hours each night in flight at a cost of about 17.7 kJ. It spends about one hour per night in a night roost digesting its food and the remaining 18 hours in its day roost. They estimated that the total daily energy budget of a nonreproductive female is

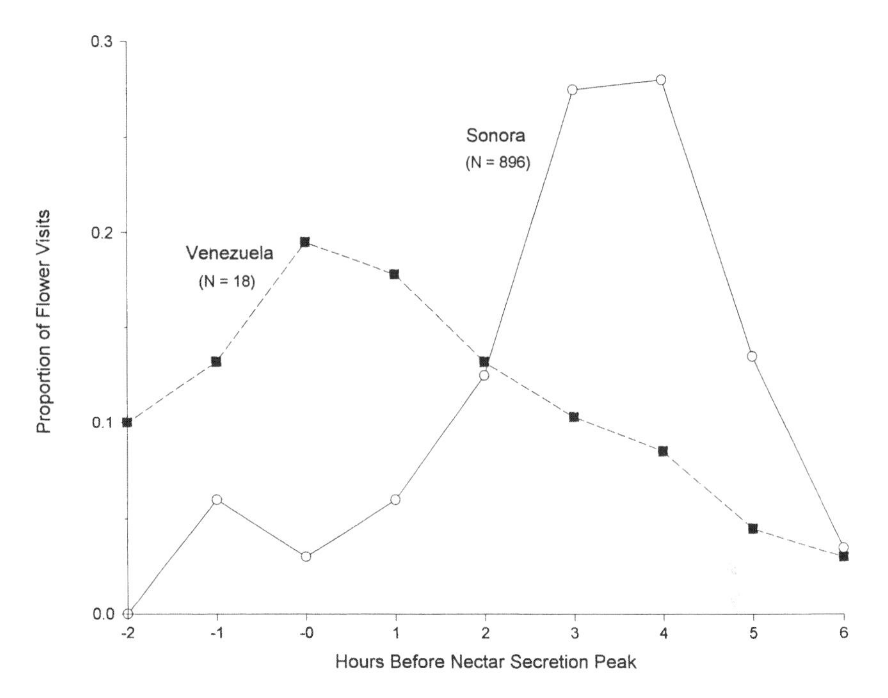

FIGURE 14.7. Timing of feeding visits by *L. curasoae* to flowers of *Pachycereus pringlei* (Sonora, Mexico) and *Pilosocereus moritzianus* (Venezuela). Data from Sonora represent mean values from 1989 and 1990. Data are from Nassar et al. (1997) and Horner et al. (1998).

about 40.2 kJ and that the bat would need to make 80–100 flower visits to acquire this energy. Because of its energy-efficient flight, only about 16–20 flower visits are needed to pay for its long-distance commute flights. A female's energy budget is 50–100% higher than 40 kJ when she is nursing a pup.

Horner et al. (1998) surveyed the feeding areas of four bats to estimate the density of open flowers and ripe fruit of columnar cacti. They found that flower density ranged from 30 to 107 flowers/ha and from 4,655 to 15,900 flowers per feeding area per night. Based on the low density of bats in their study area and the few flowers each bat needs to meet its energy requirements (as stated above, about 16–20 flowers at five visits per flower), they concluded that *L. curasoae* has an unlimited supply of energy (and protein from pollen) during the spring in this part of the Sonoran Desert. A similar situation undoubtedly occurs at Organ Pipe Cactus National Monument, where bat density is also low and cactus density high. In contrast, Petit and Pors (1996) estimated that populations of *L. curasoae* and *Glossophaga longirostris* on Curaçao are at the island's carrying capacity as determined by the density of columnar cactus flowers during the bats' lactation periods (July and August). They urged that cactus conservation be implemented on that island to prevent the extinction of these bats and other organisms feeding on cactus flowers and fruit.

Conclusions

Although the diet of *Leptonycteris curasoae* is not entirely restricted to nectar, pollen, and fruit produced by the Cactaceae, its population biology nonetheless is strongly influenced by the flowering and fruiting biology of columnar cacti. Conversely, although *Leptonycteris* bats are not necessarily the exclusive pollinators of the flowers of columnar cacti, fruit set is often strongly influenced by this bat (see chapter 10, this volume). In both an evolutionary and an ecological sense, then, these two sets of organisms are intimately linked. Cactus phenology (sensu lato) has directly influenced the nutritional physiology as well as the reproductive and migratory biology of this bat. Strong geographic and seasonal variation in the availability of flowers and fruit have selected for seasonal reproductive cycles and long-distance migration. Cactus phenology has also influenced the roosting ecology, mating system, and genetic structure of the bat. On the cactus side of the interaction, bat behavior (sensu lato) directly influences fruit set and, to a lesser extent, seed dispersal (see chapter 15, this volume). It also influences plant-plant competition for pollinators as well as plant density and genetic structure.

The heart of this interaction involves the brief (ca. 400 millisecond) contact between a cactus plant and a bat as it plunges its face into a flower for a tongueful of nectar and a faceful of pollen. During this contact, the bat gains energy and nutrients while the flower gains conspecific (and sometimes heterospecific) pollen grains, which sets the fertilization process in motion. Later in the season, bats return to the ripe fruits they helped pollinate and again make brief contact with the plant to gain energy and nutrients, initiating the seed-dispersal process. From these two fleeting contacts, repeated numerous times a night and for countless seasons, the current patterns of abundance and distribution of both columnar cacti and bats have emerged. Without these interactions, the arid regions of North and South America would be very different in appearance and function.

Summary

In this chapter, we review major features of the evolution and population biology of the lesser long-nosed bat, *Leptonycteris curasoae* (Phyllostomidae, Glossophaginae), one of the most important pollinators of columnar cacti in arid regions of the United States, Mexico, and northern South America. This bat is a relatively derived member of its subfamily and is notable for its large size and fast, energy-efficient flight. Unlike most of its relatives, *L. curasoae* is highly gregarious and annually migrates substantial distances, at least in western Mexico. It lives in widely scattered colonies containing tens of thousands to over 100,000 adults. In Mexico and northern South America, the size and sexual composition of colonies change seasonally as bats of both sexes move among caves. Certain caves serve as mating sites. Thousands of pregnant females migrate over 1,000 km from south-central Mexico into the Sonoran Desert each spring to form maternity colonies. Births occur in mid-May, during the peak flowering times of columnar cacti; lactation occurs when cactus fruits and flowers are available. Similar timing of reproductive events occurs in Venezuela, where this species also appears to be migratory, but over a smaller spatial scale. In contrast, females living in southern Mexico give birth in December or January, times of peak flowering in the tropical dry forest. Two reproductive demes thus exist in this species in Mexico. The diet of *L. curasoae* varies geographically. Bats specialize on flowers and fruits of Cactaceae and flowers of Agavaceae in arid parts of their range but have a broader diet in more mesic habitats. Foraging behavior of this bat includes long commutes (up to 30 km) between day roosts and foraging areas. In the Sonoran Desert, these bats visit flowers on many plants throughout their large (up to 2.5 km^2) foraging areas and have the potential to move pollen relatively long distances within and between populations. Individuals of this species sometimes forage in groups, but they apparently display no socially cohesive behavior (e.g., allogrooming) in their day roosts. Columnar cacti and *Leptonycteris* bats have mutually influenced each other's evolution and ecology.

Resumen

En este capítulo revisamos aspectos fundamentales de la evolución y biología poblacional del murciélago magueyero, *Leptonycteris curasoae* (Phyllostomidae, Glossophaginae), uno de los polinizadores más importantes de los cactus columnares en regiones áridas de los Estados Unidos, México y el norte de Sur América. Este murciélago es un miembro relativamente derivado dentro de su subfamilia, notable, por su gran tamaño y vuelo rápido y eficiente desde el punto de vista energético. A diferencia de la mayoría de las especies relacionadas, *L. curasoae* es altamente gregario y migra a distancias considerables anualmente; al menos en el oeste de México. Este murciélago vive en colonias ampliamente dispersas que contienen desde decenas de miles a más de 100,000 adultos. En México y el norte de Sur América, el tamaño y composición sexual de las colonias cambia estacionalmente. A medida que murciélagos de ambos sexos se mueven entre las cuevas, éstas sirven como sitios de apareamiento. Miles de hembras preñadas migran, por encima de los 1,000 km desde el sur y centro de México hasta el desierto de Sonora cada primavera para formar colonias de maternidad. Los nacimientos ocurren a mitad del mes de Mayo durante los máximos de floración de los cactus columnares; el período de lactancia coincide, con la época cuando las frutas y flores de los cactus están disponibles. Una periodicidad similar de eventos reproductivos occurre en Venezuela donde esta especie también pareciera ser migratoria, pero posiblemente sobre una escala geográfica menor. En contraste, las hembras qué viven en el sur de México, paren sus crías en Diciembre o Enero durante máximos de floración en el bosque seco tropical. Se pueden distinguir, por lo tanto, las dos poblaciones reproductivas de esta especie en México. La dieta de *L. curasoae* varía geograficamente. Estos murciélagos se especializan en flores y frutos de Cactaceae y las flores de Agavaceae en partes áridas de su distribución, pero tienen una dieta más amplia en habitats mésicos. La búsqueda de alimentos por éste murciélago incluye vuelos largos (hasta 30 km); entre refugios diurnos y áreas de alimentación. En el desierto de Sonora los murciélagos visitan flores de muchas plantas sobre una amplia área de búsqueda (hasta 2.5 km^2). Tienen el potencial de mover polen sobre distancias relativamente largas dentro y entre poblaciones. Algunos individuos de esta especie pueden realizar búsquedas de alimento en grupos, pero aparentemente estos murciélagos no despliegan comportamientos de cohesión social (ej. acicalamiento entre individuos) en sus refugios diurnos. *Leptonycteris* ha establecido una fuerte interacción con las cactáceas columnares y ha influído recíprocamente su evolución y ecología.

ACKNOWLEDGMENTS

Fleming thanks several landowners for permission to study bats and cacti on their property at Bahia Kino and the Mexican government for research permits. The U.S. National Park Service provided permission to work at Organ Pipe Cactus National Monument. He was assisted by and collaborated with many people, including M. Tuttle, M. Horner, C. Sahley, V. and D. Dalton, J. Wilkinson, and L. Sternberg, in his field and laboratory studies. Financial support for his studies was provided by the National Geographic Society, National Fish and Wildlife Foundation, Arizona Fish and Game Department, and the U.S. National Science Foundation. Nassar thanks PROVITA for logistic support on Margarita Island and PROFAUNA and Fundación Instituto Botánico de Venezuela for research permits. Assistance in the field was provided by J. Castro, M. Muñoz, J. Molinari, and local residents. Financial support was provided by Wildlife Conservation Society, the University of Miami, and Sigma Xi. We thank N. Simmons for her comments on this chapter. Mexican research in this chapter has benefited from the Programa para Conservacion de Murcielagos Migratorios (PCMM) administered by Bat Conservation International and Universidad Nacional Autónoma de México.

REFERENCES

Altringham, J. D. 1996. *Bats: Biology and Behaviour.* Oxford: Oxford University Press.

Alvarez, T., and Q. L. Gonzalez. 1970. Analisis polinico del contenido gastrico de murcielagos Glossophaginae de Mexico. *Anales del Escuela Nacional Ciencias Biologia, Mexico* 18:137–65.

Arends, A., F. J. Bonaccorso, and M. Genoud. 1995. Basal rates of metabolism of nectarivorous bats (Phyllostomidae) from a semiarid thorn forest in Venezuela. *Journal of Mammalogy* 76:947–56.

Arita, H. T. 1991. Spatial segregation in long-nosed bats, *Leptonycteris nivalis* and *Leptonycteris curasoae,* in Mexico. *Journal of Mammalogy* 79:706–14.

Arita, H. T., and S. R. Humphrey. 1988. Revision taxonomica de los murcielagos magueyeros del genero *Leptonycteris* (Chiroptera: Phyllostomidae). *Acta Zoologica Mexico* 29:1–60.

Bullock, S. H., and J. A. Solis-Magallanes. 1990. Phenology of canopy trees of a tropical deciduous forest in Mexico. *Biotropica* 22:22–35.

Carpenter, R. E. 1969. Structure and function of the kidney and the water balance of desert bats. *Physiological Zoology* 42:288–302.

Carpenter, R. E., and J. B. Graham. 1967. Physiological responses to temperature in the long-nosed bat, *Leptonycteris sanborni. Comparative Biochemistry and Physiology* 22:709–22.

Ceballos, G., T. H. Fleming, C. Chavez, and J. Nassar. 1997. Population dynamics of *Leptonycteris curasoae* (Chiroptera: Phyllostomidae) in Jalisco, Mexico. *Journal of Mammalogy* 78:1220–30.

Cockrum, E. L. 1991. Seasonal distribution of northwestern populations of the long-nosed bats, *Leptonycteris sanborni* Family Phyllostomidae. *Anales del Instito Biologia Universidad Nacional Autonoma de Mexico, Serie Zoologica* 62:181–202.

Cuervo-Diaz, A., J. Hernandez-Camacho, and A. Cadena. 1986. Lista actualizada de los mamiferos de Colombia. Anotaciones sobre su distribucion. *Caldasia* 15:471–501.

Easterla, D. A. 1972. Status of *Leptonycteris nivalis* (Phyllostomatidae) in Big Bend National Park, Texas. *Southwestern Naturalist* 17:287–92.

Feinsinger, P. 1987. Approaches to nectarivore-plant interactions in the New World. *Revista Chilena Historia Natural:* 60:285–319.

Fleming, T. H. 1988. *The Short-Tailed Fruit Bat.* Chicago: University of Chicago Press.

———. 1993. Plant-visiting bats. *American Scientist* 81:460–67.

———. 1995a. The use of stable isotopes to study the diets of plant-visiting bats. *Symposium of the Zoological Society of London* 67:99–110.

———. 1995b. Pollination and frugivory in phyllostomid bats of arid regions. *Marmosiana* 1:87–93.

Fleming, T. H., R. A. Nuñez, and L.S.L. Sternberg. 1993. Seasonal changes in the diets of migrant and non-migrant nectarivorous bats as revealed by carbon stable isotope analysis. *Oecologia* 94:72–75.

Fleming, T. H., M. D. Tuttle, and M. A. Horner. 1996. Pollination biology and the relative importance of nocturnal and diurnal pollinators in three species of Sonoran Desert columnar cacti. *Southwestern Naturalist* 41:257–69.

Fleming, T. H., A. A. Nelson, and V. M. Dalton. 1998. Roosting behaviors of the lesser long-nosed bat, *Leptonycteris curasoae. Journal of Mammalogy* 79:147–55.

Fleming. T. H., C. T. Sahley, J. N. Holland, J. Nason, and J. L. Hamrick. 2001. Sonoran Desert columnar cacti and the evolution of generalized pollination systems. *Ecological Monographs* 71:511–30.

Frankie, G. W., H. G. Baker, and P. A. Opler. 1974. Comparative phenological studies of trees in tropical wet and dry forests in the lowlands of Costa Rica. *Journal of Ecology* 62:881–919.

Freeman, P. W. 1995. Nectarivorous feeding mechanisms in bats. *Biological Journal of the Linnean Society* 56:439–63.

Gardner, A. L. 1977. Feeding habits. In *Biology of Bats of the New World Family Phyllostomatidae.* Part II, eds. R. J. Baker, J. K. Jones, Jr., and D. C. Carter, 293–350. Texas Tech University Museum Special Publication No. 13. Lubbock, Texas: Texas Tech University.

Gentry, H. S. 1982. *Agaves of Continental North America.* Tucson: University of Arizona Press.

Griffiths, T. A. 1982. Systematics of the New World nectar feeding bats (Mammalia, Phyllostomidae), based on the morphology of the hyoid and lingual regions. *American Museum Novitates* 2742:1–45.

Haiduk, M. W., and R. J. Baker. 1982. Cladistical analysis of G-banded chromosomes of nectar-feeding bats (Glossophaginae: Phyllostomidae). *Systematic Zoology* 31:252–65.

Hayward, B., and E. L. Cockrum. 1971. The natural history of the western long-nosed bat, *Leptonycteris sanborni. Western New Mexico University Research Science* 1:75–123.

Herrera, L.G.M. 1997. Evidence of altitudinal movements of *Leptonycteris curasoae* (Chiroptera: Phyllostomidae) in central Mexico. *Revista Mexicana de Mastozoologia* 2:116–18.

Horner, M. A., T. H. Fleming, and C. T. Sahley. 1998. Foraging behaviour and energetics of a nectar-feeding bat *Leptonycteris curasoae* (Chiroptera: Phyllostomidae). *Journal of Zoology* 244:575–86.

Howell, D. J. 1974. Bats and pollen: physiological aspects of the syndrome of chiropterophily. *Comparative Biochemistry and Physiology* 48:263–76.

Howell, D. J., and B. S. Roth. 1981. Sexual reproduction in agaves: the benefits of bats: cost of semelparous advertising. *Ecology* 62:3–7.

Hoyt, R. A., J. S. Altenbach, and D. J. Hafner. 1994. Observations on long-nosed bats (*Leptonycteris*) in New Mexico. *Southwestern Naturalist* 39:175–79.

Koopman, K. F. 1981. The distributional patterns of New World nectar-feeding bats. *Annals of the Missouri Botanical Gardens* 68:352–69.

Linares, O. 1998. *Mamiferos de Venezuela.* Caracas: Sociedad Concervacionista Audubon de Venezuela.

Martino, A., J. Aranguren, and A. Arends. 1997. Los quiropteros asociados a la cueva de Piedra Honda (Peninsula de Paraguaná, Venezuela): su importancia como reserva biologica. *Acta Cientifica Venezolana* 48:182–87.

Martino, A., A. Arends, and J. Aranguren. 1998. Reproductive pattern of *Leptonycteris curasaoe* Miller (Chiroptera: Phyllostomidae) in northern Venezuela. *Mammalia* 62:69–76.

McCracken, G. F., and M. F. Gassel. 1997. Genetic structure in migratory and nonmigratory populations of Brazilian free-tailed bats. *Journal of Mammalogy* 78:348–57.

McCracken, G. F., M. K. McCracken, and A. T. Vawter. 1994. Genetic structure in migratory populations of the bat *Tadarida brasiliensis mexicana. Journal of Mammalogy* 75:500–514.

Moreno, A. 2000. Ecological studies of the Mexican long-nosed bat (*Leptonycteris nivalis*). Ph.D. diss., Texas A&M University, College Station.

Nassar, J., N. Ramírez, and O. Linares. 1997. Comparative pollination biology of Venezuelan columnar cacti and the role of nectar-feeding bats in their sexual reproduction. *American Journal of Botany* 84:918–27.

Norberg, U., and J. M. Rayner. 1987. Ecological morphology and flight in bats (Mammalia: Chiroptera): wing adaptations, flight performance, foraging strategy and echolocation. *Philosophical Transactions of the Royal Society of London* 316B:335–427.

Petit, S. 1995. The pollinators of two species of columnar cacti on Curaçao, Netherlands Antilles. *Biotropica* 27:538–41.

———. 1997. The diet and reproductive schedules of *Leptonycteris curasoae curasoae* and *Glossophaga longirostris elongata* (Chiroptera: Glossophaginae) on Curaçao. *Biotropica* 29: 214–23.

Petit, S., and L. Pors. 1996. Survey of columnar cacti and carrying capacity for nectar-feeding bats on Curaçao. *Conservation Biology* 10:769–75.

Quiroz, D. L., M. S. Xelhuantzi, and M. C. Zamora. 1986. Analisis palinologico del contenido gastrointestinal de los murcielagos *Glossophaga soricina* y *Leptonycteris yerbabuena* de las Grutas de Juxtlahuaca, Guerrero. *Instituto Nacional de Antropologia Historia Serie Prehistoria* 1–62.

Rojas-Martinez, A., A. Valiente-Banuet, M. C. Arizmendi, A. Alcantara-Eguren, and H. T. Arita. 1999. Seasonal distribution of the long-nosed bat (*Leptonycteris curasoae*) in North America: Does a generalized migration pattern really exist? *Journal of Biogeography* 26:1065–77.

Sahley, C. T. 1996. Bat and hummingbird pollination of an autotetraploid columnar cactus, *Weberbauerocereus weberbaueri* (Cactaceae). *American Journal of Botany* 83:1329–36.

Sahley, C. T., M. A. Horner, and T. H. Fleming. 1993. Flight speeds and mechanical power outputs of the nectar-feeding bat, *Leptonycteris curasoae* (Phyllostomidae: Glossophaginae). *Journal of Mammalogy* 74:594–600.

Smith, J. D., and H. H. Genoways. 1974. Bats of Margarita Island, Venezuela, with zoogeographic comments. *Bulletin of the Southern California Academy of Sciences* 73:64–79.

Sosa, V. 1997. Dispersal and recruitment ecology of columnar cacti in the Sonoran Desert. Ph.D. diss., University of Miami.

Sosa, M., and P. J. Soriano. 1993. Solapamiento de dieta entre *Leptonycteris curasoae* y *Glossophaga longirostris* (Mammalia: Chiroptera). *Revista Biologia Tropical* 41:529–32.

Steenbergh, W. F., and C. H. Lowe. 1977. Ecology of the saguaro: II. *Reproduction, Germination, Establishment, Growth, and Survival of the Young Plant.* National Park Service Science Monographs Series No. 8. Washington, D.C.: Government Printing Office.

Turner, R. M., J. E. Bowers, and T. L. Burgess. 1995. *Sonoran Desert Plants: an Ecological Atlas.* Tucson: University of Arizona Press.

Van den Bussche, R. A. 1992. Restriction-site variation and molecular systematics of New World leaf-nosed bats. *Journal of Mammalogy* 73:29–42.

Valiente-Banuet, A., M. C. Arizmendi, A. Martinez-Rojas, and L. Dominquez-Canesco. 1996. Geographical and ecological correlates between columnar cacti and nectar-feeding bats in Mexico. *Journal of Tropical Ecology* 12:103–19.

Valiente-Banuet, A., M. C. Arizmendi, A. Martinez-Rojas, and P. Davila. 1997. Pollination of two columnar cacti (*Neobuxbaumia mezcalaensis* and *Neobuxbaumia macrocephala*) in the Tehuacán Valley, central Mexico. *American Journal of Botany* 84:452–55.

Wilkinson, G. S., and T. H. Fleming. 1996. Migration and evolution of lesser long-nosed bats, *Leptonycteris curasoae,* inferred from mitochondrial DNA. *Molecular Ecology* 5:329–39.

Wilson, D. E., and D. M. Reeder. 1993. *Mammal Species of the World.* 2nd ed. Washington, D.C.: Smithsonian Institution Press.

Woloszyn, D., and B. W. Woloszyn. 1982. *Los Mammiferos de la Sierra de la Laguna, Baja California Sur.* Mexico City: CONACYT.

Zamacona, M.C.H. 1991. Los quiropteros de "Las Grutas" Ciudad Hidalgo, Michoacán, Mexico. Tesis Profesional, Universidad Michoacána de San Nicolas de Hidalgo, Morelia, Mexico.

CHAPTER 15

Why Are Columnar Cacti Associated with Nurse Plants?

Vinicio J. Sosa
Theodore H. Fleming

Introduction

The answer to the question posed in our chapter title is closely related to the requirements of columnar and, very likely, other cacti for successful establishment and the factors that affect their establishment. Until recently, the recruitment ecology of columnar cacti has been neglected except for studies conducted during the 1960s and 1970s on the saguaro (*Carnegiea gigantea*) at its northernmost distribution limit (Niering et al., 1963; Turner et al., 1966, 1969; Steenbergh and Lowe, 1977). Such a narrow focus was unjustified, as there exist about 70 species of columnar cacti (Tribe Pachycereeae; Gibson, 1982) distributed throughout arid parts of the North American continent. Some of these species are dominant plants in their communities, where they are important sources of food (fruit, seeds, pollen, nectar, and plant pulp) and shelter for other organisms (Alcorn et al., 1961; Steenbergh and Lowe, 1977, 1983; Fleming et al., 1996; chapter 11, this volume). Fortunately, interest in the population and community ecology of other columnar cacti is currently increasing. As a result, we now have a reasonable understanding of the establishment ecology of columnar cacti.

In this chapter, we will first examine the generality of the association between cacti and perennial plants that act as nurse plants in arid and semiarid plant communities of North America. We then review evidence supporting each of the three main mechanisms proposed to explain the origin or maintenance of the cactus–nurse-plant association. Finally, we will briefly comment on future lines of research that will result in an even better understanding of the columnar cactus–nurse-plant interaction.

Association Patterns

Most columnar cacti need the shelter of a non-conspecific perennial plant for seedling establishment and growth. This dependence was initially inferred from the frequent observation of saguaros growing beneath or near the canopies of trees, usually palo verde (*Cercidium microphyllum*), and shrubs in the Sonoran desert (Shreve, 1917, 1931). This pattern of a positive spatial association between seedlings of one species and sheltering adults of another species has been called "the nurse-plant syndrome" (Niering et al., 1963; Turner et al., 1966, 1969; Steenbergh and Lowe, 1969, 1977).

Here we will refer to the spatial pattern in which a columnar cactus is growing under the canopy of a tree or shrub as a "cactus–nurse-plant" association or syndrome, although we realize that: (1) The association can occur randomly with a probability equal to the proportion of total plant cover; and (2) before implying a nursing effect, a benefit due to the association must be proved for the cactus, usually by experimentation.

Many studies report the presence of young cacti growing beneath the cover of shrubs and trees, which supposedly provide protection or a more suitable microenvironment for cacti than the environment beyond the limits of their canopies (Table 15.1). The putative nurse plants with which cacti are associated belong to many different genera and families, but there is an apparent preference for legumes (38% of the total of perennial species reported as nurse plants of cacti), which are frequently the dominant trees or shrubs in arid and semiarid environments. Only nine out of 70 species of extant columnar cacti have been examined regarding their association with nurse plants. Eight of them are usually associated with perennials, at least during their seedling or juvenile stages. However, this pattern might not be general: Recent studies indicate that some species may or may not be associated, depending on several factors, such as the land form or the vegetation type in which they occur.

For example, near Bahia Kino, Sonora, the saguaro was significantly associated in lower bajadas but not in floodplains, where it reaches its maximum density. Cardón (*Pachycereus pringlei*) was associated in floodplains and lower bajadas, but not in upper bajadas. Senita (*Lophocereus schottii*), in contrast, was always associated wherever it was present (V. Sosa and A. Contreras, unpubl. data). The occurrence of the cactus–nurse-plant association might also vary between vegetation types, as illustrated by organ pipe (*Stenocereus thurberi*), which is not associated in the Lower Sonoran Desert but is associated with trees or shrubs in tropical dry forest in Baja California Sur (Arriaga et al., 1993). This suggests that soil texture or other characteristics linked to topography, as well as biotic factors (e.g., plant density, abundance of cactus predators, nurse-plant quality), might affect the dependency of some cacti on perennials for establishment.

Three main mechanisms have been proposed to explain the origin and maintenance of the columnar cactus–nurse-plant association (McAuliffe, 1988):

1. *Nonrandom seed dispersal.* More seeds are found in sites under shrub and tree canopies than in open sites away from shrubs and trees. This pattern may be produced by (a) nonrandom deposition of seeds by animal dispersers, wind, or lam-

TABLE 15.1

Cactus Species that Have Been Reported Associated with Putative Nurse Plants in Arid or Semiarid Regions of North America[a]

Cactus Species	*GF[b]*	*Nurse Species*	*Region, Country*	*Reference*
Carnegiea gigantea	c	*Cercidium microphyllum, Ambrosia deltoidea,* and several shrubs and trees	Sonoran Desert, U.S.A.	Shreve (1931); Niering et al. (1963); Turner et al. (1966, 1969); Stenbergh and Lowe (1969, 1977); McAulliffe (1984a); Hutto et al. (1986); Franco and Nobel (1989)
		Bursera microphylla, Olneya tesota, Hymenoclea monogyra	Sonoran Desert, U.S.A.	Sosa and Contreras, unpubl.
Cephalocereus columnatrajani	c, g	*Eupatoriun spinosorum*	Tehuacán Valley, Mexico	Valiente-Banuet et al. (1991b)
Coryphanta pallida		*Caesalpinia melanadenia, Castela tortuosa*		
Echinocereus conglomeratus	c	*Euphorbia antisyphilitica, Hamatocactus hamatacanthus, Opuntia leptocaulis*	Chihuahuan Desert, Mexico	Silvertown and Wilson (1994)
Echinocereus engelmannii	c	*Opuntia fulgida*	Sonoran Desert, U.S.A.	McAuliffe (1984b)
Ferocactus acanthodes	g	*Pleuraphis (Hilaria) rigida,* shrubs	Sonoran Desert, U.S.A.	Jordan and Nobel (1981); Franco and Nobel (1989)
Ferocactus peninsulae	g	*Haematoxylon brasiletto*	Tropical dry forest, B.C.S., Mexico	Arriaga et al. (1993)
Hamatocactus hamatacanthus	g	*Opuntia leptocaulis, Euphorbia antisyphilitica*	Chihuahuan Desert, Mexico	Silvertown and Wilson (1994)
Lophocereus schottii	c	*Bursera microphylla, Olneya tesota,* shrubs	Sonoran Desert, Mexico	Sosa and Contreras, unpubl.
Mammillaria casoi	g	*Eupatoriun spinosorum*	Tehuacán Valley, Mexico	Valiente-Banuet et al. (1991b)
Mammillaria colina	g	*Caesalpinia melanadenia, Castela tortuosa*		
Mammillaria dioica	g	Shrubs	Sonoran Desert, B.C., Mexico	Shainsky (1978), cited by Cody (1993)
Mammillaria grahamii	g	*Olneya tesota, Hymenoclea monogyra*	Sonoran Desert, Mexico	Sosa and Contreras, unpubl.
Mammillaria microcarpa	g	*Opuntia fulgida*	Sonoran Desert, U.S.A.	McAuliffe (1984b)
Neobuxbaumia tetetzo	c	*Caesalpinea melanadenia, Cordia cylindrostachya, Verbesina* sp., *Mimosa luisana*	Tehuacán Valley, Mexico	Valiente-Banuet et al. (1991a,b), Flores-Martínez et al. (1994)
Opuntia acanthocarpa	a	*Pleuraphis rigida, Thamnosma montana, Salazaria mexicana, Muhlenbergia porteri, Stipa speciosa, Menodora spinescens*	Mohave Desert, U.S.A.	Cody (1993)
Opuntia echinocarpa	a	*Pleuraphis rigida, Krameria parvifolia, K. grayii*	Mohave Desert, U.S.A.	Cody (1993)

(continued)

TABLE 15.1 *(continued)*

Cactus Species	*GF*[b]	*Nurse Species*	*Region, Country*	*Reference*
Opuntia leptocaulis	a	*Larrea tridentata*	Chihuahuan Desert, U.S.A.	Yeaton (1978)
		Euphorbia antisyphilitica	Chihuahuan Desert, Mexico	Silvertown and Wilson (1994)
Opuntia ramosissima	a	*Pleuraphis rigida*	Mohave Desert, U.S.A.	Cody (1993)
Opuntia streptacantha	a	*Acacia schaffneri*	Chihuahuan Desert, Mexico	Yeaton and Romero-Manzanares (1986)
Pachycereus pringlei	c	*Bursera microphylla, Olneya tesota, Jatropha cinerea, J. cuneata, Hymenoclea spp.*	Sonoran Desert, Mexico	Sosa and Contreras, unpubl.
Stenocereus thurberii	c	*Haematoxylon brasiletto, Tecoma stans, Jathropha vernicosa*	Tropical dry forest, Baja California Sur, Mexico	Arriaga et al. (1993)
		Bursera microphylla, Olneya tesota	Sonoran Desert, Mexico	Sosa and Contreras, unpubl.

[a] Entries are in alphabetical order for each cactus species, then references in chronological order.

[b] GF = growth form: a = arbuscular; c = columnar; g = globose.

inar water flow; (b) nonrandom seed predation; or (c) the combined result of differential dispersal and predation of seeds.

2. *Nonrandom seedling predation.* Seedlings may escape predation more often under nurse-plant canopies than in the open.
3. *Nonrandom seedling establishment.* Seedling establishment may be higher under nurse-plant canopies than in the open because of the positive effects of nutrient accumulation and shading on microclimatic conditions.

The first two mechanisms can be referred to as biotic effects of nurse plants and the third as abiotic or environmental effects of nurse plants.

Differential Seed Dispersal

Dispersal is very important for columnar cacti that lack vegetative reproduction, because it is the only way that their seeds will be carried to favorable establishment sites. Perennial shrubs and trees may contribute to the nonrandom distribution pattern of columnar cacti by providing dispersal sites for seeds. An important structural difference between North American deserts and mesic habitats is the sparseness of plants providing perches and shade for birds and other animals. By providing shade and nesting and roosting sites, woody desert plants could serve as recruitment foci for bird-disseminated seeds (Archer et al., 1988; Milton et al., 1998; personal obs.).

Differential dispersal of cactus seeds by animals to sites under perennial canopies has received little attention compared with the other proposed mechanisms. Olin et al. (1989) reported that white-winged doves (*Zenaida asiatica*), an important consumer of cactus seeds, deposited saguaro seeds under mesquite canopies when feeding their chicks. To explore differential dispersal, Sosa (1997) studied cardón, organ pipe, and saguaro at a locality near Bahia Kino, Sonora, in the Gulf Coast zone of the Sonoran Desert. At this site, a diverse guild of facultative frugivores disperses and/or destroys columnar cactus seeds. Based on visitation frequency, treatment of seeds in their digestive tract, and deposition site, five species of birds and one lizard were the most important dispersers: verdin (*Auriparus flaviceps*), ash-throated flycatcher (*Myiarchus cinerascens*), curve-billed thrasher (*Toxostoma curvirostre*), cactus wren (*Campylorhyncus brunneicapillus*), Lloyd's bushtit (*Psaltriparus minimus*), and the desert iguana (*Dipsosaurus dorsalis*). Medium-sized mammals such as the coyote (*Canis latrans*), gray fox (*Urocyon cinereoargenteus*), and black-tailed jackrabbit (*Lepus californicus*), as well as the lesser long-nosed bat (*Leptonycteris curasoae*) also dispersed cactus seeds, but their efficiency was low because they often deposited seeds in unfavorable sites. Deposition microhabitats used by avian dispersers were largely areas under the crowns of trees or large shrubs (Table 15.2); other dispersers deposited seeds on open ground, on rocks, in ant middens, or in caves.

Seeds on the ground around fruiting cacti came from several sources. Some were scattered by birds and bats when pecking on, shaking, or pulling out fruits; some came from fruits after falling to the ground; and others were removed from fallen fruits by vertebrates. Most of these seeds were quickly removed (perhaps eaten) by bird and rodent granivores within a few hours or days after falling to the ground (see

TABLE 15.2

Percentage of Locations Where Common Frugivores Landed after Eating Columnar-Cactus Fruit Pulp and Seeds[a]

			Landing Site		
Bird Species	*Function[b]*	*N[c]*	*Non-Cactus Perenial*	*Other Columnar Cactus*	*Ground*
Zenaida macroura (white-winged dove)	P	28	14	79	7
Callipepla gambelli (Gambel's quail)	P	27	22	8	70
Melanerpes uropygialis (gila woodpecker)	D	18	0	100	0
Auriparus flaviceps (Verdin)	D	12	75	25	0
Myiarchus cinerascens (ash-throated flycatcher)	D	15	73	27	0
Others[d]	D, P	15	71	29	0

[a] Data from the Sonoran Desert near Bahia Kino, Sonora, Mexico.

[b] D = disperser, P = seed predator.

[c] *N* = sample size.

[d] Includes the dispersers *Campylorhynchus brunneicapillus* (cactus wren) and *Toxostoma curvirostre* (curved-bill thrasher), and the seed smasher *Carpodacus mexicanus* (house finch).

TABLE 15.3
Standing Seed Crop in the Soil after Fruiting of Columnar Cacti in Different Microhabitats[a]

	Number of Cactus Seeds					
Microhabitat	*Lophocereus schottii (Senita)*	*Carnegiea gigantea (Saguaro)*	*Pachycereus pringlei (Cardón)*	*Stenocereus thurberi (Organ Pipe)*	*Subtotal*	*Average*[b]
Base of cardón	46	6	12	6	70	44
Base of organ pipe	0	1	1	44	46	29
Base of saguaro	0	64	0	4	68	42
Under mesquite canopy	13	84	85	415	597	373
In open ground	2	15	0	1	18	11
Species subtotals	61	170	98	470	799	
Germination (% from species subtotals)	0	25	5	4		

[a] Data from a Sonoran Desert flatland at Bahía Kino. Soil samples taken to a depth of 3.5 cm. Cells are totals of 16 samples of 1,500 cc of soil collected in 1991 ($n = 10$) and in 1992 ($n = 6$) in each microhabitat.

[b] Average number of seeds per 1,500 cc of soil.

below). Seed density under the closest mesquite canopies averaged 4% of the seed density around saguaros and 32% of the seed density around cardóns. Conversely, seeds remaining after the fruiting peak of each species of cactus were 11 times more abundant in the soil under mesquite canopies than around parent cacti or in the open (Table 15.3). Seeds of the four species of columnar cacti present in the study area were found in these three microhabitats, but rarely in the open, in two years of study. Similarly, Hutto et al. (1986) reported that the abundance of saguaro seeds found in the open was only 3% of the abundance of seeds collected beneath palo verde canopies. Based on these studies, disproportionate seed dispersal to trees and shrubs suffices to explain, at least for species with no vegetative propagation, the association between columnar cacti and perennials.

Differential Seed Predation

Nonrandom seed and seedling predation may be another factor determining the association between juvenile columnar cacti and nurse plants (Niering et al., 1963; Turner et al., 1969). In North American deserts, rodents (predominantly heteromyids), birds, and ants are the main seed eaters (Reichman and Price, 1993). We know of only two experiments comparing cactus seed survival with predation in the open and beneath perennial canopies. In the Zapotitlán Valley of south-central Mexico, Valiente-Banuet and Ezcurra (1991) placed seeds of tetetzo (*Neobuxbaumia tetetzo*) in the open and in the shade of *Mimosa luisana* under two conditions: exposed to or protected from predators. They reported that a significantly higher number of seeds survived in the predator-excluded conditions, but found no difference in seed survivorship between the shaded and open microhabitats.

Similar results were reported for a series of experiments investigating differential removal of cardón, organ pipe, and saguaro seeds by predators at Bahia Kino (V. Sosa and A. Hernández, unpubl. data). Seed survival in three microhabitats (open space, mesquite shade, and the base of fruiting cacti) was either comparable or was higher in the open than in the other microhabitats, contrary to the hypothesis of lower seed predation under plant cover (Fig. 15.1). Tracking and live-trapping studies indicated that shaded microhabitats were used more frequently than open ground by granivorous vertebrates. However, in these experiments, seeds were placed in plots in which litter or pebbles had been removed to control for confounding factors and to facilitate recording of the tracks of seed predators. Previous workers have proposed that columnar cactus seeds can escape predation through the physical protection or concealment offered by debris, rocks, and other objects found under tree cover (Shreve, 1910; Niering et al., 1963; Turner et al., 1969; Steenbergh and Lowe, 1969, 1977). V. Sosa and A. Hernández (unpubl. data) explicitly tested this hypothesis by placing cardón seeds against two contrasting substrates (mesquite leaf litter and sand) in each of three microhabitats (in the open, at the base of fruiting cardóns, and under mesquite canopies). Cardón seeds supposedly should be more cryptic in the leaf litter than on the sand. Their results clearly supported this hypothesis. Almost all seeds were removed (only 5.2% were left in the open) when placed on the sandy substrate, compared with 19.4%, 11.8%, and 12.2% of the seeds left in the open, shaded, and cardón-base microhabitats, respectively, when placed on mesquite litter (Fig. 15.2). Under natural conditions, however, leaf litter is usually deposited only under tree or shrub canopies, which explains how seeds could escape predation under perennial canopies and which helps to account for the cactus–nurse-plant association.

DIFFERENTIAL SEEDLING PREDATION

Two studies conducted in the Sonoran Desert comparing predation on columnar cactus seedlings (defined here as cactus plants less than one year old) report no difference in seedling predation between open or shaded microhabitats. All saguaro seedlings transplanted in the open and under the canopy of a palo verde tree were gone after one year of exposure to predators at Saguaro National Monument, Arizona (Turner et al., 1969). Similarly, except for one individual of cardón, no seedlings of organ pipe, cardón, and saguaro exposed to predators (mainly jackrabbits, desert cotton-tails, and ground squirrels) survived one year after being transplanted into either the open or under mesquite canopies at Bahia Kino (Fig. 15.3; Sosa, 1997). The only cardón survivor was apparently protected by twigs that fell from the mesquite canopy after transplanting.

This observation led to another experiment with cardón seedlings in which three levels of protection against predators were provided under the canopies of mesquites: no protection, partial protection with mesquite twigs, and complete protection with mesh cages. After one year, percentages of surviving cardón seedlings were 3%, 58%, and 83%, respectively, which indicates that a protective layer of twigs can increase

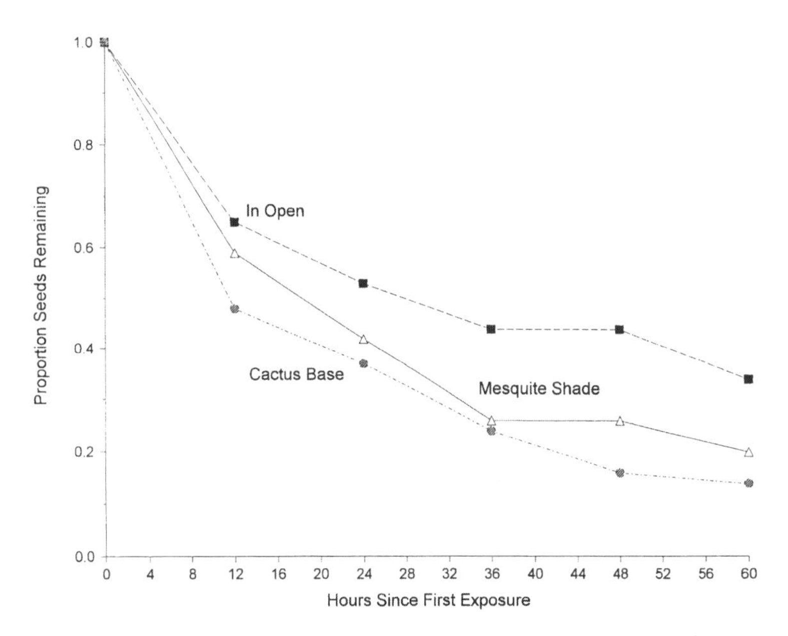

FIGURE 15.1. Comparison of the proportion of cardón (*Pachycereus pringlei*) seeds remaining after three days in different microhabitats. Data were lumped by seed density (40 or 120 seeds m^{-2}), because no significant effect of this treatment was found; $n = 30$ plots.

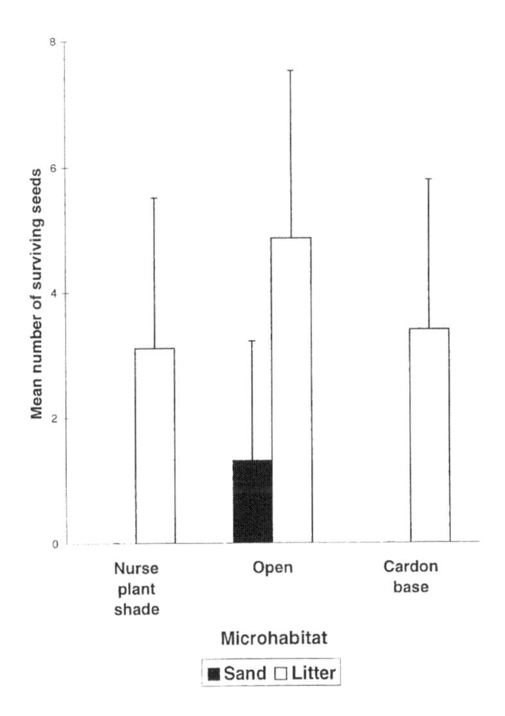

FIGURE 15.2. Survivorship of cardón seeds in three microhabitats differing in plant cover 7 km N of Bahía Kino, Sonora Mexico. Bars are 95% confidence intervals for mean proportions under two treatments of soil background: sand and litter. Bars show 95% confidence intervals around the mean; $n = 20$ plots containing 25 seeds.

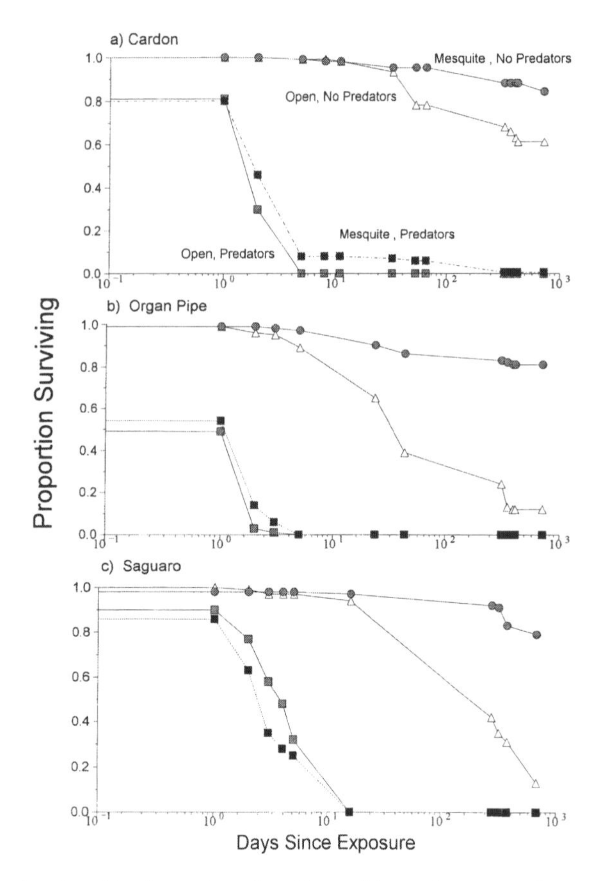

FIGURE 15.3. Survivorship of transplanted seedlings of three columnar cacti under two treatments of cover (beneath mesquite canopies or in the open) and two conditions of exposure to predators (exposed or not exposed to predators); (a) *Pachycereus pringlei* seedlings ($n = 130$ per treatment); (b) *Stenocereus thurberi* seedlings ($n = 190$ per treatment); (c) *Carnegiea gigantea* seedlings ($n = 120$ per treatment). Notice log-scale on the abscissa.

seedling survivorship nearly 20-fold (Fig. 15.4). Thus, in the Sonoran Desert, differential predation of seedlings further contributes to the association between columnar cacti and shrubs or trees. However, differential predation is not related to microhabitat use by predators but to the concealment and protection provided by debris produced by nurse plants.

In the Zapotitlán Valley, the effect of predation on tetetzo seedlings depended on type of shade and aspect of the hill where the experimental plots were set (Valiente-Banuet and Ezcurra, 1991). Survivorship of seedlings exposed to predators was significantly higher under artificial shade than in the open in north-, south-, and west-facing plots. In east-facing plots, seedling survivorship was higher beneath the canopy of *Mimosa luisana* than in the open. Thus differential seedling predation could also be contributing to the cactus-perennial association in the semiarid region of south-

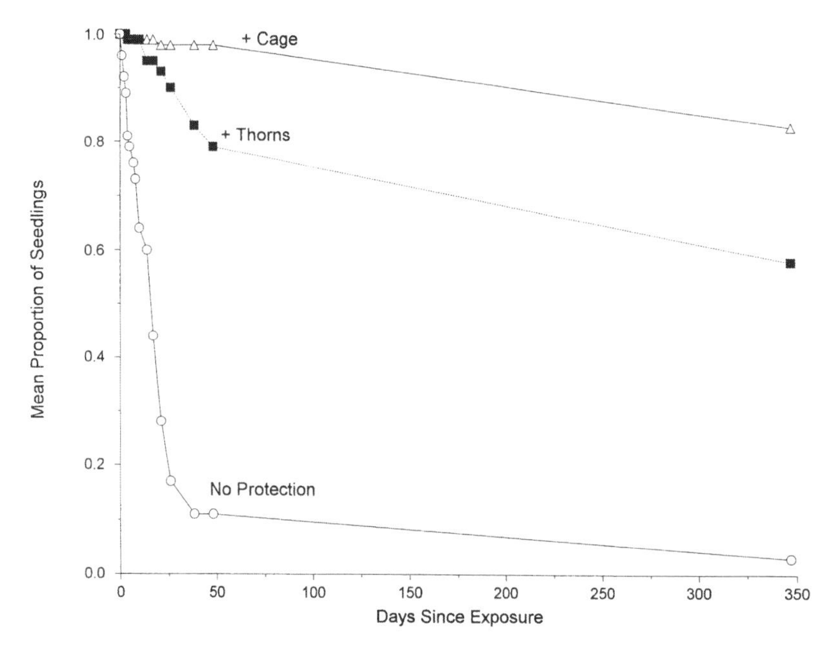

FIGURE 15.4. Survivorship of transplanted seedlings of *Pachycereus pringlei* exposed to different levels of protection against herbivores near Bahía Kino, Sonora. Treatments were: no protection = exposed to herbivores; thorns = partially protected from herbivores by a bed of thorny twigs; cage = protected from herbivores by a wire cage. Replicate (n = 16) plots of 10 seedlings per treatment. Seedlings were transplanted under mesquite canopies to ensure proper temperature conditions for seedling survival.

central Mexico. However, because predation explains only 3% of the variation in survivorship, shade (as a buffer against temperature) was considered to be the major cause of the association (see below).

Abiotic Environmental Effects of Perennial Canopies

The effects of tree and shrub canopies on the physical microenvironment of associated cacti have been the subject of most of the observational, computer simulation, and experimental studies of the cactus–nurse-plant association. Nurse plants may provide an environment of reduced light intensity, which would reduce the probability of seedling death due to heat stress and desiccation (Shreve, 1931; Turner et al., 1966; Lowe and Hinds, 1971; Despain, 1974; Nobel, 1988; Belsky et al., 1989; Valiente-Banuet and Ezcurra, 1991; Arriaga et al., 1993; Sosa, 1997). Nurse-plant foliage also buffers low temperature extremes, thus protecting seedlings from frost (Niering et al., 1963; Steenbergh and Lowe, 1976, 1977, 1983; Nobel, 1980; Franco and Nobel, 1989). Nutrient levels beneath plant canopies may be higher because of the accumulation of litterfall, animal droppings or feces, and of fine, windblown organic debris or because of nitrogen fixation (Turner et al., 1966; García-Moya and McKell, 1970; Tiedemann and Klemmedson, 1973; Charley and West, 1975; Felkler and Clark, 1981; Virginia and Jarrell, 1983). Finally, although not yet studied, nurse canopies might reduce wind speed and consequently soil erosion around their associated plants.

Evidence for increased establishment and survivorship of columnar cacti under nurse plants as a result of ameliorated abiotic conditions is definitive only for temperature buffering. In all experimental studies, seed germination and seedling survivorship of saguaro, cardón, organ pipe, and tetetzo columnar cacti have been higher under nurse-plant cover than in the open, when predators were excluded (see Fig. 15.3; Turner et al., 1966; Steenbergh and Lowe, 1969; Valiente-Banuet and Ezcurra, 1991; Sosa, 1997; V. Sosa and A. Hernández, unpubl. data). Dessication due to high summer temperatures or damage suffered from winter frosts has been used to explain high cactus mortality away from cover. In general for both agaves and cacti, the smallest seedlings are the most vulnerable, both in terms of temperature tolerance and temperatures experienced (Nobel, 1984, 1988). Thus one major indirect effect of shade is to reduce cactus seed and seedling mortality due to drought, sun scalding, and tissue freezing.

In contrast, characteristics of the soil under tree or shrub canopies seem to contribute to a lesser extent or to have no effect on seedling establishment and growth. Saguaro seeds survived better in soil collected under palo verde trees and from the open than in soils from under either mesquite or ironwood; however, seedlings grown under mesquite and ironwood trees (with significantly higher soil nutrient content) reached larger sizes than in the other treatments; albedo apparently contributed to such confounding results (Turner et al., 1966). In experiments on survivorship of seedlings of cardón, organ pipe, and saguaro near Bahia Kino, Sosa (1997) found that levels of nitrogen and organic matter were significantly higher under mesquite canopies than in the open. However, the lack of reciprocal treatments of seedlings transplanted to soil collected from the open—both shaded and unshaded—precluded any definitive conclusion about the effect of soil fertility on seedling survivorship. In the Tehuacán Valley, nitrogen concentration was slightly higher ($p = 0.06$) in soils shaded by *Mimosa luisana*—the main nurse plant of tetetzo—than in the open; it was significantly lower in soil collected under other nurse species than in the open (Valiente-Banuet et al., 1991a,b). Experiments on seedling growth yielded contrasting results: *Neobuxbaumia tetetzo* seedlings grew taller in open-space soil than *Pachycereus hollianus* seedlings, which grew taller in nurse-plant soil (Godínez-Alvarez and Valiente-Banuet, 1998). In a tropical deciduous forest, no differences in nitrogen levels and other fertility-related variables were found between soils from the open and soils shaded by plants associated with organ pipe (Arriaga et al., 1993). Recent work suggests that fertility islands might not be restricted to live perennial plants in arid or semiarid ecosystems (Barnes and Archer, 1996).

The reasons why studies of the role of soil in the nurse-plant syndrome are inconclusive include inappropriate experimental design as well as an interaction between soil and shade that results in greatest recruitment of columnar cacti under nurse plants. More factorial experiments that include as treatments artificial shade over soils both from open sites and below nurse-plant canopies are needed to clarify the effects of soil on cactus seedling performance (see, e.g., Godínez-Alvarez and Valiente-Banuet, 1998).

Levels of photosynthetically active radiation (PAR) are lower under nurse plants than in the open (Nobel, 1988; Franco and Nobel, 1989; Valiente-Banuet and Ezcurra, 1991; Sosa, 1997). Both competition for water with nurse plants and PAR limitation reduces predicted growth of *Carnegiea gigantea* and other succulent seedlings under nurse-plant canopies (Franco and Nobel, 1988, 1989, 1990). However, this cost does not necessarily result in increased seedling mortality, because it is offset by the beneficial effects of thermal buffering (which reduces drought- or freezing-related mortality) and by the increased nitrogen content of most shaded soils.

In summary, although several mechanisms are involved in the origin and maintenance of the columnar cactus–nurse-plant syndrome, most studies indicate that the main influences of nurse plants on columnar-cactus establishment and survival are biotic (via directed dispersal and protection from herbivory), rather than abiotic (via protection from desiccation or frost and increased levels of soil nitrogen). This is similar to the explanation of the association of three species of chollas (subgenus *Cylindropuntia*) with their nurse plants in the Mojave Desert (Cody, 1993). However, the importance of each mechanism may vary with climate, the abundance of predators, and the amount of clumping of dispersed seeds. For example, in the Sonoran Desert near Bahia Kino, seed clumping beneath mesquite canopies is the combined result of directed dispersal of seeds to nurse plants and escape from predation. At Bahia Kino and in Arizona, escape from strong predation on columnar cactus seeds and seedlings probably contributes more to the nurse-plant syndrome than does any beneficial effect of shade. In the dense forests of columnar cacti in south-central Mexico, however, seed dispersal is nearly uniform, and abiotic mechanisms (e.g., shade) appear to be the main cause of the cactus–nurse-plant association (Valiente-Banuet and Ezcurra, 1991).

The importance of shade as a temperature buffer for young columnar cacti may depend on the climate prevailing in a specific region. In regions where extreme high temperatures are below the desiccation threshold or extreme low temperatures are above the freezing threshold of columnar cacti, cactus establishment could occur in the open, provided that seed and seedling mortality due to predators is not excessive. However, buffering against high temperatures during the summer may be an important effect of shade in the Bahia Kino region, which is one of the hottest parts of the Sonoran Desert. Similarly, buffering against low temperatures may be the crucial benefit provided by shade at the northernmost limits of cacti, where frost risk is higher than at lower latitudes (Nobel 1980, 1988; Cody, 1984). Additionally, topography (slope and orientation of slope aspect), soil characteristics such as albedo and rockiness, density and structure of vegetation, and the tolerance of each cactus species to extreme environmental conditions are surely interacting with climate to determine the importance of nurse canopies as temperature buffers (Steenbergh and Lowe, 1977).

It is important to note that although the three mechanisms reviewed here might explain the recruitment of columnar cacti under potential nurse-plant canopies, the presence of young columnar cacti growing away from such cover suggests that other mechanisms are also likely to be involved in cactus recruitment. The possibility exists, for example, that rock crevices or outcroppings and animal and plant remains can offer seeds and seedlings the same kinds of protection and mitigation of environmental stress as nurse plants. Nonbasaltic rocks can locally enhance water availability through runoff and temporary puddles (M. L. Cody, pers. comm.). One of these processes may explain the occurrence of isolated young individuals of organ pipe, the most common columnar cactus at Bahia Kino (V. Sosa and A. Contreras, unpubl. data). Crevices in rocks and hollows among rocks, however, might offer poor edaphic or light conditions for the establishment of seedlings.

A more general hypothesis, currently awaiting careful examination, is that columnar cacti establish in the shade of another plant or any object (e.g., rock, plant remains, bones) that offers shade, moisture, and protection from predators to seeds and seedlings. If this hypothesis proves to be true, the cactus–nurse-plant syndrome should be extended to a "cactus–nurse-object" syndrome.

Promising Lines of Research

After reviewing evidence for the main mechanisms that have been proposed to explain the cactus–nurse-plant association, it appears to us that more than one mechanism or perhaps all three of those discussed above are involved in the association. Thus, one important goal in the future is to determine which mechanism, under what conditions, and at what life stage of which species is most important for cactus fitness. Sensitivity analysis will allow us to better understand cactus demography and to recommend effective measures for the conservation of columnar cacti. Experimental long-term studies that compare the effect of factors such as grazers, damage to the canopies of nurse plants, and abundance of nurse plants, on cactus recruitment will certainly provide important insights into the management of the cactus–nurse-plant interaction.

Other aspects of the cactus–nurse-plant association requiring further study include seed dispersal patterns, quality of different species as nurse plants, and cost/benefit analysis of the association in different arid or semiarid habitats. Seed dispersal due to abiotic factors has barely been investigated, yet it can account for the nurse-plant syndrome. For example, the association of cholla cacti with some nurse-plant species is initiated by the trapping of windblown fruits of chollas by the dense vegetation of those nurse species (Cody, 1993). Determination of disperser guilds, dispersal quality, seed rain, and soil-seed banks created by dispersers or modified by predators and sheet-water flow is lacking for most species of columnar cacti. We currently know little about the longevity of cactus seeds in soil and whether they have delayed germination. We also need comparative studies of nurse-plant quality as perch or roost sites for frugivorous birds or bats, as producers of concealing/protective litter, and as modifiers of the microenvironment under their canopies and around their rhizospheres. There is evidence that some nurse species, such as *Olneya tesota* (ironwood), are not only important for the recruitment of columnar cacti but are key species for the dynamics of plant communities in arid environments (Búrquez and Quintana, 1994; Suzán et al., 1996). Some species may also be involved in reciprocal nursing (sensu Cody, 1993), for example, *Lophocereus schottii* and *Olneya tesota.*

Finally, research on the competitive or inhibitory outcomes of the cactus–nurse-plant interaction seems promising. Because columnar cacti are long-lived plants, they can often outlive their nurse plants (McAuliffe, 1988). Still, during the time that columnar cacti grow together with their nurse plants, both may be subject to competition, usually for water (Vandermeer, 1980; McAuliffe, 1984a; Franco and Nobel, 1988, 1989; Flores-Martínez et al., 1994). Degree of overlap between root systems, soil type (depth and water potential), and amount and temporal distribution of rainfall may determine the intensity of this competition (Franco and Nobel, 1990). We need more studies to determine whether the replacement of nurse plants by their associated cacti is the result of competition or simply reflects intrinsically different lifespans. Linking detailed measurements of root growth, soil water potential, and nutrient and water uptake of both nurse and associated plants to field experiments on cactus recruitment will hopefully answer this question. Whenever possible, studies should be performed throughout the geographical range of a cactus species to identify general patterns.

Conclusions

We shall now succinctly answer the question posed in our chapter title. Columnar cacti are associated with nurse plants because of three mechanisms, whose relative importance is related to (1) the life-cycle stage of the cactus and (2) variation among species, sites, and local abiotic and biotic conditions. First, trees and shrubs function as perches for bird dispersers, resulting in light but crucial seed rain under their canopies. Second, because of the accumulation of litter beneath their canopies, nurse plants offer concealment and protection from predators to seeds and seedlings. Finally, nurse plants provide a favorable microenvironment for seedling performance by mitigating extreme temperatures and perhaps by contributing to relatively fertile soil in their shade.

Summary

We review the generality of the nurse-plant–columnar-cactus association (a columnar cactus growing under the canopy of a tree or shrub) and the mechanisms that explain it. Each of the nine species (out of 70) of columnar cacti investigated have been reported associated with nurse plants at some stage of its life cycle. This association is not constant across landforms or localities; it is apparently due to topographically related traits, surrounding vegetation, and the nursing quality of plants. We found evidence for three mechanisms that in sequence or simultaneously explain the association. First, seed dispersal by birds directed to tree and shrubs concentrates a small but crucial number of seeds under their canopies. Second, some of these seeds and seedlings escape predation while concealed by leaf and twig litter dropped by the nurse plant. Finally, surviving seedlings benefit from temperature buffering provided by the shade under nurse plants. Currently, most studies indicate that the main benefit that columnar cacti obtain from nurse plants is protection against predators; however, the importance of each mechanism may vary with the climate, predator abundance, and amount of clumping of dispersed seeds. Finally, we propose some lines of research that would allow us to better understand the nurse-plant–cactus interaction, such as comparing seed rain among different species of nurse plants and competition between cacti and their nurse plants.

Resumen

Revisamos la generalidad de la asociación cacto columnar-planta nodriza y los mecanismos que la producen y mantienen. Consideramos como planta nodriza, cualquier planta leñosa bajo cuya copa se desarrolle un cacto columnar. Nueve especies investigadas de cactos columnares (de las setenta existentes) se han registrado asociadas a plantas nodrizas. Sin embargo, la asociación no es constante para cada especie ni entre geoformas, ni entre localidades. Esto es debido aparentemente, a características asociadas a la topografía, vegetación circundante y calidad de la planta como nodriza. Encontramos que la asociación es causada, por al menos tres mecanismos que actúan en secuencia, ó simultáneamente. Primero, la dispersión de semillas hacia los árboles y arbustos por aves y otros grupos de vertebrados que concentran una cantidad mínima, pero crucial de semillas bajo sus copas. Después algunas de estas semillas y plántulas que se establecen escapan a los depredadores al quedar ocultas en la hojarasca de la planta nodriza o entre las ramillas caídas. Finalmente, las plántulas sobrevivientes se benefician del amortiguamiento termal de la sombra producida, por la nodriza. Hasta ahora la mayoría de los estudios indican que el principal beneficio que reciben los cactos columnares de la planta nodriza es el escape a la depredación sin embargo, la importancia de cada mecanismo podría variar con el clima, la abundancia de depredadores y la agregación de las semillas dispersadas. Para finalizar, proponemos algunas líneas de investigación que permitirían entender mejor la ecología de la asociación cacto-nodriza, tales como: comparación de lluvia de semillas bajo diferentes especies de nodrizas y la competencia entre cacto y nodriza.

ACKNOWLEDGMENTS

We thank M. L. Cody, P. S. Nobel, and A. Valiente-Banuet for their valuable comments on the manuscript. Funding was provided by the Instituto de Ecología, A.C. (account 902-19), the National Geographic Society, and the National Science Foundation.

REFERENCES

Alcorn, S. M., S. E. McGregor, and G. Olin. 1961. Pollination of saguaro cactus by doves, nectar-feeding bats, and honey bees. *Science* 133:1594–95.

Archer, S., C. Scifres, C. R. Bassham, and R. Maggio. 1988. Autogenic succession in a subtropical savanna: conversion of grassland to thorn woodland. *Ecological Monographs* 58:11–127.

Arriaga, L., Y. Maya, S. Díaz, and J. Cancino. 1993. Association between cacti and nurse perennials in a heterogeneous tropical dry forest in northwestern Mexico. *Journal of Vegetation Science* 4:349–56.

Barnes, P. W., and S. Archer. 1996. Influence of an overstory tree (*Prosopis glandulosa*) on associated shrubs in a savanna parkland: implications for patch dynamics. *Oecologia* 105:493–500.

Belsky, A. J., R. G. Amundson, J. M. Duxbury, S. J. Riha, A. R. Ali, and S. M. Mwonga. 1989. The effects of trees on their physical, chemical, and biological environments in a semi-arid savanna in Kenya. *Journal of Applied Ecology* 26:1005–24.

Búrquez, A., and M. A. Quintana. 1994. Islands of diversity: ironwood ecology and the richness of perennials in a Sonoran Desert biological reserve. In *Ironwood: an Ecological and Cultural Keystone of the Sonoran Desert,* eds. G. P. Nabhan and J. L. Carr, 9–27. Occasional Papers in Conservation Biology No. 1. Washington, D.C.: Conservation International.

Charley, J. L., and N. E. West. 1975. Plant-induced soil chemical patterns in some shrub-dominated semi-desert ecosystems of Utah. *Journal of Ecology* 63:945–63.

Cody, M. L. 1984. Branching patterns in columnar cacti. In *Being Alive on Land,* eds. N. Margaris, M. Arianoutsou-Faraggitaki, and W. C. Oechel, 201–36. The Hague: Junk.

———. 1993. Do cholla cacti (*Opuntia* spp., Subgenus *Cylindropuntia*) use or need nurse plants in the Mojave Desert? *Journal of Arid Environments* 24:139–54.

Despain, D. G. 1974. The survival of saguaro (*Carnegiea gigantea*) seedlings on soils of differing albedo and cover. *Journal of Arizona Academy of Science* 9:102–7.

Felkler, P., and P. R. Clark. 1981. Nodulation and nitrogen fixation (acetylene reduction) in desert ironwood. *Oecologia* 48:292–93.

Fleming, T. H., M. D. Tuttle, and M. A. Horner. 1996. Pollination biology and the relative importance of nocturnal and diurnal pollinators in three species of Sonoran desert columnar cacti. *Southwestern Naturalist* 41:257–69.

Flores-Martínez, A., E. Ezcurra, and S. Sánchez-Colón. 1994. Effect of *Neobuxbaumia tetetzo* on growth and fecundity of its nurse plant *Mimosa luisana. Journal of Ecology* 82:325–30.

Franco, A. C., and P. S. Nobel. 1988. Interactions between seedlings of *Agave deserti* and the nurse plant *Hilaria rigida. Ecology* 69:1731–40.

———. 1989. Effect of nurse plants on the microhabitat and growth of cacti. *Journal of Ecology* 77:870–86.

———. 1990. Influences of root distribution and growth on predicted water uptake and interspecific competition. *Oecologia* 82:151–57.

García-Moya, E., and C. M. McKell. 1970. Contribution of shrubs to the nitrogen economy of a desert-wash plant community. *Ecology* 51:81–88.

Gibson, A. C. 1982. Phylogenetic relationships of Pachycereeae. In *Ecological Genetics and Evolution. The Cactus-Yeast-Drosophila Model Ecosystem,* eds. J.S.F. Baker and W. T. Starmer, 3–16. Sydney: Academic Press.

Godínez-Alvarez, H., and A. Valiente-Banuet. 1998. Germination and early seedling growth of Tehuacán Valley cacti species: the role of soils and seed ingestion by dispersers on seedling growth. *Journal of Arid Environments* 39:21–31.

Hutto, R. L., J. R. McAuliffe, and L. Hogan. 1986. Distributional associates of the saguaro (*Carnegiea gigantea*). *Southwestern Naturalist* 31:469–76.

Jordan, P. W., and P. S. Nobel. 1981. Seedling establishment of *Ferocactus acanthodes* in relation to drought. *Ecology* 62:901–6.

Lowe, C. H., and D. S. Hinds. 1971. Effect of paloverde (*Cercidium*) trees on the radiation flux at ground level in the Sonoran Desert in winter. *Ecology* 52:916–22.

McAuliffe, J. R. 1984a. Sahuaro-nurse tree associations in the Sonoran Desert: competitive effects of sahuaros. *Oecologia* 64:319–21.

———. 1984b. Prey refugia and the distributions of two Sonoran Desert cacti. *Oecologia* 65:82–85.

———. 1988. Markovian dynamics of simple and complex desert plant communities. *American Naturalist* 131:459–90.

Niering, W. A., R. H. Whittaker, and C. H. Lowe. 1963. The saguaro: a population in relation to environment. *Science* 142:15–23.

Milton, S. J., W.R.J. Dean, G.I.H. Kerley, M. T. Hoffman, and W. G. Whitford. 1998. Dispersal of seeds as nest material by the cactus wren. *Southwestern Naturalist* 43:449–52.

Nobel, P. S. 1980. Morphology, surface temperatures, and northern limits of columnar cacti in the Sonoran Desert. *Ecology* 61:1–7.

———. 1984. Extreme temperatures and thermal tolerances for seedlings of desert succulents. *Oecologia* 62:310–7.

———. 1988. *Environmental Biology of Agaves and Cacti.* Cambridge: Cambridge University Press.

Olin, G., S. M. Alcorn, and J. M. Alcorn. 1989. Dispersal of viable saguaro seeds by white-winged doves (*Zenaida asiatica*). *Southwestern Naturalist* 34:281–84.

Reichman, O. J., and M. V. Price. 1993. Ecological aspects of heteromyid foraging. In *Biology of the Heteromyidae,* eds. H. H. Genoways and J. H. Brown, 539–73. Special Publication No. 10. American Society of Mammalogists.

Shreve, F. 1910. The rate of establishment of the giant cactus. *Plant World* 13:235–40.

———. 1917. The establishment of desert perennials. *Journal of Ecology* 5:210–16.

———. 1931. Physical conditions in sun and shade. *Ecology* 12:96–104.

Silvertown, J., and J. B. Wilson. 1994. Community structure in a desert perennial community. *Ecology* 75:409–17.

Sosa, V. J. 1997. Dispersal and recruitment ecology of columnar cacti in the Sonoran Desert. Ph.D. diss., University of Miami, Coral Gables, Florida.

Steenbergh, W. F., and C. H. Lowe. 1969. Critical factors during the first years of life of the saguaro (*Cereus giganteus*) at Saguaro National Monument, Arizona. *Ecology* 50:825–34.

———. 1976. *Ecology of the Saguaro I. The Role of Freezing Weather in a Warm-Desert Plant Population.* National Park Service Science Monographs Series No. 1. Washington, D.C.: National Park Service.

———. 1977. *Ecology of the Saguaro: II. Reproduction, Germination, Establishment, Growth, and Survival of the Young Plant.* National Park Service Science Monographs Series No. 8. Washington, D.C.: National Park Service.

———. 1983. *Ecology of the Saguaro: III. Growth and Demography.* National Park Service Science Monographs Series No. 17. Washington, D.C.: National Park Service.

Suzán, H., G. P. Nabhan, and D. T. Patten. 1996. The importance of *Olneya tesota* as a nurse plant in the Sonoran Desert. *Journal of Vegetation Science* 7:635–44.

Tiedemann, A. R., and J. O. Klemmedson. 1973. Nutrient availability in desert grassland soils under Mesquite (*Prosopis juliflora*) trees and adjacent open areas. *Proceedings of the Soil Science Society of America* 37:107–11.

Turner, R. M., S. M. Alcorn, G. Olin, and J. A. Booth. 1966. The influence of shade, soil and water on saguaro seedling establishment. *Botanical Gazette* 127:95–102.

Turner, R. M., S. M. Alcorn, and G. Olin. 1969. Mortality of transplanted saguaro seedlings. *Ecology* 50:835–44.

Valiente-Banuet, A., and E. Ezcurra. 1991. Shade as a cause of the association between the cactus *Neobuxbaumia tetetzo* and the nurse plant *Mimosa luisana* in the Tehuacán Valley, Mexico. *Journal of Ecology* 79:961–71.

Valiente-Banuet, A., A. Bolongaro-Crevenna, O. Briones, E. Ezcurra, M. Rosas, H. Nuñez, and G. Barnard. 1991a. Spatial relationships between cacti and nurse shrubs in a semi-arid environment in central Mexico. *Journal of Vegetation Science* 2:15–20.

Valiente-Banuet, A., F. Vite, and J. A. Zavala-Hurtado. 1991b. Interaction between the cactus *Neobuxbaumia tetetzo* and the nurse shrub *Mimosa luisana. Journal of Vegetation Science* 2:11–14.

Vandermeer, J. H. 1980. Saguaros and nurse trees: a new hypothesis to account for population fluctuations. *Southwestern Naturalist* 25:357–60.

Virginia, R. A., and W. M. Jarrell. 1983. Soil properties in a mesquite dominated Sonoran Desert ecosystem. *Soil Science Society of America Journal* 47:138–44.

Yeaton, R. I. 1978. A cyclical relationship between *Larrea tridentata* and *Opuntia leptocaulis* in the northern Chihuahuan Desert. *Journal of Ecology* 66:651–56.

Yeaton, R. I., and A. Romero-Manzanares. 1986. Organization of vegetation mosaics in the *Acacia schaffneri-Opuntia streptacantha* associations, Southern Chihuahuan desert, Mexico. *Journal of Ecology* 74:211–17.

CHAPTER 16

Cacti in the Dry Formations of Colombia

Adriana Ruiz
Jaime Cavelier
Mery Santos
Pascual J. Soriano

Introduction

Located in the northwestern corner of South America, Colombia has a variety of vegetation types, including lowland wet and moist forest, montane rainforest, paramos and savanna as well as dry formations such as dry deciduous forest, thorn woodland and desert shrub (Etter, 1993). These dry formations are located in the lowlands along the Caribbean coast, within inter-Andean valleys below 1,000 m, and in the highlands of the Eastern Andes at altitudes of about 2,500 m. The flora of the southern Caribbean dry zone and inter-Andean dry valleys shows greater floristic affinities with the dry deciduous forests that extend from southern Mexico to Peru than with the other three genetic stocks from which the dry flora of South America is derived. These three stocks are: the Caatingas of northeast Brazil, the Chaco, forests and the shrublands and forests of Mediterranean climate of central and southern Chile (Sarmiento, 1975). The dry formations in Colombia were at their largest during the Miocene; since then, they have undergone expansions and contractions as a result of the climatic changes associated with glaciations. During the Pleistocene (ca. 13,000 years ago), the inter-Andean dry valleys were much larger and joined with the Caribbean dry zone. Today they are isolated and greatly reduced in area, representing the Holocene refuges of flora and fauna of dry climates (Hernández et al., 1992). Dry formations also occur in other regions of tropical South America, including the inter-Andean dry valleys of Venezuela and central Peru, along the dry Pacific coast of Ecuador and Peru, on the western flank of the Peruvian Andes, and in the Puna of Bolivia and Peru (Sarmiento, 1975).

Some of the most conspicuous plants in these dry formations are Cactaceae of different growth forms, including columnar, barrel, decumbent, epiphytic, and cladodes (Marshall and Bock, 1941). Although most Cactaceae are very conspicuous (except for understory species of the genus *Mammillaria*), we currently lack a complete account of the species present in Colombia, and little has been done to update the taxonomic status of the species apparently present in the country. The literature on Colombian cacti ranges from lists of species presumed to occur in the country (e.g., Croizat, 1944) to regional (e.g., Rieger, 1976) and local vegetation surveys in which cacti are present (e.g., Schnetter, 1968). The CITES list of Colombian cacti includes 38 species, of which 11 are epiphytes that occur in wet formations. Nevertheless, this list is of limited use because there are no descriptions, references to herbarium specimens, or bibliographies for any of the species (Hunt, 1992). In a review of the xerophytic and subxerophytic formations of Colombia (Hernández et al., 1995), several species of cacti are reported that require further taxonomic work. Thus the biology and ecology of Colombian cacti are almost unknown, with the exception of recent work on phenology (Santos, 1995; Ruiz et al., 2000) and pollination and dispersal of columnar cacti in the upper valley of Río Magdalena (Ruiz et al., 1997).

In this chapter, we summarize taxonomic and ecological information available on the cacti present in the dry formations of Colombia. We base this review on (1) our own collections and work with cacti of the ten dry inter-Andean valleys; (2) on the study of specimens deposited in the Herbario Nacional Colombiano (COL), Universidad del Valle and Universidad de Antioquia; and (3) on critical analysis of the literature and databases (e.g., Tropicos at www.mobot.org). This review should be considered preliminary and subject to change when more collections become available.

The Distribution of Dry Formations

Lowland dry formations are located in the southern Caribbean dry zone and in the inter-Andean valleys. In the Caribbean, they are located in the upper and middle Guajira Peninsula (Rieger, 1976; Sugden, 1982), in the lower part of the northwest flank of the Sierra Nevada de Santa Marta (Cleef, 1984; Lozano, 1984), and along a narrow band on the coast of the Departments of Atlántico, Bolivar, Córdoba, and Sucre (Hernández et al., 1995). There are also extensive dry forests and woodlands in the Departments of Bolivar, Cesar, Córdoba, and Sucre. The Caribbean areas and inter-Andean valleys where dry formations occur are summarized in Table 16.1 and Fig. 16.1.

Highland dry formations are located in the Eastern Cordillera of the Andes, especially southwest of Bogotá (Soacha, Fute, and Laguna de La Herrera), in the southern part of Ubaté Valley, Boyacá (Ramiriquí, Paipa, "La Candelaria" at Villa de Leyva), and along Río Negro in Cáqueza (Cundinamarca). There are also dry formations in the southern Andes around the towns of Ipiales and Pasto (Table 16.1, Fig. 16.1). The canyons of Río Chicamocha and La Tatacoa are included in the lowland formations but have a great altitudinal gradient, extending above 2,000 m.

TABLE 16.1

Dry Formations of Colombia[a]

Formation	*Vegetation Type*[b]
Lowland dry formations	sensu Holdridge
Caribbean coast[c]	
1. Upper and middle Guajira Península	ST-s, ST-t
2. Santa Marta region	ST-t,
3. Caribbean coast of Atlántico, Bolivar and Sucre	T-vd
4. Dry forests and woodlands	T-d
Inter-Andean dry valleys	
5. Canyon of Cauca Río (Santafé de Antioquia)	T-d
6. Upper valley of Sucio Río (Dabeiba)	T-d
7. Upper valley of Cauca Río (Yumbo and Anaime Río)	TP-d
8. Canyon of Dagua Río	T-vd, T-d
9. Canyon of Patía Río and Juanambú	T-vd, TP-d
10. Aguachica	T-d
11. Ocaña	TP-d
12. Cúcuta	T-vd, T-d, TP-d
13. Canyon of Chicamocha River	TP-t, T-vd, bs-P
14. Upper valley of the Magdalena River (Colombia and Tatacoa)	ST-t, T-D, TP-d
Upland dry formations	
15. Southwest of Bogotá (Fute, Soacha and Lag. La Herera)	TP-d
16. Ubaté Valley (Checua region)	TP-d
17. Villa de Leyva	TP-d
18. Negro Río (Cáqueza)	TP-d
19. Ipiales and Pasto	TP-d, TLM-d

[a] Based on Espinal and Montenegro (1977); Hernández et al. (1995); and Cavelier et al. (1996). The dry pocket of "Abrego" (sensu Hernández et al. [1995]) is not included, because it is a tropical premontane moist forest. The dry pockets of southwest Bogotá, Ubaté Valley, and Villa de Leyva sensu Hernández et al. (1995), are all part of a larger area of tropical lower Montane dry Forest that runs from Bogotá to Chicamocha but have a distinct flora (i.e., cacti), probably as a result of local drier conditions and/or edaphic factors.

[b] st-s = subtropical desert shrub, ST-t = subtropical thorn woodland, T-d = tropical dry forest, T-t = tropical thorn woodland, T-vd = tropical very dry forest, TLM-d = tropical lower montane dry forest, TP-d = tropical premontane dry forest, TP-df = tropical premontane dry forest, TP-t = tropical premontane thorn woodland.

[c] Numbers correspond to those in Fig. 16.1.

Physical Environment

Along the Caribbean coast of South America, between 10° and 12° north, there is an area where rainfall is significantly lower than in other locations at the same latitude around the world. This southern Caribbean dry zone extends between Cartagena, Colombia, and Margarita Island in Venezuela (Johnston, 1909), including the peninsulas of Guajira (Rieger, 1976) and Paraguaná (Tamayo, 1941, 1967). This zone is characterized by a mean annual rainfall of less than 1,000 mm (Lahey, 1958), a marked seasonal rainfall with a single peak in the second half of the year, and constant northeastern trade winds (Snow, 1976). This dry climate has occurred in this area for at least the past 20,000 years (Ochsenius, 1981). The dry formations of the inter-Andean valleys are defined as those areas below 1,000 m and with mean annual rainfall less than 2,000 mm. These areas include tropical dry forest (1,000–2,000

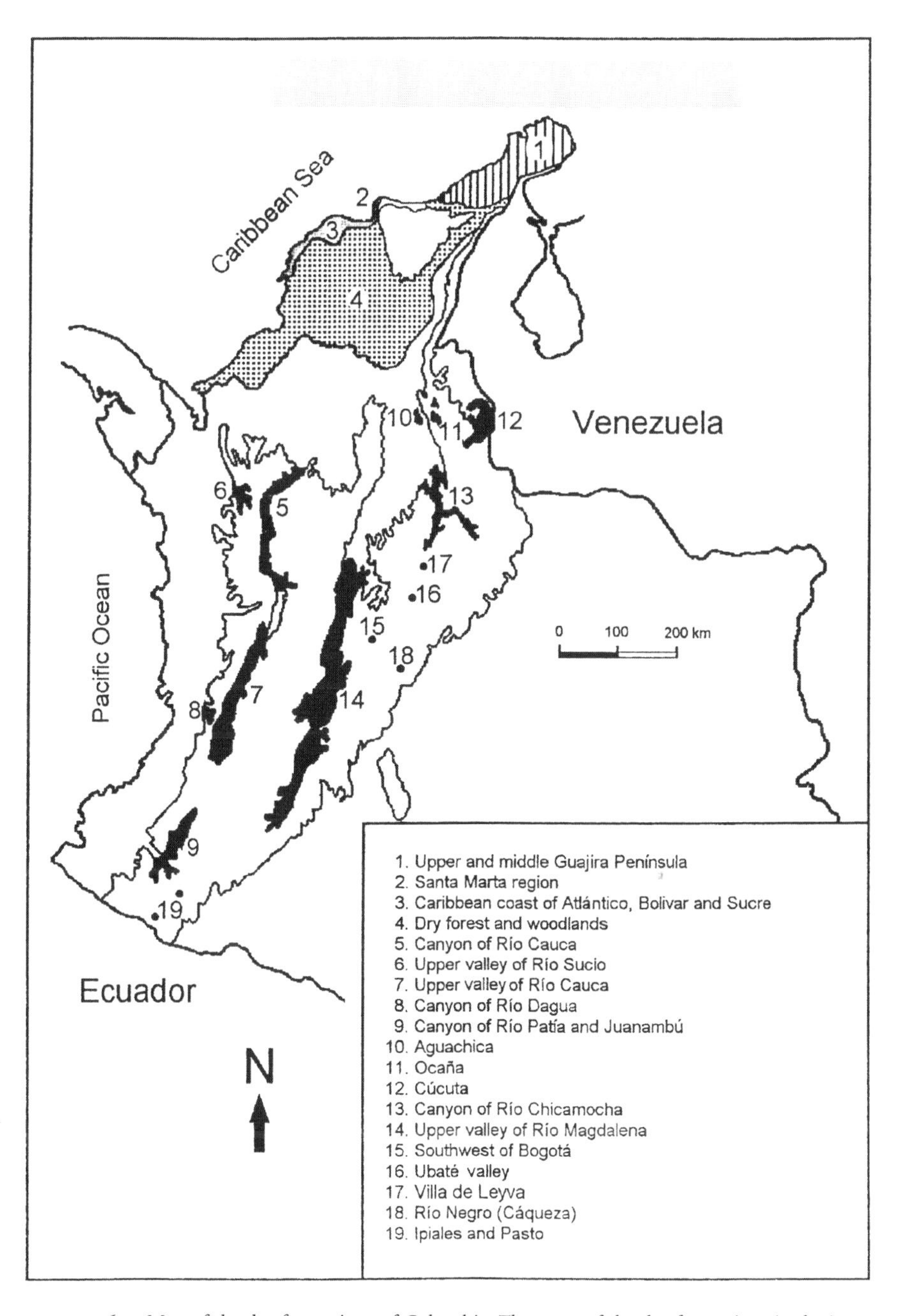

FIGURE 16.1. Map of the dry formations of Colombia. The areas of the dry formations in the inter-Andean valleys are somewhat larger than in Fig. 12.1 of this volume because in that figure only the areas where columnar cacti occur are shown. In chapter 12, the dry forest and woodlands of the Caribbean coast are not represented as they are in the present figure.

mm/year), tropical very dry forest (500–1,000 mm), tropical thorn woodland (250–500 mm), and tropical desert scrub (125–250 mm) sensu Holdridge (Espinal and Montenegro, 1977).

The highland dry formations are located on the plateau of the Departments of Cundinamarca and Boyacá (ca. 2,500 m), with mean annual temperatures around 13°C. Mean annual rainfall varies between 500 mm and 1,000 mm as in the case of Sutamarchán (658 mm), La Candelaria (760 mm), Tinjacá (830 mm), and Villa de Leyva (1,050 mm) (Hernández et al., 1995). These dry formations grow on shallow soils (10–15 cm deep), which frequently have a clay pan horizon, low organic matter, and are easily eroded (Hernández et al., 1995; van der Hammen, 1996).

Physiognomy and Floristics of the Dry Formations

Physiognomy

The lowland dry (seasonal) formations in tropical America were first described by Beard (1944, 1955) using the names "deciduous seasonal forest," "thorn woodland," "cactus shrub," and "desert" (Fig. 16.2). In this review, we consider mainly the drier formations, where columnar cacti occur, including some deciduous seasonal forests (Cuatrecasas, 1958; Dugand, 1973). The thorn woodland, cactus shrub, and desert of Beard's classification system correspond to the thorn woodland, desert scrub, and desert, respectively, sensu Holdridge (Espinal and Montenegro, 1977). Thorn woodlands are dominated by shrubs with a mean height of 3–10 m. These shrubs are mostly evergreen with small leaves (microphylls), some of them pubescent. Spines are also present in some species. These shrubs are mixed with cacti of different growth forms, including columnar, cladodes, barrel, epiphytic, and decumbent. Along a rainfall gradient, these thorn woodlands are in contact with dry forests (sensu lato), where cacti co-occur with seasonal deciduous and evergreen trees. Cactus shrubs are much more open than thorn woodlands and are dominated by columnar and cladode cacti 3–5 m tall. There are a few small shrubs with smaller leaves (microphylls, notophylls, and nanophylls sensu Raunkaier, 1934). Deserts are even more open than cactus shrubs. They occupy drier areas and are dominated by shrubs and cacti less than 1 m in height. These are not true deserts when compared with those in the temperate zone (e.g., Atacama in South America and Mohave in North America) because rainfall (ca. 1,000 mm/year) is much higher.

The upland dry formations of Colombia occur in the drier parts of the plateau of Cundinamarca and Boyacá, where, instead of an upper montane rainforest (sensu Grubb, 1977), the vegetation is 2–3 m in height and contains microphyllous shrubs and succulent plants such as Cactaceae (i.e., *Opuntia* spp. and *Mammillaria* spp.) and Agavaceae (*Furcraea* sp. and *Agave* sp.). In the area of Mondoñedo-La Herrera, on the southwest side of the Bogotá plateau, the short vegetation is dominated by *Opuntia* cacti and the shrub *Dodonea* sp. (Vink and Wijninga, 1987; van der Hammen, 1996).

FIGURE 16.2. Photographs of the dry formations of Colombia. (a) Deciduous seasonal forest with *Acanthocereus tetragonus;* (b) thorn woodland with *Stenocereus griseus;* (c) cactus shrub with *Stenocereus griseus* and *Opuntia* sp.; and (d) "desert" with *Melocactus* sp.

In the Checua region between Bogotá and Ubaté, the thorn woodland is 3–10 m tall and is dominated by *Durantha mutisii* and *Condalia thomasiana* (van der Hammen, 1997; Fernández-Alonso, 1997). In some of the lowland dry formations (e.g., the canyons of Río Chicamocha and La Tatacoa), Cactaceae are replaced by Agavaceae (*Furcraea* and *Agave*).

FLORISTICS

Lowland formations. Several floristic studies have been conducted in the southern Caribbean dry zone of Colombia, in the Guajira Peninsula (Rieger, 1976; Sugden, 1982; Sugden and Forero, 1982) around Santa Marta (Schnetter, 1968), Tayrona (Bastidas and Corredor, 1977; Lozano, 1984), and Barranquilla (Dugand, 1941, 1970). The drier and larger areas of xerophytic vegetation of the lowlands of Colombia occur in the upper and middle Guajira Peninsula. Floristic inventories of these lowlands include a total of 316 species (Sugden and Forero, 1982). This list includes the survey of Rieger (1976) as well as the collections of Romero-Castañeda in 1953 and Saravia in 1962 (Sugden and Forero, 1982). In the upper and middle Guajira, Rieger (1976) recognized four formations where the following cacti occur: *Cereus hexagonus, Cereus margaritensis, Pilosocereus lanuginosus,* and *Stenocereus griseus* (Table 16.2). Besides these cactus species, *Subpilocereus repandus* occurs in the upper and middle Guajira Peninsula (Hernández et al., 1995). In his study of the flora of the hills of Fernando and La Llorona, southwest of Santa Marta, Schnetter (1968) reported a vegetation gradient from dry deciduous forest (e.g., *Pseudobombax*) to very open and short vegetation dominated by shrub (e.g., *Haematoxylon, Melochia*), passing through thorn woodlands (e.g., *Mimosa, Stenocereus* [*Lemaireocereus* in Schnetter's paper]) and cactus shrub (e.g., *Mimosa, Opuntia, Pereskia, Stenocereus*) (see Cleef, 1984). Tayrona National Park is somewhat wetter, with seasonally deciduous forests and thorn woodlands. Bastidas and Corredor (1977) described four associations: (1) *Capparis odoratissima* and *Platymiscium polystachium,* (2) *Anacardium excelsum* and *Hura crepitans,* (3) *Pereskia colombiana* and *Stenocereus* (*Lemaireocereus*) *griseus,* and (4) *Prosopis juliflora* and *Adipera bicapsularis* (*P. colombiana* = *P. guamacho;* Trujillo and Ponce [1988]). The flora around Barranquilla (Dugand, 1941, 1970) is characterized by dry deciduous forest, rather than formations dominated by shrubs and cacti. The largest and least-known areas are the dry forests and woodlands of the Caribbean plains (IAVH, 1997), because they have been almost totally transformed to extensive pastures for cattle ranching. It is likely that the columnar cactus *Cereus hexagonus* is present on these plains, as it also occurs along the coast (Table 16.2).

Floristic studies of the dry inter-Andean valleys are scarce. The most complete survey was carried out in the upper valley of Río Magdalena (Santos, 1995; Cavelier et al., 1996a). This area includes the following formations, some of which contain columnar and other cacti: dry evergreen forest with no Cactaceae; dry deciduous forest with *Acanthocereus pentagonus* and *Cereus hexagonus;* thorn woodlands with *A. pentagonus, C. hexagonus, Melocactus* sp., *Monvillea smithiana, Opuntia pubescens,*

Opuntia sp., and *Stenocereus griseus;* and cactus shrub with *A. pentagonus, C. hexagonus, Mammillaria* sp., *Melocactus* sp., *Monvillea smithiana, O. pubescens, Opuntia* sp., *Pilosocereus* sp., and *S. griseus.* The canyon of Patia River has been subject to preliminary floristic surveys that do not include Cactaceae (Martinez and Ordoñez, 1991; Fernández and Fernández, 1992), but the canyon is now known to contain cacti (Cavelier et al., 1996b).

TABLE 16.2

Cactus Species of the Dry Formations in the Caribbean Coast and in the Uplands of the Cordillera Oriental of the Andes of Colombia[a]

Species	*Reference*
Caribbean lowlands[b]	
Acanthocereus pentagonus (L.) Britton and Rose	Lozano (1986); Dugand (1970) (= *A. tetragonus*)
Acanthocereus sicariguensis Croizat and Tamayo	Dugand (1966)
Acanthocereus sp.	Rieger (1976); Sugden and Forero (1982)
*Cereus hexagonus** (L.) J. S. Muell	Rieger (1976); Sugden and Forero (1982)
*Cereus margaritensis** (J .R. Johnst.) Backeb.	Rieger (1976); Lozano (1986); Sugden and Forero (1982); Bastidas and Corredor (1977)
Epiphyllum cf. *phyllanthus* (L.) Haw.	Lozano (1986)
Hylocereus polyrhizus (F. A. C. Weber) Britton and Rose	Lozano (1986); Sugden and Forero (1982)
Mammilaria simplex Haw.	J. Hernández (pers. com.)
Melocactus curvispinus subsp. caesius (Wendland)	Taylor (1991); Schnetter (1968)
Melocactus sp. 1 (Saravia 2073)	Sugden and Forero (1982)
Melocactus sp. 2 (Saravia 2162)	Sugden and Forero (1982)
Opuntia caribea Britton and Rose	J. Hernández (pers. com.)
Opuntia elatior Mill.	Sugden and Forero (1982)
Opuntia pennellii Britton and Rose	Britton and Rose (1963); Croizat (1944)
Opuntia wentiana Britton and Rose	Rieger (1976); Lozano (1986); Sugden and Forero (1982); Schnetter (1968); Bastidas and Corredor (1977); Dugand (1970)
Opuntia sp. (Saravia 2196)	Sugden and Forero (1982)
Pereskia aculeata Mill.	Sugden and Forero (1982)
Pereskia bleo (Kunth) DC	Britton and Rose (1963); Croizat (1944)
Pereskia colombiana Britton and Rose	Lozano (1986); Sugden and Forero (1982) Schnetter (1968); Bastidas and Corredor (1977); Rieger (1976) + Dugand (1970) = (*P. guamacho* F.A.C. Weber)
*Pilosocereus lanuginosus** (L.) Byles and G.D. Rowley	Rieger (1976); Lozano (1986); Sugden and Forero (1982) Bastidas and Corredor (1977)
*Pilosocereus remolinensis** Backeb.	Croizat (1943c, 1944)
Rhipsalis baccifera (J.S. Muell.) Stearn	Lozano (1986)
*Stenocereus griseus** (Haw.) Buxb. ex Bravo	Rieger (1976); Lozano (1986); Schnetter (1968); Bastidas and Corredor (1977)
*Subpilocereus horrispinus** Backeb.	Ponce (1989)

(continued)

TABLE 16.2 *(continued)*

Species	Reference
*Subpilocereus russelianus** (Otto in Salm-Dyck)Backeb.	Lozano (1986); Bastidas and Corredor (1977)
*Subpilocereus repandus** (L.) Backeb.	Hernández et al. (1995)
Uplands	
Browningia sp. nov ?	Hernández et al. (1995)
Cylindropuntia sp. (Engelmann) F. M. Knuth	Hernández et al. (1995)
Mammillaria bogotensis Werderm.	Dugand (1954), Hernández et al. (1995); A. Ruiz, pers. obs.
Melocactus curvispinus subsp. *curvispinus* Pfeffer	Taylor (1991)
Melocactus obtusipetalus Lem.	Taylor (1991)
Melocactus andinus R. Gruber	Taylor (1991)
Opuntia ficus-indica (L.) Mill. (introduced sp.)	Vink and Wijninga (1987)
Opuntia pittieri	Vink and Wijninga (1987)
Opuntia schumanni Speg.	Vink and Wijninga (1987)
Wigginsia vorwerckiana.	Hernández et al. (1995)

[a] *Stenocereus griseus* was originally reported as *Lemaireocereus griseus* (Rieger, 1976). *Subpilocereus horrispinus* was originally reported as *Pilocereus horrispinus* (Croizat, 1943c, 1944).

[b] Columnar cacti are marked with a *.

In contrast to the lack of general floristic surveys, we now have lists of Cactaceae that are present in each of the ten dry inter-Andean valleys (Table 16.3). This list includes six genera of columnar cacti: *Armathocereus, Cereus, Monvillea, Pilosocereus, Stenocereus,* and *Subpilocereus.* The genus *Armathocereus* is represented by one species (*A. humilis*) in five inter-Andean valleys. One species of *Cereus* (*C. hexagonus*) is present in three inter-Andean valleys (Ocaña, Chicamocha, and the upper valley of Río Magdalena), and one species of *Stenocereus* (*S. griseus*) is present at four sites (the canyons of Río Patía and Juananbú, Cúcuta, the canyon of Río Chicamocha, and upper valley of Río Magdalena). The *S. griseus* reported for the Caribbean coast may be a different species or a subspecies (Hernández et al., 1995). The genus *Pilosocereus* may be represented by up to four species: *P. colombianus* in the canyon of Río Dagua, *P.* cf. *lanuginosus* in Cúcuta, *Pilosocereus* sp.1 (possibly *P. colombianus*) in Río Patía and Juananbú, and *Pilosocereus* sp. 2 in La Tatacoa (upper valley of Río Magdalena). The latter species has very different vegetative (yellow spines and scarce pseudocephalium) and fruit characteristics (red pulp) that suggest a new and endemic species (Ruiz et al., 2000). The genus *Subpilocereus* is represented by one species (*S.* cf. *horrispinus*) in Río Chicamocha; the genus *Monvillea* includes one species (*M. smithiana*) common to La Tatacoa in the upper valley of Río Magdalena, Cúcuta, and the canyon of Río Chicamocha.

Upland dry formations. These are the least-known dry formations in Colombia. Although no columnar cacti occur in these formations, there are other growth forms

(Table 16.2). Southwest of Bogotá at Laguna de la Herrera, the vegetation is short and dense and includes shrubs mixed with *Furcraea* sp., *Opuntia schumanni, O. ficus-indica* (an introduced species), and *O. pittieri* (Vink and Wijninga, 1987). These and other dry pockets of the Bogotá plateau also contain cacti of the genus *Cylindropuntia* (Hernández et al., 1995). Dry formations in the Ubaté Valley (Checua region) are composed of relatively dense woodlands (4–8 m tall) dominated by *Condalia* sp. nov. (25% cover), *Myrsine guianensis* (15%), and *Dodonea viscosa* (5%) (van der Hammen, 1997). These formations contain no Cactaceae. At La Candelaria in Villa de Leyva (Boyacá), thorn woodlands are dominated by the shrubs *Acacia farnesiana, Opuntia pittieri,* and *Poponax flexuosa* (Molano, 1990). *Mammillaria bogotensis* occurs in the understory (Dugand 1954; Hernández et al., 1995; A. Ruiz, pers. obs.). *Wigginsia*

TABLE 16.3
Species of Cactaceae and Their Abundances in the Ten Inter-Andean Dry Valleys below 1,000 m[a]

	Inter-Andean Dry Formations Valley Abundance[b,c]									
Species	*1*	*2*	*3*	*4*	*5*	*6*	*7*	*8*	*9*	*10*
Acanthocereus pentagonus (L.) Britton and Rose					O	A				
Acanthocereus sp. 1								F		
*Armatocereus humilis** (Britton and Rose) Backeb	O		F	F	F				R	
*Cereus hexagonus** (L.) J.S. Muell							R		O	O
Cleistocactus sepium (Kunt) F.A.C. (6)					U					
Frailea colombiana (1) and (2) (Werd.) Backeb and Knuth				(E)						
Hylocereus c.f. polyrhizus (F.A.C. Weber) Britton & Rose									O	
Hylocereus sp. 2	O								O	
Hylocereus sp. 3			O							
Hylocereus sp. 4					O					
Hylocereus sp. 5						O	O			
Hylocereus sp. 6										R
Hylocereus sp. 7					O					
Mammillaria colombiana (1) and (5) Salm-Dyck									F	R
Mammillaria simplex (1) Haw.								VA	F	
Melocactus curvispinus curvispinus Pfeffer				O						
Melocactus schatzlii Till and R. Gruber									U[(4)]	
Melocactus sp. 1 (sp. nov ?)									F	
Melocactus sp. 2								VA		
Melocactus sp. 3										VA
*Monvillea smithiana** (Britton and Rose) Backeb.								F	F	A
Nyctocereus kalbreyerianus (2) (Werckle) Croizat										U
Opuntia bella				A						
Opuntia cf. *elatior* Mill.									F	
Opuntia cf. *pittieri*			A	VA	A		O			

(continued)

TABLE 16.3 *(continued)*

Species	*Inter-Andean Dry Formations Valley Abundance*[b,c]									
	1	*2*	*3*	*4*	*5*	*6*	*7*	*8*	*9*	*10*
Opuntia cf. *schumannii* Speg.									F	
Opuntia sp. 1 (*O. dillenii* (1) (Ker Gawl.) Haw.?)	O								O	
Opuntia sp. 2								VA		
Opuntia sp. 3										VA
O. pubescens J.C. Wendl. ex Pfeiff.					O			A	VA	F
O. caribaea Britton and Rose								U[(1)]	R	
Pereskia colombiana (3) Britton and Rose										U
Pilosocereus colombianus * (Rose) Byles & G.D. Rowley				VA						
Pilosocereus sp. 1* (*P. colombianus* ?)					VA					
P. cf. *lanuginosus**								VA		
P. cf. *colombianus* (1)*									VA	
Pilosocereus sp. 2 (sp. nov ?)										R
Rhipsalis baccifera (J.S. Muell.) Stearn				F					F	
Rhipsalis sp.1								O		
Stenocereus griseus * (Haw.) Buxb. ex Bravo					VA			R	VA	VA
Subpilocereus cf. *horrispinus* *									R	
Selenicereus inermis (Otto) Britton and Rose	0									
Total number of species	4	0	3	7	8	1	3	10	19	12

[a] VA = very abundant, A = abundant, F = frequent, O = occasional, R = rare, E = extinct ?, U = unknown.

[b] 1 = Canyon of Río Cauca at Santafé de Antioquia, 2 = Upper Valley of Río Sucio at Dabeiba, 3 = Upper valley of Río Cauca, 4 = Canyon of Río Dagua, 5 =Canyon of Río Patía and Juanambú, 6 =Aguachica, 7 = Ocaña, 8 = Cúcuta, 9 =Canyon of Río Chicamocha, 10 = Upper valley of Río Magdalena.

[c] Modified from Cavelier et al., (1996b) using names and information on well- documented species in Hernández et al., (1995)[(1)], Croizat (1943a,b, 1944)[(2)], Espinal and Montenegro (1977)[(3)], Taylor (1991)[(4)], Hunt (1983)[(5)], Madsen (1989)[(6)]. Columnar cacti are marked with a *.

vorwerckiana, a small understory cactus, is apparently present in some of the dry pockets of the plateau of Cundinamarca-Boyacá around Sogamoso and Bogotá (Hernández et al., 1992), as well as in the highest part of the Río Chicamocha watershed, where *Mammillaria bogotensis* and *Browningia* sp. also occur (Hernández et al., 1995). Lastly, in the dry pocket of Río Negro (Cáqueza), *Opuntia* sp. and Agavaceae (*Furcraea* sp. and *Agave* sp.) co-occur with short woodlands. Not much information is available for the flora of the dry formations around Ipiales and Pasto, except that the species *Cleistocactus sepium* (*Brozicactus sepium*) occurs in the nearby canyon of Río Juananbú (Madsen, 1989).

In summary, Cactaceae occur both in the lowland and upland dry formations of Colombia, but columnar cacti are restricted to the lowlands of the Caribbean (eight species) and inter-Andean valleys (eight to ten species). Only four columnar cacti are apparently restricted to the Caribbean coast (*Cereus margaritensis, Pilosocereus remolinensis, Subpilosocereus repandus,* and *Subpilosocereus russelianus*) and four to the inter-Andean valleys (*Armathocereus humilis, Monvillea smithiana, Pilosocereus colombianus,* and *Pilosocereus* sp. nov. in the upper valley of Rio Magdalena).

Phenology of Columnar Cacti and Mutualistic Relations

Of the fourteen species of columnar cacti of Colombia, only four have been the subject of biological and ecological studies. Flowering and fruiting phenology have been studied in *Cereus hexagonus, Monvillea smithiana, Pilosocereus* sp., and *Stenocereus griseus* at La Tatacoa in the upper valley of the Río Magdalena (Ruiz et al., 2000). The results of this study showed that these species have bimodal, multimodal, or irregular patterns of flower and fruit production without significant temporal overlap in the flowering or fruiting peaks among species. In *S. griseus,* flowering was concentrated during the dry season and fruiting during the wet season (Fig. 16.3). Flowering in *Pilosocereus* sp. was also concentrated during the dry season with no marked seasonality for fruiting. In contrast, flowering and fruiting in *C. hexagonus* occurred during the wet season, and *M. smithiana* showed no seasonal pattern (Fig. 16.3). Although *C. hexagonus* is likely to be pollinated by moths (Sphingidae), the other three species are pollinated by bats (*Glossophaga longirostris*) and dispersed by both bats and birds. In a detail study of the mutualistic relationships between cacti and bats at the same site, the resident bat *G. longirostris* depends on flowers and fruits produced by these cacti, particularly those of *S. griseus,* for its diet and reproduction (Ruiz et al., 1997). Furthermore, seasonal variation in the diet of this glossophagine bat was correlated with the availability of cactus resources, including pollen, nectar, and fruits. The presence of lactating females coincided with the peak in fruit production of *S. griseus* (Ruiz et al., 1997). Current research on mutualistic relationships between columnar cacti and glossophagine bats at the Cúcuta and Chicamocha inter-Andean valleys has documented the occurrence of the migrant bat *Leptonycteris curasoae,* which pollinates and disperses cacti (see chapter 12, this volume).

Conservation Status

Colombia currently lacks national parks created for the protection of the fauna and flora of dry formations. Only Tayrona and Macuira National Parks and the Sanctuary of Fauna and Flora of Los Colorados (1,000 ha on the Caribbean coast) contain areas of dry formations. None of the inter-Andean valleys or upland dry formations are protected. As a result, they are in an advanced state of degradation. Whereas the vegetation of the flat inter-Andean valleys has been almost completely replaced by ir-

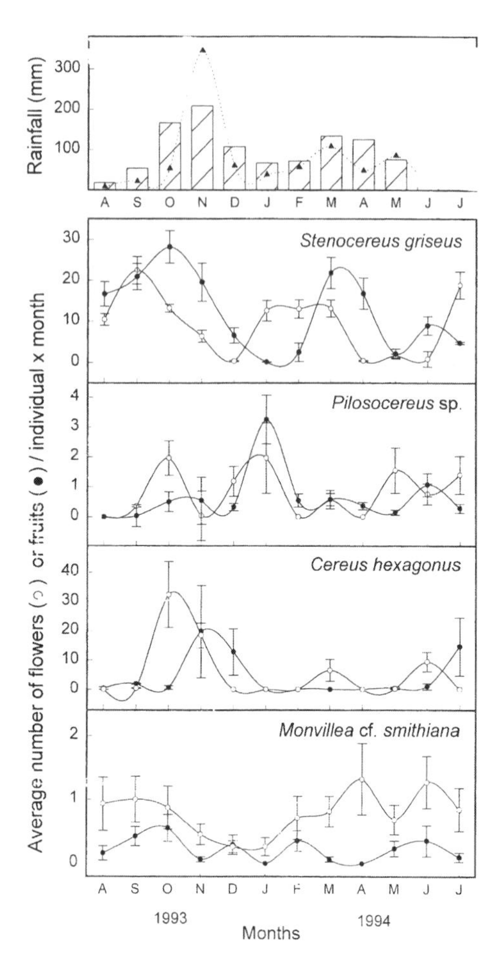

FIGURE 16.3. Flowering and fruiting patterns of four columnar cacti studied in the inter-Andean valley of the Magdalena River (August 1993–July 1994). The vertical lines represent the standard error of the mean. In the upper panel, th bars represent the mean monthly rainfall (1964–1994) and the triangles, the monthly rainfall during the study period (1993–1994). Modified from Ruiz et al. (2000).

rigated fields for crops such as rice and sugarcane, the dry slopes have been transformed by cattle grazing since the first half of the 16th century (Cavelier et al., 1996a). A similar situation occurs in the dry formations of the Caribbean lowlands. Of the cactus species, only one is apparently extinct: *Frailea colombiana,* an endemic of the canyon of Río Dagua (Hernández et al., 1992, 1995). Although there is a CITES list of Colombian cacti with a total of 38 species, little is known about their population sizes and their actual conservation status. Species like *Pilosocereus* sp. 2 (sp. nov. ?) in the Tatacoa region of the upper valley of Río Magdalena have a very restricted distribution and are likely to become threatened or endangered, due to the conversion of their habitat to pastures for cattle ranching.

Research and Conservation Priorities

The cacti of the dry formations of Colombia are relatively well known. Nevertheless, the knowledge on the taxonomy of cacti will greatly benefit from detailed work on the genus *Hylocereus* (also abundant in wet formations), as well as on the columnar cacti *Stenocereus griseus* and *Pilosocereus* spp. *S. griseus* in the inter-Andean valleys of the canyons of Río Patía and Juanambú, the canyon of Río Chicamocha, Cúcuta, and the upper valley of Río Magdalena that exhibit several differences from the populations of the Caribbean, including the number and length of spines in the areoles and fruits, size and position of the flowers, and length of the branches. No information has been included on the cacti of the savannas of Río Orinoco, where *Melocactus* sp. (*M. mazelianus;* see Taylor [1991]) and *Cereus* spp. are known to occur. The only two genera that have been recently reviewed are *Melocactus* (Taylor, 1991) and *Mammillaria* (Hunt 1983, 1984, 1985, 1987).

We need much more information on the biology and ecology of almost all cacti. In particular, we need data on phenology, pollination, dispersal, germination, and growth requirements, as well as population densities and dynamics. This information is needed to design management or restoration programs in degraded dry habitats where cacti and their mutualists still exist. Degradation of populations of columnar cacti may have effects on the stability of pollinators and dispersers, particularly glossophagine bats (e.g., *Glossophaga longirostris*).

Although all inter-Andean valleys and the dry formations in the Caribbean have been subject to disturbance for the past 400 years, there are still intact patches of vegetation that can be preserved by the creation of national parks. These areas could be used as "seeds" for restoration programs in nearby areas that are badly degraded. Based on the number of cactus species, the most important sites would be the canyon of Río Chicamocha (19 species), the upper valley of Río Magdalena at La Tatacoa and the Municipality of Colombia (12 species), and Cúcuta (ten species). Ex-situ conservation programs may be needed for the *Pilosocereus* sp. at La Tatacoa, because it has a very narrow distribution range within the large upper valley of Río Magdalena and because this area is under increasing pressure from agriculture and cattle ranching.

Summary

Dry formations occur both in the lowlands (below 1,000 m) and in the uplands of Colombia (ca. 2,500 m). In the lowlands, they occur along the Caribbean coast and in ten dry inter-Andean valleys; in the uplands, they are restricted to the Cordillera Oriental of the Andes. At least 26 species of cacti occur in the Caribbean lowlands, 42 species in the inter-Andean valleys, and ten species in the uplands. Columnar cacti of the genera *Cereus* (two species), *Monvillea* (one species), *Pilosocereus* (four or five species), *Stenocereus* (one species), and *Subpilocereus* (three species) occur only in the lowlands (Caribbean plus inter-Andean), whereas in the uplands, the most common species are the cladode *Opuntia* (three species). *Furcraea* and *Agave* (Agavaceae) are more important than Cactaceae in the upper Andes of Colombia. Total richness of Cactaceae in Colombia is around 60 species and 20 genera. This diversity is higher than that of Panama (17 species and eight genera; Woodson and Schery [1958]) and Ecuador (41 species and 17 genera; Madsen [1989]). In Venezuela, the total number of species is unknown (B. Trujillo, pers. com.). Cactus diversity in Colombia is significantly lower than in the region of the southwestern United States and Mexico. The latter region is the center of diversification of this family, with around 2,000 species present (Bravo-Hollis and Sanchez-Mejorada, 1978).

Resumen

En Colombia, las formaciones vegetales secas se presentan en las tierras bajas (<1,000 m) y en las tierras altas (ca. 2,500 m). En las tierras bajas se presentan a lo largo de la costa del Caribe y en diez valles interandinos. Mientras que en las tierras altas, están restringidos a la Cordillera Oriental de los Andes. Existen 26 especies de cactus en las tierras bajas del Caribe. En los valles secos interandinos estan 42 especies y en las tierras altas 10 especies. En las tierras bajas (Caribe y valles interandinos) se encuentran cactáceas columnares de los géneros: *Cereus* (dos especies), *Monvillea* (una especie), *Pilosocereus* (cuatro a cinco especies), *Stenocereus* (una especie), y *Subpilocereus* (tres especies). Mientras que en las tierras altas las especies más comunes pertenecen al género de cladodios *Opuntia* (tres especies). *Furcraea* y *Agave* (Agavaceae) son más importantes que las cactáceas en las tierras altas de Colombia. La riqueza total de especies de Cactaceae en Colombia es de aproximadamente 60 especies y 20 géneros. Esta riqueza es más alta que en Panamá (17 especies y 8 generos; Woodson and Schery, 1958) y Ecuador (41 especies y 17 generos; Madsen, 1989). En Venezuela, el número total de especies es todavía desconocido (B. Trujillo, pers. com.). La riqueza de Cactáceas en Colombia es significativamente más baja que en el sur y occidente de los Estados Unidos y México. La última región es el centro de diversificación de esta familia con aproximadamente 2,000 especies (Bravo-Hollis and Sanchez-Mejorada, 1978).

ACKNOWLEDGMENTS

Thanks to Dr. Pablo Leyva, director of Instituto de Hidrología, Meteorología y Estudios Ambientales, for the financial support that made the field work in the inter-Andean dry valleys of Colombia possible. Thanks also to Dr. Ricardo Callejas (Universidad de Antioquia) and Dr. Michael Alberico (Universidad del Valle) for allowing us access to the herbarium of these universities, and to Diana Alvira, Jorge Hernández, Diego Lizcano, and Johanna Santamaría for assistance during the preparation of the final versions of this chapter. Special thanks to Theodore H. Fleming for editorial comments and for encouraging us to prepare this chapter.

REFERENCES

Bastidas, N., and H. Corredor, H. 1977. Contribución al estudio fitosociológico del Parque Nacional Natural Tayrona (Ensenadas de Chegue y parte este de Nenguage). Tesis de pregado, Universidad Nacional de Colombia, Bogotá.

Beard, J. S. 1944. Climax vegetation in tropical America. *Ecology* 25:127–58.

———. 1955. The classification of tropical American vegetation types. *Ecology* 36:89–100.

Bravo-Hollis, H., and H. Sanchez-Mejorada. 1978. *Las Cactáceas de México.* Vol. I, II, and III. Mexico: Universidad Nacional Autónoma de México.

Britton, N. L., and J. N. Rose. 1963. *The Cactaceae.* Vol. I–IV. New York: Dover Publications.

Cavelier, J., A. Ruiz, M. Santos, M. Quiñones, and P. Soriano. 1996a. *El proceso de Degradación y Sabanización del Valle alto del Magdalena.* Colombia: Fundación Alto del Magdalena, Neiva, Huila.

Cavelier, J., A. Ruiz, and M. Santos. 1996b. *El Uso de Cactáceas como Bioindicadores del Proceso de Desertificación en Enclaves Secos Internadinos.* Bogotá: Instituto de Hidrología, Meteorología y Estudios Ambientales, IDEAM.

Cleef, A. M. 1984. Synopsis of the coastal vegetation of the Santa Marta area. In *La Sierra Nevada de Santa Marta (Colombia), Transecto Buritaca–La Cumbre,* ed. T. van der Hammen and P. M. Ruiz, 423–37. Studies on Tropical Andean Ecosystems, Vol. 2. Berlin: J. Cramer.

Croizat, L. 1943a. Notes on *Cereus* and *Acanthocereus. Caldasia* II(7):116–22.

———. 1943b. Euphorbiaceae Cactaceae que novae vel criticae Colombianae. *Caldasia* 2:123–39.

———. 1943c. Notes on *Pilocereus, Monvillea* and *Malacocarpus* with special reference to Colombia and Venezuela. *Caldasia* 2:251–60.

———. 1944. A check list of Colombian and presumed Colombian Cactaceae. *Caldasia* 2(11):337–55.

Cuatrecasas, J. 1958. Aspectos de la vegetación natural de Colombia. *Revista de la Academia Colombiana de Ciencias Exactas, Físicas y Naturales* 10:221–68.

Dugand, A. 1941. Estudios geobotánicos colombianos. Descripción de una sinesia típica en la subxerofitia del litoral Caribe. *Revista de la Academia Colombiana de Ciencias Exactas, Físicas y Naturales* 4:135–41.

———. 1954. Tres cactáceas colombianas poco conocidas. *Mutisia* 1:7–11.

———. 1966. Notas sobre la flora de Colombia y paises vecinos. *Phytologia* 13:379–81.

———. 1970. Observaciones botánicas y geobotánicas en la costa colombiana del Caribe. *Revista de la Academia Colombiana de Ciencias Exactas, Físicas y Naturales* 13:415–65.

———. 1973. Elementos para un curso de geobotánica en Colombia. *Cespedecia* 2:139–481.

Espinal, L. S., and E. Montenegro. 1977. *Formaciones Vegetales de Colombia. Memoria explicativa sobre el mapa ecológico de Colombia.* Bogotá: Instituto Geográfico "Agustín Codazzi" (IGAC), Subdirección Agrológica.

Etter, A. 1993. Diversidad ecosistémica en Colombia hoy. In *Nuestra Diversidad Biológica,* 43–61. Bogotá: CEREC and Fundación Alejandro Angel Escobar.

Fernández, A., and S. I. Fernández. 1992. Contribución al estudio florístico de la hoya hidrográfica del Río Patía. *Novedades colombianas* 5:27–44.

Fernández-Alonso, J. L. 1997. Nueva especie de *Condalia* Cav. (Rhamnaceae) y notas sobre géneros de la familia en la flora de Colombia. *Caldasia* 19:101–8.

Grubb, P. J. 1977. Control of forest growth and distribution on wet tropical mountains: with special reference to mineral nutrition. *Annual Review of Ecology and Systematics* 8:83–107.

Hernández C, J., T. Walschburger, R. Ortiz, and A. Hurtado. 1992. Origen y distribución de la biota Suramericana y colombiana. In *La Diversidad Biológica de Iberoamérica,* ed. G. Halffter,

55–104. Volumen especial de Acta Zoológica Mexicana, CYTED-D. Xalapa, Mexico: Acta Zoológica Mexicana.

Hernández, J., V. R. Almonacid, and H. Sánchez. 1995. Zonas áridas y semiáridas de Colombia. In *Desiertos. Zonas Áridas y Semiáridas de Colombia,* ed. C. Hernández, 111–62. Bogotá: Diego Samper Ediciones.

Hunt, D. 1983. A new review of *Mammillaria* names A–C. *Bradleya* 1:105–28.

———. 1984. A new review of *Mammillaria* names D–K. *Bradleya* 2:65–96.

———. 1985. A new review of *Mammillaria* names L–M. *Bradleya* 3:53–66.

———. 1987. A new review of *Mammillaria* names S–Z. *Bradleya* 5:16–48.

———. 1992. Cactaceae checklist. CITES, Royal Botanical Gardens, Kew.

Instituto Alexander von Humboldt 1997. *Caracterización Ecológica de Cuatro Remanentes de Bosque Seco de la Región Caribe Colombiana.* Grupo de Exploraciones y Monitoreo Ambiental—GEMA No. 3.

Johnston, J. R. 1909. Flora of the Islands of Margarita and Coche, Venezuela. *Contributions of Gray Herbarium* 37.

Lahey, J. F. 1958. On the origin of the dry climate in northern South America and the southern Caribbean. *Department of Meteorology, University of Wisconsin Science Report* 10:1–290.

Lozano, G. 1984. Comunidades vegetales del flanco norte del Cerro "El Cielo", y la flora vascular del Parque Nacional Tairona (Magdalena, Colombia). In *La Sierra Nevada de Santa Marta (Colombia), Transecto Buritaca–La Cumbre,* ed. T. van der Hammen and P. M. Ruiz, 407–22. Studies on Tropical Andean Ecosystems, Vol. 2. Berlin: J. Cramer.

———. 1986. Comparacion florística del Parque Nacional Natural Tayrona, La Guajira y la Macuira-Colombia y los medanos de Coro-Venezuela. *Mutisia* 67:1–26.

Madsen, J. E. 1989. Cactaceae. *Flora de Ecuador* 35:3–78

Marshall, W. T., and T. M. Bock. 1941. *Cactaceae.* Pasadena, Calif.: Abbey Garden Press.

Martinez, A., and P. Ordoñez. 1991. Levantamiento ecológico de la Quebrada Las Tallas, Valle del Patía, Cauca. Tesis de pregado, Fundación Universitaria de Popayán, Popayán, Colombia.

Molano, J. 1990. *Villa de Leyva. Ensayo de Interpretación Social de una Catástrofe Ecológica.* Bogotá: Fondo FEN.

Ochsenius, C. 1981. Ecología del Pleistoceno tardío en el cinturón árido pericaribeño. *Revista CIAF* 6:365–72.

Ponce, M. 1989. Distribución de las cactáceas en Venezuela y su ámbito mundial. Trabajo Especial de Ascenso a Profesor Agregado. Universidad Central de Venezuela, Facultad de Agronomía, Maracay.

Raunkaier, C. 1934. *The Life Forms of Plants and Statistical Plant Geography.* Oxford: Oxford University Press.

Rieger, W. 1976. Vegetationskundliche Untersuchungen auf der Guajira-Halbinsel (Nordost-Kolumbien). *Giessener Geographische Schriften* 40:1–142.

Ruiz, A., M. Santos, P. Soriano, J. Cavelier, and A. Cadena. 1997. Relaciones mutualísticas del murciélago *Glossophaga longirostris* Miller con las cactáceas columnares en la zona árida de La Tatacoa, Huila. *Biotropica* 29:469–79.

Ruiz, A., M. Santos, J. Cavelier, and P. Soriano. 2000. Estudio fenológico de cactáceas en el enclave seco de La Tatacoa, Colombia. *Biotropica* 32:397–407.

Santos, M. 1995. Fenología de cuatro cactáceas y su relación con polinizadores y dispersores en el enclave seco de La Tatacoa, Huila, Colombia. Tesis de pregado, Departamento de Ciencias Biológícas, Universidad de Los Andes, Bogotá, Colombia.

Sarmiento, G. 1975. The dry plant formations of South America and their floristic connections. *Journal of Biogeography* 2:233–51.

Schnetter, R. 1968. Die Vegetation des Cerro San Francisco und des Cerro La Llorona im Trockengebiet bei Santa Marta, Kolumbien. *Berichte der Deutschen Botanischen Gesellschaft* 81:289–302.

Snow, J. W. 1976. The climate of northern South America. In *World Survey of Climatology*. Vol. 12, ed. W. Schwerdtfeger, 245–379. Amsterdam: Elsevier.

Sugden, A. M. 1982. The vegetation of the Serranía de Macuira, Guajira, Colombia: a contrast of arid lowlands and an isolated cloud forest. *Journal of the Arnold Arboretum* 63:1–30.

Sugden, A. M., and E. Forero. 1982. Catálogo de las plantas vasculares de la Guajira con comentarios sobre la vegetación de la Serranía de Macuira. *Colombia Geográfica* 10:23–77.

Tamayo, F. 1941. Exploraciones botánicas en la Península de Paraguaná, Estado Falcón. *Boletín de la Sociedad Venezolana de Ciencias Naturales* 7:1–90.

———. 1967. El espinar costero. *Boletín de la Sociedad Venezolana de Ciencias Naturales* 111:163–68.

Taylor, N. P. 1991. The genus *Melocactus* (Cactaceae) in Central and South America. *Bradleya* 9:1–80.

Trujillo, B., and M. Ponce, M. 1988. Lista-inventario de Cactáceas silvestres en Venezuela con sinonimia y otros aspectos relacionados. *Ernstia* 47:1–20.

van der Hammen, T. 1996. *Plan Ambiental de la Cuenca Alta del Río Bogotá. Análisis de la Problemática y Soluciones Recomendadas.* Bogotá: Corporación Autónoma Regional de Cundinamarca.

———. 1997. El bosque de Condalia. *Caldasia* 19:355–59.

Vink, R., and V. Wijninga. 1987. *The Vegetation of the Semi-Arid Region of La Herrera (Cundinamarca, Colombia).* Bogotá: Informe Laboratorio Hugo de Vries, Universidad de Amsterdam e Instituto de Ciencias Naturales, Universidad Nacional de Colombia.

Woodson, R. E., and R. W. Schery. 1958. Cactaceae. *Annals of the Missouri Botanical Garden* 45:68–91.

CHAPTER 17

Priority Areas for the Conservation of New World Nectar-Feeding Bats

Mery Santos
Héctor T. Arita

Introduction

Bats (order Chiroptera) are the dominant group of mammals in most neotropical lowland forests, both in terms of number of individuals and number of species (Emmons, 1997). Tropical bats exploit a wide variety of food resources, including fruit, nectar, pollen, insects, fish, small vertebrates, and blood (Wilson, 1973; Gardner, 1977; Fleming, 1988). In tropical and subtropical areas of the world, some bat species have specialized to feed primarily on nectar and pollen from flowers. In the Old World, a dozen species of the subfamily Macroglossinae show different degrees of morphological and physiological adaptation to a diet based on flower products. In the New World, 30 species of the family Phyllostomidae show similar patterns of adaptation. Depending on the author, these bats are classified within the tribe Glossophagini, subfamily Phyllostominae (Baker et al., 1989) or within the subfamilies Glossophaginae and Lonchophyllinae (Griffiths, 1982; Koopman, 1993).

Although the natural history of most species of nectar-feeding bats has not been studied in detail, it is known that these bats are very important or essential pollinators of at least 130 plant genera worldwide (Howell and Hodgkin, 1976; Marshall, 1983). Several of these chiropterophilous plants show traits that suggest a close evolutionary relationship with bats. Most chiropterophilous flowers, for example, show nocturnal anthesis, emit strong musky aromas, have white or pale corollas, produce abundant pollen exposed in elongated anthers, and secrete copious amounts of dilute nectar (Faegri and van der Pijl, 1979; Grant and Grant, 1979; Fleming, 1988). Furthermore, pollen produced by chiropterophilous plants tends to have much higher concentrations of protein and to contain higher proportions of some amino acids that are essential for bats (Howell, 1974; Arita and Martínez del Río, 1990).

In a complementary way, anthophilous bats show a series of morphological, physiological, and behavioral adaptations to their ecological relationship with plants

(Howell, 1974; Koopman, 1982; chapter 5, this volume). For example, nectar-feeding bats are comparatively small (7–30 g in the case of glossophagines; Table 17.1), allowing the animals to hover close to the flowers or to hang from them to gain access to nectar and pollen. In most species the muzzle is elongated, with the lower jaw protruding farther than the upper one, and the tongue is narrow and very long, with long hair-like papillae at the tip. Teeth are reduced, particularly the lower incisors, which in some species are absent altogether. Some species have a groove in the lower jaw that allows fast movements of the tongue (Koopman, 1982).

Some bat-pollinated trees are of economic importance, including balsa (*Ochroma pyramidale*), several species of *Ceiba* and *Bombax* (from whose fruits the kapok fiber is obtained), southeast Asian durian (*Durio zibethinus*), jícaro or calabash (*Crescen-*

TABLE 17.1

Distribution and Conservation Status of the 30 Species of Nectar-Feeding Bats of the New World

Species	*Number of Quadrats*[a]	*Categories of Distribution*[b]	*Categories of Risk of Extinction*[c]
Anoura caudifer	39	D	LR: lc
A. cultrata	19	C	LR: lc
A. geoffroyi	50	D	LR: lc
A. latidens	18	C	LR: nt
A. luismanueli	2	A	DD
Choeroniscus godmani	23	C	LR: nt
C. intermedius	20	C	LR: nt
C. minor	30	D	LR: lc
C. periosus	3	A	VU
C. mexicana	16	C	LR: nt
Glossophaga longirostris	12	B	LR: lc
G. commissarisi	25	D	LR: lc
G. leachii	5	A	LR: lc
G. morenoi	3	A	LR: nt
G. soricina	71	D	LR: lc
Hylonycteris underwoodi	12	B	LR: nt
Leptonycteris curasoae	18	C	VU
L. nivalis	8	B	EN
Lichonycteris obscura	34	D	LR: lc
L. spurrelli	31	D	LR: lc
Lonchophylla bokermanni	4	A	VU
L. dekeyseri	11	B	VU
L. handleyi	11	B	VU
L. hesperia	4	A	VU
L. mordax	19	C	LR: lc
L. robusta	11	B	LR: lc
L. thomasi	36	D	LR: lc
Musonycteris harrisoni	2	A	VU
Platalina genovensium	6	A	VU
Scleronycteris ega	6	A	VU

[a] Potential number of quadrats in which the species is present.

[b] A = very restricted distribution, 1–7 quadrats; B = restricted distribution, 8–15 quadrats; C = widely distributed, 16–23 quadrats; D = very widely distributed, 24 or more quadrats.

[c] DD = data deficient; EN = endangered; LR: lc = lower risk: least concern; LR: nt = lower risk: near threatened; VU = vulnerable.

tia alata), and several species of agaves and cacti (Janzen, 1983; Nobel, 1994). In the dry zones of Mexico and southwestern United States, the geographic range of nectar-feeding bats and some species of agaves and columnar cacti coincide, and the presence and abundance of bats in a particular area is closely correlated with the availability of blooming plants (Alvarez and González-Q., 1970; Quiroz et al., 1986; Arita, 1991; Valiente-Banuet et al., 1996). A close mutualistic relationship has been documented in the Sonoran desert for the Mexican cardón (*Pachycereus pringlei*) and its pollinator, the lesser long-nosed bat (*Leptonycteris curasoae*) (Horner et al., 1998). Several species of bat-pollinated cacti are threatened both in North and South America, mainly due to habitat loss and overexploitation of natural resources (Nobel, 1994; M. Santos, pers. obs.).

Some nectar-feeding bats are considered endangered or threatened because of their specialized diets, restricted distributions, low population levels, or recent population declines (Arita, 1993; SEDESOL, 1994; IUCN, 1996; Hutson et al., 1998). Several of these species roost in caves (e.g., *Anoura* spp., *Choeroniscus godmani, Leptonycteris* spp., *Lionycteris spurrelli*) (Molinari, 1994; Muñoz, 1995), making them particularly vulnerable to people who destroy or disturb caves or who exterminate all kinds of bats in misdirected campaigns against vampire bats.

In this chapter, we analyze the distribution of the 30 species of New World nectar-feeding bats to determine geographic areas that deserve particular attention in management plans aimed at the conservation of these animals. Traditionally, species richness has been used as the sole criterion for diversity at large scales. Alternatively, measures such as phylogenetic indices (Cousins, 1991; Pressey et al., 1993; Humphries et al., 1995), the presence of threatened, endangered, or endemic species (Ceballos and Brown, 1995), or indices of rarity (Arita, 1993) have been used. For the analyses presented here, we used species richness and an index of rarity based on area of geographic range to identify those areas that should have priority for a conservation plan for nectar-feeding bats. We based our analysis on the concept of complementarity of species among sites (Faith, 1992; Arita and Santos del Prado, 1999). To construct a set of complementary areas for conservation, a first site is chosen on the basis of a given criterion (e.g., highest species richness). Then the second area is chosen as the one with the highest number of species that do not occur in the first area (the complementary set); the third area is the one with the most species not occurring in sites one or two, and so on until all species are accounted for.

Methods

We compiled a list of nectar-feeding bats for the entire American continent, excluding insular species. Thus, our list included all glossophagines with distributions in the neotropical region and adjacent parts of northern Mexico and southwestern United States (Table 17.1; Koopman, 1981, 1982, 1993; Molinari, 1994). We drew distributional maps for all species (Fig. 17.1) based on the data and maps of Koopman (1981, 1982), Eisenberg (1989), Redford and Eisenberg (1992), and Marinho-Filho and Sazima (1998), updating the information for particular species, such as *Anoura luismanueli,* recently described by Molinari (1994).

We constructed a grid of 96 squares of 5-by-5 degrees longitude and latitude covering the continents and placed it on the distribution maps to determine the potential presence or absence of each species in each quadrat. Sixteen squares contained no species, so our final database included 80 squares (Fig 17.2). We calculated the species richness in each quadrat simply as the sum of all species potentially present there (see Fig 17.3). According to their geographic range, species were assigned to one of four categories of rarity: very restricted range (one to seven quadrats), restricted range (eight to 15 quadrats), wide range (16 to 23 quadrats), and very wide range (24 or more quadrats). The limits to these categories were determined by the quartiles and the median of the frequency distribution of the areas of distribution, as suggested by Arita (1993) and Gaston (1994).

We used an iterative process of complementarity analysis (Pressey et al., 1993) to determine sets of priority areas for the conservation of nectar-feeding bats in the New World. In the first step, the quadrat (or quadrats) with the highest species richness was selected as the most important area. In the second step, all species present in this area were eliminated from the database, and the quadrat with the highest number of species belonging to this complementary set was selected and added to the list of priority areas. To select the third area, all species present in areas one and two were excluded. The process was iterated until all species were accounted for in the list of priority areas.

A variation of this analysis of priority areas was performed giving higher weights to species with restricted distribution. The iterative process was repeated using only the 15 species with restricted or very restricted ranges (Table 17.1). This analysis identified priority areas for the conservation of endemic and microendemic species. In a final analysis, the areas of distribution of all species were included by using an index that measures the average area of distribution of species present in a given area (Kershaw et al., 1994; Arita et al., 1997; Arita and Ortega, 1998). We used the formula

$$I_c = \frac{\sum_{i=1}^{n_c} 1/A_i}{n_c}$$

where I_c is the value of the index for area *c*, A_i is the area of distribution of species *i*, and n_c is the number of species in area *c*. The index for a given area is the average value of the inverse of the distributional ranges of the species that occur in that area. Thus, the index measures the degree of "restrictedness" of the species potentially occurring at a site (Arita et al., 1997). We repeated the iterative complementary analysis described above, using the value of this index as the criterion instead of species richness.

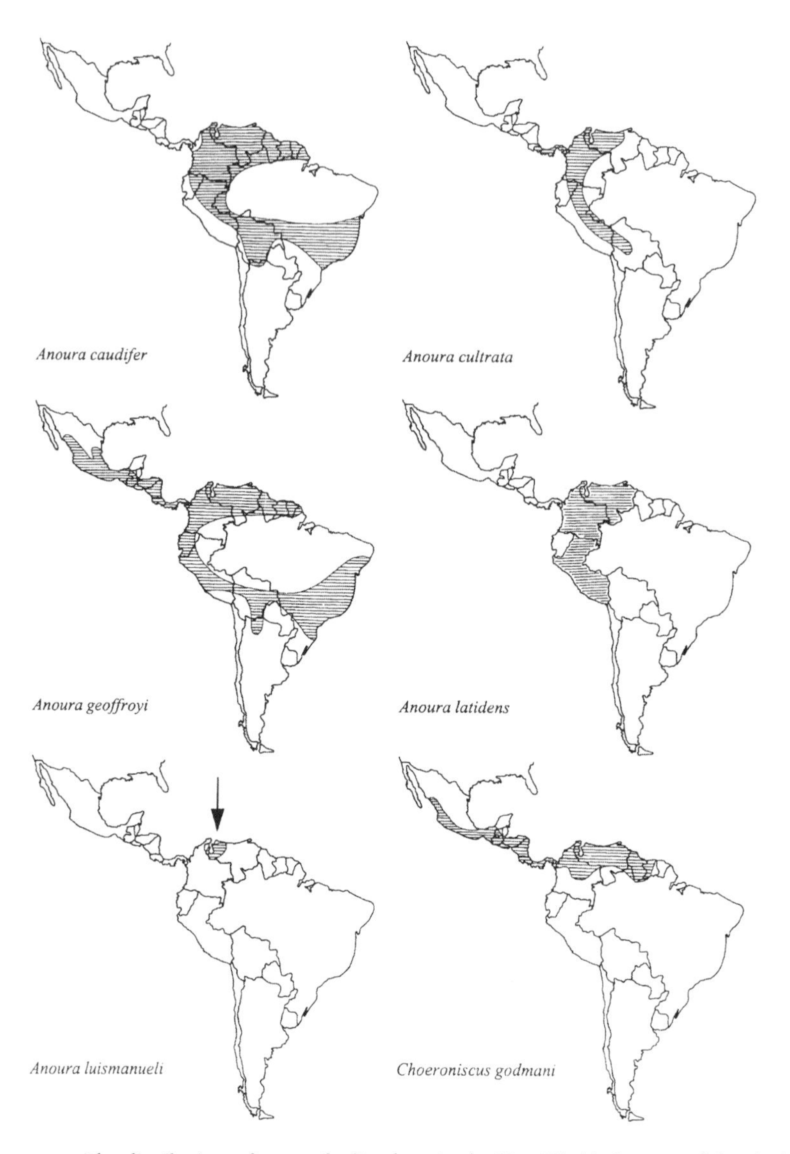

FIGURE 17.1. The distribution of nectar-feeding bats in the New World. Sources of data include Alvarez et al. (1991); Anderson et al. (1982); Arita (1991); Cockrum and Petryszyn (1991); Easterla (1972); Eisenberg (1989); Emmons (1997); Webster et al. (1998); Hall (1981); Koopman (1981, 1982, 1993); Linares (1987); Marinho-Filho and Sazima (1998); Medellín et al. (1997); Molinari (1994); Muñoz (1995); Redford and Eisenberg (1992); Reid (1997); Téllez and Ortega (1999); Webster and Handley (1986).

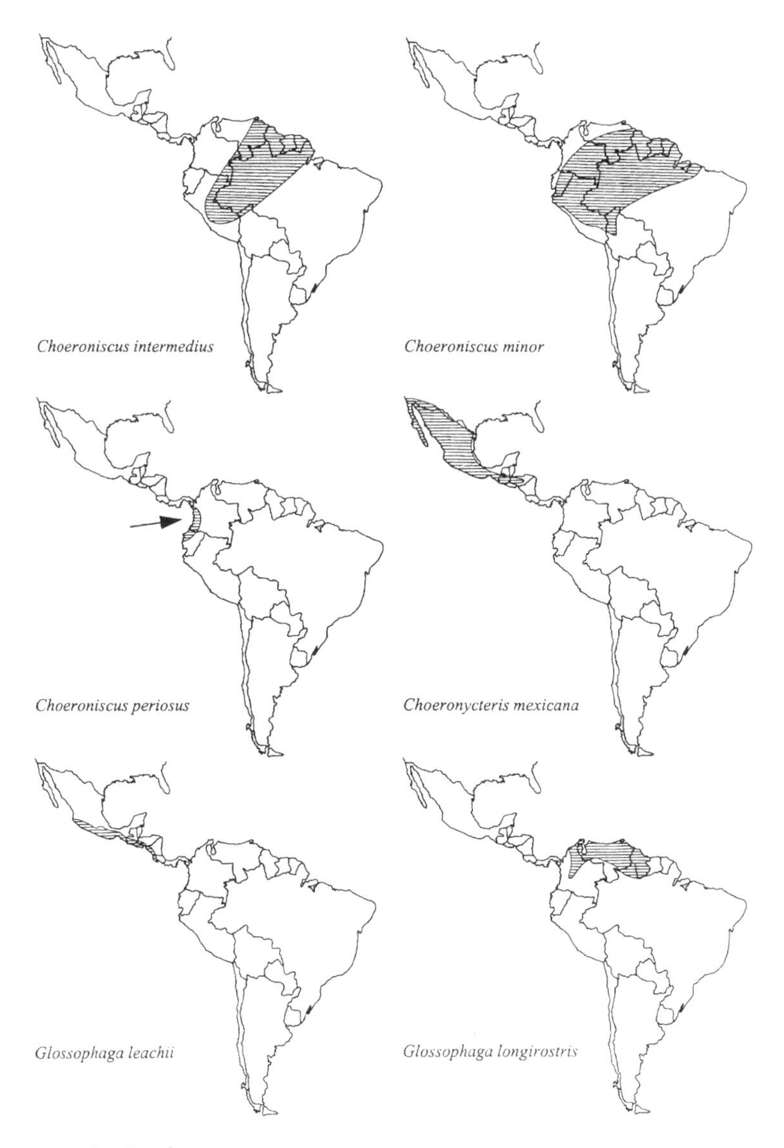

FIGURE 17.1. *Continued*

Results and Discussion

Conservation at the Species Level

As with other phyllostomids, there is a latitudinal trend in the number of species of New World nectar-feeding bats, with sites at lower latitudes harboring more species. Only three species reach the southwestern extreme of the United States, whereas 12 species occur in Mexico, 14 in Central America, and 19 in Colombia (Table 17.2).

FIGURE 17.1. *Continued*

Similarly, although 14 species occur in Ecuador, only three reach the northern part of Argentina and none has been reported from Chile (Koopman, 1981, 1982, 1993; Anderson et al., 1982; Linares, 1987; Eisenberg, 1989; Molinari, 1994; Muñoz, 1995; Medellín et al., 1997; Reid, 1997; Arita and Ortega, 1998).

Several species of nectar-feeding bats have restricted ranges. Most of these bats are associated with dry tropical and subtropical areas (Koopman, 1981). Compared with other neotropical bats, nectar-feeding species have relatively small geographic

FIGURE 17.1. *Continued*

ranges (Arita, 1993). Four species are endemic to Central America (*Glossophaga leachii, G. morenoi, Hylonycteris underwoodi,* and *Musonycteris harrisoni*) (Arita and Ortega, 1998). The former three occur mostly in dry tropical areas of western Mexico. Another two species (*Choeronycteris mexicana* and *Leptonycteris nivalis*) are quasi-endemic to Central America; their ranges extend to the southwestern extreme of the United States. In a similar pattern, some South American species are restricted to tropical dry areas. *G. longirostris,* for example, is found only in the arid regions of

FIGURE 17.1. *Continued*

Colombia, Venezuela, and adjacent areas (Koopman, 1982; Webster and Handley, 1986). *Platalina genovensium* is restricted to a few dry areas of the Pacific versant of Peru (Koopman, 1981).

These arid formations of North and South America are characterized by the presence of several species of agaves and columnar cacti that are pollinated by bats (Fleming, 1993; Fleming et al., 1993). For example, in Arizona, Texas, and northern and central Mexico, long-nosed bats (*Leptonycteris curasoae* and *L. nivalis*) are the main

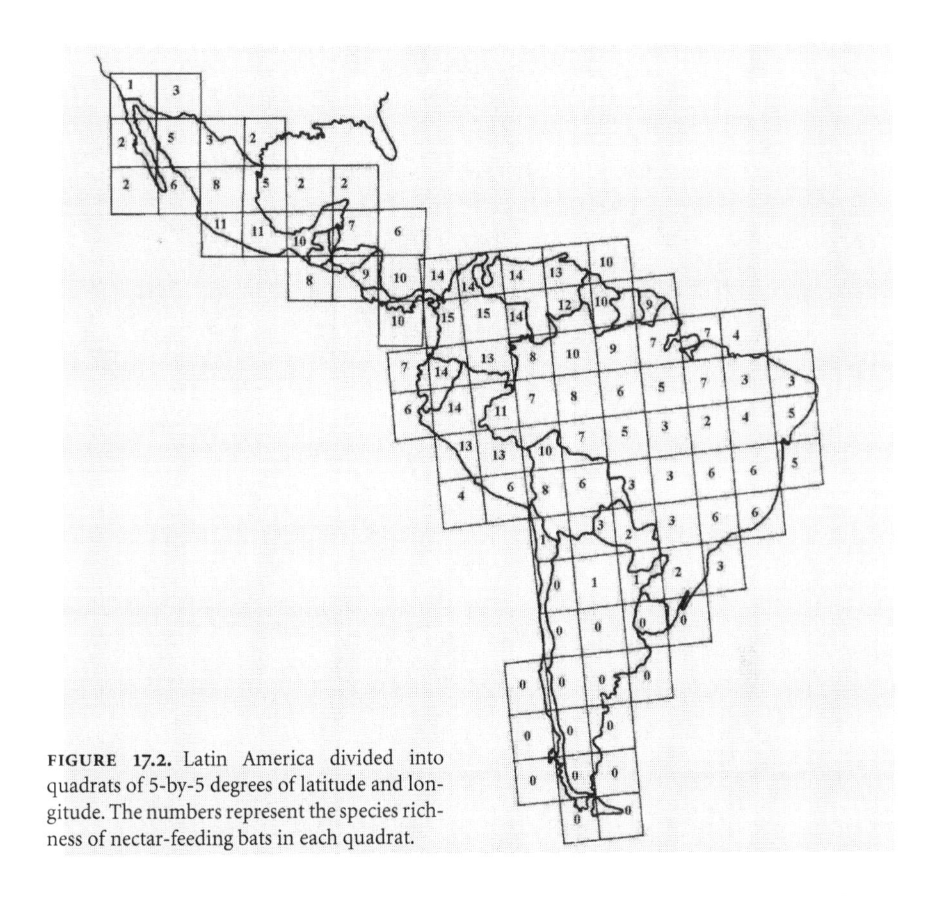

FIGURE 17.2. Latin America divided into quadrats of 5-by-5 degrees of latitude and longitude. The numbers represent the species richness of nectar-feeding bats in each quadrat.

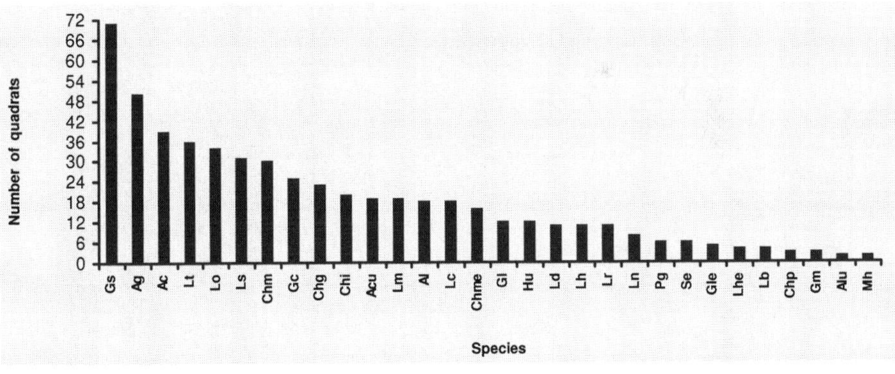

FIGURE 17.3. The distributional ranges of nectar-feeding bats in the New World. The bars represent the number of quadrats included in the potential distribution of each species. Gs: *Glosssophaga soricina* (71 quadrats); Ag: *Anoura geoffroyi* (50); Ac: *Anoura caudifer* (39); Lt: *Lonchophylla thomasi* (36); Lo: *Lichonycteris obscura* (34); Ls: *Lionycteris spurrelli* (31); Chm: *Choeroniscus minor* (30); Gc: *Glossophaga commissarisi* (25); Chg: *Choeroniscus godmani* (23); Chi: *Choeroniscus intermedius* (20); Acu: *Anoura cultrata* (19); Lm: *Lonchophylla mordax* (19); Al: *Anoura latidens* (18); Lc: *Leptonycteris curasoae* (18); Chme: *Choeronycteris mexicana* (16); Gl: *Glossophaga longirostris* (12); Hu: *Hylonycteris underwoodi* (12); Ld: *Lonchophylla dekeyseri* (11); Lh: *Lonchophylla handleyi* (11); Lr: *Lonchophylla robusta* (11); Ln: *Leptonycteris nivalis* (8); Pg: *Platalina genovensium* (6); Se: *Scleronycteris ega* (6); Gle: *Glossophaga leachii* (5); Lhe: *Lonchophylla hesperia* (4); Lb: *Lonchophylla bokermanni* (4); Chp: *Choeroniscus periosus* (3); Gm: *Glossophaga morenoi* (3); Alu: *Anoura luismanueli* (2); Mh: *Musonycteris harrisoni* (2).

TABLE 17.2

Species of Nectarivorous Bats that Have Been Reported or Are Potentially Present in the United States and Latin America[a]

Species	*Country[b]*												*References[c]*
	US	*MX*	*CA*	*CO*	*VE*	*GU*	*EC*	*PE*	*BO*	*BR*	*PA*	*AR*	
A. caudifer			X	X	X	X	X	X	X	X	X?	X	1,5, 6, 7,12
A. cultrata			X	X	X		X	X	X	X?			1, 3, 7,14
A. geoffroyi		X	X	X	X	X	X	X	X	X	X?	X	1, 5, 7, 9
A. latidens				X	X			X					7
A. luismanueli				X									10
C. godmani		X	X	X	X	X				X?			5, 6, 7, 9,14
C. intermedius			X	X	X		X	X	X				5, 6, 7
C. minor				X	X	X	X	X	X	X			1, 5, 6, 7
C. periosus				X			X						6, 7
C. mexicana	X	X	X										9
G. longirostris			X	X	X				X				2, 4, 7
G. commissarisi	X	X	X				X		X				2, 3, 7, 9
G. leachii		X	X	X									3, 9
G. morenoi		X											9
G. soricina		X	X	X	X	X	X	X	X	X	X	X	2,3,5,6,7,9
H. underwoodi		X	X										5
L. curasoae	X	X		X	X								7, 8, 9, 11
L. nivalis	X	X											7, 9
L. obscura		X	X	X	X	X	X	X	X	X			1, 6, 7, 9
L. spurrelli			X	X	X	X	X	X	X	X			5, 6, 7
L. bokermanni										X			7
L. dekeyseri										X			7
L. handleyi				x			X	X					7
L. hesperia							X	X					6, 7
L. mordax			X	X			X	?	?	X			2, 6, 7
L. robusta			X	X	X		X						6, 7
L. thomasi			X	X	X	X	X	X	X	X			1, 2, 6, 7
M. harrisoni		X											9, 13
P. genovensium						X	X						6, 7
S. ega				X	X					X			5, 6, 7, 11
Total	3	12	14	19	16	10	14	15?	10?	16?	3?	3	

[a] Full names of bat species are listed in Table 17.1.

[b] US (United States); MX (Mexico); CA (Guatemala through Panama); CO (Colombia); VE (Venezuela); GU (Guianas and Surinam); EC (Ecuador); PE (Peru); BO (Bolivia); BR (Brazil); PA (Paraguay); AR (Argentina). No nectar-feeding bats are known from Chile and Uruguay.

[c] 1 = Anderson et al. (1982); 2 = Eisenberg (1989); 3 = Hall (1981); 4 = Webster et al. (1998); 5 = Koopman (1981); 6 = Koopman (1982); 7 = Koopman (1993); 8 = Linares (1987); 9 = Medellín et al. (1997); 10 = Molinari (1994); 11 = Muñoz (1995); 12 = Redford and Eisenberg (1992); 13 = Téllez and Ortega (1999); 14 = Marinho-Filho and Sazima (1998).

pollinators of several species of *Agave* (Easterla, 1972; Howell, 1974; Martínez del Río and Eguiarte, 1987), the saguaro (*Carnegiea gigantea*) and the Mexican cardón (*Pachycereus pringlei*) (Horner et al., 1998). In the Tehuacán Valley of central Mexico, the tetetzo cactus (*Neobuxbaumia tetetzo*) is pollinated by *L. curasoae* and *Choeronycteris mexicana,* which show patterns of local migration associated with the flowering phenology of these plants (Valiente-Banuet et al., 1996). In the Andean arid regions of Colombia and Venezuela, *L. curasoae* and *Glossophaga longirostris* pollinate several species of columnar cacti, including *Pilosocereus tillianus, Stenocereus griseus,* and *S. repandus* (Soriano et al., 1991; Ruiz et al., 1997).

In a pattern that contrasts with that of species from dry areas, nectar-feeding bats associated with tropical rainforests tend to be widely distributed in the New World. The species with the widest ranges (e.g., *Anoura caudifer, A. geoffroyi, Glossophaga soricina, Lichonycteris obscura, Lionycteris spurrelli, Lonchophylla thomasi*) are found in tropical rainforests (Emmons, 1997). These forests were connected with almost no major barrier from southern Mexico to southern Peru, Bolivia, and Brazil before the existence of humans in the continent. In contrast, dry tropical areas form geographic patches of smaller size than the extensions of tropical rainforest, fostering the development of isolated populations of endemic species. Thus we speculate that the local-level pattern of close relationships between nectar-feeding bats and agaves and cacti is associated with a geographic pattern of isolated populations restricted to dry tropical and subtropical areas. Even migratory species, such as both species of *Leptonycteris,* travel over relatively small areas and tend to be restricted to dry areas throughout the year (Arita, 1991).

In general, species with the widest ranges are also the most abundant nectar bats at local levels. For example, *Glossophaga soricina,* the nectar-feeding bat with the widest range in the New World, is also one of the most abundant bats in most neotropical sites (Arita, 1993); it is particularly common in disturbed areas (Alvarez et al., 1991). Exceptions to this pattern do exist, however, due to particular situations. *Lichonycteris obscura,* for example, is common in South America (Emmons, 1997), but is considered extremely rare in the northern limit of its distribution in Mexico (Arita and Santos del Prado, 1999).

New World nectar-feeding bats are relatively small phyllostomids. On average, they have body masses of less than 30 g and forearm lengths of less than 55 mm. The small size of these bats is associated with energetic and aerodynamic constraints related to their diet and foraging behavior (Norberg et al., 1993; Norberg, 1994). For mammals, body mass correlates positively with geographic range and negatively with local abundance (Gaston, 1994). Thus one would expect nectar-feeding bats to be, when compared with other bats, restricted in range and locally abundant. In an analysis of 150 New World bat species, Arita (1993) found the first prediction to hold, but nectar-feeding bats did not differ from other bats in terms of local abundance.

Because we currently have little information on the natural history and population trends of most of species of nectar-feeding bats, it is difficult to establish a sound conservation strategy for these animals. However, with the consensus of several specialists, the International Union for the Conservation of Nature (IUCN) has produced a list of ten glossophagines that are considered endangered or vulnerable, and another six species are considered to be at lower risk but near threatened (Table 17.1; Hutson et al., 1998). Based on field surveys, the government of the United States considered both species of long-nosed bats (*Leptonycteris curasoae* and *L. nivalis*) to be threatened (but see Cockrum and Petryszyn [1991] for arguments that challenge this assessment). In Mexico, long-nosed bats are also included in the official list of threatened species (SEDESOL, 1994).

Because of the scarcity of data on nectar-feeding bats, it is difficult to implement effective conservation actions focused on particular species. An alternative strategy is to select sets of areas, which, if protected, can provide adequate habitat for most species. Strategies based on the protection of geographic areas would seem to be a better choice than species-oriented conservation plans in the case of the New-World nectar-feeding bats.

PRIORITY-AREAS ANALYSIS

The stepwise priority-areas analysis produced a group of six regions that include all the species of the neotropical nectar-feeding bat fauna. This group constitutes a possible system of areas that should be given special attention in a plan aimed at the conservation of these bats (Fig. 17.4, Tables 17.3, 17.4). Because of its iterative design, the procedure identified first the area with the highest species richness, and subsequently produced a set of complementary areas that are rich in species not included in the first region. Completing the sequence, the algorithm added areas that are relatively poor in species but that harbor endemic species not found in the richest areas. The six priority areas were, in order of urgency, the Andean region and eastern plains of Colombia, the Transvolcanic belt and the Balsas basin in Mexico, southern Ecuador and northwestern Peru, southeastern Brazil, the Amazon region of southern Venezuela and northern Brazil, and the Mérida Mountains in Venezuela (Fig. 17.4).

The iterative analyses including the species with restricted distributions and using the index of restrictedness yielded the same set of six priority areas but in a different order (Fig. 17.5). In this analysis, the descending order of importance was the Transvolcanic belt and the Balsas basin in Mexico, the coastal area of Ecuador, the coastal area of northern Peru, southeastern Brazil, the Mérida Mountains in Venezuela, and the northwestern part of the Brazilian Amazon region.

From the results of analyses using different criteria, we have identified the following sites as priority areas for the conservation of the nectar-feeding bats of the New World.

Andean region and eastern plains of Colombia. (Fig. 17.4, region 1.) This area provides habitat for 17 of the 30 neotropical nectar-feeding bats, being the zone with the highest species richness. However, this area has a low value of the index of restrictedness, because 14 of the 17 species have wide distributional ranges. Thus the fauna of this area is particularly rich in species because of the overlap of species' ranges, not because of the presence of endemic species. In fact, the only endemic species is *Choeroniscus periosus,* classified as vulnerable by the IUCN (Hutson et al., 1998). Another vulnerable species that inhabits this area is *Lonchophylla handleyi* (IUCN, 1996); *Glossophaga longirostris* and *L. robusta* are bats with relatively restricted distributions but they also occur in other areas of the continent.

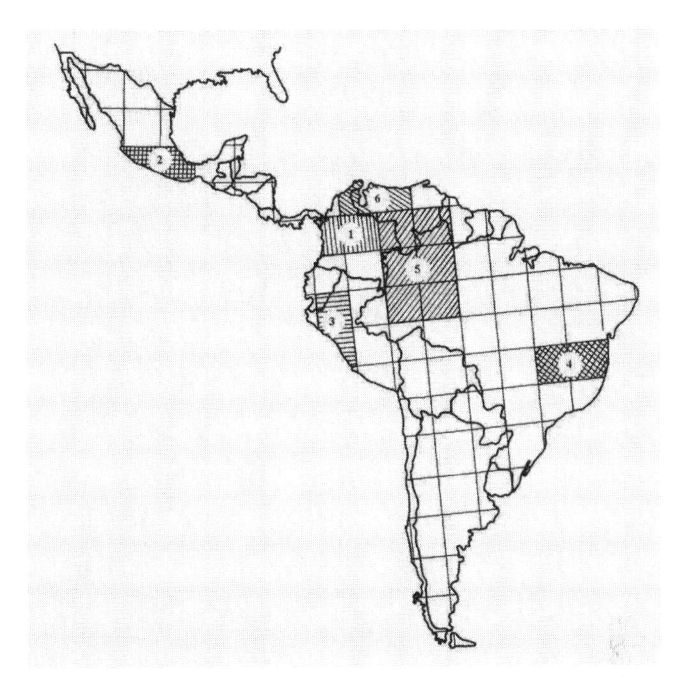

FIGURE 17.4. Priority areas for the conservation of glossophagine bats based on species richness and a stepwise complementarity selection of areas. Major areas, in order of importance, are (1) Andean region and eastern plains of Colombia; (2) Transvolcanic belt and the Balsas basin in Mexico; (3) western Ecuador and northwestern Peru; (4) southeastern Brazil; (5) Amazon region of southern Venezuela and northern Brazil; and (6) northeastern Colombia and Mérida Mountains, Venezuela.

TABLE 17.3

Potential List of Nectarivorous Bats inside the Areas with Highest Species Richness[a]

Area 1	*Area 2*	*Area 3*	*Area 4*	*Area 5*	*Area 6*
*A. caudifer**	*A. geoffroyi*	*A. caudifer*	*A. caudifer*	*A. caudifer*	*A. caudifer*
*A. cultrata**	*C. godmani*	*A. cultrata*	*A. geoffroyi*	*A. cultrata*	*A. cultrata*
*A. geoffroyi**	*C. mexicana**	*A. geoffroyi*	*G. soricina*	*A. geoffroyi*	*A. geoffroyi*
*A. latidens**	*G. morenoi**	*A. latidens*	*L. mordax*	*A. latidens*	*A. latidens*
*C. godmani**	*G. commissarisi*	*C. intermedius**	*L. bokermanni**	*C. godmani*	*A. luismanueli**
*C. minor**	*G. leachii**	*C. minor*	*L. dekeyseri**	*C. intermedius*	*C. godmani*
*C. periosus**	*G. soricina*	*G. commissarisi*		*C. minor*	*C. minor*
*G. commissarisi**	*H. underwoodi**	*G. soricina*		*G. commissarisi*	*G. longirostris*
*G. longirostris**	*L. curasoae*	*L. obscura*		*G. longirostris*	*G. soricina*
*G. soricina**	*L. nivalis**	*L. spurrelli*		*G. soricina*	*L. curasoae*
*L. curasoae**	*M. harrisoni**	*L. thomasi*		*L. curasoae*	*L. obscura*
*L. obscura**		*L. handleyi*		*L. obscura*	*L. spurrelli*
*L. spurrelli**		*L. hesperia*		*L. spurrelli*	*L. mordax*
*L. mordax**		*P. genovensium*		*L. thomasi*	*L. thomasi*
*L. robusta**				*S. ega**	*L. robusta*
*L. handleyi**					
*L. thomasi**					

[a] * Species in the complementary set of each area, that is, species not found in previous regions. Area 1 = Andean region and eastern plains of Colombia; Area 2 = Transvolcanic belt and the Balsas basin in México; Area 3 = western Ecuador and northwestern Perú; Area 4 = southeastern Brazil; Area 5 = Amazon region of southern Venezuela and northern Brazil; Area 6 = northeaestern Colombia and Mérida Mountains, Venezuela (see Fig. 17.4). Full names of bat species are listed in Table 17.1.

TABLE 17.4
Potential List of Rare Species or Those with Restricted Distributions[a]

Area 1 (Ic = 0.124)	*Area 2 (Ic = 0.152)*	*Area 3 (Ic = 0.129)*	*Area 4 (Ic = 0.122)*	*Area 5 (Ic = 0.112)*	*Area 6 (Ic = 0.108)*
A. geoffroyi	*C. minor*	*L. handleyi*	*A. caudifer*	*A. cultrata*	*C. intermedius*
C. godmani	*C. periosus*	*P. genovensium*	*L. bokermanni*	*A. latidens*	*S. ega*
C. mexicana	*L. hesperia*		*L. dekeyseri*	*A. luismanueli*	
G. morenoi	*L. mordax*			*G. longirostris*	
G. commissarisi	*L. robusta*			*L. obscura*	
G. leachii				*L. spurrelli*	
G. soricina				*L. thomasi*	
H. underwoodi					
L. curasoae					
L. nivalis					
M. harrisoni					

[a] According to the highest values of the restrictedness index (Ic), the most important regions for the conservation of nectarivorous bats are: Area 1 = Transvolcanic belt and the Balsas basin in Mexico; Area 2 = Pacific coast in Ecuador; Area 3 = north Pacific coast of Peru; Area 4 = southeastern Brazil; Area 5 = northeastern Colombia and Mérida Mountains, Venezuela; Area 6 = northwestern Amazon region in Brazil (see Fig. 17.5). Full names of bat species are listed in Table 17.1.

Transvolcanic belt and the Balsas basin in Mexico. (Fig. 17.4, region 2; Fig. 17.5, region 1.) This region is particularly rich in species with restricted distributions, including four Central American endemics (*Glossophaga leachii, G. morenoi, Hylonycteris underwoodi,* and *Musonycteris harrisoni*) (Arita and Ortega, 1998). The high concentration of endemic species makes this area the second most important in terms of species richness and the first priority for species with restricted distributions. The area is part of the transitional zone between the nearctic and neotropical realms and includes most of the terrestrial biomes of Mexico (Toledo and Ordóñez, 1998). The Tehuacán valley, part of the Balsas basin, is the area with the highest diversity of columnar cacti (Valiente-Banuet et al., 1996).

Northeastern Colombia and Mérida Mountains, Venezuela. (Fig. 17.4, region 6; Fig. 17.5, region 5.) The main biomes in this area of northern South America are xerophytic scrub, tropical dry forest, savannas, and, to a lesser extent, tropical rainforest (Linares, 1987). *Anoura luismanueli,* a recently described species, is restricted to the mid-altitude cloud forests of the Mérida Mountains of Venezuela (Molinari, 1994). Although the presence of this bat in Colombia is highly likely, it is not yet known from that country (Molinari, 1994).

Southeastern Brazil. (Fig. 17.4, region 4; Fig. 17.5, region 4.) The main biomes of this area are cerrado (a tropical savanna), Atlantic rainforest, and a small area of caatinga (semiarid brushland). With only six nectar-feeding bats, this area is comparatively poor in species, but two of them are endemic. *Lonchophylla dekeyseri* is endemic to the Cerrado, and *L. bokermannni* occurs both in cerrado and Atlantic forest. Both

FIGURE 17.5. Priority areas for the conservation of glossophagine bats based on the presence of endemic species and species with restricted distributions. The areas, in order of importance, are (1) Transvolcanic belt and the Balsas basin in Mexico; (2) Pacific coast of Ecuador; (3) north Pacific coast of Peru; (4) southeastern Brazil; (5) northeastern Colombia and Mérida Mountains, Venezuela; and (6) northwestern Amazon region in Brazil.

species are considered vulnerable because of the high rate of destruction of their natural habitat. In cerrado, for example, 50% of the natural vegetation has been lost, and only 1% of the area is protected by federal regulation. Similarly, in the Atlantic forest, probably the most threatened environment of Brazil, only 2% of the area is protected (Marinho-Filho and Sazima, 1998).

Amazon region of southern Venezuela and northern Brazil. (Fig. 17.4, region 5; Fig. 17.5, region 6.) This area has a rich fauna of 14 nectar-feeding bats, but only one species (*Scleronycteris ega*) is endemic. This species is considered vulnerable (IUCN, 1996) because of its narrow distributional range (Fig. 17.1) and because of the high rate of habitat loss in the Amazon basin.

Western Ecuador and northwestern Peru. (Fig. 17.4, region 3; Fig. 17.5, regions 2 and 3.) Among the 14 species of nectar-feeding bats of this area, four are of particular concern because of their restricted distributions (*Choeroniscus intermedius, C. periosus, Lonchophylla hesperia,* and *Platalina genovensium*). These bats occur chiefly in the arid areas of this region, including the central Andean valleys and the Atacama and Pacific deserts in Peru, the dry coastal formations of Ecuador, and the northern Andean valleys of Ecuador and Colombia. These South American arid formations are the natural habitat for numerous species of columnar cacti (Sarmiento, 1975), which almost surely have developed mutualistic relationships with bats. However, almost nothing is known about the natural history of the plants and the bats of this region (but see Sahley, 1996; Sahley and Baraybar, 1996).

From the analysis of the priority areas described above, a general pattern is evident. Areas with high species richness tend to be regions of convergence in the distribution of species with wide ranges. These areas generally contain high habitat heterogeneity and include, at least partially, zones covered with rainforest. Conversely, areas of low or intermediate species richness but with high percentages of endemic species tend to be dry zones, isolated from the more continuously distributed humid forests. Our combination of analyses produced a set of priority areas that takes into account both the richest regions and those with highest percentages of endemic nectar-feeding bats.

A shortcoming of all stepwise procedures for identifying priority areas for conservation is that they produce a set of regions that is biased, first, to species-rich areas and second, to regions with endemic species. These sets of areas are not necessarily the most efficient in terms of encompassing the entire fauna with the minimum number of areas, a number that depends on the degree of complementarity among sites. However, algorithms to optimize the efficiency of a system of priority areas are not yet fully developed, and the computer power required increases exponentially with the number of possible combinations that have to be analyzed, so this kind of analysis cannot be performed on personal computers (Pressey et al., 1993). The procedure used in this chapter has the advantage of providing concrete and fast results that constitute a first approach to the conservation of New World nectar-feeding bats.

Because of the methods we used, our results show a static view of the distributions of nectar-feeding bats. In fact, several species of nectarivorous bats are known to be migratory. The two species of *Leptonycteris,* for example, perform long-distance latitudinal movements in northern Mexico and southwestern United States and altitudinal migrations in central Mexico (Alvarez and González-Q., 1970; Cockrum, 1991; Wilkinson and Fleming, 1996; but see Rojas-Martínez et al., 1999 for an opposing view). Our system of priority areas for conservation is designed for the protection of overall species diversity but might be inadequate for the conservation of particular taxa, especially migratory bats. Because of their particular habits, migratory nectar-feeding bats are more susceptible to extinction than other bats, and require special strategies to guarantee their preservation (Arita and Santos del Prado, 1999).

After determining appropriate priority areas, the next step is to determine their suitablility for establishing real conservation strategies. Unfortunately, the rapid growth of human populations all over Latin America threatens the future of most of the areas that are critical for nectar-feeding bats. In Colombia, for example, which includes areas with the highest species richness of nectar-feeding bats, none of the ten Andean arid regions that constitute the genetic reservoir of the flora and fauna of the northern South American dry ecosystems are included in the system of national parks (Hernández-Camacho, 1992; Cavelier et al., 1996). Similarly, the tropical dry forests of Mexico, the prime habitat for many nectar-feeding bats, are poorly represented in the national system of protected areas (Flores-V. and Gerez, 1994; Gómez-Pompa and Dirzo, 1995). In particular, the basin of the Balsas River, which includes populations of many of the endemic nectar-feeding bats and which is the richest area of columnar cacti in the world, is almost totally unprotected (Valiente-Banuet et al., 1996). The conservation of nectar-feeding bats, of columnar cacti and other bat-pollinated plants, and of the mutualistic relationship between plants and bats will require the combined efforts of scientists and conservationists as well as the determination of governments to protect the critical areas identified in this chapter.

Summary

Nectar-feeding bats show traits that make them more vulnerable to extinction than other chiropteran species. Several of these characteristics are determined by the close mutualistic association that exists between the bats and the plants they pollinate: high dietary specialization, association with dry tropical and subtropical areas, small body size, and restricted distributional ranges. Thirty species of nectar-feeding bats (tribe Glossophagini) occur in the New World, with ten of them considered endangered or threatened. Countries with the highest species richness of nectar-feeding bats include Colombia, Venezuela, Brazil, and Peru. We performed priority-areas analyses based on species richness and on the presence of endemic species, and identified six areas of particular importance for the conservation of the nectar-feeding bats of the New World: the Andean region and the eastern plains of Colombia; the Transvolcanic belt and the Balsas basin in Mexico; northeastern Colombia and the Mérida Mountains, Venezuela; southeastern Brazil; the Amazon region of southern Venezuela and northern Brazil; and western Ecuador and northwestern Peru. Many of these areas are being transformed at an accelerated pace by the activity of humans; few protected areas have been established in these regions. To ensure the survival of nectar-feeding bats and their food plants, including many species of columnar cacti, it is necessary to implement conservation programs in these regions.

Resumen

Los murciélagos nectarívoros reúnen características qué los hacen vulnerables a la extinción en mayor grado que otros quirópteros. Varias de esas características están determinadas, por la cercana asociación mutualística de los murciélagos con las plantas que polinizan: alta especialización alimentaria, asociación con zonas secas tropicales y subtropicales; tamaño corporal pequeño y área de distribución restringida. Treinta especies de murciélagos nectarívoros (glosofaginos) habitan el continente americano siendo diez de ellas consideradas amenazadas ó en peligro de extinción. Los países con mayor número de especies de glosofaginos son Colombia, Venezuela, Brasil y Perú. Mediante análisis de áreas prioritarias, para la conservación, basados tanto en riqueza de especies como en presencia de especies endémicas identificamos seis áreas de particular importancia, para la conservación de los murciélagos nectarívoros del Nuevo Mundo: la región andina y parte de los llanos orientales de Colombia; el eje volcánico transversal y la cuenca del río Balsas en México; la región nor-oriental de Colombia y la cordillera de Mérida en Venezuela; la región sur-oriental de Brasil; la región amazónica de Venezuela y Brasil y la región occidental de Ecuador y el noroeste de Perú. Muchas de estas regiones están siendo transformadas rápidamente, por la acción del hombre. Se han establecido pocas áreas protegidas en ellas, por lo que es urgente implementar estrategias para la conservación de los murciélagos nectarívoros que las habitan y las plantas de las que se alimentan, incluyendo varias especies de cactáceas columnares.

ACKNOWLEDGMENTS

We thank J. Uribe for helping us in the search for information on the distribution and ecology of nectar-feeding bats. G. Guerrero and L. B. Vázquez collaborated in the production of the chapter.

REFERENCES

Alvarez, J., M. R. Willig, J. K. Jones, Jr., and W. D. Webster. 1991. *Glossophaga soricina. Mammalian Species* 379:1–7.

Alvarez, T., and L. González-Q. 1970. Análisis polínico del contenido gástrico de murciélagos Glossophaginae de México. *Anales de la Escuela Nacional de Ciencias Biológicas, México* 18:137–65.

Anderson, S., K. F. Koopman, and G. K. Creighton. 1982. Bats of Bolivia: an annotated checklist. *American Museum of Natural History Novitates* 2750:1–24.

Arita, H. T. 1991. Spatial segregation in long-nosed bats, *Leptonycteris nivalis* and *Leptonycteris curasoae,* in Mexico. *Journal of Mammalogy* 72:706–14.

———. 1993. Rarity in neotropical bats: correlations with phylogeny, diet, and body mass. *Ecological Applications* 3:506–17.

Arita, H. T., and C. Martínez del Río. 1990. Interacciones flor-murciélago: un enfoque zoocéntrico. *Publicaciones especiales del Instituto de Biología, Universidad Nacional Autónoma de México, México* 4:1–35.

Arita, H. T., and J. Ortega. 1998. The Middle American bat fauna: conservation in the neotropical-nearctic border. In *Bat Biology and Conservation,* eds. T. H. Kunz and P. A. Racey, 295–308. Washington, D.C.: Smithsonian Institution Press.

Arita, H. T., and K. Santos del Prado. 1999. Conservation biology of nectar-feeding bats in Mexico. *Journal of Mammalogy* 80:31–41.

Arita, H. T., F. Figueroa, A. Frisch, P. Rodríguez, and K. Santos del Prado. 1997. Geographical range size and the conservation of Mexican mammals. *Conservation Biology* 11:92–100.

Baker, R. J., C. S. Hood, and R. L. Honeycutt. 1989. Phylogenetic relationships and classification of the higher categories of the New World bat family Phyllostomidae. *Systematic Zoology* 38:228–38.

Cavelier, J., A. Ruiz, and M. Santos. 1996. *El Uso de Cactáceas como Bioindicadores del Proceso de Desertificación en Enclaves Secos Interandinos de Colombia.* Bogotá: Instituto de Hidrología, Meteorología y Estudios ambientales, IDEAM.

Ceballos, G., and J. H. Brown. 1995. Global patterns of mammalian diversity, endemism, and endangerment. *Conservation Biology* 9:559–68.

Cockrum, E. L. 1991. Seasonal distribution of northwestern populations of the long-nosed bat, *Leptonycteris sanborni,* family Phyllostomidae. *Anales del Instituto de Biología, Universidad Nacional Autónoma de México, México, serie Zoología* 62:181–202.

Cockrum, E. L., and Y. Petryszyn. 1991. The long-nosed bat, *Leptonycteris:* an endangered species in the Southwest? *Occasional Papers, The Museum, Texas Tech University* 142:1–32.

Cousins, S. H. 1991. Species diversity measurement: choosing the right index. *Trends in Ecology and Evolution* 6:190–92.

Easterla, D. A. 1972. Status of *Leptonycteris nivalis* (Phyllostomidae) in Big Bend National Park, Texas. *Southwestern Naturalist* 17:287–92.

Eisenberg, J. F. 1989. *Mammals of the Neotropics. The Northern Neotropics. Panamá, Colombia, Venezuela, Guyana, Suriname, French Guyana.* Vol. 1. Chicago: University of Chicago Press.

Emmons, L. H. 1997. *Neotropical Rainforest Mammals. A Field Guide.* 2nd ed. Chicago: University of Chicago Press.

Faegri, K., and L. van der Pijl. 1979. *The Principles of Pollination Biology.* Oxford: Pergamon Press.

Faith, D. P. 1992. Conservation evaluation and phylogenetic diversity. *Biological Conservation* 61:1–10.

Fleming, T. H. 1988. *The Short-Tailed Fruit Bat. A Study in Plant-Animal Interactions.* Chicago: University of Chicago Press.

———. 1993. Plant-visiting bats. *American Scientist* 81:460–67.

Fleming, T. H., R. A. Núñez, and L.S.L. da Silveira. 1993. Seasonal changes in the diets of migrant and non-migrant nectarivorous bats as revealed by carbon stable isotope analysis. *Oecologia* 94:72–75.

Flores-V. O., and P. Gerez. 1994. *Biodiversidad y Conservación en México: Vertebrados, Vegetación y Uso de Suelo.* 2nd ed. Mexico: Universidad Nacional Autónoma de México.

Gardner, A. L. 1977. Feeding habits. In *Biology of Bats of the New World Family Phyllostomidae.* Part II, eds. R. J. Baker, J. K. Jones, Jr., and D. C. Carter, 293–350. *Special Publications, The Museum, Texas Tech University* 13:1–364.

Gaston, K. J. 1994. *Rarity.* London: Chapman and Hall.

Gómez-Pompa, A., and R. Dirzo. 1995. *Reservas de la Biósfera y Otras Áreas Naturales Protegidas de México.* Ciudad de México, Mexico: Secretaria de Medio Ambiente, Recursos Humanos y Pesca.

Grant, V., and K. A. Grant. 1979. The pollination spectrum in the southwestern American cactus flora. *Plant Systematics and Evolution* 133:29–37.

Griffiths, T. A. 1982. Systematics of the New World nectar-feeding bats (Mammalia, Phyllostomidae), based on the morphology of the hyoid and lingual regions. *American Museum Novitates* 2742:1–45.

Hall, E. L. 1981. *The Mammals of North America.* Vol. I. New York: John Wiley and Sons.

Hernández-Camacho, J. 1992. Vulnerabilidad y estrategias para la conservación de algunos biomas de Colombia. In *La Diversidad Biológica de Iberoamérica,* ed. G. Halffter, 191–202. Volumen Especial de Acta Zoológica Mexicana. Xalapa, Mexico: Acta Zoológica Mexicana.

Horner, M. A., T. H. Fleming, and C. T. Sahley. 1998. Foraging behaviour and energetics of a nectar-feeding bat, *Leptonycteris curasoae* (Chiroptera: Phyllostomidae). *Journal of Zoology* 244:575–86.

Howell, D. J. 1974. Bats and pollen: Physiological aspects of the syndrome of chiropterophily. *Comparative Biochemistry and Physiology* 48A:263–76.

Howell, D. J., and N. Hodgkin. 1976. Feeding adaptations in the hairs and tongues of nectar-feeding bats. *Journal of Morphology* 148:329–36.

Humphries, C. J., P. H. Williams, and R. I. Vane-Wright. 1995. Measuring biodiversity value for conservation. *Annual Review of Ecology and Systematics* 26:93–111.

Hutson, T., S. Mickleburgh, and P. Racey. 1998. *Global Action Plan for Microchiropteran Bats.* Final draft. Gland, Switzerland: IUCN.

IUCN. 1996. *The 1996 IUCN Red List of Threatened Animals.* Gland, Switzerland: IUCN.

Janzen, D. H. 1983. *Costa Rica Natural History.* Chicago: University of Chicago Press.

Kershaw, M., P. H. Williams, and G. M. Mace. 1994. Conservation of Afrotropical antelopes: consequences and efficiency of using different site selection methods and diversity criteria. *Biodiversity and Conservation* 3:354–72.

Koopman, K. F. 1981. The distributional patterns of New World nectar-feeding bats. *Annals of the Missouri Botanical Garden* 68:352–69.

———. 1982. Biogeography of the bats of South America. In *Mammalian Biology in South America,* eds. M. A. Mares and H. H. Genoways. *Special Publication Pymatuning Laboratory of Ecology, University of Pittsburgh* 6:273–302.

———. 1993. Order Chiroptera. In *Mammal Species of the World. A Taxonomic and Geographic Reference.* 2nd ed., eds. D. E. Wilson and D. M. Reeder. Washington, D.C.: Smithsonian Institution Press.

Linares, O. J. 1987. *Murciélagos de Venezuela.* Caracas, Venezuela: Cuardernos Lagoven.

Marinho-Filho, J., and I. Sazima, I. 1998. Brazilian bats and conservation biology. In *Bats, Biology and Conservation,* eds. T. H. Kunz and P. A. Racey. Washington, D.C.: Smithsonian Institution Press.

Marshall, A. G. 1983. Bats, flowers and fruit: evolutionary relationships in the Old World. *Biological Journal of the Linnean Society* 20:115–35.

Martínez del Río, C., and L. E. Eguiarte. 1987. Bird visitation to *Agave salmiana:* comparisons among hummingbirds and perching birds. *Condor* 89:357–63.

Medellín, R., H. T. Arita, and O. Sánchez. 1997. *Identificación de los Murciélagos de México. Clave de campo. Asociación Mexicana de Mastozoología, A. C. Publicaciones Especiales* 2:1–37.

Molinari, J. 1994. A new species of *Anoura* (Mammalia Chiroptera Phyllostomidae) from the Andes of northern South America. *Tropical Zoology* 7:73–86.

Muñoz, J. A. 1995. *Clave de Murciélagos Vivientes en Colombia.* Medellín, Colombia: Editorial Universidad de Antioquia.

Nobel, P. S. 1994. *Remarkable Agaves and Cacti.* New York: Oxford University Press.

Norberg, U. M. 1994. Wing design, flight performance and habitat use in bats. In *Ecological Morphology, Integrative Organismal Biology,* eds. P. C. Wainwright and S. M. Reilly, 205–39. Chicago: University of Chicago Press.

Norberg, U. M., T. H. Kunz, J. F. Steffensen, Y. Winter, and O. von Helversen. 1993. The cost of hovering flight in a nectar feeding bat, *Glossophaga soricina,* estimated from aerodynamic theory. *Journal of Experimental Biology* 182:207–27.

Pressey, R. L., C. J. Humphries, C. R. Margules, R. I. Vane-Wright, and P. H. Williams. 1993. Beyond opportunism: key principles for systematic reserve selection. *Trends in Ecology and Systematics* 8:124–28.

Quiroz, D. L., M. S. Xelhuantzi, and M. C. Zamora. 1986. Análisis palinológico del contenido gastrointestinal de los murciélagos *Glossophaga soricina* y *Leptonycteris yerbabuenae* de las grutas de Juxtlahuaca, Guerrero. Ciudad de México, Mexico: Instituto Nacional de Antropología e Historia.

Redford, K. H., and J. F. Eisenberg. 1992. *Mammals of the Neotropics. The Southern Cone. Chile, Argentina, Uruguay.* Vol. 2. Chicago: University of Chicago Press.

Reid, F. 1997. *A Field Guide to the Mammals of Central America and Southeast Mexico.* Oxford: Oxford University Press.

Rojas-Martínez, A. E., A. Valiente-Banuet, M. C. Arizmendi, A. Alcántara-Eguren, and H. T. Arita. 1999. Seasonal distribution of the long-nosed bat (*Leptonycteris curasoae*) in North America: Does a generalized migration pattern really exist? *Journal of Biogeography* 26:1065–77.

Ruiz, A., M. Santos, P. Soriano, J. Cavelier, and A. Cadena. 1997. Relaciones mutualísticas entre el murciélago *Glossophaga longirostris* y las cactáceas columnares de la zona árida de La Tatacoa, Colombia. *Biotropica* 29:469–79.

Sahley, C. T. 1996. Bat and hummingbird pollination of an autotetraploid columnar cactus, *Weberbauerocereus weberbaueri* (Cactaceae). *American Journal of Botany* 83:1329–36.

Sahley, C. T., and L. Baraybar. 1996. The natural history of the long-snouted bat, *Platalina genovensium* (Phyllostomidae: Glossophaginae), in southwestern Peru. *Vida Silvestre Neotropical* 5:101–9.

Sarmiento, G. 1975. The dry plant formations of South America and their floristic connections. *Journal of Biogeography* 2:233–51.

SEDESOL (Secretaria de Desarrollo Social). 1994. Norma Oficial Mexicana NOM-059-ECOL-1994, que determina las especies y subespecies de flora y fauna silvestres terrestres y acuáticas en peligro de extinción, amenazadas, raras y las sujetas a protección especial, y que establece especificaciones para su protección. *Diario Oficial de la Federación* 438:2–60.

Soriano, P. J., M. Sosa, and O. Rossell. 1991. Hábitos alimentarios de *Glossophaga longirostris* Miller (Chiroptera: Phyllostomidae) en una zona árida de los Andes venezolanos. *Revista de Botánica Tropical* 39:263–68.

Téllez, G., and J. Ortega. 1999. *Musonycteris harrisoni. Mammalian Species* 622:1–3.

Toledo, V. M., and M. Ordóñez. 1998. El panorama de la biodiversidad de México: una revisión de los hábitats terrestres. In *Diversidad Biológica de México: Orígenes y Distribución,* eds. T. P. Ramamoorthy, R. Bye, A. Lot, and J. Fa. Mexico: Instituto de Biología, Universidad Nacional Autónoma de México.

Valiente-Banuet, A., M. C. Arizmendi, A. Rojas-Martínez, and L. Domínguez-Canseco. 1996. Ecological relationships between columnar cacti and nectar-feeding bats in Mexico. *Journal of Tropical Ecology* 12:103–19.

Webster, W. D., and C. O. Handley, Jr. 1986. Systematics of Miller's long tongued bat, *Glossophaga longirostris,* with description of two new subspecies. *Occasional Papers, The Museum, Texas Tech University* 100:1–22.

Webster, D., C. O. Handley, Jr. W., and P. Soriano. 1998. *Glossophaga longirostris. Mammalian Species* 576:1–5.

Wilkinson, G. S., and T. H. Fleming. 1996. Migration and evolution of lesser long-nosed bats, *Leptonycteris curasoae,* inferred from mitochondrial DNA. *Molecular Ecology* 5:329–39.

Wilson, D. E. 1973. Bat faunas: A trophic comparison. *Systematic Zoology* 22:14–29.

Epilogue

Plants of the family Cactaceae are among the most evocative organisms living in arid regions throughout the New World. According to Taylor (1997), Mexico and the southwestern United States currently contain the greatest diversity of cacti. Fully 27% of all cactus genera are endemic to this region and at least 570 species are native. Arborescent columnar cacti, the subjects of this book, dominate the landscapes in many parts of south-central, central, and western Mexico as far north as the Sonoran Desert. Although lower in diversity, these plants nonetheless are critical elements in the arid regions of northern and western South America. In each of these areas, columnar cacti function as "keystone" species by interacting with a host of other organisms. They have also had considerable economic significance for populations of ancient and modern people.

As described in this book, columnar cacti and their biotic interactions are important at a number of ecological levels. At the population level, demographically critical biotic interactions occur during the early stages of cactus life cycles. These interactions, which include pollination and seed dispersal, significantly affect the finite rates of increase of populations of both cacti and their mutualists. At the community level, the maintenance of biological diversity also depends on these same interspecific interactions. In the case of columnar cacti, seed dispersal is often directed to the canopies of perennial plants that serve as recruitment foci for cactus seedlings. From an evolutionary viewpoint, the foraging behavior of pollinators and seed dispersers also affects the genetic structure of cactus populations. And the reproductive phenology and morphology of flowers and fruit have undoubtedly influenced the evolution of cactus-visiting vertebrates. For example, the morphology, physiology, foraging and migratory behavior, and reproductive biology of the phyllostomid bat *Leptonycteris curasoae* all show strong influences of a dependence on cactus flower and fruit resources.

Such positive interactions as commensalism and mutualism among plants and between plants and animals are common in arid habitats, and these interactions have important conservation implications. Numerous studies indicate that columnar cacti need "nurse plants" for establishment. In the Sonoran Desert, for example, these plants include mesquite and ironwood—species that are frequently harvested for charcoal production and commercial carving, respectively. The legume *Mimosa luisana* plays a similar role as a nurse plant in the Tehuacán Valley of south-central Mexico. Conservation and management of populations of columnar cacti therefore requires preservation of perennial plants in their habitats. Any process that has a negative impact on populations of nurse plants also negatively affects columnar cacti. Likewise, conservation of cactus-visiting birds and bats—the primary pollinators and dispersers of columnar cacti—is critical for the continued reproductive success of these plants. Roost or nest sites, which include caves and mines in the case of bats, and alternate food plants are critical resources that need to be conserved for these organisms. Furthermore, because such important cactus pollinators as the bat *Leptonycteris curasoae* and white-winged doves are migratory in parts of their ranges, conservation plans require a large-scale geographic perspective. In western Mexico, migrant birds and bats move from their winter grounds to their spring/summer breeding grounds and back again through a variety of habitats that provide them with food and shelter. Preservation of wildland habitats along migration routes is as important as preserving habitats at the ends of migration routes. National—and in some cases, international—conservation strategies are needed to prevent the extinction of migratory animals.

Any plan of conservation and sustainable development in forests of columnar cacti needs to consider the role of humans and how they interact with natural resources. Information provided in this volume indicates that humans living in centers of diversity and endemism of columnar cacti (e.g., in central Mexico) have been using these resources for thousands of years. Indeed, ethnic and campesino populations in central Mexico today continue to conduct different forms of in situ and ex situ management of cactus populations. These practices provide excellent examples of artificial selection that have produced significantly higher frequencies of favored phenotypes (e.g., plants with larger and sweeter fruits) in ex situ populations. Such studies also provide an excellent example of a "mutualistic" association between humans and cacti that results in the conservation of these plants under controlled and wild conditions. Other human practices, such as forestry and goat raising, negatively affect natural plant communities and undoubtedly affect recruitment in cactus populations. However, in certain populated areas in Mexico (e.g., on ejidos or Propiedad Communal), people have developed strict rules for the use of their natural resources under state or federal guidelines. In other areas (e.g., in the Balsas River basin), cacti are protected by local people in the absence of state or federal legislation.

Ultimately, however, we need to increase the amount of arid land under legal protection to preserve columnar cacti and their mutualists. Mexico and the United States have led the way in setting aside national parks, national monuments, and biosphere reserves containing substantial populations of these plants and animals. Countries such as Venezuela, Colombia, and Peru lag far behind in protecting their arid lands. The recent establishment of a biosphere reserve containing about 460,000 ha in the Tehuacán-Cuicatlán Valley in Mexico can perhaps serve as a model of how to protect arid lands in Latin America. This valley is a major center of diversity for cacti of tribe

Pachycereeae, and its biosphere reserve includes nine different vegetation types dominated by these cacti. Many of these species are endemic to the valley and some are the ancestors of species found elsewhere in North America. In addition to habitat protection, activities involved in the establishment of the reserve include: (1) an intensive educational campaign about the importance of cacti and their mutualists, as well as information about the overall importance of biological diversity, in local municipalities; and (2) ecological restoration of degraded areas. In both cases, positive interactions between scientists, land managers, and local communities have been key processes in the maintenance of these unique, highly diverse, and very fragile ecosystems. Without such cooperative interactions, columnar cacti and their mutualists will largely disappear from the earth. It is hard to imagine what New World arid lands would look like if many of their most evocative elements were gone.

REFERENCE

Taylor, N. P. 1997. Cactaceae. In *Cactus and Succulent Plants—Status Survey and Conservation Action Plan,* comp. S. Oldfield, 17–20. International Union for the Conservation of Nature/ Cactus and Succulent Specialist Group (IUCN/SSC). Gland, Switzerland: IUCN.

INDEX